Soil Health Series: Volume 4 Soil Health and Sustainability in India

SERIES EDITOR
Douglas L. Karlen

EDITOR
Ranjan Bhattacharyya

CONTRIBUTORS

CHAPTER 1
Shrila Das
ICAR- Indian Agricultural Research Institute, Pusa, New Delhi, 110012
Dibyendu Mukhopadhyay
Uttar Banga Krishi Viswavidyalaya, Pundibari, Cooch Bihar, West Bengal, 736165
Ranjan Bhattacharyya
ICAR- Indian Agricultural Research Institute, Pusa, New Delhi, 110012

CHAPTER 2
D. Mandal
ICAR- Indian Agricultural Research Institute, Pusa, New Delhi, 110012
P. Jha
ICAR- Indian Institute of Soil Science, Nabibagh, Berasia Road, Bhopal, 462038, India
Raman Jeet Singh
ICAR-Indian Institute of Soil and Water Conservation. Dehradun, India 248195
M. Madhu
ICAR-Indian Institute of Soil and Water Conservation. Dehradun, India 248195
S. Barman
ICAR- Indian Agricultural Research Institute, Pusa, New Delhi, 110012
Ranjan Bhattacharyya
ICAR- Indian Agricultural Research Institute, Pusa, New Delhi, 110012
Rajesh Kaushal
ICAR-Indian Institute of Soil and Water Conservation. Dehradun, India 248195

CHAPTER 3
Sumanta Kundu
ICAR- Central Research Institute for Dryland Agriculture, Hyderabad, 500 059, India
Ch. Srinivasarao
ICAR- Indian Agricultural Research Institute, Pusa Campus, New Delhi, 110012, India
V.K. Singh
ICAR- Central Research Institute for Dryland Agriculture, Hyderabad, 500 059, India
J. Naveen
ICAR- Central Research Institute for Dryland Agriculture, Hyderabad, 500 059, India

CHAPTER 4
Avijit Ghosh
ICAR- Indian Grassland and Fodder Research Institute, Jhansi, India
Sukanya Misra
RLBCAU, Jhansi, India
Ranjan Bhattacharyya
ICAR- Indian Agricultural Research Institute, Pusa, New Delhi, 110012

CHAPTER 5
Dibyendu Chatterjee
ICAR National Rice Research Institute, Cuttack
Rajendiran Selladurai
Central Horticulture Exp Station, ICAR llHR
Mohammad Shahid
ICAR National Rice Research Institute, Cuttack
Dibyendu Sarkar
Bidhan Chandra Krishi Viswavidyalaya Nadia, W.B.
Saikat Ranjan Das
ICAR National Rice Research Institute

CHAPTER 6
Uttam Kumar Mandal
ICAR- Central Soil Salinity Research Institute, Regional Research Station, Canning Town, West Bengal, 743329, India
V. Ramesh
ICAR- Central Tuber Crops Research Institute, Thiruvananthapuram, India
Sudipa Mal
School of Agriculture, Sanskriti University, Mathura, Uttar Pradesh, 281401, India
T. D. Lama
ICAR- Central Soil Salinity Research Institute, Regional Research Station, Canning Town, West Bengal, 743329, India
S. K. Sarangi
ICAR-Central Institute for Women in Agriculture, Bhubaneswar, Odisha, 751003, India
Sourav Mullick
The Neotia University, West Bengal, 743368, India
Amit Ghosh
ICAR- Central Soil Salinity Research Institute, Regional Research Station, Canning Town, West Bengal, 743329, India
Dibyendu Bikas Nayak
ICAR- Central Soil Salinity Research Institute, Regional Research Station, Canning Town, West Bengal, 743329, India
Dhiman Burman
ICAR- Central Soil Salinity Research Institute, Regional Research Station, Canning Town, West Bengal, 743329, India

CHAPTER 7
K. S. Anil Kumar
ICAR- National Bureau of Soil Survey and Land Use Planning, Hebbal, Bangalore, 560024
M. Lalitha
ICAR- National Bureau of Soil Survey and Land Use Planning, Hebbal, Bangalore, 560024
K. M. Nair
ICAR- National Bureau of Soil Survey and Land Use Planning, Hebbal, Bangalore, 560024
Rajendra Hegde
ICAR- National Bureau of Soil Survey and Land Use Planning, Hebbal, Bangalore, 560024

B. S. Dwivedi
ICAR- National Bureau of Soil Survey and Land Use Planning, Amravati Road, Nagpur

CHAPTER 8

Pramod Jha
ICAR- Indian Institute of Soil Science, Nabi Bagh, Berasia Road, Bhopal, MP, India, 462038
K. M. Hati
ICAR- Indian Institute of Soil Science, Nabi Bagh, Berasia Road, Bhopal, MP, India, 462038
N. K. Lenka
ICAR- Indian Institute of Soil Science, Nabi Bagh, Berasia Road, Bhopal, MP, India, 462038
B. L. Lakaria
ICAR- Indian Institute of Soil Science, Nabi Bagh, Berasia Road, Bhopal, MP, India, 462038
M. V. Coumar
ICAR- Indian Institute of Soil Science, Nabi Bagh, Berasia Road, Bhopal, MP, India, 462038
B. P. Meena
ICAR- Indian Institute of Soil Science, Nabi Bagh, Berasia Road, Bhopal, MP, India, 462038
J. K. Thakur
ICAR- Indian Institute of Soil Science, Nabi Bagh, Berasia Road, Bhopal, MP, India, 462038
A. K. Biswas
ICAR- Indian Institute of Soil Science, Nabi Bagh, Berasia Road, Bhopal, MP, India, 462038
A.K. Patra
ICAR- Indian Institute of Soil Science, Nabi Bagh, Berasia Road, Bhopal, MP, India, 462038

CHAPTER 9

Anup Das
ICAR Research Complex for Eastern Region, Patna, 800014, Bihar, India
Christi B. K. Sangma
ICAR Research Complex for NEH Region, Umiam, Umroi road, Meghalaya, 793 103
Mahasweta Chakraborty
ICAR Research Complex for NEH Region, Umiam, Umroi road, Meghalaya, 793103, India
Saurav Saha
ICAR Research Complex for Northeastern Hill Region, Sikkim center, Tadong, 737102, Sikkim, India
Amit Kumar
CAR Research Complex for Northeastern Hill Region, Sikkim center, Tadong, 737102, Sikkim, India
Bappa Das
ICAR- Central Coastal Agricultural Research Institute, Old Goa, 403402, Goa, India
Jaynata Layek
ICAR Research Complex for NEH Region, Umiam, Umroi road, Meghalaya, 793 103
Subhash Babu
ICAR Indian Agricultural Research Institute, Pusa, New Delhi, India
Prashant Pandey
ICAR Research Complex for Northeastern Hill Region, Sikkim center, Tadong, 737102, Sikkim, India

CHAPTER 10

Avijit Ghosh
ICAR- Indian Grassland and Fodder Research Institute, Jhansi, India
Pramod Jha
ICAR- Indian Institute of Soil Science, Nabibagh, Berasia Road, Bhopal, 462038, India
Sukanya Misra
RLBCAU, Jhansi, India
Ranjan Bhattacharyya
ICAR- Indian Agricultural Research Institute, Pusa, New Delhi, 110012

CHAPTER 11

Ashim Datta
ICAR-Central Soil Salinity Research Institute, Kamal, Haryana, India
Nirmalendu Basak
ICAR-Central Soil Salinity Research Institute, Kamal, Haryana, India
Bholanath Saha
Dr. Kalam Agricultural College, Bihar Agricultural University, Kishanganj, Bihar, India
Piu Basak
Deportment of Agricultural Chemistry & Soil Science, Bidhan Chandra Krishi Viswavidyalaya, Mahanpur, West Bengal, India
Arvind Kumar Rai
ICAR-Central Soil Salinity Research Institute, Kamal, Haryana, India
Rajender Kumar Yadav
ICAR-Central Soil Salinity Research Institute, Kamal, Haryana, India

CHAPTER 12

Ruma Das
ICAR- National Bureau of Soil Survey and Land Use Planning, RC, Kolkata, West Bengal-700091, India
Samrat Ghosh
Climate and Sustainability, Emergent Ventures India Pvt. Ltd., Gurgaon, Haryana, 122011, India
Khushboo Rani
ICAR- Indian Institute of Soil Science, Nabibagh, Bhopal, 462038, India
Amit Phonglosa
Directorate of Extension Education, Odisha University of Agriculture & Technology, Bhubaneswar, Odisha, 751003, India
B. N. Ghosh
ICAR- National Bureau of Soil Survey and Land Use Planning, RC, Kolkata, West Bengal-700091, India

CHAPTER 13

Amit K. Dash
ICAR-Indian Institute of Seed Science, Mau, 275103, India
ICAR-Indian Agricultural Research Institute, New Delhi, 110012, India
Abir Dey
ICAR-Indian Agricultural Research Institute, New Delhi, 110012, India

Ranjan Bhattacharyya
ICAR-Indian Agricultural Research Institute,
New Delhi, 110012, India
T.K. Das
ICAR-Indian Agricultural Research Institute,
New Delhi, 110012, India
Arti Bhatia
ICAR-Indian Agricultural Research Institute,
New Delhi, 110012, India
Shrila Das
ICAR-Indian Agricultural Research Institute,
New Delhi, 110012, India

CHAPTER 14
Dibyendu Sarkar
Bidhan Chandra Krishi Viswavidyalaya, Mohanpur,
741252, West Bengal
Siddhartha Mukherjee
Ramakrishna Mission Vivekananda Educational and
Research Institute, Ranchi, 834008, Jharkhand
G. Pratibha
ICAR- Central Research Institute for Dryland
Agriculture, Hyderabad, 500059, Telangana
Sumanta Kundu
ICAR- Central Research Institute for Dryland
Agriculture, Hyderabad, 500059, Telangana

CHAPTER 15
Kuntal M. Hati
ICAR- National Bureau of Soil Survey and Land Use
Planning, Regional Centre Kolkata, Salt Lake,
Kolkata, 700091, India
Ranjan Bhattacharyya
ICAR- Indian Agricultural Research Institute, Pusa,
New Delhi, 110012, India
Pramod Jha
ICAR- Indian Institute of Soil Science, Nabibagh,
Berasia Road, Bhopal, 462038, India
J. Somasundaram
ICAR- Indian Institute of Soil and Water
Conservation, Research Centre, Fernhill,
Udhagamandalam, Tamil Nadu, 643004, India
B. N. Ghosh
ICAR- National Bureau of Soil Survey and Land Use
Planning, Regional Centre Kolkata, Salt Lake,
Kolkata, 700091, India
Avijit Ghosh
ICAR- Indian Grassland and Fodder Research
Institute, Jhansi, 284003, India
Douglas L. Karlen
DL Karlen Consulting LLC, St. Paul, MN, 55102

Soil Health Series: Volume 4 Soil Health and Sustainability in India

Edited by Ranjan Bhattacharyya

SSSA
Soil Science Society of America

WILEY

Editorial Correspondence:
Soil Science Society of America, Inc.
5585 Guilford Road, Madison, WI 53711-58011, USA
soils.org

Registered Offices:
John Wiley & Sons, Inc., 111 River Street, Hoboken, NJ 07030, USA

For details of our global editorial offices, customer services, and more information about Wiley products, visit us at www.wiley.com.

Wiley also publishes its books in a variety of electronic formats and by print-on-demand. Some content that appears in standard print versions of this book may not be available in other formats.

Library of Congress Cataloging-in-Publication Data Applied for:
Paperback: 9780891187394

Cover Design: Wiley
Cover Image: © Getty Images for Unsplash, © Equalstock/Unsplash, © Arjun MJ/ Unsplash

Set in 9.5/12.5pt STIXTwoText by Straive, Pondicherry, India

Contents

Soil Health Insights from India

Douglas L. Karlen, Soil Health Series Editor

Soil Health and Sustainable Agriculture in India is the fourth volume in the Soil Health Series being published by the Soil Science Society of America (SSSA) and Wiley. Volumes one and two provide the background and general methods for assessing soil biological, chemical, and physical properties and processes. Building upon that soil health foundation, a series of volumes are being added to show how the concept is being applied to protect and improve soil resources around the globe. Volume four is edited by Dr. Ranjan Bhattacharyya, an excellent scientist from the Division of Environment Science, ICAR-Indian Agricultural Research Institute. The size and regional agroecological diversity of India make this volume an important contribution to the Series.

The subcontinent of India has heterogeneous landforms and a variety of climatic conditions that create subtle but very important differences in critical soil health indicators and management practices needed to sustain the soil resources. Lofty mountains, river deltas, high altitude forests, and peninsular plateaus have a variety of geological formations, endowed with temperature regimes ranging from arctic cold to equatorial heat, and rainfall extremes from aridity (<10 cm) to sites receiving the world's maximum average annual rainfall (1120 cm).

Soil health and sustainable agricultural practices in this wide variety of agroecological regions are discussed by highly respected authors in fifteen chapters. Beginning with a broad overview of India's agricultural ecoregions and soil health challenges, specific soil health properties, processes, and indicators are discussed for Himalayan, Arid, Semi-Arid, Sub-Humid, Coastal, and Southern Peninsular Ecoregions. Indian Vertisols are also examined in depth. Soil management strategies to maintain or improve soil health are then examined for those ecoregions, saline, acidic, and waterlogged soils. Finally, conservation agriculture strategies for enhancing soil health and mitigating greenhouse gas (GHG) emissions and crop residue management strategies for dryland and irrigated areas are explored. A series of recommendations for long-term soil health enhancement throughout the subcontinent are presented in Chapter 15. The editor, authors, and I hope you find this volume useful in helping guide sustainable agricultural management practices that enhance productivity while protecting the fragile soil, water, and air resources within India or wherever you may be located in the world.

Preface

Land degradation in India is a pervasive problem and a major soil health challenge. Despite years of study and substantial investment in remediation and prevention, soil erosion continues to be a major environmental problem. Changing climate and weather patterns are further exacerbating the situation. India's soil degradation affects approximately 120 million hectares of land. Issues include water erosion, soil acidification, flooding or waterlogged soils, wind erosion, salinity, and other combined factors. This is extremely serious, as India supports 18% of the world's population. Major causes of soil degradation are both natural and human induced. Natural causes mainly include droughts, floods, and tornadoes. Soil degradation caused by human activities arises due to deforestation, suboptimal agricultural practices, inadequate forest management, surface mining, urban expansion, and accelerated development. Poor farming practices involve excessive tillage, unbalanced use of inorganic fertilizers, improper irrigation techniques, insufficient organic inputs, and poor crop cycle planning.

Managing agricultural land to achieve soil health can help optimize agricultural inputs to soil (including fertilizer, water, energy, etc.) thereby reducing costs while maintaining or improving yields. Improving soil health improves adaptation to the effects of climate change, and will also contribute to carbon neutrality objectives.

This book examines the challenges and opportunities related to restoring and enhancing soil health in degraded farmland in India. It includes information on various soil restoration practices such as tree-based ecological restoration, conservation agriculture, integrated nutrient management, soil management in dryland agriculture, grassland management, and organic farming systems. The book discusses the importance of new fertilizer formulations for better input efficiency, as well as quick soil assessment and monitoring. You will find recommendations for sustainable soil management to support India's growing population. These include ensuring food safety, protecting biodiversity, adapting to climate change, managing waste, addressing soil pollution, and comprehensive soil management.

Ranjan Bhattacharyya

1

Agricultural Ecoregions and Soil Health Challenges Within the Indian Sub-Continent

Shrila Das, Dibyendu Mukhopadhyay, and Ranjan Bhattacharyya

Chapter Overview

Every agro-ecological zone in India faces various soil health challenges, including deforestation, overgrazing, changes in land use (such as converting forests to farms), cultivation of steep slopes, and degradation of marginal lands due to human activity. Agriculture-specific activities, such as overexploitation of vegetation, removal of crop residues or in situ burning, and minimal or no addition of organic manures coupled with intensive cultivation, are leading to the depletion of soil organic carbon (SOC) and degradation of soil quality. Major challenges to the conservation of soil resources and the maintenance of soil health include water and wind erosion, salinity, decline in SOC, nutrient imbalance, and soil pollution and contamination. This chapter aims to discuss these soil health challenges encountered in different agro-ecological zones.

Introduction

India is known for its diverse landforms, ranging from lofty mountains and riverine deltas to high-altitude forests and peninsular plateaus, and the variety of climatic conditions associated with them. The range of geological formations are accompanied by temperatures varying from arctic cold to equatorial heat and rainfall varying from extreme aridity (less than 10 cm annually) to the world's highest rainfall (1,120 cm per annum). To develop long-term land use strategies, India has been divided into 20 agro-ecological regions (AERs) with 60 agro-ecological subregions (AESRs) based on diverse soil, bioclimatic, and physiographic conditions. AER s of India with their dominant soil characteristics are described in Table 1.1.

Soil Health Series: Volume 4 Soil Health and Sustainability in India, First Edition.
Edited by Ranjan Bhattacharyya.

Table 1.1 Agro-ecological Regions of India with their Dominant Soil Characteristics.

	AER	% of total geographic area	Soil characteristics	Moisture and temperature regimes
1	**Western Himalayas, cold arid ecoregion**	4.7	Desert and calcareous soils (Cryorthents and Cryorthids)	Aridic soil moisture and cryic soil temperature regimes
2	**Western Plain, Kachchh and part of Kathiawar Peninsula, hot arid ecoregion**	9.78	Desert and saline soils (Torripsamments, Camborthids, Calciorthids, Salorthids, and Natrargids)	Aridic soil moisture and hyperthermic soil temperature regimes
3	**Deccan plateau, hot arid ecoregion**	1.5	Red and black soils (Rhodustalfs, Haplustalfs, Ustropepts, and Pellusterts)	Aridic-ustic soil moisture and isohyperthermic soil temperature regimes
4	**Northern Plain (and Central Highlands) including Aravallis, hot semi-arid ecoregion**	9.8	Alluvium-derived soils (Ustochrepts, Natrustalfs, Ustipsamments, and Ustifluvents)	Typic-ustic soil moisture with hyperthermic soil temperature regime
5	**Central (Malwa) Highlands, Gujarat plains, and Kathiawar Peninsula ecoregion**	5.4	Medium and deep black soils (Ustochrepts and Chromusterts)	Typic-ustic soil moisture with hyperthermic and isohyperthermic soil temperature regime.
6	**Deccan Plateau, hot semi-arid ecoregion**	9.5	Shallow and medium black soils (Ustorthents, Ustropepts, and Chromusterts/Pellusterts)	Ustic soil moisture with (iso) hyperthermic soil temperature regime
7	**Deccan Plateau (Telangana) and Eastern Ghats, hot semi-arid ecoregion**	9.5	Shallow and medium black soils (Ustorthents, Ustropepts, and Chromusterts/Pellusterts)	ustic soil moisture with (iso) hyperthermic soil temperature regime

8	**Eastern Ghats and Tamil Nadu Uplands and Deccan (Karnataka) Plateau, hot semi-arid ecoregion**	5.2	Red and black soils (Ustorthents, Rhodustalfs, Ustropepts, and Chromusters)	Moisture soil regime is ustic and temperature regime is the isohyperthermic
9	**Northern Plain, hot subhumid (dry) ecoregion**	3.7	Alluvium-derived soils (Ustochrepts, Haplustalfs, Eutrochrepts, and Ustifluvents)	Ustic soil moisture with hyperthermic soil temperature regime
10	**Central Highlands (Malwa and Bundelkhand), hot subhumid (dry) ecoregion**	5.8	Red and black soils (Ustorthents, Ustochrepts, Haplustalfs, and Chromusterts)	Ustic soil moisture with hyperthermic soil temperature regime
11	**Chattisgarh/Mahanadi Basin agro-ecoregion**	4.3	Red and yellow soils (Ustorthents, Ustochrepts, Haplustalfs, Rhodulstalfs, and Plinthustalfs)	Ustic soil moisture with hyperthermic soil temperature regime
12	**Eastern Plateau (Chhotanagpur) and Eastern Ghats, hot subhumid ecoregion**	8.2	Red and lateritic soils (Ustochrepts, Haplustalfs, Plinthustalfs, Paleustalfs, Haplustults, and Rhodustalfs)	Ustic soil moisture with hyperthermic soil temperature regime
13	**Eastern Plain, hot subhumid (moist) ecoregion**	3.4	Alluvium-derived soils (Ustifluvents, Ustochrepts, Eutrochrepts, Aquic Hapludalfs, and Haplustalfs)	Udic and ustic soil moisture regime
14	**Western Himalayas, warm subhumid (to humid with inclusion of perhumid) ecoregion**	6.3	Brown forest and podzolic soils (Eutrochrepts, Hapludalfs, Haplustalfs, Hapludolls, Haplohumults, and Haplaquepts)	Mesic, thermic, and cryic-mesic soil temperature regime
15	**Assam and Bengal Plain, hot subhumid to humid (inclusion of perhumid) ecoregion**	3.7	Alluvium-derived soils (Haplaquepts, Haplaqualfs, Dystrochrepts, Eutrochrepts, Fluvaquents, Hapludalfs)	Udic-ustic (to Udic) and Hyperthermic

(Continued)

Table 1.1 (Continued)

	AER	% of total geographic area	Soil characteristics	Moisture and temperature regimes
16	**Eastern Himalayas, warm perhumid ecoregion**	2.9	Brown and red hill soils (Haplumbrepts, Paleudults, Hapludolls, and Argiudolls)	Udic soil moisture regime and mesic, thermic, and hyperthermic soil temperature regime
17	**Northeastern Hills (Purvachal), warm perhumid ecoregion**	3.3	Red and lateritic soils (Hapludults, Paleudults, Dystrochrepts, Hapludalfs, and Rhodustalfs)	Udic soil moisture with hyperthermic, thermic soil temperature regime
18	**Eastern Coastal Plain, hot subhumid to semiarid ecoregion**	2.6	Coastal alluvium-derived soils (Haplaquents, Halaquepts, Ustifluvents, Pellusterts, and Ustropepts)	Udic and ustic soil moisture regime and mesic, thermic soil temperature regime
19	**Western Ghats and Coastal Plain, hot humid-perhumid ecoregion**	3.6	Red, lateritic, and alluvium-derived soils (Dystropepts, Eutropepts, Hapludults, and Haplaquepts with localized Haplorthorx)	Udic and ustic soil moisture regime and isohyperthermic temperature regime
20	**Islands of Andaman-Nicobar and Lakshadweep, hot humid to perhumid island ecoregion**	0.3	Red loamy soils including marine alluviumderived soils along the coast. They qualify for the Great Groups of Hapludalfs, Dystropepts, Eutropepts and Sulfaquents (along the coast).	Udic soil moisture and isohyperthermic soil temperature regime

1) **Western Himalayas, cold arid ecoregion**
   - Covers 4.7% of the total geographical area (TGA) of the country
   - Desert and calcareous soils (Cryorthents and Cryorthids)
   - Represents aridic soil moisture and cryic soil temperature regimes
2) **Western Plain, Kachchh and part of Kathiawar Peninsula, hot arid ecoregion**
   - Covers 9.78% of the TGA of the country
   - Desert and saline soils (Torripsamments, Camborthids, Calciorthids, Salorthids, and Natrargids)
   - Represents aridic soil moisture and hyperthermic soil temperature regimes
3) **Deccan plateau, hot arid ecoregion**
   - Covers 1.5% of the TGA of India
   - Red and black soils (Rhodustalfs, Haplustalfs, Ustropepts, and Pellusterts)
   - Represents the aridic-ustic soil moisture and isohyperthermic soil temperature regimes
4) **Northern Plain (and Central Highlands) including Aravallis, hot semi-arid ecoregion**
   - Covers 9.8% of the TGA of the country
   - Alluvium-derived soils (Ustochrepts, Natrustalfs, Ustipsamments, and Ustifluvents)
   - Represents typic-ustic soil moisture with hyperthermic soil temperature regime
5) **Central (Malwa) Highlands, Gujarat plains, and Kathiawar Peninsula ecoregion**
   - Covers 5.4% of the TGA of the country
   - Medium and deep black soils (Ustochrepts and Chromusterts)
   - Represents typic-ustic soil moisture with hyperthermic and isohyperthermic soil temperature regime.
6) **Deccan Plateau, hot semi-arid ecoregion**
   - Covers 9.5% of the TGA of the country
   - Shallow and medium black soils (Ustorthents, Ustropepts, and Chromusterts/Pellusterts)
   - Represents ustic soil moisture with (iso) hyperthermic soil temperature regime
7) **Deccan Plateau (Telangana) and Eastern Ghats, hot semi-arid ecoregion**
   - Covers 9.5% of the TGA of the country
   - Shallow and medium black soils (Ustorthents, Ustropepts, and Chromusterts/Pellusterts)
   - Represents ustic soil moisture with (iso) hyperthermic soil temperature regime

8) **Eastern Ghats and Tamil Nadu Uplands and Deccan (Karnataka) Plateau, hot semi-arid ecoregion**
   - Covers 5.2% of the TGA of the country
   - Red and black soils (Ustorthents, Rhodustalfs, Ustropepts, and Chromusters)
9) **Northern Plain, hot subhumid (dry) ecoregion**
   - Covers 3.7% of the TGA of the country
   - Alluvium-derived soils (Ustochrepts, Haplustalfs, Eutrochrepts, and Ustifluvents)
   - Represents ustic soil moisture with hyperthermic soil temperature regime
10) **Central Highlands (Malwa and Bundelkhand), hot subhumid (dry) ecoregion**
   - Covers 5.8% of the TGA of the country
   - Red and black soils (Ustorthents, Ustochrepts, Haplustalfs, and Chromusterts)
   - Represents ustic soil moisture with hyperthermic soil temperature regime
11) **Chattisgarh/Mahanadi Basin agro-ecoregion**
   - Covers 4.3% of the TGA of the country
   - Red and yellow soils (Ustorthents, Ustochrepts, Haplustalfs, Rhodulstalfs, and Plinthustalfs)
12) **Eastern Plateau (Chhotanagpur) and Eastern Ghats, hot subhumid ecoregion**
   - Covers 8.2% of the TGA of the country
   - Red and lateritic soils (Ustochrepts, Haplustalfs, Plinthustalfs, Paleustalfs, Haplustults, and Rhodustalfs)
   - Represents ustic soil moisture with hyperthermic soil temperature regime
13) **Eastern Plain, hot subhumid (moist) ecoregion**
   - Covers 3.4% of the TGA of the country
   - Alluvium-derived soils (Ustifluvents, Ustochrepts, Eutrochrepts, Aquic Hapludalfs, and Haplustalfs)
   - Represents udic and ustic soil moisture regime
14) **Western Himalayas, warm subhumid (to humid with inclusion of perhumid) ecoregion**
   - Covers 6.3% of the TGA of the country
   - Brown forest and podzolic soils (Eutrochrepts, Hapludalfs, Haplustalfs, Hapludolls, Haplohumults, and Haplaquepts)
   - Represents mesic, thermic, and cryic-mesic soil temperature regime
15) **Assam and Bengal Plain, hot subhumid to humid (inclusion of perhumid) ecoregion**
   - Covers 3.7% of the TGA of the country
   - Alluvium-derived soils (Haplaquepts, Haplaqualfs, Dystrochrepts, Eutrochrepts, Fluvaquents, Hapludalfs)

16) **Eastern Himalayas, warm perhumid ecoregion**
    - Covers 2.9% of the TGA of the country
    - Brown and red hill soils (Haplumbrepts, Paleudults, Hapludolls, and Argiudolls)
    - Represents udic soil moisture regime
17) **Northeastern Hills (Purvachal), warm perhumid ecoregion**
    - Covers 3.3% of the TGA of the country
    - Red and lateritic soils (Hapludults, Paleudults, Dystrochrepts, Hapludalfs, and Rhodustalfs)
    - Represents udic soil moisture with hyperthermic, thermic soil temperature regime
18) **Eastern Coastal Plain, hot subhumid to semiarid ecoregion**
    - Covers 2.6% of the TGA of the country
    - Coastal alluvium-derived soils (Haplaquents, Halaquepts, Ustifluvents, Pellusterts, and Ustropepts)
19) **Western Ghats and Coastal Plain, hot humid-perhumid ecoregion**
    - Covers 3.6% of the TGA of the country
    - Red, lateritic, and alluvium-derived soils (Dystropepts, Eutropepts, Hapludults, and Haplaquepts with localized Haplorthorx)
20) **Islands of Andaman-Nicobar and Lakshadweep, hot humid to perhumid island ecoregion**
    - Covers 0.3% of the TGA of the country.
    - Medium to very deep, Red loamy soils including marine alluvium derived soils along the coast, hot perhumid (Andaman and Nicobar Islands) to highly calcareous and sandy in nature, hot humid (Lakshadweep Group of Islands).

Soil Health Challenges in India

Deforestation and Removal of Natural Vegetation, Overgrazing, Change in Land Use Plan

Soil degradation in India affects an estimated 120 million hectares (Mha) of land, encompassing issues such as water erosion, acidification, flooding, wind erosion, salinity, and various other factors. This situation is particularly critical given that India is home to 18% of the global human population and 15% of the world's livestock population, yet it only possesses 2.4% of the world's land area (Bhattacharyya et al., 2015). Deforestation, removal of natural vegetation, overgrazing, and changes in land-use plans are some of the human-induced forms of soil degradation. The UN Food and Agricultural Organization (FAO) defines deforestation as the conversion of forested land to another land use or a long-term reduction in tree canopy cover to below 10%.

A new study by Utility Bidder, a UK-based comparison site for energy and utility costs, ranks the countries contributing the most to global deforestation rates in 2023. India ranks second, having destroyed an average of 668,400 ha of forestry between 2015 and 2020. Deforestation and overgrazing have led to land degradation in eight Indian states, where over 20% of land area is now wasteland (NRSA, 2019). Vegetation loss is driven by excessive cutting beyond the silvicultural permissible limit, unsustainable fuelwood and fodder extraction, agricultural encroachment into forest lands, forest fires, and overgrazing, all of which contribute to land degradation. Additionally, the clearing of forest areas due to shifting cultivation results in the loss of primary vegetation and the formation of secondary vegetation (Ray et al., 2021). Moreover, owing to shortening of shifting cultivation cycles, the secondary vegetation does not have adequate time to regenerate. The repeated short-cycle use of land eventually converts the area under degraded fallow. Therefore, shifting cultivation is considered the single most important factor causing forest degradation in the northeastern Himalayan region. After a short jhum cycle (shifting cultivation), weedy species such as *Eupatorium odoratum*, *Imperata cylindrica*, *Mikania micrantha*, and *Saccharum spontaneum* colonize the affected areas (Saxena & Ramakrishnan, 1983). A cattle population of 467 million grazes on 11 Mha of pastures, averaging 42 head per hectare, which far exceeds the sustainable threshold of five animals per hectare (Sahay, 2015). This high livestock density in arid regions leads to overgrazing, which decreases infiltration and accelerates runoff and soil erosion. As a result, soil loss due to overgrazing is 5–41 times greater than normal at the mesoscale and 3–18 times greater at the macroscale (Sharma, 1997).

Change in a land use system is a process, by which human activities transform the natural landscape or existing land use to another land use in which the functional roles of land for economic activities are emphasized. A report of meta-analysis (more than 1,786 paired datasets) on land uses from Indian (1990–2019) indicated that the maximum loss of SOC was under grasslands at 36.1%, followed by forest lands, plantation lands (35.5%), cultivated lands (31.1%), barren lands (27.3%), and horticulture lands (11.5%; Padbhushan et al., 2022). Land use changes from anthropogenic activities also modify the land surface temperature and surface energy and finally cause changes within microclimates at the regional level and at local level. Kaliraj et al. (2017) reported that about 18% of the fallow lands and barren lands were converted into settlements and about 5.15% and 5.86% of the land uses were converted into barren and fallow lands, respectively, in coastal plains of southern Tamil Nadu, India, between 2000 and 2010. Therefore, any change in land use systems (cultivation or fallow land) will have an impact on soil health and its resilience. The transformation of tea garden/forest land area into the field crop production and vice versa definitely have an effect on the soil biota and the surrounding ecosystem. The changes in mineralizable carbon (Figure 1.1)

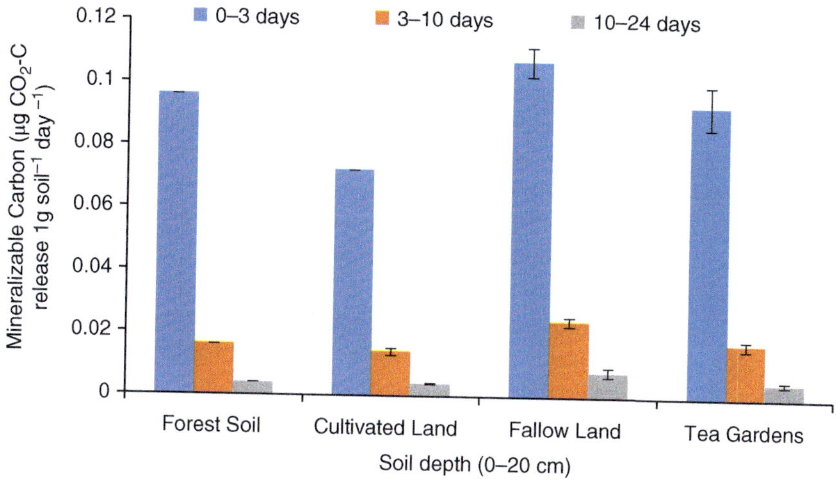

Figure 1.1 Mineralizable carbon (µg of CO_2-C released per gram of soil per day) fractions in soils under different land use systems. Error bars indicate the standard deviation ($P = 0.05$).

in soil under different land use patterns show the need for effective technological intervention to build up carbon sequestration in soil as well as to improve the microbial biomass carbon (Pal et al., 2020). Dharumarajan et al. (2017) found that erratic rainfall distribution, poor soil fertility, and ground water deterioration are the factors responsible for the area of fallow lands in Tamil Nadu expanding with increasing costs of cultivation. Moreover, there is increase in population growth (from 11.42% in 1991 to 14.20% in 2014) in the Indian coastal tracts, with rapid transformation of built-ups, settlements, and recreational uses directly causing degradation of the coasts (Kaliraj et al., 2017).

The changes in mineralizable carbon at different incubation periods (0–3, 3–10, and 10–24 days) at (0–20) cm soil depth are shown in Figure 1.1. Significant differences were observed in some of the ecosystems (tea gardens, fallow land) with the given period of incubation, while the changes were not significant in most of the cases at the initial period of incubation (Pal et al., 2020).

Cultivating Steep Slopes and Degrading Marginal Lands

According to land capability classification, steep land (with shallow depth, coarse texture with >30% slope) is classed as unsuitable for farming as it may cause water erosion. In addition, low moisture availability in arid climates aggravates drought. Cultivation on steep slopes leads to severe soil health challenges such as soil erosion up to 100 ton ha^{-1}, which can remove top soil in 20 years (based on the universal soil loss equation; Verma et al., 2018); rapid drying of the upper part of soil facing the sun, leading to sparse or nil vegetative cover during dry season; soil fertility impoverishment because of leaching loss of nutrients, which leads to soil acidity, particularly in permeable soils (Entisols, Inceptisols, Oxisols); and drought, owing to wind erosion being more intense on steep land. Marginal lands are defined as being severely limited by various soil physical, chemical, and biological properties (texture, drainage, fertility, slope or climate; Esch et al., 2018). Marginal lands are found in different part of India. Soil attributes that are indicators of marginal land (salt content, electrical conductivity, drainage, base saturation, salinity, sodicity, rooting depth, organic carbon, cation exchange capacity, etc.) should be addressed; otherwise, soil quality, as well as sustainability of the land, will be hampered (Csikós & Tóth, 2023). Important indicators corresponding to soil processes are given in Table 1.2.

Table 1.2 Soil Health Indicators Affecting the Soil Processes.

Soil health indicators	Soil processes
Physical	
Soil texture	All the physico-chemical processes
Aggregate stability of soil	Rate of infiltration, aeration, root density
Water holding capacity of soil	Characteristics of soil moisture retention
Surface and subsurface hardness	Seed germination, root growth, internal drainage
Chemical	
pH	Nutrient availability, toxicity
Available N, P, K	Availability of N-P-K from soil to plants
Other minor elements	Micronutrients, nutrient transformation
Biological	
Stock of organic matter in soil	Carbon storage, water and nutrient reserves
Carbon fractions	Supporting biological activities
Mineralizable nitrogen	Ability to supply nitrogen
Residue decomposition	Microbial population, nutrient mobilization to plants

Threats to Conservation of Soil Resources

Erosion by Water, Wind, and Cyclonic Storms

Climate change, particularly the changing trends in precipitation, have a severe impact on soil erosion, especially in hilly regions. The effect of increased precipitation and/or increased intensity may vary between different slope categories. For moderate slopes, there will be proportionately higher increase of soil erosion mainly because of increased erosive power and the availability of erodible material. On very steep slopes, the proportional increase in erosion may be less because of limiting erodible materials, such as hard rock outcrops. Another aspect of climate change is conversion of snowfall to rainfall. The frequency of extreme rain events (>100 mm intensity) has increased over central India from 1951 to 2015 (Goswami et al., 2006). Such extreme precipitation results in severe soil erosion, which affects the productive capacity of the soils (Mandal et al., 2010). The critical permissible limit of soil erosion is considered as 1000 t km^{-2} yr^{-1}. The soil loss map (Figure 1.2) shows that in the northeastern states (Nagaland, Meghalaya, Arunachal Pradesh, Assam, Chhattisgarh, and Jharkhand), erosion rates are above the critical permissible limits. Other states such as Uttar Pradesh, Uttarakhand, Madhya Pradesh, and Manipur also face erosion rates above the critical limit (NAAS, 2017). Therefore, site-specific best management practices should be identified to bring the prevailing erosion rates within the permissible limits in these areas.

In addition, wind erosion occurs over 14.28 M ha of the land, contributing to 15.66% of the total land degradation in India. Wind erosion occurs in the arid and semi-arid regions of India, especially in Rajasthan (hot desert), Ladakh (cold desert), and Haryana. Wind erosion also occurs in Tamil Nadu, Punjab, Andhra Pradesh, Gujarat, and Odisha to a lesser extent. It causes considerable loss of surface soils rich in nutrients equivalent to 4 g C kg^{-1} and 0.37 g N kg^{-1} of soil in Rajasthan (Santra et al., 2013). Therefore, it is necessary to consider effective management practices that can help to retain soil health and also support the crop/grassland productivity in the arid ecosystem.

Tropical cyclones (high winds, intense rainfall, terrestrial flooding, and storm surges) are natural disasters that causes devastating hazards in the coastal regions of world. Effects vary depending on the origins of the storms and the locations of landfall. About 5% of global tropical cyclones originate in the Bay of Bengal, but adjacent countries, including Bangladesh, India, and Myanmar, experience more than 75% of the global casualties (Alam & Dominey-Howes, 2014; Burman et al., 2013). The Bay of Bengal is more turbulent and the epicenter of frequent devastating cyclones, sea surges, and so forth. A huge amount of damage occurs when thousands of hectares of fertile agricultural land and adjoining areas become a wasteland owing to ingress of saline sea water. The sea water increases soil pH and soil salinity and stagnates for a long time, making the land unsuitable for cultivation (Debnath, 2013; Mitra et al., 2020). Analysis of cyclone impacts in coastal areas in India has been reported by many researchers (Thakur et al., 2018; Parida et al., 2018, Mandal et al., 2019). Mishra et al. (2022) reported that the cyclone Nisarga in 2020 caused the shoreline along the coastal zone of Maharashtra to erode by 56.3%.

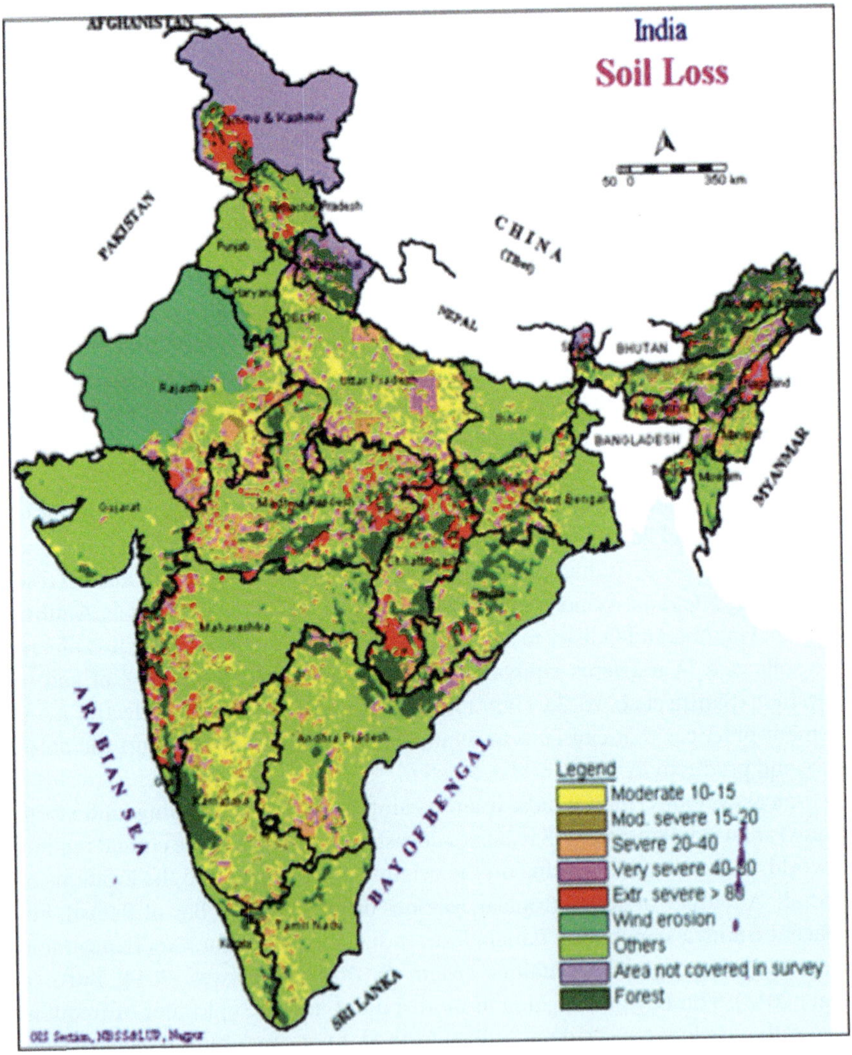

Figure 1.2 Average annual gross erosion rate in different states of India.

Development of Salt-Affected Soils

Salt-affected soils cause major constraints in sustainable food production system of the world. In India, approximately 6.727 million ha of land (2.956 million ha under saline and 3.771 million ha is sodic) has salt-affected soil (Arora et al., 2016; Arora & Sharma, 2017). This land is in the states of Gujarat (2.23 million ha), Uttar Pradesh (1.37 million ha), Maharashtra (0.61 million ha), West Bengal (0.44 million ha), and Rajasthan (0.38 million ha; Mandal et al., 2018) and in four major agriculturally important ecological regions (semi-arid Indo-Gangetic alluvial tract, arid and semi-arid tracts, peninsular regions, and coastal-alluvial region). In addition, approximately 0.3 Mha is characterized as acid sulfate soils (in Kerala and the Andaman and Nicobar group of islands), which have a pH of <4.

Table 1.3 shows the distribution of salt-affected soils in India.

Table 1.3 Distribution of Salt-Affected Soils in India.

State	Sodic soils	Saline soil	Coastal saline soils	Total
Gujarat	14.3	71.2	37.1	32.9
Uttar Pradesh	35.6	1.3	—	20.3
Maharashtra	11.2	10.4	0.6	9.0
West Bengal	—	—	35.4	6.5
Rajasthan	4.7	11.4	—	5.6
Tamil Nadu	9.4	—	1.1	5.5
Andhra Pradesh	5.2	—	6.2	4.1
Haryana	4.8	2.9	—	3.4
Bihar	2.8	2.8	—	2.3
Punjab	4.0	—	—	2.2
Karnataka	3.9	0.1	—	2.2
Orissa	—	—	11.8	2.2
Madhya Pradesh	3.7	—	—	2.1
Andaman & Nicobar Islands	—	—	6.2	1.1
Kerala	—	—	1.6	0.3
Jammu and Kashmir	0.5	—	—	0.3
Total[a]	100 (3.78)	100 (1.71)	100 (1.25)	100 (6.74)

[a] Values in parentheses indicate total area in millions of hectares. *Source:* Adapted from Mandal et al. (2018).

Another major degradational problem in semi-arid tropical areas under rain-fed agriculture is regressive pedogenic processes that lead to the formation of pedogenic $CaCO_3$ and concomitant development of subsoil sodicity. These problems affect major soil types in India in the Indo-Gangetic plains, arid and semi-arid regions of western and central India, and the peninsular region in the southern India (Pal, 2019). The largest area under sodic soils (35.6%) occurs in the state of Gujarat. The fundamental mechanism in the precipitation of pedogenic $CaCO_3$ in the semi-arid tropical environment is water loss through evapotranspiration. In clay-rich soils of semi-arid central India, pedogenic $CaCO_3$ formation results in the simultaneous development of sodicity throughout the soil depth. The accumulation of soil inorganic carbon in terms of pedogenic $CaCO_3$ production in soils has been seen as a curse since it reduces soil productivity (Srivastava et al., 2002; Bhattacharyya et al., 2004; Pal, 2019). The creation of pedogenic $CaCO_3$ in an arid climate raises the pH and the relative quantity of Na^+ ions in both soil exchange sites and solution, causing the fine clay particles to disperse. Pedogenic $CaCO_3$ development generates a chemical condition that allows clay particles to deflocculate and travel downward. Therefore, due to these two simultaneously occurring pedogenic processes, the relative amount of sodium and the soil pH increase with depth.

Soil salinity commonly occurs in arid environments, particularly in desert soils where intense evaporation leads to high salinity levels. This is especially true in inland river basins with relatively high water tables. Despite salinity effects not always being immediately apparent, crop yields can still suffer. Desert soils, primarily composed of sandy soil (90%–95%), are found in low-rainfall regions and are characterized by low nitrogen and organic matter content, but high levels of calcium carbonate and phosphate, making them infertile (Moharana et al., 2021). The calcium content in the lower soil layers is 10 times higher than in the topsoil. In India, desert soils cover about 4% of the total area and are found in regions such as Rajasthan, parts of Punjab and Haryana between the Indus and the Aravallis, the Rann of Kuchchh in Gujarat, and coastal areas of Orissa, Tamil Nadu, and Kerala (https://en.wikipedia.org/wiki/Desert). These soils are prone to wind erosion and support low population densities.

The largest area of saline soils (71.2%) is found in the state of Uttar Pradesh. Over 72% of coastal saline soils are located in the states of Gujarat and West Bengal. Increased soil salinization is due to (a) increased demand for water, (b) increased evaporation (c) increased salt accumulation in crops via increased transpiration, and (d) rising sea water level causing increased saline water intrusion. Moharana et al. (2019) have reported the development of secondary salinity and salt migration in the irrigated areas of upland, medium land, lowland, and wetlands of Indira Gandhi Nahar Pariyojna (IGNP) that have also caused changes in the land use from dry lands crops to irrigated crops.

Waterlogged Soils

Excess soil water leading to waterlogging is one of the major causes of land degradation in India, rendering agricultural land sub productive. Factors contributing to waterlogging are faulty water management in the agricultural sector, disparity between the inflow and outflow of irrigated water, and improper drainage systems, among others. The expansion of irrigated areas over the years has exacerbated the problem, with inadequate drainage having led to increased waterlogging in agricultural lands. Additionally, excess monsoon rains cause seasonal waterlogging in low-lying areas. Soil is considered waterlogged when surplus water stagnates due to poor drainage or when a shallow water table rises, saturating soil pores in the crop root zone. This saturation restricts normal air circulation, decreases oxygen levels, and increases carbon dioxide levels, altering the physical, chemical, and microbiological properties of the soil. Approximately 11.6 Mha, or 8.3% of India's net sown area, are waterlogged (Planning Commission, 2011), with over 20% of this area in the eastern region, where surface waterlogging is a major issue. Eastern India, particularly the 15 eastern districts of Uttar Pradesh, all districts of Bihar and West Bengal, and parts of Odisha, constitutes a significant portion of the Ganga Meghna-Brahmaputra basin, home to 500 million of the world's poorest people (Saxena NC, Planning Commission). State-wise break up of the areas affected by water is presented in Table 1.3.

The introduction of canal irrigation through the Indo-Gangetic Plains has brought about significant improvements, including enhanced micro-climate conditions, changes in land use and cropping patterns, and better soil moisture conditions. However, prolonged irrigation using surface flow methods has also led to several issues such as rising water table levels, waterlogging, marshland formation, increased soil salinity, and reduced biodiversity. The expansion of canal irrigation has been linked to widespread waterlogging and salinity problems in various areas (Pandey, 2013). Effective regulation of waterlogging can be achieved through methods such as raised bed farming, mechanical drainage (both surface and sub-surface), bio-drainage, and controlled irrigation measures. Examples of effective bio-drainage plants include *Syzygium cumini*, *Pongamia pinnata*, *Terminalia arjuna*, *Casuarina glauca*, and *Eucalyptus tereticornis*. Additionally, using tolerant or resistant crop varieties and proper nutrient management can enhance the survival of agricultural crops in waterlogged soils.

Acid Soils and Acid Sulfate Soils

Globally, 3,950 Mha of arable land are affected by soil acidity (Ultisols or Oxisols with a pH of 5.5 or lower), including approximately 1,043 Mha of farmland in Asia. This includes 5 Mha in the Near East, 212 Mha in the Far East, 314 Mha in Southeast Asia and the Pacific, and 512 Mha in North and Central Asia (Sumner & Noble, 2003; Osman, 2018). In India, about 28% of the TGA is affected by acidity to varying degrees (Maji et al., 2012). Highly acidic soils in India are primarily found in the Himalayan ecosystem, the red and lateritic regions of the southern and eastern plateau, and some pockets in the greater plains. Soils with pH values below 4.5 cover only 1.9% of India's TGA, amounting to approximately 6.24 Mha. Moderately acidic soils with pH values ranging from 4.5 to 5.5 cover 24.4146 Mha, or 7.4% of the TGA. As noted by Mandal et al., (2023), acid soil occurrence is higher in the Himalayan ecosystem owing to rainfall conditions. The Eastern Himalayan ecosystem, which includes the Northeastern region and parts of sub-Himalayan West Bengal and Jharkhand, has more extensive acid soil areas compared with the Western Himalayan region, covering Uttarakhand, Himachal Pradesh, and Jammu and Kashmir. Peninsular India and some coastal areas, particularly in Tamil Nadu, Kerala, Karnataka, and Goa, also have acid soils. Analysis of the acid soil map of India shows that the total area of highly acidic soils (strongly and moderately acidic) amounts to 30.6572 Mha, which is 9.3% of the TGA. Arunachal Pradesh has the largest extent of strongly acidic soils, followed by Manipur, Sikkim, Tamil Nadu, Kerala, Chhattisgarh, Nagaland, Tripura, and Assam. Chhattisgarh has the largest area of moderately acidic soils (5.93 Mha), followed by Kerala (2.78 Mha), Assam (2.33 Mha), Arunachal Pradesh (1.74 Mha), Nagaland (1.48Mha), Manipur (1.43 Mha), Mizoram (1.26 Mha), and Meghalaya (1.18 Mha). Other states have less than 1,100 hectares of moderately acidic soils. Acidic soils are not found in Haryana, Punjab, Rajasthan, Gujarat, and Delhi.

If left unmanaged, soil acidity can negatively affect soil quality and plant growth. Acid soils are generally low in available phosphorus and exchangeable bases such as potassium and calcium. They also have low base saturation, low cation exchange capacity, and high levels of potentially toxic elements such as aluminum, iron, and manganese. Consequently, the productivity of acid soils is very low ($<1\,t\,ha^{-1}$) due to deficiencies in phosphorus, calcium, magnesium, molybdenum, and boron and the toxicity of aluminum and iron. The availability of boron decreases in acid soil due to its adsorption on sesquioxide, iron, and aluminum hydroxy compounds. Kaolinitic and illitic clay minerals dominate acid soils. Soil pH affects the bioavailability of plant nutrients and indirectly influences crop growth. Therefore, managing acid soil should focus on improving its production potential through amendments or adjustments in agricultural practices to achieve optimum yields under acidic conditions. Even within crops, variety performance can vary under acidic conditions. Acid soils can be managed by growing crops suitable for particular soil pH (acid-tolerant crops), ameliorating soils with inorganic amendments (lime, phosphogypsum) or organic amendments (farmyard manure, poultry manure, pig manure, green manure, biochar), and adopting good agronomic practices (crop scheduling, crop diversification, lime application techniques, seed priming, integrated nutrient management, growing multi-purpose tree species, agro-forestry systems, adding crop residues) to counteract soil acidity (Majumdar et al., 2022).

Additionally, the patchy occurrence of acid sulfate soils in the region has likely limited research attention and complicated the detection of potential and/or active acid sulfate soils (Lindgren et al., 2022). Acid sulfate soils, a group of soils with high acidity and other limitations, were called "Cat Clays" by Dutch farmers in the 17th century due to their yellowish mottles and crusts composed of jarosite and related sulfates. Similar clays in northern Germany were known as Maibolt (Prokopovich, 1988). According to Dent (1986), acid sulfate soils are defined as soils with a pH (1:2.5 in 0.01 M $CaCl_2$) below 4.0, containing jarosite mottles or a soluble SO_4-S content over 0.05% within 0.60 m of the soil depth. These soils are characterized by a low pH (<3.5) and the presence of sulfidic materials and/or a sulfuric horizon (Shamshuddin et al., 2014), which contain oxidizable sulfur compounds such as pyrite (FeS_2). Acid sulfate soils are associated with toxic levels of water-soluble iron, aluminum, and manganese. Major soil-related problems include poor physical properties, nutrient deficiencies and toxicity, soil salinity, variable acidity at different depths, and adverse impacts on ecosystems and biodiversity. The application of lime, rock phosphate, and single superphosphate has been beneficial in improving soil properties and crop yields in the acid sulfate soils of India (Burman et al., 2010; Maji & Bandyopadhyay, 1996). Additionally, adding locally available calcium-rich oyster shell powder along with green or leaf manures has proven to be an effective ameliorative measure (Das, 2022). Management measures for potential and active acid sulfate soils include a range of broad strategies aimed at mitigating their adverse effects.

Depletion of Organic Matter

Intensive agriculture, inadequate management practices, and improper soil conservation measures cause soil and environmental degradation and lead to a decline in soil organic matter (SOM), resulting in poor soil health. Intensive agriculture has increased from 19% to 38% of the cropped area under irrigation over the past 40 years, with incorrect extraction from limited ground water resources. In addition, poor management and maintenance of canal irrigation has led to waterlogging and salinization problems. Further, excessive use of nitrogen fertilizers has led to nitrate leaching into groundwater. The main contributors to greenhouse gas emissions from agricultural systems include unnecessary tillage for land preparation and planting, indiscriminate irrigation, and excessive fertilizer applications.

The burning of crop residues for cooking, heating, or disposal is a widespread issue in India, leading to not only environmental pollution but also loss of SOM. The Ministry of New and Renewable Energy reported that around 125 million tonnes (Mt) of crop residues are burned, among the approximately 500 Mt of residue generated. Uttar Pradesh generates the highest amount of crop residues (60 Mt), followed by Punjab (51 Mt) and Maharashtra (46 Mt). Cereals contribute the most residues (352 Mt), followed by fiber crops (66 Mt), oilseeds (29 Mt), pulses (13 Mt), and sugarcane (12 Mt). Rice and wheat are the primary cereals responsible for crop residue production, contributing 34% and 22%, respectively (NAAS, 2012).

Diversification of crops, proper crop rotation, and adherence to adequate soil and water conservation practices are needed to prevent soil erosion. Moreover, farming in marginal areas with low SOM have significantly accelerated wind and water erosion. Wind erosion is a particularly serious issue in arid and semi-arid regions, coastal areas with sandy soils, and the cold desert regions of Leh in northern India (Bhattacharyya et al., 2015).

Nutrient Imbalance

Soils have immense potential to support a healthy environment that sustains the production of food, fodder, biomass energy, and fiber and thereby meets basic human needs. Healthy soils contribute to achieving several Sustainable Development Goals (SDGs), including SDG 2 (Zero Hunger), SDG 3 (Good Health and Well-being), SDG 6 (Clean Water and Sanitation), SDG 12 (Responsible Production and Consumption), SDG 13 (Climate Action), and SDG 15 (Life on Land). The natural biogeochemical cycles of elements operating in the lithosphere and atmosphere are the key components of earth's climatic system. Deforestation, fossil fuel emission, land use change, and greenhouse gas emissions are factors that alter the elements' biogeochemical cycles. For example, the conversion of forest systems to coffee plantations resulted in considerable build-up of available phosphorous in the lateritic surface soils of the Western Ghats region. The practice was rooted in the belief that low-activity clay soils rich in oxides of iron and aluminum irreversibly fix large quantities of the element (Nair et al., 2016). Thus, the heavy input of phosphatic fertilizers increased the phosphorous content in surface soils of coffee-growing areas from 6.12 kg ha^{-1} to 37.07 kg ha^{-1} in forests.

Proper nutrient budgeting is one of the important aspects for maintaining soil fertility and balanced and sustainable nutrient management. Soil nutrient imbalance is a major cause of soil degradation, leading to some soils becoming nutrient depleted and losing their capacity to support crops, while others accumulate such high nutrient concentrations that they become toxic to plants and animals. Depletion is not limited to macronutrients; micronutrient loss is also widespread. Despite the prevalence of micronutrient deficiencies, there are only a few global studies addressing the issue (Bouwman et al., 2017; Vitousek et al., 2009). Nutrient-deficient soils produce nutrient-deficient plants, forages, and crops, resulting in nutrient deficiencies in animals and humans, with implication for the One Health concept. For example, owing to improper soil management, vegetables and cereals do not contain the same levels of vitamins and nutrients as they did 70 years ago, leading to the deterioration of human health (Colino, 2022). Intensive farming practices, particularly with wheat (*Triticum aestivum* L.) and rice (*Oryza sativa* L.) in India, have severely depleted soil nutrients. The imbalanced consumption ratio of nitrogen, phosphorus, and potassium widened to as much as 5:2:1 in 2009–2010, compared with the target ratio of 4:2:1. With the Green Revolution and increased grain production, there is also increased the number of nutrient deficiencies from one (nitrogen) in 1950 to nine (nitrogen, phosphorus, potassium, sulfur, boron, copper, iron, manganese, and zinc) in 2005–2006 in Indian soils. Despite a significant increase in fertilizer use, overall consumption remains low in most parts of the country. Widespread deficiencies of zinc, sulfur, iron, copper, manganese, and boron are common. Annually, around 20 Mt of the three major nutrients are removed by crops (Tandon, 1992), but the additions through inorganic fertilizers and organic manures fall short. Another estimate suggests an annual nutrient gap of 8–10 Mt of nitrogen, P_2O_5, and K_2O over the past 50 years (Tandon, 2004). Additionally, soil erosion contributes to nutrient loss, with an annual loss of 8 Mt of plant nutrients due to 5.3 billion tonnes of soil loss (Prasad & Biswas, 2000). The detrimental effects of residue burning (90–140 Mt of crop residues are burned annually; NAAS, 2012) include greenhouse gas emissions, loss of plant nutrients (nitrogen, phosphorus, potassium, and sulfur), and soil and environmental pollution (Samra et al., 2003). Nutrient depletion is a significant concern in the middle and lower Indo-Gangetic Plains. Biswas et al. (2006) reported that more depletion of nutrient occurred under two rice crops in a rice-jute-rice system than the traditional jute-wheat system, which only grows a single crop of wheat. In rice-wheat and rice-rice-rapeseed (*Brassica napus*) systems, available phosphorus has declined despite a slightly positive phosphorus balance (Yadvinder-Singh et al., 2000). Mohanty et al. (2020) stated that even continuous cropping without fertilizer alters soil microbial metabolism, affecting nitrogen acquisition from mineralization. Therefore, long-term crop residue retention combined with external optimum dose of nutrient inputs is crucial for maintaining soil health.

Soil Pollution

Soil pollution as a threat to human life is largely ignored at the national level in India owing to the lack of comprehensive information, as various researchers have sporadically generated information on different aspects of pollution affecting soil quality. Indian soils face threats from the deliberate use of contaminated organic matter, amendment materials, and irrigation water, as well as from atmospheric depositions and effluent spillages. Pollutants range from salts and toxic metals to metalloids and persistent organics, with varying degrees of toxicity originating from both industrial and geogenic sources. Moreover, soil quality is often linked to human health (Parr et al., 2009). The Central Pollution Control Board (CPCB, 2009) identified 43 critically polluted zones within 16 states, with most zones located in just four states: Gujarat, Uttar Pradesh, Maharashtra, and Tamil Nadu. Among the industries responsible for the production of several aforesaid pollutants, leading to groundwater pollution and consequent deterioration of the quality of drinking and irrigation water, are chlor-alkali, textiles, glass, and rubber production; animal hide processing and leather tanning; metal processing; pharmaceuticals; oil and gas drilling; pigment manufacture; ceramic manufacture; and soap and detergent production (Table 1.4). Buildup of heavy metals in soils and groundwater depends on the nature of contamination, which varies according to the type of industries operating. Atmospheric deposition also plays a significant role in the accumulation of heavy metals in vegetables, and the consumption of these crops can lead to biomagnification that poses health risks. Sharma et al. (2008) have reported that copper and cadmium accumulation in vegetables and lead only through cauliflower caused health risk.

Another source of soil pollutant is untreated municipal solid wastes that contain hazardous materials. The unorganized, indiscriminate, and unscientific dumping and disposal method of municipal wastes causes severe environmental consequences such as runoff of toxic compounds into nearby land areas, as well as surface water and groundwater, eventually leading to water pollution. For example, increased in total metal concentrations of chromium, copper, iron, nickel, lead, and zinc from 32.46 to 108.85 $mg\,kg^{-1}$ was reported from three municipal waste dumpsites of Allahabad, Uttar Pradesh, by Tripathi and Misra (2012).

Other sources of soil pollution are excessive use of fertilizers and pesticides, antibiotics and hormones in livestock, and irrigation of farms having heavy metals. Adhikari et al. (2012) reported that there was build-up of heavy metals such as cadmium, lead, cobalt, chromium, and nickel after application of 150% and/or 100% recommended doses of NPK for >39 years, although the contamination level was within safe limits. Another important pollutant in this aspect is arsenic. In nine districts of West Bengal, have extensive arsenic-contaminated soil and groundwater (CGWB, 2013). Other parts of India such as Uttar Pradesh, Jharkhand, Bihar, Madhya Pradesh, Chhattisgarh, Assam, Manipur, Tripura, Arunachal Pradesh, Punjab, and Andhra Pradesh have also suffered from arsenic contamination, but to a lesser extent (Chakraborti et al., 2003; Chetia et al., 2011; Singh et al., 2011; Sharma et al., 2016). Another source of soil pollution is application of large amount of pesticides that ultimately enter soil and cause serious health hazards. In India, the presence of pesticide residues (parathion, endosulfan, DDT) was reported in several foods, including vegetables, cereals, pulses, grains, wheat flour, meat, fish, poultry, and bovine milk, among others. (Kulshrestha, 2013). Therefore, real assessment on the status of these toxic pollutants is needed for effective remediation and thus clean soil and environment.

Table 1.4 Impact of Industrial Activities on Soil Groundwater Quality.

Industrial area	Soil quality parameters affected	Surface water quality parameters affected	Groundwater quality parameters affected
Ratlam industrial area, Madhya Pradesh	EC, ESP	—	TDS, SAR, coloration, Pb, Cd
Nagda industrial area, Madhya Pradesh	EC, ESP	TDS, high Na, Cl, SO_4^{-2}	
Pithampur (Dhar) industrial area, Madhya Pradesh	EC, SAR, Co, Cr	—	EC, SAR, Cr
Patancheru industrial area, Medak district, Andhra Pradesh	—	EC, As	High pH, EC, Ni, As
Zinc smelting area in Udaipur, Rajasthan	Zn, Cd	Zn, Cd, F	Zn, Cd
Textile industries in Pali, Rajasthan	EC	—	EC, Na, Cu, Pb, Cr, As
Korba industrial area (Chhattisgarh)	Acidic pH, EC, Cr, Cu	—	Cd, Co, Cr, Ni, Zn
Tiruppur industrial area	EC, ESP	EC	Pb, Cr
Coimbatore industrial area, Coimbatore	Ni, Pb, Cd, Cr	—	EC, Na, Cl, SO_4^{-2}

Note. Adopted from Panwar et al. (2010) and Saha and Sharma (2006). EC: Electrical conductivity; ESP: Exchangeable sodium percentage; SAR: Sodium adsorption ratio; TDS: Total dissolved solids.

Conclusions

Healthy soil is essential for human health as well as achieving sustainable development goals (SDGs). While India has reached self-sufficiency in food production, challenges remain in achieving zero hunger, good health, and no poverty. Addressing the nearly 147 Mha of degraded land in India is crucial. Preventing further land degradation and investing in soil conservation are vital for sustainable development and nature conservation. The ongoing nutrient depletion in agricultural fields poses a significant threat to soil quality. Insufficient nutrient application and poor-quality water lead to sparse plant cover and low vegetative input into soils. In arid and rainfed areas, nutrient depletion by crops far exceeds the benefits from added fertilizers. Alfisols, Ultisols, and Oxisols, with low cation exchange capacity, are the most affected. SOC loss ranges from 0.22% to 6.0% in different cropping sequences in India due to inadequate fertilization, while balanced fertilization results in lower SOC depletion (0.22%–2.92%). Many soils in the tropics, particularly in densely populated areas of Asia, have lost significant portions of their original SOC pool due to soil fertility mining practices. Best management practices, such as restoring degraded soils and ecosystems and adopting recommended agricultural practices on prime land while retiring marginal agricultural lands, can help sequester soil carbon. The restoration of degraded soils and ecosystems, especially in regions with rapidly increasing populations, is crucial. This restoration strategy can enhance biomass production, improve soil quality, and increase the SOC pool. In Southeast Asia, the potential for restoring degraded soils ranges from 18.3 to 35.0 $\mathrm{Tg\,C\,yr^{-1}}$. These estimates are achievable if regional governments adopt appropriate policies and implement plans to restore degraded soils through afforestation, establishing planted fallows, and improving grazing lands. The significant challenges require a coordinated and planned approach to achieve successful solutions.

References

Adhikari, T., Wanjari, R. H., Biswas, A. K., et al. (2012). *Final report of the project entitled "Impact assessment of continuous fertilization on heavy metals and microbial diversity in soils under References 309 long-term fertilizer experiment."*. Submitted to Ministry of Forest and Environment.

Alam, E., & Dominey-Howes, D. (2014). A new catalogue of tropical cyclones of the northern Bay of Bengal and the distribution and effects of selected landfalling events in Bangladesh. *International Journal of Climatology*, *35*, 801–835.

Arora, S., & Sharma, V. (2017). Reclamation and management of salt-affected soils for safeguarding agricultural productivity. *Journal of Safe Agriculture*, *1*, 1–10.

Arora, S., Singh, Y. P., Vanza, M., & Sahni, D. (2016). Bioremediation of saline and sodic soils through halophilic bacteria to enhance agricultural production. *Journal of Soil and Water Conservation*, *15*, 302–305.

Bhattacharyya, R., Ghosh, B. N., Mishra, P. K., Mandal, B., Rao, C. S., Sarkar, D., Das, K., Anil, K. S., Lalitha, M., Hati, K. M., & Franzluebbers, A. (2015). Soil degradation in India: Challenges and potential solutions. *Sustainability*, *7*(4), 3528–3570.

Bhattacharyya, T., Pal, D. K., Chandran, P., Mandal, C., Ray, S. K., Gupta, R. K., & Gajbhiye, K. S. (2004). *Managing soil carbon stocks in the Indo-Gangetic Plains, India*. Rice–Wheat Consortium for the Indo-Gangetic Plains.

Biswas, B., Ghosh, Dasgupta, M. K., Trivedi, N., Timsina, J., & Dobermann, A. (2006). Integrated assessment of cropping systems in the Eastern Indo-Gangetic plain. *Field Crops Research*, *99*(1), 35–47.

Bouwman, A. F., Beusen, A. H. W., Lassaletta, L., Van Apeldoorn, D. F., Van Grinsven, H. J. M., & Zhang, J. (2017). Lessons from temporal and spatial patterns in global use of N and P fertilizer on cropland. *Scientific Reports*, *7*(1), Article 40366

Burman, D., Bandyopadhyay, B. K., & Mahanta, K. K. (2010). Management of acid sulphate soil of coastal Sundarbans region: Observations under on-farm trial. *J. Indian Soc. Coast. Agric. Res.*, *28*, 8–11.

Burman, D., Bandyopadhyay, B. K., Mandal, S., Mandal, U. K., Mahanta, K. K., Sarangi, S. K., Maji, B., Rout, S., Bal, A. R., Gupta, S. K., & Sharma, D. K. (2013). *Land shaping—A unique technology for improving productivity of coastal land* (Bulletin No. CSSRI/Canning Town/Bulletin/2013/02). Central Soil Salinity Research Institute, Regional Research Station.

Central Ground Water Board (CWGB). (2013). Annual Report. CWGB, Government of India. Accessed January 16, 2013. http://gis2.nic.in/cgwb/Gemsdata.aspx

Central Pollution Control Board (CPCB). (2009). *Comprehensive environmental assessment of industrial clusters* (Ecological Impact Assessment Series: EIAS/5/2009–2010). CPCB, Ministry of Environment and Forest, Government of India.

Chakraborti, D., Mukherjee, S. C., Pati, S., et al. (2003). Arsenic groundwater contamination in Middle Ganga Plain, Bihar, India: A future danger? *Environmental Health Perspectives*, *111*, 1194–1201.

Chetia, M., Chatterjee, S., Banerjee, S., et al. (2011). Groundwater arsenic contamination in Brahmaputra river basin: A water quality assessment in Golaghat (Assam), India. *Environmental Monitoring and Assessment*, *173*, 371–385.

Colino, S. (2022, June 26). Fruits and vegetables are less nutritious than they used to be. *National Geographic*.

Csikós, N., & Tóth, G. (2023). Concepts of agricultural marginal lands and their utilisation: A review. *Agricultural Systems*, *204*, Article 103560

Das, M. (2022). Soil Management Intervention in Cyclone Affected Coastal Areas. Available online: http://www.iiwm.res.in/trainings/Short_Course/chapters/8.pdf.

Debnath, A. (2013). Condition of agricultural productivity of Gosaba C.D. Block, South24 Parganas, West Bengal, India after severe cyclonic Aila. *International Journal of Scientific and Research Publications*, *3*(7), 97–100.

Dent, D. (1986). *Acid Sulphate Soils: A Baseline for Research and Development*. ILRI Publication.

Dharumarajan, S., Lalitha, M., Natarajan, A., Naidu, L. G. K., Balasubramanian, R., Hegde, R., Vasundhara, R., Anil Kumar, K. S., & Singh, S. K. (2017). Biophysical and socio-economic causes for increasing fallow land in Tamil Nadu. *Soil Use and Management*, *33*, 487–498.

Esch, E., MacDougall, A. S., Esch, E., & MacDougall, A. S. (2018). *More at the margin: Leveraging ecosystem services on marginal lands to improve agricultural sustainability and slow trends of farming costs outpacing yield gains* (abstract #B33E-2714). American Geophysical Union, Fall Meeting 2018.

Goswami, B. N., Venugopal, V., Sengupta, D., Madhusoodanan, M. S., & Xavier, P. K. (2006). Increasing trend of extreme rain events over India in a warming environment. *Science*, *314*(5804), 1442–1445.

Kaliraj, S., Chandrasekar, N., Ramachandran, K., Srinivas, Y., & Saravanan, S. (2017). Coastal landuse and land cover change and transformations of Kanyakumari coast, India using remote sensing and GIS. *The Egyptian Journal of Remote Sensing and Space Science*, *20*, 169–185.

Kulshrestha, S. (2013). *Report of the expert committee to frame a policy for monitoring of pesticide residues in fruits & vegetables*. Ministry of Health & Family Welfare.

Lindgren, A., Jonasson, J. K., Ohrling, C., & Giese, M. (2022). Acid sulfate soils and their impact on surface water quality on the Swedish west coast. *J. Hydrol. Reg. Stud.*, *40*, 101019.

Maji, A.K.; Reddy, G.P.O.; Sarkar, D. (2012). Acid Soils of India—Their Extent and Spatial Variability; NBSS Publication No. 145; National Bureau of Soil Survey and Land Use Planning: Nagpur, India; p. 138.

Maji, B., & Bandyopadhyay, B. K. (1996). Effect of liming on yield of safflower and nutrient availability in coastal acid saline soils of Sundarbans, West Bengal. *J. Indian Soc. Coast. Agric. Res.*, *14*, 47–51.

Majumdar, S., Samborlang, Wanniang, K., & Bhera, U. K. (2022). Acid soil management. *Indian Farming*, *72*, 12–17.

Mandal, D., Patra, S., Sharma, N. K., Alam, N. M., Jana, C., & Lal, R. (2023). Impacts of Soil Erosion on Soil Quality and Agricultural Sustainability in the North-Western Himalayan Region of India. *Sustainability*, *15*(6), 5430. https://doi.org/10.3390/su15065430.

Mandal, D., Singh, R., Dhyani, S., & Dhyani, B. (2010). Landscape and land use effects on soil resources in a Himalayan watershed. *Catena*, *81*, 203–208.

Mandal, S., Raju, R., Kumar, A., Kumar, P., & Sharma, P. C. (2018). Current status of research, technology response and policy needs of salt-affected soils in India—A review. *Journal of the Indian Society of Coastal Agricultural Research, 36*, 40–53.

Mandal, U. K., Burman, D., Bhardwaj, A. K., Nayak, D., Samui, A., Mahanta, K. K., Lama, T. D., Maji, B., Mandal, S., Raut, S., & Sarangi, S. K. (2019). Waterlogging and coastal salinity management through land shaping and cropping intensification in climatically vulnerable Indian Sundarbans. *Agricultural Water Management, 216*, 12–26.

Mishra, M., Kar, D., Santos, C. A. G., da Silva, R. M., & Das, P. P. (2022). Assessment of impacts to the sequence of the tropical cyclone Nisarga and monsoon events in shoreline changes and vegetation damage in the coastal zone of Maharashtra, India. *Marine Pollution Bulletin, 174*, Article 113262

Mitra, A., Dutta, J., Mitra, A., & Thakur, T. (2020). Amphan supercyclone: A death knell for Indian Sundarbans. *Journal of Applied Forest Ecology, 8*(1), 41–48.

Mohanty, S., Parihar, M., Dubey, G., Parmar, R., Jain, D., Singh, M., Patra, A., Chaudhari, S., & Kollah, B. (2020). Nitrification under the influence of long-term fertilizer application in a tropical vertisol. *Archives of Agronomy and Soil Science, 68*(4), 516–530.

Moharana, P., Singh, R., Singh, S., Tailor, B., Jena, R., & Meena, M. (2019). Development of secondary salinity and salt migration in the irrigated landscape of hot arid India. *Environmental Earth Sciences, 78*, Article 454

Moharana, P. C., Jena, R. K., Kumar, N., Singh, R. S., & Rao, S. S. (2021). Assessment of soil organic and inorganic carbon stock at different soil depths after conversion of desert into arable land in the hot arid regions of India. *Carbon Management, 12*(2), 153–165.

Nair, K. M., Anil Kumar, K. S., Srinivas, S., Nagaraj, J. S., Violet D'Souza, M., Raghuramulu, Y., Hegde, R., & Singh, S. K. (2016). *Soil quality monitoring sites (SQMS) for traditional coffee-growing areas of India* (NBSS publ. no. 1095). National Bureau of Soil Survey and Land Use Planning.

National Academy of Agricultural Sciences (NAAS). (2012). *Management of crop residues in the context of conservation agriculture* (Policy Paper No. 58). NAAS.

National Academy of Agricultural Sciences (NAAS). (2017). *Mitigating land degradation due to water erosion* (Policy paper 88). NAAS.

National Remote Sensing Agency. (2019). Wasteland Atlas of India.

Osman, K. T. (2018). Acid soils and acid sulfate soils. In *Management of Soil Problems*. Springer.

Padbhushan, R., Kumar, U., Sharma, S., Rana, D. S., Kumar, R., Kohli, A., et al. (2022). Impact of land-use changes on soil properties and carbon pools in India: A meta-analysis. *Frontiers in Environmental. Science, 9*, Article 794866

Pal, D., Patra, P. K., Deb, D., Bhattacharyya, R., & Mukhopadhyay, D. (2020). Effect of land use on soil carbon fractions. *Journal of the Indian Society of Soil Science, 68*, 392–399.

Pal, D. K. (2019). Is soil inorganic carbon ($CaCO_3$, SIC) sequestration a bane or a hidden treasure in soil ecosystem services? In *Ecosystem Services and Tropical Soils of India*. Springer International Publishing.

Pandey, A. (2013). Environmental impacts of canal irrigation in India. *Mediterranean Journal of Social Sciences*, *4*(11), 138–143.

Panwar, N. R., Saha, J. K., Adhikari, T., et al. (2010). *Soil and water pollution in India: Some case studies* (IISS Technical Bulletin). Indian Institute of Soil Science.

Parida, B. R., Behera, S. N., Oinam, B., Patel, N. R., & Sahoo, R. N. (2018). Investigating the effects of episodic Super-cyclone 1999 and Phailin 2013 on hydro-meteorological parameters and agriculture: An application of remote sensing. *Remote Sensing Applications: Society and Environment*, *10*, 128–137.

Parr, J. F., Papendick, R. I., Hornick, S. B., & Meyer, R. E. (2009). Soil quality: Attributes and relationship to alternative and sustainable agriculture. *American Journal of Alternative Agriculture*, *7*, 5–11.

Planning Commission. (2011). Mid-term appraisal of eleventh five year plan 2007–2012. Oxford University Press., 500p.

Prasad, R. N., & Biswas, P. P. (2000). Soil resources of India. In G. B. Singh & B. R. Sharma (Eds.), *50 Years of natural resource management*. Indian Council of Agricultural Research.

Prokopovich, N. P. (1988). Cat clays. In *General Geology* (pp. 65–69). Encyclopedia of Earth Science; Springer.

Ray, P., Chattaraj, S., Bandyopadhyay, S., Jena, R. K., Singh, S. K., & Ray, S. K. (2021). Shifting cultivation, soil degradation, and agricultural land-use planning in the northeastern hill region of India using geo-spatial techniques. *Land Degradation & Development*, *32*(14), 3870–3892.

Saha, J. K., & Sharma, A. K. (2006). *Impact of the use of polluted irrigation water on soil quality and crop productivity near Ratlam and Nagda industrial area* (Agricultural Bulletin, IISS-1). Indian Institute of Soil Science.

Sahay, K. B. (2015). Problems of livestock population. *The Tribune*. Accessed on 20 March 2015. http://www.tribuneindia.com/2000/20000411/edit.html

Samra, J. S., Singh, B., & Kumar, K. (2003). Managing crop residues in the rice-wheat system of Indo Gangetic Plain. In J. K. Ladha, J. E. Hill, J. M. Duxbury, R. K. Gupta, & R. J. Buresh (Eds.), *Improving the productivity and sustainability of rice-wheat systems: Issues and impacts* (ASA Special Publication No. 65 (pp. 173–195). ASA Inc., CSSA Inc., and SSSA Inc.

Santra, P., Mertia, R. S., Kumawat, R. N., Sinha, N. K., & Mahla, H. R. (2013). Loss of soil carbon and nitrogen through wind erosion in the Indian Thar Desert. *Journal of Agricultural Physics*, *13*(1), 13–21.

Saxena, K. G., & Ramakrishnan, P. S. (1983). Growth and allocation strategies of some perennial weeds of slash and burn agriculture (Jhum) in northeastern India. *Canadian Journal of Botany*, *61*(4), 1300–1306.

Saxena, N.C. in 'Wells and Welfare in the Ganga Basin: Essay on Public Policy and Private Initiative: Shah Tushaar http://planningcommission.nic.in/hindi/reports/articles/ncsxna/index.php?repts=wells.htm#_ftn1 accessed on 5.5.2017 at 12.43 p.m.

Shamshuddin, J., Azura, A. E., Shazana, M. A. R. S., Fauziah, C. I., Panhwar, Q. A., & Naher, U. A. (2014). Properties and management of acid sulfate soils in Southeast Asia for sustainable cultivation of rice, oil palm, and cocoa. *Adv. Agron.*, *124*, 91–142.

Sharma, K. D. (1997, April 23–May 3). Assessing the impact of overgrazing on soil erosion in arid regions at a range of spatial scales. Human impact on erosion and sedimentation. In Proceedings of the International Symposium of the Fifth Scientific Assembly of the International Association of Hydrological Sciences (IAHS), Rabat, Morocco, pp. 119–123.

Sharma, R. K., Agrawal, M., & Marshall, F. M. (2008). Atmospheric deposition of heavy metal (copper, zinc, cadmium and lead) in Varanasi city, India. *Environmental Monitoring and Assessment*, *142*, 269–278.

Sharma, S., Kaur, J., Nagpal, A. K., & Kaur, I. (2016). Quantitative assessment of possible human health risk associated with consumption of arsenic contaminated groundwater and wheat grains from Ropar Wetland and its environs. *Environmental Monitoring and Assessment*, *188*, Article 506

Singh, V., Brar, M. S., Sharma, P., & Brar, B. S. (2011). Distribution of arsenic in groundwater and surface soils in south western districts of Punjab. *Journal of the Indian Society of Soil Science*, *59*, 376–380.

Srivastava, P., Bhattacharyya, T., & Pal, D. K. (2002). Significance of the formation of calcium carbonate minerals in the pedogenesis and management of cracking clay soils (Vertisols) of India. *Clays and Clay Minerals*, *50*, 111–126.

Sumner, M. E., & Noble, A. D. (2003). Soil acidification: The world story. In Z. Rengel (Ed.), *Handbook of Soil Acidity*. Marcel Dekker Inc.

Tandon, H. L. S. (1992, September 7–11). *Assessment of soil nutrient depletion*. In Proceedings of the FADINAP Regional Seminar on Fertilization and the Environment, Chiangmai, Thailand.

Tandon, H. L. S. (2004). *Fertilizers in Indian agriculture—From 20th to 21st century*. FDCO.

Thakur, M. K., Lakshmi Kumar, T. V., Dwivedi, S., & Narayanan, M. S. (2018). On the rainfall asymmetry and distribution in tropical cyclones over Bay of Bengal using TMPA and GPM rainfall products. *Natural Hazards*, *94*, 819–832.

Tripathi, A., & Misra, D. R. (2012). A study of physico-chemical properties and heavy metals in contaminated soils of municipal waste dumpsites at Allahabad, India. *International Journal of Environmental Science, 2*, 2024–2033.

Verma, G. P., Singh, Y. P., & Singh, A. K. (2018). Management of steep land. *Indian Farming, 68*(05), 16–19.

Vitousek, P. M., Naylor, R., Crews, T., David, M. B., Drinkwater, L. E., Holland, E., Johnes, P. J., et al. (2009). Nutrient imbalances in agricultural development. *Science, 324*(5934), 1519–1520.

Yadvinder-Singh, Dobermann, A., Bijay-Singh, Bronson, K. F., & Khind, C. S. (2000). Optimal phosphorus management strategies for wheat–rice cropping on loamy sand. *Soil Science Society of America Journal, 64*, 1413–1422.

2

Soil Health in the Indian Himalayan Region

D. Mandal, P. Jha, Raman Jeet Singh, Rajesh Kaushal, M. Madhu, S. Barman, and Ranjan Bhattacharyya

Chapter Overview

Since the early days of civilization, natural resources such as land, soil, water, biodiversity and climate have remained the basis of supporting and sustaining life for all plants and animals on Earth. However, in recent times intensive use and over-exploitation of resources have robbed them of their natural resilience. To effectively tackle the complex problems and to fulfil basic needs for livelihood, food security, poverty, employment, equity and environmental services, efficient and judicious utilization of natural resources for enhanced and sustained productivity is a matter of serious concern for policy makers, planners, scientists, conservationists and environmentalists.

Varied geological, physiographic, climatic, and vegetation characteristics have given rise to multiple soil types in the Himalayan region. Most Himalayan soils belong to Entisols and Inceptisols. While Entisols are dominant in the Western Himalayas, the latter are found extensively in the northeast. The Ultisols also occupy a large area in the northeastern region. According to the Government of India's harmonized database, about 120.7 million hectares are degraded in India, 70% of which is caused by water erosion. The extent of soil loss caused by water erosion varies from $<5\,t\cdot ha^{-1}\cdot yr^{-1}$ for dense forest areas to more than $80\,t\cdot ha^{-1}\cdot yr^{-1}$ in the Shiwalik region. Land degradation due to acid soils is more severe in Indian Himalayas than anywhere else in India

The data indicates that land degradation due to open forest is about six times more in the northeastern Himalayas (NEHs) (5.54% of the TGA) than in the northwestern Himalayas (NWHs) (0.95% of the TGA). In the NEHs, a high

Soil Health Series: Volume 4 Soil Health and Sustainability in India, First Edition.
Edited by Ranjan Bhattacharyya.

percentage of the area falls under shifting cultivation and indiscriminate forest clearance. Comparatively more land in the Indian Himalaya (0.49% of the TGA) os waterlogged and marshy than in India as a whole (0.27% of the TGA). About 0.94% of the TGA in the NEHs region is occupied by wetlands compared with 0.11% in the NWHs. Although only 1.79% of the TGA of India is affected by barren and stony wastelands, about 6.69% of the TGA in the IHR is reported to be stony wasteland. Data further reveals that this type of land degradation is about 11 times more severe in the NWHs (11.55% of the TGA) than in the NEHs (1.02% of the TGA).

In this chapter we will explore various strategies to Improve soil health and soil conservation including conservation agriculture, grassed buffer strips for soil conservation in the Indian Himalayas. We will also examine residue management and diversified cropping, site-specific management of land uses for highland sustainability, agroforestry and grass-based systems, improvement of shifting agriculture, ameliorating acid soils, location-specific farming systems for conservation and production as a comprehensive package for different categories of farmers and farming situations.

Introduction

Land is one of the most precious natural resources in India, and it is experiencing enormous pressure owing to several factors. The increasing anthropogenic pressures on the Himalayan resources to meet the ever-increasing demands for material supplies are also leading to severe land degradation problems. The signs of degradation are evident in terms of declining land productivity, loss of biodiversity, increased sedimentation of rivers and reservoirs, drying up of water sources, recurring droughts and floods, and the deteriorating environment. The continued degradation of the fragile Himalayan region would adversely affect the socioeconomic and environmental stability not only of the highlands but also the adjoining Indo-Gangetic plains.

The Indian Himalayan region (IHR) varies between 250 and 300 km in width and extends approximately 2,500 km from Jammu and Kashmir in the west to Arunachal Pradesh, ranging from 21°57′ to 37°5′ N and from 72°40′ to 97°25′ E. This great chain of mountains in Indian territory runs along the northern border of the country from the eastern border of Pakistan in the west to the frontiers of Myanmar in the east, covering 11 states of India (Figure 2.1). The region has a total geographical area of about 533,604 km^2, representing about 16.2% of the country's total area, and is inhabited by 39,628,311 people, or 4% of the total population of India, per the 2001 census. The region is vast, rugged, and versatile. It

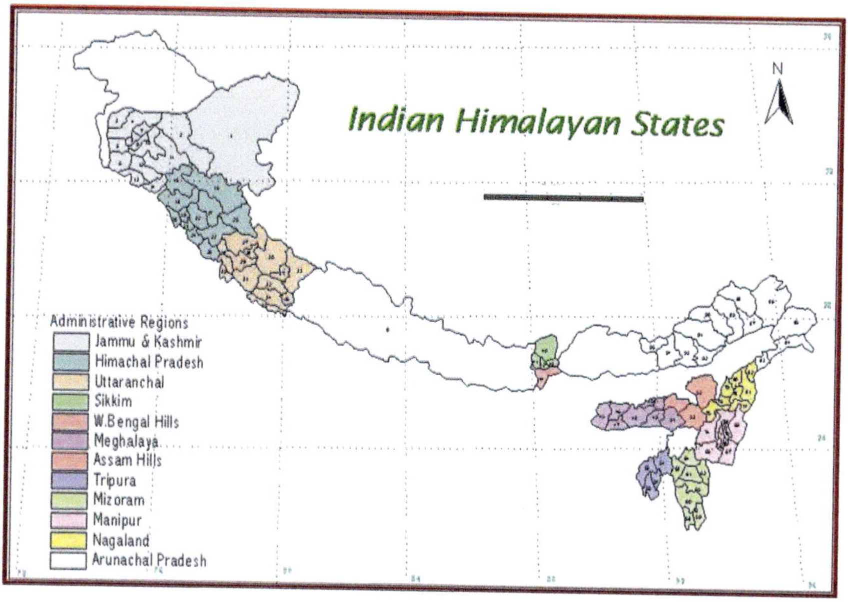

Figure 2.1 Distribution of Indian Himalayan Region.

supports remarkable cultural, ethnic, and biological diversity. It consists of two distinct geopolitical subregions, namely, the north eastern hills (NEHs) and the north western hills (NWHs). In total, there are eight states in the NEHs and three states in NWHs. The NWHs include the states of Jammu and Kashmir, Himachal Pradesh, and Uttarakhand, while the NEHs include Arunachal Pradesh, Manipur, Meghalaya, Mizoram, Nagaland, Sikkim, Tripura, and Assam. Climatically, the regions differ from one another. Annual rainfall in the NEHs is very high (2,800–12,000 $mm \cdot yr^{-1}$) in comparison with the NWHs (350–3,000 $mm \cdot yr^{-1}$). Very low precipitation (~300 mm) occurs in the cold arid region of the NWHs region. The cold arid region is a part of greater and inner Himalayan chains, which are characterized by lofty ranges, massive rockfalls, bare ridges, and extensive glacier and snow fields. The climate is therefore extremely wet and cold in the NEHs, whereas it is dry and cold in the NWHs.

The Western Himalayas occupy a major part (about 62%) of the total Himalayan range. About half of the area in the Himalayan region is degraded. The severity of land degradation by state is in the following order: Mizoram > Himachal Pradesh > Uttarakhand > Nagaland > Tripura. The major causes of degradation are soil erosion owing to water and some adverse soil conditions. The rates are quite high for the Shiwalik Hills (>80 $t \cdot ha^{-1} \cdot yr^{-1}$) and areas with shifting cultivation (40 $t \cdot ha^{-1} \cdot yr^{-1}$).

Status of Soil Resources and State of Degradation in the Indian Himalayan Region

Varied geological, physiographic, climatic, and vegetational characteristics have given rise to multiple soil types in the Himalayan region. The majority of Himalayan soils belong to Entisols and Inceptisols (Table 2.1). While Entisols are dominant in the Western Himalayas, the later are found extensively in north eastern part. The Ultisols too occupy large area in the NEHs region.

Approximately eight taxonomy soil orders are reported to occur in the rest of India, with the maximum area being occupied by Inceptisols followed by Entisols, Alfisols, Vertisols, and Aridisols (Velayutham & Bhattacharyya, 2000). Based on physiography, climate, growing period, and soil characteristics such as soil depth, texture, and available water capacity, the country has been delineated into 60 agro-ecological subregions on a 1:4 million scale. The Himalayan region includes 13 of these 60 agro-ecological subregions.

Extent of Land Degradation in Himalayan Region

According to the Government of India harmonized database, about 120.7 million ha of land area is degraded in India (NAAS, 2010), 70% of which is due to water erosion. The extent of soil loss caused by water erosion varies from $<5\,t \cdot ha^{-1} \cdot yr^{-1}$ for dense forest areas to more than $80\,t \cdot ha^{-1} \cdot yr^{-1}$ in the Shiwalik region (Singh et al., 1992). The extent of land degradation in various states of the IHR is presented in Table 2.2.

Figures 2.2 and 2.3 show comparisons of areas affected by various forms of degradation in the IHR and India as a whole. An analysis of soil erosion data revealed that the erosion of $>10\,t \cdot ha^{-1} \cdot yr^{-1}$ is more severe at the country level than in the

Table 2.1 Characteristics of Northern Hilly Regions.

Attribute	Northwestern hills	Northeastern hills
Area, million ha	32.13	18.70
Annual average rainfall, mm	1,740	2,500
Soils	Brown forest podozolic Entisols > Inceptisols > Alfisols	Brown Red and Laterite Inceptisols > Entisols > Ultisols
Natural vegetation	Himalayan moist temperate subtropical pine and subalpine forests, subhumid tropical broad-leaved forest	Subtropical pine forest, temperate wet evergreen forest, and tropical moist deciduous forest

Table 2.2 Percentage of Land Degradation Area in Various States of Indian Himalayas.

States	Classes[a]										
	1	2	3	4	5	6	7	8	9	10	11
Jammu and Kashmir	5.88	3.10	0.16	0.16	0.00	0.62	0.00	13.97	10.43	59.63	6.04
Himachal Pradesh	16.50	0.63	0.63	0.63	0.00	1.21	0.00	11.87	21.57	0.00	46.95
Uttarakhand	14.57	3.31	0.00	3.31	3.31	1.99	0.66	1.32	18.54	12.58	40.40
Northwestern Himalayas	**9.13**	**2.72**	**0.22**	**0.76**	**0.54**	**0.95**	**0.11**	**11.55**	**13.68**	**41.74**	**18.69**
Arunachal Pradesh	2.05	2.63	3.29	6.16	11.67	1.48	0.00	1.23	11.91	0.00	59.57
Assam	24.42	5.47	5.05	16.84	3.44	1.64	2.70	0.00	0.00	0.00	40.43
Manipur	1.42	5.68	4.26	4.26	61.65	4.26	1.42	0.00	0.00	0.00	17.05
Meghalaya	5.90	25.66	2.95	7.37	35.99	10.32	0.00	0.00	0.00	0.00	11.80
Mizoram	0.00	0.00	7.43	0.00	47.99	34.18	0.00	0.00	0.00	0.00	10.40
Nagaland	0.00	2.00	2.00	2.00	88.00	0.00	0.00	0.00	0.00	0.00	6.00
Sikkim	0.00	0.00	0.00	4.76	0.00	0.00	0.00	23.81	23.81	23.81	23.81
Tripura	3.23	3.23	9.68	6.45	51.61	9.68	0.00	0.00	0.00	0.00	16.13
Northeastern Himalayas	**8.77**	**5.44**	**4.32**	**8.51**	**25.04**	**5.54**	**0.94**	**1.02**	**4.33**	**0.64**	**35.71**
Himalayas hill	**8.96**	**3.97**	**2.11**	**4.34**	**11.84**	**3.06**	**0.49**	**6.69**	**9.37**	**22.79**	**26.54**
All India Total	**22.29**	**2.83**	**1.55**	**1.74**	**2.17**	**2.52**	**0.27**	**1.79**	**1.60**	**3.89**	**59.35**

[a] Classes: 1 = exclusively water erosion (>10 $t \cdot ha^{-1} \cdot yr^{-1}$); 2 = water erosion under open forest (<40% canopy); 3 = exclusively acid soils (pH < 5.5); 4 = acid soils under water erosion; 5 = acid soils under open forest; 6 = exclusively open forest; 7 = waterlogged and marshy (permanent); 8 = barren/stony waste; 9 = snow covered and glacial area; 10 = area not surveyed; 11 = others.

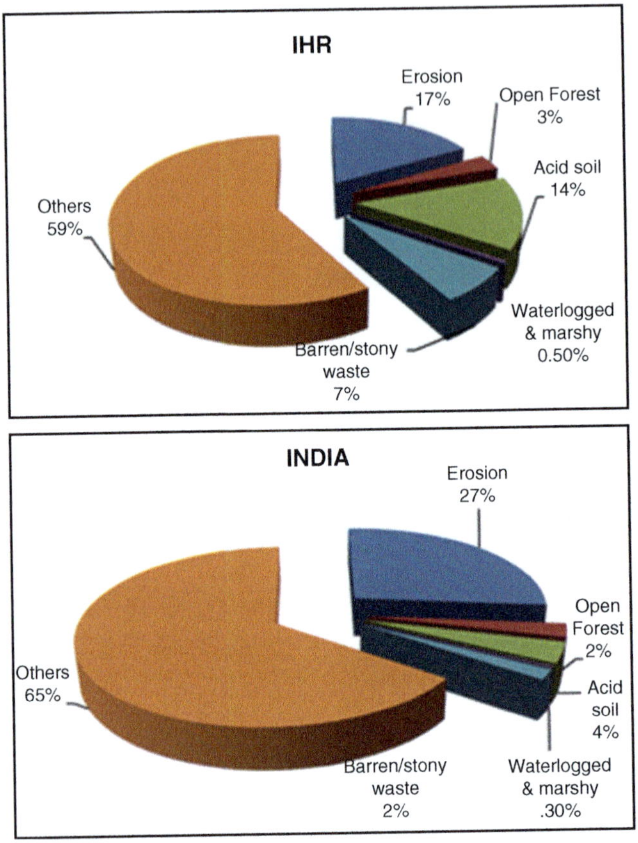

Figure 2.2 Percentage of areas under different categories of land degradation in Indian Himalayas and India.

IHR. Compared with 26.67% of the total geographical area (TGA) affected by water erosion in India, only about 17.27% of the TGA is affected in the IHR. The extent of water erosion is more severe in the NEHs (22.27% of the TGA) than in the NWHs (12.61% of the TGA). According to NAAS (2010) within the northwestern Himalayan region, Uttarakhand has highest area (21.19% of the TGA) under water erosion, followed by Himachal Pradesh (17.76% of the TGA), with the least being in Jammu and Kashmir (9.14% of the TGA; Table 2.2). Among the north-eastern states, Assam has highest area affected by water erosion (46.73% of the TGA), followed by Meghalaya (38.93% of the TGA), Tripura (12.91% of the TGA), Manipur (11.36% of the TGA), Arunachal Pradesh (10.84% of the TGA), Sikkim (4.76% of the TGA), and Nagaland (4.0% of the TGA).

Land degradation due to acid soils is more severe in Indian Himalayas (13.95% of the TGA) than in the country as a whole (3.72% of the TGA). The extent of the

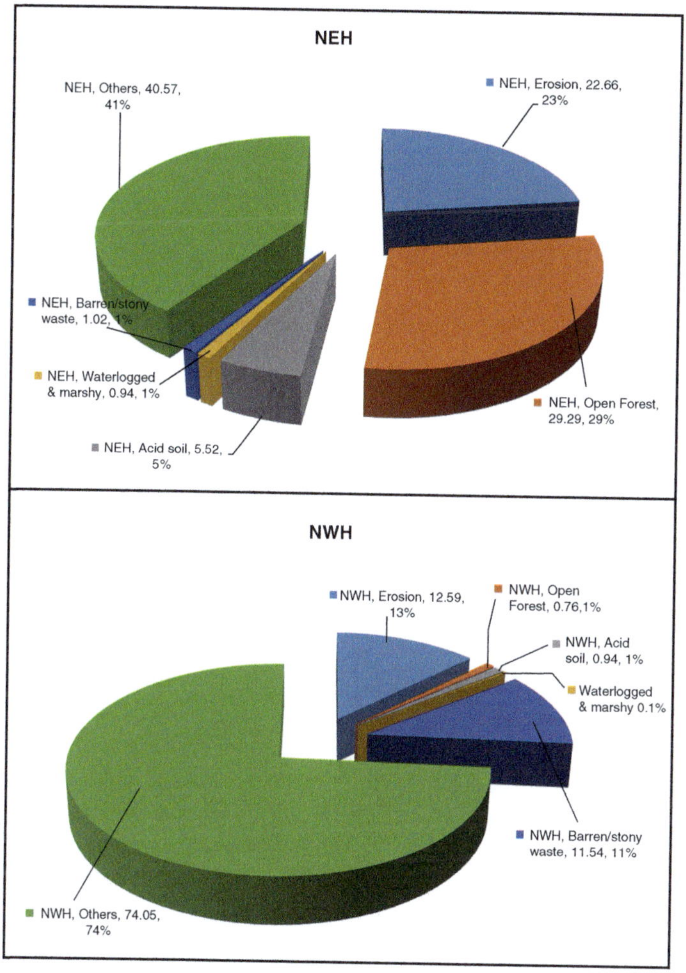

Figure 2.3 Percentage of areas under different categories categories of land degradation in North East and North West Himalayas in India.

area affected by acid soils is much greater in the NEHs (29.36% of the TGA) than that in the NWHs (0.76% of the TGA). Among the northeastern states, Nagaland has the greatest area (90.0% of the TGA) under acid soils, followed by Manipur (65.91% of the TGA), Tripura (61.29% of the TGA), Mizoram (55.42% of the TGA), Arunachal Pradesh (14.96% of the TGA), and Assam (8.49% of the TGA). Sikkim is reported to not have a soil acidity problem.

Land degradation in forest areas is an acute problem in Indian Himalayas. The proportion of the area under open forest with <40% canopy is more in the Himalayan region (3.06% of the TGA) than the nation overall (2.52% of the TGA).

The data analysis further indicated that land degradation due to open forest is about six times more in the NEHs (5.54% of the TGA) than in the NWHs (0.95% of the TGA). In the NEHs, a high percentage of the area falls under shifting cultivation and indiscriminate forest felling. Comparatively more area in the IHR (0.49% of the TGA) is occupied by waterlogged and marshy land than in India as a whole (0.27% of the TGA). About 0.94% of the TGA in the NEHs region is occupied by wetlands compared with 0.11% in the NWHs. Although only 1.79% of the TGA of India is affected by barren and stony wastelands, about 6.69% of the TGA in the IHR is reported to be stony wasteland. Data further revealed that this type of land degradation is about 11 times more severe in the NWHs (11.55% of the TGA) than in the NEHs (1.02% of the TGA).

Extent of Soil Erosion and Priority Areas in the Indian Himalayas

Over-exploitation of natural resources has resulted in severe land degradation problems, depletion of water resources, decrease in productivity of arable and nonarable lands, and recession in the snow-covered area, which is also a consequence of global warming. Recent estimates indicate that nearly 39% area of the Indian Himalayas has potential erosion rate of more than $40\,t \cdot ha^{-1} \cdot yr^{-1}$, which is alarming. In the northwestern Himalayas, on average, 17% of the area falls in the very severe category with erosion rates exceeding $40\,t \cdot ha^{-1} \cdot yr^{-1}$, while about 25% of the area has an erosion rate of more than $10\,t \cdot ha^{-1} \cdot yr^{-1}$ (Sharda & Mandal, 2018). The trend is similar in the northeastern Himalayan states. A maximum area of 34% and 63.5% in the state of Uttrakhand in the western Himalayas and Nagaland in the NEHs region, respectively, are in the very severe category with erosion rates greater than $40\,t \cdot ha^{-1} \cdot yr^{-1}$. Overall, about 23% of the area in the Himalayan states has potential erosion rates exceeding $40\,t \cdot ha^{-1} \cdot yr^{-1}$. The situation calls for serious efforts to employ appropriate conservation measures to reduce and mitigate land degradation problems.

Table 2.3 presents soil erosion data under different categories and the corresponding soil loss tolerance limit (T) values in the western Himalayan region. About 19.5% of the area in the region falls under the category of severe and very severe erosion, whereas the percentage of the area within the very low, low, and moderate erosion classes is 5.17%, 7.48%, and 9.82%, respectively. These values clearly reflect that soil erosion by water is a major problem in one-third of the area in the region, with a soil erosion rate of more than $10.0\,t \cdot ha^{-1} \cdot yr^{-1}$. An analysis of tolerance limits data (Table 2.3) showed that T-values varied between 2.5 and $12.5\,t \cdot ha^{-1} \cdot yr^{-1}$ depending on soil depth and soil resistivity to erosion. About 1.17% of the TGA can tolerate a soil loss of only $2.5\,t \cdot ha^{-1} \cdot yr^{-1}$, while about 3.89% of the area has a T-value of $12.5\,t \cdot ha^{-1} \cdot yr^{-1}$. Based on the analysis by Sharda and Mandal (2018), only about 0.85% of the TGA of the region requires different

Table 2.3 Area Under Different Potential Erosion Rates, Tolerance Limits, and Priority Classes in Western Himalayan Region[a].

Area under different erosion categories	**Erosion categories (t·ha^{-1}·yr^{-1})**						
	<5 (very low)	5–10 (low)	10–20 (moderate)	20–40 (severe)	>40 (very severe)	Others	
	1.71 (5.17)	2.48 (7.48)	3.25 (9.82)	1.92 (5.80)	4.54 (13.70)	19.23 (58.02)	
Area under different T-value	**T-value (t·ha^{-1}·yr^{-1})**						
	2.5	5.0	7.5	10.0	12.5	Rock/unreported	
	0.39 (1.17)	0.25 (0.75)	3.52 (10.63)	9.01 (27.21)	1.28 (3.89)	18.67 (56.34)	
Area under different priority classes	**Priority class (t·ha^{-1}·yr^{-1})**						
	1	2	3	4	5	6	Others
	>35	25–35	15–25	5–15	0–5	<0	—
	0.28 (0.85)	3.97 (11.99)	1.53 (4.63)	2.12 (6.42)	0.36 (1.09)	10.08 (30.42)	14.77 (44.59)

[a] Area is measured in millions of hectares. Values in the parentheses are percentages of area.

degrees of erosion management, while 30.42% of the TGA falls under no treatment category due to higher values of permissible erosion limits (Table 2.3). As evident from Table 2.3, although 12.65% area of the region falls under very low and low erosion categories, about 39.76% has high priority for conservation treatment as the soil loss tolerance limit varies only between 2.5 and $10.0\,t \cdot ha^{-1} \cdot yr^{-1}$. Similarly, although about 19.5% of the area has prevailing erosion rates of more than $20\,t \cdot ha^{-1} \cdot yr^{-1}$, only about 0.85% of the area falls under very high erosion risk priority class 1; however, about 12.84% area falls in priority class 2. The areas where the prevailing erosion rate exceeds the permissible limits need to be prioritized with regard to applying appropriate conservation measures to bring the erosion rate to the permissible level. The acute problem of soil erosion in Shiwalik and the high Himalayas is quite evident through large-scale deformation of the terrain. Increased soil erosion affects the soil productivity and causes considerable off-site damage in terms of rapid siltation of multipurpose reservoirs and lakes. Therefore, these areas are relatively more sensitive and minimal soil disturbance is particularly important. The spatial distribution of priority areas under different classes is presented in Figure 2.4.

Similarly, soil erosion data under different categories, corresponding T-values, and priority risk areas for the northeastern Himalayan region are given in Table 2.4 (Mandal & Sharda, 2013). Analysis of these data revealed that soil erosion rates vary enormously across the region, ranging from less than $5\,t \cdot ha^{-1} \cdot yr^{-1}$ to more than $40\,t \cdot ha^{-1} \cdot yr^{-1}$. About 29.82% of the area in the region falls in the very severe erosion category, whereas the percentage of the region in the very low, low, moderate, and severe erosion classes is 4.47%, 21.16%, 16.79%, and 12.97%, respectively. The results further reveal that owing to strong resistance capacity, T-values were more or less consistent at around 7.5 and $12.5\,t \cdot ha^{-1} \cdot yr^{-1}$, except for about 0.26% area under a T-value of $5.0\,t \cdot ha^{-1} \cdot yr^{-1}$, which confirms a greater role of soil depth in deciding the T-values. The areas of major concern are soils on ridges and side slopes that cannot tolerate soil loss exceeding $7.5\,Mg \cdot ha^{-1} \cdot yr^{-1}$. West Tripura (N), Dhalai (N), and Darjeeling hills are areas of major concern because they can withstand a soil loss of only $5.0\,t \cdot ha^{-1} \cdot yr^{-1}$. Therefore, these areas are the most sensitive and need careful attention during cultivation.

Based on the analysis, about 58.94% of the region's TGA is projected to require different degrees of erosion management, with 28.38% of the TGA falling under the no treatment category owing to higher values of permissible erosion limits (Table 2.4). As evident from Table 2.4, although 42.79% of the area of the region falls under severe and very severe erosion categories, 65.09% of the area has a high priority for conservation treatment because the tolerance limit for soil loss is less than $10.0\,t \cdot ha^{-1} \cdot yr^{-1}$ (Mandal & Sharda, 2013). However, an assessment based on a comparison of the prevailing and permissible rates of erosion indicates that nearly 29.70% of the region's area is very critical, as it is

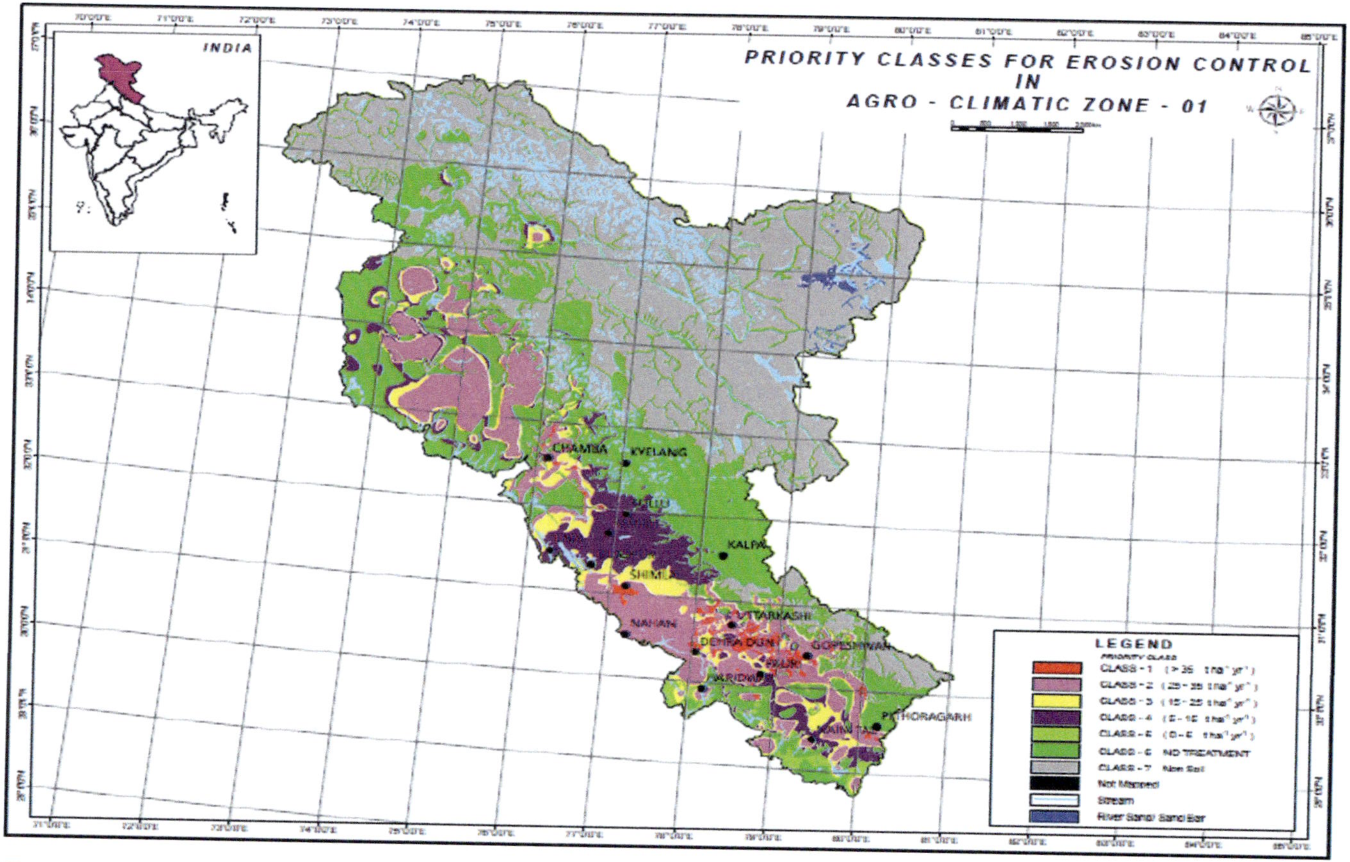

Figure 2.4 Priority classes for soil erosion control in western Himalayan region of India.

Table 2.4 Area Under Different Potential Erosion Rates, Tolerance Limits, and Priority Classes in the Northeastern Himalayan Region[a].

Area under different erosion categories	**Erosion categories (t·ha^{-1}·yr^{-1})**						
	<5 (very low)	5–10 (low)	10–20 (moderate)	20–40 (severe)	>40 (very severe)	Others	
	1.23 (4.47)	5.82 (21.16)	4.62 (16.79)	3.57 (12.97)	8.20 (29.82)	4.07 (14.79)	
Area under different T value	**T-value (t·ha^{-1}·yr^{-1})**						
	2.5	5.0	7.5	10.0	12.5	Rock/unreported	
	–	0.07 (0.26)	4.71 (17.13)	13.11 (47.70)	5.77 (21.00)	3.82 (13.91)	
Area under different priority classes	**Priority class (t·ha^{-1}·yr^{-1})**						
	1	2	3	4	5	6	Others
	>35	25–35	15–25	5–15	0–5	<0	–
	0.004 (0.01)	8.16 (29.69)	3.48 (12.69)	3.09 (11.27)	1.45 (5.28)	7.80 (28.38)	3.48 (12.68)

[a] Area is measured in millions of hectares. Values in the parentheses are percentages of area.

classified up to priority class 2. In general, the areas where the prevailing erosion rate is more than the permissible limits need to be prioritized through appropriate conservation measures to reduce the erosion rate and lower it to the permissible level.

The spatial distribution of priority areas delineated under different classes is presented in Figure 2.5. Effective planning and execution of soil conservation programs is absolutely necessary as cultivation has been extended to marginal lands through extensive deforestation, resulting in high soil erosion and low productivity. Indiscriminate deforestation and the practice of *jhum* (slash and burn) cultivation have led to accelerated erosion for which proper conservation measures need to be adopted, especially on very steep slopes.

In the northwestern Himalayas, the highest area under priority class 1 is recorded in Uttarakhand state (4.34% of the TGA) followed by Himachal Pradesh (0.91% of the TGA). In the state of Jammu and Kashmir, no area falls under

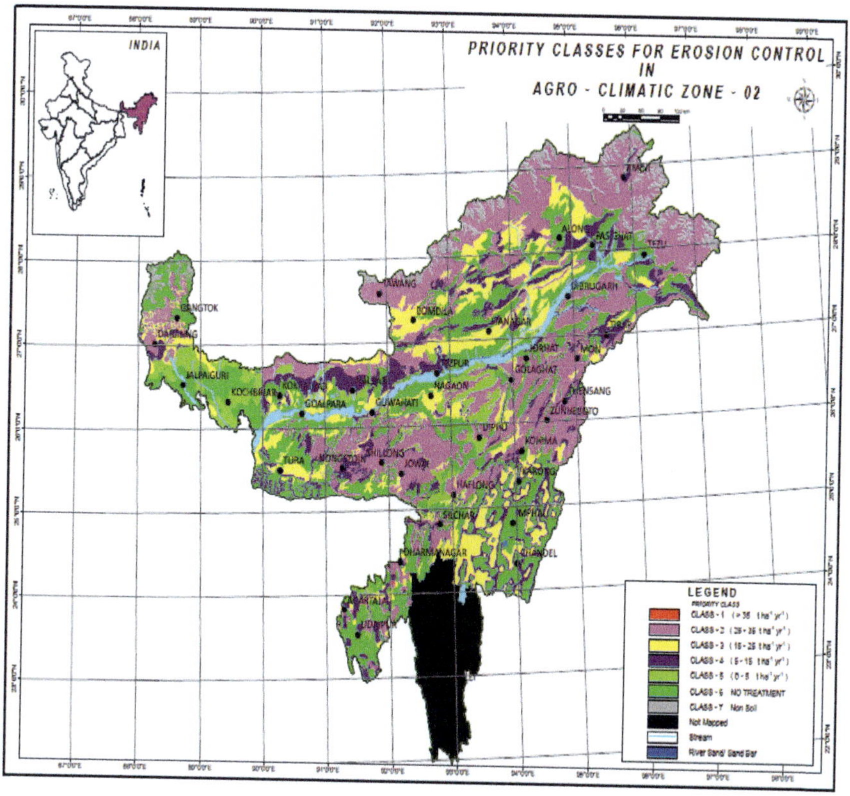

Figure 2.5 Priority classes for soil erosion control in eastern Himalayan region of India.

priority class 1 (Figure 2.4). Similarly, the highest area under priority class 2 also falls in Uttarakhand (29.30% of the TGA), followed by Himachal Pradesh (15.21% of the TGA) and Jammu and Kashmir (7.01% of the TGA) (Figure 2.5).

Erosion Control Practices

Arable Lands

Alley cropping with different leguminous trees for erosion control, which improves moisture and nutrient availability for increased crop production and maintains productive capacity of the soil, has been developed for different agro-ecological regions. Subabul (*Leucaena leucocephala*) has been most widely used as hedgerow on field bunds to produce mulch material for moisture conservation and nutrient recycling. Alley cropping with *L. leucocephala* is effective for erosion control on sloping lands up to 30%. Contour-paired rows of *Leucaena* hedge, *Leucaena* and *Eucalyptus* trees, and 0.75-m-wide grass barriers at 1.0-m intervals in maize reduced runoff from 40% to 30% of rainfall and soil loss from 21 to $8\,t \cdot ha^{-1}$ on 4% sloping land (Narain et al., 1992). Any reduction in crop yield could be minimized by shifting the management of trees to contour hedgerows. The sediment deposition along the hedgerows and tree rows increased considerably, with consequent reduction in soil loss. The hedgerows of perennial pigeonpea enabled higher maize-equivalent yield and effectively controlled erosional losses. Application of *Leucaena* before or at harvesting of maize increased moisture conservation and yield of the following wheat crop under Doon Valley conditions.

Degraded Lands

Suitable agri-horticultural systems have been developed for degraded lands (Figure 2.1) of different agro-ecological regions. These systems allow normal cultivation of agricultural crops in the initial 4–5 years, after which the horticultural crops provide adequate cover to check erosion and provide continuous returns. Fruit plants, including mango, litchi, peach, and *Citrus* species such as kinnow, lemon, and sweet orange, were identified for horticultural development in the degraded lands of Doon Valley (Central Soil and Water Conservation Research and Training Institute [CSWCRTI], 2000). Intercropping of turmeric with kinnow or peach and intercropping of cowpea, blackgram, and toria with mango and litchi resulted in the highest net returns, without any adverse effect on the performance of the fruit crops. The yields of green pods of cowpea and seeds of blackgram and toria were 1.6–2.1, 0.5–0.6, and $0.4–0.5\,t \cdot ha^{-1}$, respectively, in association with mango and litchi in these gravelly lands.

In marginal and degraded lands of Doon Valley, ameliorative effects of agroforestry have been demonstrated (Dadhwal & Tomar, 1999). Adopting agroforestry systems led to decreases in soil pH by 0.4 units and bulk density by 0.1 $Mg \cdot m^{-3}$ and improvements in organic carbon (OC) content by 0.16% and water holding capacity by 1%–5%.

About 10 million ha in the states of Jammu and Kashmir and Himachal Pradesh are classified as cold desert, which is an unstable and ecologically fragile region. The biotic pressures on the desert vegetation owing to fuel, fodder, and grazing needs have led to serious ecological degradation. Mid to high altitude zones of Jammu and Kashmir have suffered because of faulty land-use practices such as cultivation on steep slopes, over-grazing, and exploitation of natural tree cover that have threatened the sustainability of the ecosystem. Agroforestry species found to be suitable for this area include *Grewia optiva*, *Morus serrata*, *Celtis australis*, *Robinia pseudoacacia*, *Ulmus wallichiana*, *Quercus* spp., and *Bauhinia variegata, among others*. Erosion-prone areas can be managed by silvipastoral land use. Fodder trees are preferred as they can be trimmed during lean periods and the roughages supplemented with palatable and nutritious leaves to feed livestock. *Hippophae rhamnoides*, an important indigenous multipurpose shrub of the region, is used as fuel and fodder by the local inhabitants. A study conducted in Spiti Valley of Himachal Pradesh revealed that a 1-ha plantation of this shrub can meet the fuelwood needs of about 20 families (ICFRE, 1993). Recently, some *Eucalyptus* and *Populus* clones have been introduced as agroforestry species for re-vegetation of cold desert areas in Kashmir Valley.

Bouldery Wastelands

About one-third of the lands in the Himalayan region are bouldery wastelands (Figure 2.2), which are formed due to excessive erosion and deforestation on the hill slopes. These lands are virtually devoid of any economic vegetation but can be rehabilitated by establishing silvipastures for production of fodder, fuel wood, and fiber to supplement livelihood gathering of resource-poor farmers. Long-term studies have been conducted on silvipasture systems in the highly degraded lands of the Shiwalik foothills, ravinous areas, and bouldery riverbed lands. In the Shiwalik foothills, silvipasture systems involving *Eulaliopsis binata*, *Saccharum munja*, or *Vetiveria zizanioides* with *Acacia nilotica* were found to be promising. Further, a system based on *Eucalyptus tereticornis* + *E. binata* has been identified as economically viable and eco-friendly for rehabilitation of these lands.

Studies were undertaken to identify suitable tree and grass species for developing suitable silvipasture systems on the relatively flat, gravelly, and bouldery riverine land for biomass production and environmental protection under Doon Valley conditions. These technologies were aimed at attaining more profitable use of vast areas of wastelands for varied requirements of the people and ecological rehabilitation and minimizing the impact of natural or human-related disasters. The results have indicated the potential of silvipasture systems, such as *Dalbergia sissoo* + *Chrysopogon fulvus*; *Acacia catechu* + *E. binata*; *Eucalyptus* hybrid + *C. fulvus*; and *L. leucocephala* + Napier/*Panicum maximum* for utilizing the degraded lands for productive as well as protective functions. In one study, four multi-purpose tree species (*Albizia lebbeck*, *G. optiva*, *Bauhinia purpurea*, and *L. leucocephala*) were grown in association with two grass species (*C. fulvus* and *E. binata*). The findings over a period of 14 years suggested that gravelly and bouldery lands could be effectively utilized by putting them under perennial vegetation for improving soil fertility and biomass production (Samra et al., 1999a, 1999b; Vishwanatham et al., 1999). Biomass production of *B. purpurea* and *A. lebbeck* was consistently higher than that of *G. optiva* and *L. leucocephala*, despite survival being lowest in *B. purpurea* and highest in *L. leucocephala*. *Eulaliopsis binata* showed greater survival and better growth parameters compared with *C. fulvus*. Accordingly, the biomass production of *E. binata* ($4.6\,t \cdot ha^{-1}$) was considerably higher than that of *C. fulvus* ($2.7\,t \cdot ha^{-1}$). Net returns of Indian Rupees (Rs) $3{,}500 \cdot ha^{-1} \cdot yr^{-1}$ were obtained in *G. optiva* + *E. binata* followed by Rs 3,400 in *A. lebbeck* from these bouldery lands. There was a marked decrease in soil pH and an increase in OC, available phosphorus, and available potassium in the surface layer owing to the growing of the tree species *A. lebbeck* and *L. leucocephala* compared with *G. optiva* and *B. purpurea*. *Chrysopogon fulvus* also decreased soil pH and increased OC, total nitrogen, and available potassium in the surface layer compared with *E. binata*, which had a favorable effect on available phosphorus status.

Based on these findings, a combination of *B. purpurea* or *A. lebbeck* in association with *E. binata* was identified as the most promising for establishing a silvipasture system in the gravelly and bouldery lands of Doon Valley. The area has now been completely rehabilitated, torrents and stream bank erosion are effectively checked, and land productivity has increased manifold in these otherwise unproductive wastelands.

Torrent Control

Torrents known as *choes* or *rao* are common in the foothills of the Himalayan region. The CSWCRTI conducted research on torrent training in the early 1970s. The research efforts led to the identification of the underlying causes of the problems and suggestions for possible solutions. The package of ameliorative practices included afforestation, grassland development with trenching across the land slope, and growing crops on terraced lands. The torrent banks could be effectively protected by planting *Salix*, *Populas*, and a mixture of grasses. Cutoff drains, grassed waterways, gully control structures, and other construction works to reduce runoff velocity are ideal for stream bank stabilization.

Landslide Stabilization

The young mountains of the Himalayan region are prone to landslides and associated mass movement phenomena due to tectonic activity, fragile ecosystems, torrential rainfall, and human interference. Bio-engineering, defined as the use of vegetation either alone or in conjunction with other mechanical measures, has assumed great significance in landslide control, torrent control, and rehabilitation of minespoil and bouldery lands and for slope stabilization (Dhyani et al., 1988; Juyal et al., 1991; Katiyar et al., 1988). At Nalotanala, a 4-ha landslide area on Dehradun-Mussoorie road was successfully stabilized by bio-engineering measures including agroforestry interventions, such as planting of *Ipomoea carnea*, *Vitex negundo*, and Napier with *Erythrina suberosa*, *D. sissoo*, and *A. catechu*. The area was completely revegetated and rehabilitated within a period of 10 years. The sediment load which was $320\,t \cdot ha^{-1} \cdot yr^{-1}$ before treatment reduced it to $5.5\,t \cdot ha^{-1} \cdot yr^{-1}$ with the improvement of vegetation cover from <5% to >95%. In addition, dry weather flow, which used to last for barely 100 days after the cessation of monsoon, increased to 250 days after the treatment (Sastry et al., 1981).

Minespoil Rehabilitation

About 25% of limestone mining of India is carried out in the Himalayan region. Unscientific mining and over-exploitation of limestone deposits in Doon Valley during the 1960s and 1970s following ruthless cutting of forest vegetation resulted in environmental problems and rendered mined areas and their surroundings into barren land with stony/bouldery debris. The area was carelessly quarried without guarding against detrimental effects such as denudation of the natural vegetation, pollution of air and water resources, blockade of roads, siltation of reservoirs, communication failures, spoilage of good agricultural lands, and destruction of aesthetics and perennial springs. The debris flow from these mines was as high as $550\,t \cdot ha^{-1} \cdot yr^{-1}$.

The CSWCRTI, Dehradun, undertook a research project in a mined watershed of 64 ha near Shahastradhara for developing technologies for the rehabilitation of minespoils (Figure 2.3). Biological measures with proper scientific technologies were undertaken with or without mechanical measures to restore the ecology and productivity of land and to maintain the aesthetic beauty. Multipurpose tree species (MPTS)and other plant species were selected with economical and ecological considerations in mind. The selected species were native to the area, lime-loving, and leguminous, with good soil-binding characteristics. Results indicated that planting slips of *E. binata* was ideal on the degraded, steep minespoil areas. Planting of leguminous species such as *L. leucocephala* and *Pueraria hirsuta* in the minespoil produced foliage rich in nitrogen, which served as fodder, organic manure, mulch, and so forth. Species such as *Thysanolaena maxima*, *S. munja*, *Pennisetum purpureum*, *E. binata*, *I. carnea*, and *V. negundo* performed well under geotextiles, such as GeoJute, Netlon, Geocell, Geogrid, Excelsior matting, and wire mesh, among others, by arresting soil and supporting vegetation on steep slopes.

The use of various biological measures revealed that vegetal cover increased from 10% to >80% over a period of 14 years and the debris flow decreased from $550\,t \cdot ha^{-1} \cdot yr^{-1}$ before treatment to about $8\,t \cdot ha^{-1} \cdot yr^{-1}$ afterward. Improvement in soil health was noticed during the rehabilitation of the minespoil degraded lands (Dadhwal, 1999). After 14 years of agroforestry measures and other technologies, the pH of the minespoil fell 0.6 units from pH 8.1 to pH 7.5, OC increased from 0.13% to 0.42%, while $CaCO_3$ content decreased from 54.6% to 31.0% and bulk density from 1.63 to $1.47\,Mg \cdot m^{-3}$, which showed that planting of MPTS, shrubs, and grasses helped improve soil characteristics over time. Further, there were considerable decreases in the equivalent slope of land from 38% to 19% and the resultant runoff and soil loss with adoption of mechanical and biological measures. The quality of runoff improved considerably and was within permissible limits, and the flow became perennial. The treated minespoil sites have now started generating additional reserves of water for irrigation, fuel wood, fodder, fiber, and timber on sustainable basis. These techniques have restored the ecology of the area, ameliorated soil characteristics, and improved the socioeconomic conditions of the local people.

Watershed Management Issues in the Himalayan Region

The CSWCRTI, Dehradun ICAR Research complex for the NEHs region, Barapani, and State Agricultural Universities (SAUs) have developed several model watersheds in different regions and demonstrated the usefulness of appropriate technologies for rehabilitation of arable and nonarable degraded lands, landslide and mine spoil areas, and eroding steam banks or torrents. The impacts of watershed development programs in the Himalayan region on reducing soil and water losses and improving the economic conditions of the farmers have been amply demonstrated.

The research efforts in the past four to five decades have identified a number of resource conservation techniques for the hills and valleys that reduce the risk of soil degradation, preserve the productive potential, decrease the level of inputs required, and sustain agricultural productivity in the long run. These measures include land shaping or mechanical measures, agronomic manipulations, vegetative barriers, alternate land use systems, and runoff harvesting and recycling techniques. Agronomical measures are generally recommended on mildly sloping lands with the objective of maximizing in situ rainwater conservation to ensure higher productivity and protection against erosion. The measures include contour farming, intercropping, strip cropping, mixed cropping, cover management, mulching, crop geometry, tillage practices, and diversified cropping systems. Mechanical measures are adopted to support the agronomical measures on steeper slopes or where the runoff is high by reducing the length and/or degree of slope to dissipate the energy of flowing water. They include land leveling, bunding, terracing, conservation bench terracing and contour trenching.

Over the past three decades, India has addressed the challenges posed by land degradation and attempts to enhance the productivity of rain-fed areas. The country has made major investments in the area of watershed management through an appropriate combination of technical innovations, participatory approaches, and an enabling policy environment. Ample evidence highlights the positive impacts of watershed programs in terms of reduction in soil and water losses and improved agricultural productivity in normal rainfall years in regions that were bypassed in the conventional Green Revolution era. Nevertheless, several reviews on the performance of watershed development projects (Hanumantha Rao, 2000; Joshi et al., 2000, 2004; Joy et al., 2005; Kerr et al., 2000, 2004; Palanisami et al., 2002) in India have diagnosed various limitations of watershed programs.

The participatory integrated watershed management (IWSM) approach adopted in the recent past has shown encouraging results over the previously adopted commodity-based or sectoral approaches. Operational Research Project on Watershed Management at Fakot in the outer Himalayas, which was implemented by the CSWCRTI during 1975–1986, is a successful example of this participatory approach. Similar trends in production increase and environmental benefits have been observed in other watersheds developed under National Watershed Development Project for Rainfed Areas (NWDPRA), Drought Prone Area Programme (DPAP), River Valley project (RVP), and other bilateral projects implemented during 1990s. The IWSM programs were quite effective in conserving natural resources of land, water, and vegetation for sustained productivity. Runoff reduction in the range of 1.5–2.5 times and soil loss in the range of 1.2–4.8 times was realized in the experimental watersheds (Table 2.5).

In general, the impact of IWSM projects has been analyzed through tangible benefits accrued in terms of productivity components. The environmental externalities and intangible benefits are rarely accounted for. Even by considering tangible benefits, IWSM programs were found to be economically sound with a benefit–cost

Table 2.5 Impact of Integrated Watershed Management Practices on Runoff and Soil Loss in the Himalayan Region.

Watershed management site (state)	Runoff as percentage of rainfall		Soil loss ($t \cdot ha^{-1}$)	
	Pre-treatment	Post-treatment	Pre-treatment	Post-treatment
Fakot (Uttarakhand)	42.0	14.2 (2.9)	11.9	2.5 (4.8)
Behdala (Himachal Pradesh)	30.0	15.0 (2.0)	12.0	8.0 (1.5)
Una (Himachal Pradesh)	30.0	20.0 (1.5)	12.0	10.0 (1.2)

Table 2.6 Economic Evaluation of Watershed Management Programs in Western Himalayas and Shiwaliks.

Watershed	Benefit–cost ratio	Project life (years)	Discount rate (%)
Fakot	1.92	25	10
Relmajra	1.20	20	12
Sukhomajri	2.06	25	12
Nada	1.07	30	15
Bunga	2.05	30	12
Maili	1.10	50	15
Chohal	1.12	50	15

ratio ranging from 1.10:1 to 2.94:1. Data in Table 2.6 present the benefit–cost ratio of some of the projects in the western Himalayan and Shiwalik regions.

In addition to increasing overall productivity, the IWSM projects have also helped to moderate floods downstream, mitigate the impacts of drought, improve in situ moisture conservation besides groundwater augmentation, and provide other socioeconomic benefits. These successes are largely ignored by researchers and implementers due to inherent problems from the nonavailability of tools and techniques for their quantification and valuation. An attempt was made to assess and quantify the intangible benefits by taking the Fakot watershed as a test case (Sharda & Dhyani, 2004). The results are presented in Table 2.7. It is evident that beyond the improvement in living standards and soil retention value, the intangible benefits alone work out to be worth Rs 30.54 lakhs (1 lakh = 100,000) from the 370-ha watershed. It is much higher even from the

Table 2.7 Intangible Benefits From ORP, Fakot Watershed Management Program in Middle Himalayas.

Attribute of intangible benefits	Method	Value (Rs)
Additional fodder produced	Travel cost approach	135,825
Increased availability of fuel	Travel cost approach	35,420
Land value	Existence value	1,221,750
Demonstration value	Preventive expenditure approach	351,775
Living standard or human development index	Asset or property value approach	5,740,047
Education or literacy	Shadow project approach	72,000
Livestock sector	Shadow price approach	406,385
Soil retention	Programming and functional approach	6.7 t soil loss
Reduction in out migration	Relocation approach	831,600
Total annual intangible benefits (excluding living standard and soil retention value)		3,054,755

total budget allocation of Rs 22.2 lakhs for development of a watershed of this size as per the guidelines of the Ministries of Agriculture and Rural Development applicable in 2006. Hence, investment in soil and water conservation activities following an IWSM approach is economically justified for sustained and balanced development and for ensuring environmental security.

There are many success stories of water resource development through water harvesting in watershed projects (Samra et al., 2002) in submontane Siwaliks (Sukhomajri), western Himalayas (Palampur in Himachal Pradesh and Fakot in Uttarakhand), and the northeastern Himalayan region (Barapani, Meghalaya). The created water resources have dramatically increased the productivity of agricultural lands in the areas. Less expensive ways and means to reduce seepage from the reservoirs have been worked out to enhance water storage efficiency. The small perennial water sources could also be tapped in a participatory mode in several locations. The technology has been demonstrated in villages of Kalimati, Badasi, Bhopalpani, and Sainji in Dehradun by providing irrigation water to a large number of rural families (Sharda et al., 2006).

Increased productivity through improved rainwater-use efficiency in rain-fed regions through the adoption of a holistic and participatory consortium approach is possible (Wani et al., 2003). Convergence of watershed activities such as agriculture, horticulture, livestock, fisheries, poultry, and micro-enterprises to add value to rural production provides a roadmap for improved livelihoods. In situ and ex situ conservation of rainwater through a range of innovative techniques and enhancing rainwater-use efficiency through supplemental irrigation has led to the creation of a number of bright spots in different agro-ecologies.

An integrated farming system with well-founded conservation base has been developed in the northeastern region for water resource development, minimization of erosion, protection of the environment, and production. The system comprises protected forest land on the hilltop, a well-planned water harvesting tank at the middle of the hill slope, a cattle yard, and paddy fields toward foothills. Provision of silt retention tanks in runoff water courses, compaction of pond beds for checking seepage, diverting washwater from cattle yard into paddy fields, using split bamboo channels for water conveyance, constructing terraces with care to keep topsoil on the terrace bed surface, and practicing paddy-cum-fish culture are some special features of the system (Sonwal et al., 1989).

Similarly, an integrated farming system comprising a watermill, poultry, fish cultivation, a piggery, and crop production has been successfully demonstrated in the western Himalayas within the Sainji watershed to supplement the income of small and marginal farmers. The synergistic benefits of each component ensure the best use of the available resources.

The success of interventions is further enhanced through integrated nutrient management (INM; e.g., through the use of legumes and green manures, micro and secondary nutrients), integrated pest management, micro-enterprises, village-based seed banks, and rehabilitation of common property resources. The impact of such watershed-based interventions is visible through increased productivity in benchmark watersheds, a shift in cropping patterns and diversification, improved groundwater levels, reduced runoff and soil loss, and increased incomes. This convergence model implemented with a participatory approach has potential for scaling-up (experimental models to regional coverage) in the Himalayan region.

The area treated or reclaimed in the IHR under various watershed development program up to Xth Plan (March 2007) is presented in Table 2.8. It is evident from the table that only 13.46% of the TGA of the NWHs and 15.5% of the TGA of the NEHs have been treated so far through various watershed development programs. In comparison with the situation at the national level, the treated area in the Himalayan region is only 14.36% versus 17.20% of the TGA at the country level. As the hilly region is more fragile and ecologically more sensitive, a perspective plan must be prepared and implemented following the suggested priority classes for soil conservation planning in the Himalayan region. A special centrally assisted scheme on “Watershed Development Project in Shifting Cultivation Areas (WDPSCA)” is being implemented in the northeastern states with 100% special

Table 2.8 Physical and Financial Status of Treated Area in Different States in Indian Himalaya Through Various Watershed Development Programs.

States	Total geographical area (TGA)	Area treated (million ha)	Expenditure (Lakh Rs)[a]	% of TGA treated
Jammu and Kashmir	22.22	1.19	57,609.29	5.34
Himachal Pradesh	5.57	1.45	67,688.94	25.94
Uttarakhand	5.35	1.82	76,316.63	33.93
Northwest Himalayas	**33.14**	**4.46**	**201,614.86**	**13.46**
Arunachal Pradesh	8.37	0.5	11,639.43	5.97
Assam	7.84	1.09	27,321.18	13.97
Manipur	2.23	0.44	16,227.47	19.69
Meghalaya	2.24	0.33	11,534.61	14.76
Mizoram	2.11	0.66	27,334	31.37
Nagaland	1.66	0.58	31,677.2	35.19
Sikkim	0.71	0.23	10,167.69	32.44
Tripura	1.05	0.23	11,109.52	21.57
Northeast Himalayas	**26.21**	**4.06**	**147,011.10**	**15.49**
Himalayas hill Total	**59.35**	**8.52**	**348,625.96**	**14.36**
All India Total	**328.66**	**56.54**	**1,947,057.24**	**17.20**

[a] Conversion: 1 lakh = 100,000.

assistance to the state plan. Under this scheme, an area of 3.93 lakh ha has been treated since inception up to the Xth Plan. During 2008–2009, an area of about 0.40 lakh ha was treated at an estimated cost of Rs 40.00 crore (1 crore = 10 million).

Strategies to Improve Soil Health and Soil Conservation

Conservation Agriculture

Minimum Tillage Under Rain-Fed Soybean-Based Cropping Systems in the Indian Himalayas

Minimum tillage (50% reduction in conventional tillage) under rain-fed soybean–pea and soybean–lentil cropping systems improved soil quality (including increased surface soil aggregation and carbon and nitrogen sequestration within soil aggregates) without sacrificing crop productivity, as reported by Bhattacharyya et al. (2009). They also observed that zero-tillage (ZT) increased bulk density, mean weight diameter, and the proportion of macroaggregate fractions (2.0–4.75 mm, 0.25–2.0 mm) in soil compared with conventional tillage (CT) in the surface soil layer, but not in the subsurface soil layer. Impacts of ZT included a greater proportion of macroaggregate-associated (2.0–4.75 mm size fraction) total soil nitrogen than minimum tillage and CT and higher aggregate-associated total soil nitrogen in ZT plots than CT within all aggregate-size fractions (except the <0.053-mm fraction) in the surface soil layer (0–15 cm) only. Plots under soybean–lentil and soybean–pea rotations had increased total soil nitrogen concentrations compared with soybean–wheat by ~19% and 21% for the whole soil system (<4.75 mm) in the 0- to 15-cm soil layer. Thus, short-term conservation tillage and continuous leguminous cropping under rain-fed conditions increased soil organic carbon (SOC) and total soil nitrogen storage in the soil surface in the Indian Himalayas (Bhattacharyya et al., 2009).

ZT Under Irrigated Rice–Wheat Cropping System for Higher Yields and Increased Soil Aggregation

Bhattacharyya et al. (2008) standardized tillage and residue management practices according to growing conditions and type of manure availability in the Indian Himalayas. Year-round ZT may be a more desirable practice than year-round CT under irrigated rice–wheat systems in the region. The adoption of ZT is the best management option to increase soil carbon under the rice–wheat cropping system. The management practice has the potential to increase productivity and soil aggregation, with greater accumulation of particulate organic matter carbon and SOC stabilization. Other benefits include weed control, lower cultivation costs, and higher profits. Adoption of ZT and two irrigations (at the critical crop growth stages) in a direct-seeded rice–wheat system is optimum for maximizing productivity and profitability (Bhattacharyya et al., 2008; Bhattacharyya et al., 2013).

Seasonal Tillage Alteration Under Rain-Fed Cropping Systems in the Indian Himalayas

In the subtemperate Indian Himalayas, where rainfall during the summer months is abundant and the accumulation of degree-days in winter is limited by the cooling effect of the Himalayas, conservation tillage practices designed for continental climates may be unsuitable. Hence, an attempt was made by Bhattacharyya et al. (2012) to develop an appropriate conservation tillage system, by employing seasonal tillage alteration for a rain-fed lentil–finger millet production system. The findings indicate that zero-tillage in winter (*rabi*) and CT in the rainy/summer (*kharif*) season was more desirable than CT in both seasons, in terms of crop productivity and surface (0–5 cm) SOC retention and aggregation (Bhattacharyya et al., 2012).

Manure Management Under Direct-Seeded Rice–ZT Wheat System

One hundred percent NPK application to direct-seeded rice and 100% NPK + farmyard manure (FYM)/poultry manure (PM) application to wheat may be preferred to 100% NPK application to both crops or decreased doses of NPK doses in terms of SOC retention, aggregation, and stabilization (Bhattacharyya et al., 2012). Thus, plots under PM had similar yields to FYM-treated plots under ZT, even though PM was required at 50% of FYM to obtain equivalent amounts of nitrogen (Bhattacharyya et al., 2012).

Grassed Buffer Strips for Soil Conservation in the Indian Himalayas

Vegetation strips are recommended for wider adoption to reduce runoff and soil loss in the Indian Himalayas and to increase crop yields (Ghosh et al., 2012). The long-term goal is to achieve a palmarosa+ system (palmarosa [*Cymbopogon martinii*] as a vegetation strip along with organic amendments, FYM, vermicompost, and PM) and weed mulch under minimum tillage in the region. This system decreased soil loss by ~60% over control plots on a 2% slope (Ghosh et al., 2012). After 7 years of cropping at the same site, the carbon management index (CMI) was found to be strongly positively correlated with crop productivity and negatively correlated with soil loss. Hence, this single value could potentially be used to assess soil degradation elsewhere (Ghosh et al., 2016). The net return from the plots with palmarosa+ was 85% higher, and when expressed as the net return per tonne of soil loss, it was 4.5 times higher than conventional practice (Ghosh et al., 2015).

In association with an agronomist, it was observed that hybrid Napier can be grown beneath pine forests (even on slopes of 70%). After 3 years, surface SOC was ~12% higher in plots where Napier was grown under a pine forest than plots under pine forest only (VPKAS Annual Report, 2010–2011, p. 55–57).

Residue Management and Diversified Cropping

Crop Residue Management Under a Rain-fed Maize-Wheat Cropping System

The combined NPK and FYM application had higher productivity in the maize–wheat system. In contrast, the productivity of the cropping system in the plots under maize stalk, wheat straw, and kudzu application, along with adjusted dose of NPK (50% when residues were added) to both crops, was significantly lower than both 0.5 NPK + FYM and NPK + FYM. However, depending upon resource availability, farmers can also apply kudzu (a fast-growing Asian vine of the legume family that is used for forage and erosion control), maize stalk, and wheat straw annually along with an adjusted dose of NPK to a crop and full dose of NPK to the other crop as the next option (Bhattacharyya et al., 2016).

Vegetable-Based Relay Intercropping Under Irrigated Conditions

Plots under maize–frenchbean–garden pea–frenchbean relay intercropping systems had ~10% higher SOC in the 0- to 15-cm soil layer than plots under the maize–garden pea system, after just 2 years of treatment. This finding suggests that this management practice is also very suitable for soil carbon sequestration (Ved-Prakash et al., 2007).

Intercropping System Under Colocasia-Based Cropping for Increased Profitability

The *Colocasia*–onion–frenchbean cropping system had the highest *Colocasia*-equivalent yield (52.4 Mg ha^{-1}) net cost–benefit ratio (2:31). The second highest was *Colocasia*–wheat–okra and *Colocasia*–gardenpea–frenchbean (both after 3 years of cropping). Positive residual effects of legume crops were observed with respect to the maintenance or improved soil properties (i.e., SOC, available phosphorus, and available potassium). Thus, these cropping systems are alternative vegetable-based cropping sequences that are more productive and remunerative than the rice–wheat cropping system. This finding represents a pioneering research result in the Indian Himalayas (Tuti et al., 2013a).

Innovative Intercropping System for Increasing Farm Income Under Rain-Fed Conditions

Intercropping of wheat + lentils in ratios of 2:1 and 1:1 showed an added advantage over sole and wheat + toria intercropping, after 2 years of experiments (2009–2010; Tuti et al., 2013b). The highest wheat-equivalent yield (5.24 $t \cdot ha^{-1}$), net returns (Rs 36,770 per hectare), benefit–cost ratio (1.81), Monetary Advantage Index (Rs 7,207), and land-equivalent ratio (1.19) were obtained in plots under wheat + lentils (2:1) intercropping, followed by wheat + lentils in 1:1 intercropping (Tuti et al., 2013b).

It is evident from Table 2.10, that runoff and soil loss can be decreased by ~30%–43% and 33%–54%, respectively, simply by adopting INM practices, namely, by substituting 50% of the nitrogen requirement of maize by in situ live sunn hemp mulching or by combined application of FYM (5 t ha^{-1}) + vermi-compost (1.0 $t \cdot ha^{-1}$) + PM (2.5 $t \cdot ha^{-1}$) + minimum tillage + 3 weed mulch (20, 40, and 60 days after seeding) at 2.18 $t \cdot ha^{-1}$ + palmarosa vegetation strip in the sloping crop lands of NWHR. These practices can enhance the net return by Rs 17,000 to 31,000 per hectare. In addition to their direct advantages, these practices also improve OC, resulting in more water stable aggregates and a greater infiltration rate and lowering the bulk density of eroded sloping arable lands.

Table 2.9 Mean Productivity, Rainwater, Soil Conservation, and Fertility Improvement by Integrated Nutrient Management Practice for Soil Health Improvement in Diversified Crop Rotations of the Northwestern Himalayan Region.

								Soil fertility parameters				
Location	Crop rotation	Slope (%)	Treatments	Wheat equivalent yield (kg ha^{-1})	Net return (Rs·ha^{-1})	Runoff (%)	Soil loss (t ha^{-1})	OC (g kg^{-1})	Infiltration rate (cm hr^{-1})	B.D. (g cm^{-3})	Water stable aggregates (%)	Source
Dehradun	Maize–wheat	2	Control	3,058	−1,920	48.7	18.6	5.80 (−0.80)[a]	0.88 (−0.04)	1.36 (+0.03)	40.8	Singh et al. (2017)
			50% N through urea + 50% N through FYM	6,265	17,940	36.9	12.0	6.85 (+0.25)	0.93 (+0.01)	1.32 (−0.02)	54.5	Ghosh et al. (2019)
			50% N through urea + 50% N through Green manuring	5,542	31,020	27.9	8.5	7.57 (+0.97)	0.95 (+0.03)	1.31 (−0.02)	44.7	
		2	100:60:40 kg NPK + conventional tillage (CT) + chemical weeding + *Panicum* vegetation strip (farmers' practice)	2,800	16,008	32.8	5.2	9.0	—	—	43.0	Ghosh et al. (2016)
			FYM (5 t ha^{-1}) + vermi-compost (1.0 t ha^{-1}) + poultry manure (2.5 t ha^{-1}) + MT + 3 weed mulch (20, 40 and 60 DAS) at 2.18 t ha^{-1} + palmarosa vegetation strip	3,300	17,004	22.8	3.5	11.8	—	—	60.0	

[a] Values in the parenthesis are change over initial status of the soil.

Table 2.10 Mean Productivity; Rainwater, Soil Conservation, and Nutrient Conservation by Intercropping Practice in Different Crop Rotations of the Northwestern Himalayan Region.

Location	Crop rotation	Slope	Treatments	Wheat equivalent yield	Net return	Runoff	Soil loss	Nutrient losses				Source
								OC	Total N	Avail. P	Avail. K	
		%		kg · ha^{-1}	Rs · ha^{-1}	%	t · ha^{-1}	kg · ha^{-1}				
Dehradun	Maize–wheat	4	Cultivated fallow	–	–	46.7	38.0	258.0	26.9	1.06	20.4	Sharma et al. (2017)
			Pure maize (90 cm by 20 cm)–wheat (farmers' practice)	4,358	35,346	37.5	18.3	236.0	24.3	0.88	18.0	
			Maize + cowpea (90 cm by 20 cm and 45 cm by 30 cm)-wheat	4,930[a]	41,198	27.6	10.4	109.0	19.9	0.73	10.7	
Ballowal Saunkhri (Punjab)	Maize–wheat	1–3	Pure maize 6-m-wide strip	3,200[a]	28,512	32.0	8.5	6.5	14.0	6.5	18.0	Khokhar et al. (2021)
			Maize strip 4.8 m and cowpea strip 1.2 m wide	3,300[a]	34,980	28.0	8.0	5.0	12.0	5.0	17.0	
			Maize strip 3 m and cowpea strip 3 m wide	2,700[a]	28,974	27.0	7.8	4.0	11.0	4.8	15.0	
			Maize strip 1.2 m and cowpea strip 4.8 m wide	2,000[a]	20,196	26.0	7.5	3.8	10.0	4.5	15.0	
			Pure cowpea 6 m	1,200[a]	14,124	25.0	7.0	3.2	9.0	4.0	14.0	

[a] Maize equivalent yield.

Intercropping

By following intercropping of legumes such as cowpea, either by incorporating an additional row between two maize rows or by strip intercropping in an erosion-permitting crop such as maize in the sloping lands (1%–4%) of hilly regions such as Dehradun and *Kandi* region of Punjab, the system productivity of field crops improved by 3%–13% and the net return by Rs 5,852 per hectare over farmers' practice of growing maize alone (Table 2.2). Along the same lines, reductions in the runoff (~12%–26%), soil loss (~6%–43%), OC loss (~23%–54%), total nitrogen loss (~14%–18%), available phosphorus loss(~14%–17%), and available potassium loss(6–40%) were observed compared with maize as the only crop.

Site-Specific Management of Land Uses for Highland Sustainability

Adoption of appropriate cropping resulted in ≥90% more production compared with only $2\,Mg \cdot ha^{-1} \cdot yr^{-1}$ (maximum value for different selected agro-ecological sites under rain-fed conditions) with inclusion of water harvesting and land use management. Hybrid Napier can be grown to produce a green forage yield of 4.5–$6.5\,Mg \cdot ha^{-1}$ in different pockets under pine and deodar forest floors. Growing fodder trees (e.g., mulberry [*Morus alba*], *G. optiva*, oak, and *Bauhinia* species) on sloping land can compensate for acute shortages of fodder during winter. The increased crop productivity had significant impacts on increasing total SOC and total nitrogen retention in soils after 9 years of site-specific land-use management. The appropriate cropping under high, medium, and low productivity conditions for three representative model sites (Khakal, Kanigere, and Attadhar) are reported in Table 2.11:

Table 2.11 Appropriate cropping under high, medium and low productivity conditions.

Productivity conditions (Site/Terrace)	Appropriate cropping/land use management
High productivity (terrace nos. 1, 2, 3, and 4 of Kanigere; terrace nos. 1, 2, 3, 4, and 5 of Attadhar; and terrace nos. 1 and 2 of Khakal).	Upland rice, maize, finger millet, wheat, peas, lentils, toria, hybrid Napier, and mulberry in the borders/terrace risers.
Medium productivity (terrace nos. 5 and 7 of Kanigere; terrace nos. 6, 7, and 8 of Attadhar; and terrace nos. 3, 4, 5, 6, and 7 of Khakal).	Finger millet, upland rice, horse gram, Silvi pasture systems (mulberry + hybrid Napier + winter grass), wheat, toria, lentils
Low productivity (terrace nos. 6 and 8 of Kanigere, and terrace no. 8 of Khakal).	Horsegram, silvi-pasture systems (mulberry + hybrid Napier + winter grass), finger millet, toria, lentils

These experiments represent novel attempts to develop model farming systems in the Indian Himalayas after 9 years of investigation (VPKAS Annual Report, 2010–2011, p. 46–47).

Integrated Nutrient Management and Organic Farming

Specific novel results in the Indian Himalayas indicated that INM is the best technology for income generation under both rain-fed and irrigated soybean–wheat cropping systems, and INM for improved soil health and carbon and nitrogen sequestration. Although NPK + FYM application had higher crop yields and cost–benefit ratio under the rain-fed soybean–wheat system than NPK-treated plots, the same management practice under irrigated conditions had a lower cost–benefit ratio than NPK application only if farmers had to buy FYM (Bhattacharyya et al., 2008; Bhattacharyya et al., 2010). However, NPK + FYM addition had other benefits, including lower soil bulk density values and increased water retention, soil structure development, aggregate-associated carbon and nitrogen concentrations, and total nitrogen accumulation in bulk soils at 0- to 45-cm depth after 30 years of soybean–wheat cropping (Bhattacharyya et al., 2004). The NPK + FYM treatment had higher nitrogen, phosphorus, and potassium pools in the soil; higher nutrient (NPK) balance; and improved selected soil biological properties (Kundu et al., 2007; Ved-Prakash et al., 2007).

In the irrigated system after 8 years of cropping, FYM application was associated with significantly decreased soil bulk density and increased mean weight diameter and SOC contents in different aggregate size fractions. The SOC content of micro-aggregates was significantly greater than in macro-aggregates. The response of SOC content to FYM application depended on inorganic fertilization and more balanced application of NPK than nitrogen only. Steady-state infiltration rates under NPK + FYM were significantly higher than under NPK. Soil water sorptivity (calculated using the Philip's equation) under NPK + FYM was also higher than under NPK (Bhattacharyya et al., 2007).

Appropriate Organic Farming Technology for Aromatic Rice and Vegetable-Based Cropping Systems

Extensive research was conducted to establish the appropriate organic farming package of practices for aromatic rice and garden pea-based cropping systems. After 6 years of cropping, plots with organic sources of nutrients (FYM, vermicompost, and local composts from plant wastes) were comparable to or more productive than mineral-fertilized rice in terms of grain yield and improved soil physical and biological health. SOC in plots under organic farming (FYM, vermicompost, and PM) was significantly higher than in recommended NPK-treated plots after 8 years of garden pea-based cropping systems (Saha et al., 2010). Results of another study under a garden pea–frenchbean cropping system in the Himalayas revealed that an FYM application of $20\,t \cdot ha^{-1}$ recorded higher carbon storage potential than mineral fertilization and INM. Soil cracking was drastically

Table 2.12 Measures of Climate Change Adaptation and Mitigation Through Agroforestry.

CC activity	Major CC functions	Agroforestry role
Mitigation—activities that reduce GHGs in the atmosphere or enhance the storage of GHGs stored in ecosystems	Sequester C	• Accumulate C in woody biomass and in soil
	Reduce GHG emission	• Conserve existing C stock available in the forest providing alternate source of fuel, fodder and timber through AFS • Enhance forage quality, thereby reducing CH_4 • Reduce fossil fuel consumption: – Reduce equipment runs in areas with trees – Reduce farmstead heating and cooling • Reduce CO_2 emission by C sink • Reduce N_2O emissions – By greater nutrient uptake through tress – By reducing N fertilizer consumption in tree systems
Adaptation—actions to reduce or eliminate the negative effects of CC or take advantage of the positive effects	Reduce threats and enhance resilience	• Amelioration of micro-climate – To reduce impact of extreme weather events on crop production – To maintain quality and quantity of forage to reduce livestock stress • Provide greater habitat diversity to support organisms (e.g., native pollinators and useful insects) • Provide greater structural and functional diversity to maintain and protect natural resources • Create diversified production opportunities to reduce risk under aberrant weather
	Allow species to migrate to more favorable conditions	• Provide habitat corridors for species migration

Abbreviations: AFS, agroforestry system; CC, climate change; GHG, greenhouse gas.
Note. Schoreneberger et al. (2012).

reduced with FYM application of 20 t · ha^{-1} compared with NPK and INM plots. Plots under FYM at 20 t · ha^{-1} were also more effective in moderating soil temperature and were thus more climate resilient than NPK and INM treatments (Mahanta et al., 2013).

Application of FYM 20 t · ha^{-1} + biofertilizer (BF) and combined application of organic manures (FYM 10 t · ha^{-1}, PM and vermicompost each 1.5 t · ha^{-1}, and BF) resulted in higher pod yields of garden pea compared with other treatments (control) and recommended NPK. In capsicum, however, plots under INM (NPK + FYM) had higher fruit yields compared with all other organic modules. In areas where mineral fertilizers cannot be applied, application of PM at 5 t · ha^{-1}, kudzu residues at 10 t · ha^{-1}, and FYM at 10 t · ha^{-1} to both crops under garden pea–basmati/aromatic rice system is recommended to maximize yields. Application of organic amendments, irrespective of source and rate, greatly improved soil physical properties and increased soil pH and oxidizable SOC. Thus, long-term use of local composts has the potential to contribute to offset global warming and could be sustainable in remote hill lands (Saha et al., 2007). Results of the garden pea–frenchbean cropping system showed that FYM application at 20 t · ha^{-1} provided most carbon sequestration potential (0.527 Mg·ha^{-1} · yr^{-1}) in soil and the highest sustainability index. Maximum profit (optimum yield) and turnover of invested money could be achieved with FYM application of 20.0 and 15.6 t · ha^{-1}, respectively. Application of FYM at 5.9 and 8.9 t · ha^{-1} could substitute NPK and INM, respectively (Mahanta et al., 2015).

Mahanta et al. (2017) observed that PM application at 6 Mg · ha^{-1} for each crop is recommended over inorganic fertilizer for higher soil culturable microbial diversity under gardenpea–frenchbean system in this region and similar agro-ecological zones. Although vermicompost was more effective in increasing the *Trichoderma* population and suppressing disease-causing *Fusarium* species in soil, its ability to improve the status of other culturable soil microorganisms was lower than PM. Addition of PM at 6 Mg · ha^{-1} considerably increased SOC compared with inorganic fertilization treatments.

The concept of CMI is very effective in assessing the best nutrient management practices on sloping croplands, as it showed significant correlations with yield, runoff, and soil loss (Ghosh et al., 2016). The estimation of CMI values for any fertilization management can indicate the soil degradation status quantitatively, which is valuable information for land planning and mitigating land degradation in hilly agro-ecosystems. Significant correlation between grain yield of the maize–wheat rotation and runoff and soil loss indicated that soil loss may be predicted where gauging devices cannot be used, but it requires further research in developing a mathematical model in a particular soil plant condition. Fertilization of 50% NPK + 50% PM and 50% NPK + 50% green manure was as effective as 50% NPK + 50% FYM in controlling runoff, soil erosion, increasing system sustainability and increasing soil quality and CMI (Ghosh et al., 2016).

Agroforestry and Grass-Based Systems

Agroforestry is a dynamic, ecologically based natural resources management system that, through the integration of trees on farms and in the agricultural landscape, diversifies and sustains production for increased social, economic, and environmental benefits for land users at all levels (www.icraf.cgiar.org). Agroforestry land use increases livelihood security and reduces vulnerability to climate and environmental changes, and it has an important role in increasing resilience of farming systems and buffering households against climate-related risks. Therefore, agroforestry will be required to contribute substantially to meet the demands of rising population for food, fruits, fuel wood, timber, fodder, biofuel, and bio-energy, as well as for its perceived ecological services.

Agroforestry for Resource Conservation

Agroforestry for Soil and Water Conservation

A combination of perennial and annual components in agroforestry reduces runoff, soil loss, and nutrient loss by enhancing infiltration and protecting soil aggregates by intercepting the rainfall through canopy and litter (Andreu et al., 1998; Kaushal et al., 2017). Litter and root exudates help in producing water-stable aggregates. Stem and roots trap sediments, stop their down-slope movement, and facilitate water infiltration. Deep tree roots stabilize slopes and help prevent shallow landslides. In the long term, vegetation influences the fluxes of water and sediment by increasing the soil aggregate stability and cohesion as well as by improving soil water infiltration (Zuazo and Pleguezuelo, 2008). Through the addition of soil organic matter, trees in agroforestry mitigate both excess water during and water scarcity during drought periods (Bhattacharya et al., 2007). The soil moisture in agroforestry is improved as trees lower air and soil temperatures, which in turn reduces evaporation and transpiration of crops. Evaporation from the soil is also reduced as litter fall from the trees covers soil surface and thereby creates variable micro-climates (Rhoades, 1995; Siriri et al., 2013). Soil bulk density, porosity, water holding capacity, and infiltration of the soil are improved owing to the fine roots of trees (Young, 1997). The trees in agroforestry have been reported to utilize soil moisture from deeper soil depths during the moisture stress periods owing to their deep root systems.

Plantation crop combinations, multilayer tree gardens and home gardens, hedgerow (alley) cropping, and windbreaks and shelterbelts are the common agroforestry practices that are used commonly for erosion control (Young, 1997). Hedgerows are highly beneficial in controlling the loss of nutrients and OC during the erosion processes in different land-use systems. The obstruction created by hedgerows results in the deposition of soil and nutrient near the biological barrier system. In Doon Valley, Narain et al. (1998) reported that contour tree-rows or *Leucaena* hedges reduced the runoff and soil loss by 40% and 48%, respectively, over the maize plot, reducing soil loss to about $12.5\,t \cdot ha^{-1}$. In Eastern Ghat, the efficiency of SOC and available N,P,& K increased by 44%, 63%, 56% and 33%, in *Gliricidia* + trench and *Leucaena* + trench planting systems, respectively (Hombegowda et al., 2020). The configuration of land slope also decreased by $0.41° \cdot yr^{-1}$ and $0.27° \cdot yr^{-1}$ under *Gliricidia*- and *Leucaena*-based hedgerows, respectively, indicating deposition of eroded soils on the lower slope side.

In Doon Valley, Grewal (1993) reported that the soil loss and runoff under *Eucalyptus tereticornis* + *E. binata* were reduced to $0.07\,t \cdot ha^{-1}$ and 0.05% in comparison with $5.65\,t \cdot ha^{-1}$ and 23.0% under cultivated fallow and $2.69\,t \cdot ha^{-1}$ and 20.50% under *Sesamum indicum–Brassica campestris* systems, respectively. Yadav et al. (2005) reported that the effectiveness of different inter-crops in reducing soil loss was in the order of *Chrysopogon* > Napier > pigeon pea with 81%, 56%, and 25% reduction over sole aonla. Gupta et al. (2006) compared poplar-based agroforestry system with a non-agroforestry system and found the dispersion ratio was reduced; whereas, the water-stable aggregates (>0.5 mm) increased under a poplar-based agroforestry system irrespective of the soil type as compared with a non-agroforestry system of cultivation. In Northeast Region, Saha et al. (2012) reported that hedgerows reduced soil loss by 94% and runoff by 78%. In degraded ravine land of Gujrat, a silvipasture system based on bamboo (*Dendrocalamus strictus*) and Anjan grass (*Cenchrus ciliaris*) was found to absorb more than 80% of rainfall, which was used by the plant or percolated deep to recharge the groundwater. Soil loss under the system was reduced to less than $1\,t \cdot ha^{-1} \cdot yr^{-1}$ compared with about $20\,t \cdot ha^{-1} \cdot yr^{-1}$ from degraded ravines prior to plantation.

Agroforestry for Climate Change and Carbon Sequestration

Agroforestry can pool more CO_2 from the air and can play an important role in reducing vulnerability, increasing resilience of farming systems, and buffering households against climate-related risks. The tree-based system provides a long-term strategy to increase the amount of carbon while still allowing for growing of food crops. Sequestering carbon through tree-based systems is now being considered as an attractive economic opportunity for carbon trading (Kaushal et al., 2017;

Verma et al., 2014). The average carbon sequestration potential of agroforestry has been estimated to be $25\,\mathrm{t \cdot ha^{-1}}$ over 96 million ha of land in India (Sathaye & Ravindranath, 1998). Table 2.12 provides the role of agroforestry in climate change mitigation and adaption (Schoreneberger et al., 2012).

References

Andreu, V., Rubio, J. L., & Cerni, R. (1998). Effects of Mediterranean shrub cover on water erosion. *Journal of Soil and Water Conservation, 53*, 112–120.

Bhattacharyya, R., Chandra, S., Singh, R. D., Kundu, S., Srivastva, A. K., & Gupta, H. S. (2007). Long-term farmyard manure application effects on soil properties in a silty clay loam soil under irrigated wheat-soybean rotation. *Soil & Tillage Research, 94*, 386–396.

Bhattacharyya, R., Kundu, S., Pandey, S. C., Singh, K. P., & Gupta, H. S. (2008). Tillage and irrigation effects on crop yields and soil properties under the rice–wheat system in the Indian Himalayas. *Agricultural water management, 95*(9), 993–1002.

Bhattacharyya, R., Kundu, S., Ved-Prakash, & Gupta, H. S. (2008). Sustainability under combined application of mineral and organic fertilizers in a rainfed soybean-wheat system of the Indian Himalayas. *European Journal of Agronomy, 28*, 32–46.

Bhattacharyya, R., Pandey, A. K., Gopinath, K. A., Mina, B. L., Bisht, J. K., & Bhatt, J. C. (2016). Fertilization and crop residue addition impacts on yield sustainability under a rainfed maize-wheat system in the Himalayas. *Proceedings of the National Academy of Sciences, India Section B: Biological Sciences, 86*, 21–32.

Bhattacharyya, R., Pandey, S. C., Bisht, J. K., Bhatt, J. C., Gupta, H. S., Tuti, M. D., . . . Kundu, S. (2013). Tillage and irrigation effects on soil aggregation and carbon pools in the Indian sub-Himalayas. *Agronomy Journal, 105*(1), 101–112.

Bhattacharyya, R., Pandey, S. C., Chandra, S., Kundu, S., Saha, S., Mina, B. L., Srivastva, A. K., & Gupta, H. S. (2010). Fertilization effects on yield sustainability and soil properties under irrigated wheat-soybean system of the Indian Himalayas. *Nutrient Cycling in Agro-Ecosystems, 86*, 255–268.

Bhattacharyya, R., Prakash, V., Kundu, S., Srivastva, A. K., & Gupta, H. S. (2009). Soil aggregation and organic matter in a sandy clay loam soil of the Indian Himalayas under different tillage and crop regimes. *Agriculture, Ecosystems & Environment, 132*(1–2), 126–134.

Bhattacharyya, R., Tuti, M.D., Kundu, S., Bisht, J.K. and Bhatt, J.C. (2012) Conservation tillage impacts on soil aggregation and carbon pools in a sandy clay loam soil of the Indian Himalayas. *Soil Science Society of America Journal, 76*, 617–627.

Bhattacharyya, R., Ved-Prakash, Kundu, S., Srivastva, A. K., & Gupta, H. S. (2004). Effect of long term manuring on soil organic carbon, bulk density and water retention characteristics under soybean-wheat cropping sequence in N-W Himalayas. *Journal of the Indian Society of Soil Science*, *52*, 238–242.

Central Soil & Water Conservation Research &Training Institute (CSWCRTI). (2000). *Annual report 1999–2000*. CSWCRTI.

Dadhwal, K. S. (1999). Arresting environmental degradation of limestone mined lands from north-west Himalayas through agroforestry. In K. R. Solanki, A. K. Bisaria, & A. K. Handa (Eds.), *Sustainable rehabilitation of degraded lands through agroforestry* (pp. 95–101). National Research Centre for Agroforestry.

Dadhwal, K. S., & Tomar, V. P. S. (1999). Enhancing production by managing tree roots in an agroforestry landuse system. *SAIC Newsletter*, *9*(5), 7.

Dhyani, S. K., Dadhwal, K. S., & Katiyar, V. S. (1988). Biological measures for rehabilitation of abandoned mine area—Preliminary studies. *Indian Farming*, *38*(7), 7–10.

Ghosh, B., Dogra, P., Bhattacharyya, R., Sharma, N. K., & Dadhwal, K. (2012). Effects of grass vegetation strips on soil conservation and crop yield under rainfed conditions in the Indian sub-Himalayas. *Soil Use and Management.*, *28*. https://doi.org/10.1111/j.1475-2743.2012.00454.x

Ghosh, B. N., Dogra, P., Sharma, N. K., Bhattacharyya, R., & Mishra, P. K. (2015). Conservation agriculture impact for soil conservation in maize–wheat cropping system in the Indian sub-Himalayas. *International Soil and Water Conservation Research*, *3*(2), 112–118.

Ghosh, B. N., Meena, V. S., Alam, N. M., Dogra, P., Bhattacharyya, R., Sharma, N. K., & Mishra, P. K. (2016). Impact of conservation practices on soil aggregation and the carbon management index after seven years of maize–wheat cropping system in the Indian Himalayas. *Agriculture, Ecosystems & Environment*, *216*, 247–257.

Ghosh, B. N., Meena, V. S., Singh, R. J., Alam, N. M., Patra, S., Bhattacharyya, R., Sharma, N. K., Dadhwal, K. S., & Mishra, P. K. (2019). Effects of fertilization on soil aggregation, carbon distribution and carbon management index of maize-wheat rotation in the north-western Indian Himalayas. *Ecological Indicators*, *105*, 415–424. https://doi.org/10.1016/j.ecolind.2018.02.050

Grewal, S. S. (1993). Agroforestry systems for soil and water conservation in Shivalik. In *Agroforestry in 2000 AD for semi-arid and arid tropics* (pp. 82–85). National Research Centre For Agroforestry.

Gupta, N., Kukal, S. S., & Singh, P. (2006). Soil erodibility in relation to poplar based Agroforestry system in north western India. *International Journal of Agriculture and Biology*, *8*(6), 859–861.

Hanumantha Rao, C. H. (2000). Watershed development in India: recent experiences and emerging issues. *Economic and Political Weekly*, *XXXV*(45), 3943–3947.

Hombegowda, H. C., Adhikary, P. P., Jakhar, P., Madhu, M., & Barman, D. (2020). Hedge row intercropping impact on run-off, soil erosion, carbon sequestration and millet yield. *Nutrient Cycling in Agroecosystems, 116*, 103–116.

Indian Council of Forestry Research and Education (ICFRE). (1993). *Annual research report for 1991–92*. ICFRE.

Joshi, P. K., Pangare, V., Shiferaw, B., Wani, S. P., Bouma, J., & Scott, C. A. (2004). *Socio-economic and policy research on watershed management in India: Synthesis of past experiences and needs for future research. Global Theme on Agro-ecosystem* (Report No. 7). International Crops Research Institute for Semi-Arid Tropics.

Joshi, P. K., Tewari, L., Jha, A. K., & Shiyani, R. L. (2000). Meta analysis to assess impact of watershed. Proceedings Workshop on Institutions for Greater Impact of Technologies, National Centre for Agriculture Economics and Policy Research (ICAR), New Delhi, India.

Joy, K. J., Parnjpe, S., Shah, A., Badigar, S., & Lele, S. (2005, February 24–26). *Scaling up of watershed development projects in India: Learning from the first generation projects*. Proceedings of the Fourth IWMI-Tata Annual Partners Meet, International Water Management Institute, Anand, India (pp. 133–134).

Juyal, G. P., Katiyar, V. S., Sastry, G., Singh, G., & Joshie, P. (1991). *Geojute for rehabilitation of steep mine spoil areas* (Bulletin no. T-26/D-19). Central Soil and Water Conservation Research and Training Institute.

Katiyar, V. S., Dadhwal, K. S., & Dhyani, S. K. (1988). Soil conservation measures in mine spoil stabilization. In *Wastelands development and their utilization* (pp. 226–236). Scientific Publications.

Kaushal, R., Ambrish, K., Alam, N. M., Mandal, D., Jayaparkash, J., Tomar, J. M. S., Patra, S., Gupta, A. K., Mehta, H., Panwar, P., Chaturvedi, O. P., & Mishra, P. K. (2017). Effect of different canopy management practices on rainfall partitioning in *Morus alba*. *Ecological Engineering, 102*, 374–380.

Kerr, J., Pangare, G., Pangare, V., & George, P. J. (2000). *An evaluation of dryland watershed development in India* (EPTD Discussion Paper 68). International Food Policy Research Institute.

Kerr, J., Pangare, G., & Pangare, V. (2004). *Watershed development projects in India: An evaluation* (Research report 127). International Food Policy Research Institute.

Khokhar, A., Yousuf, A., Singh, M., Sharma, V., Sandhu, P. S., & Chary, G. R. (2021). Impact of land configuration and strip-intercropping on runoff, soil loss and crop yields under rainfed conditions in the Shivalik foothills of north-west, India. *Sustainability, 13*, Article 6282. https://doi.org/10.3390/su13116282

Kundu, S., Bhattacharyya, R., Ved-Prakash, Ghosh, B. N., & Gupta, H. S. (2007). Carbon sequestration and relationship between carbon addition and storage under rainfed soybean-wheat rotation in a sandy loam soil of the Indian Himalayas. *Soil & Tillage Research, 92*, 87–95.

Mahanta, D., Bhattacharyya, R., Gopinath, K. A., Tuti, M. D., Mina, B. L., Pandey, B. M., Mishra, P. K., Bisht, J. K., Srivastva, A. K., & Bhatt, J. C. (2013). Influence of farmyard manure application and mineral fertilization on yield sustainability, carbon sequestration potential and soil property of gardenpea–french bean cropping system in the Indian Himalayas. *Scientia Horticulturae, 164*, 414–427.

Mahanta, D., Bhattacharyya, R., Mishra, P. K., Gopinath, K. A., Channakeshavaih, C., Krishnan, J., Raja, A., Tuti, M. D., Varghese, E., Pandey, B. M., Bisht, J. K., & Bhatt, J. C. (2017). Influence of a six-year organic and inorganic fertilization on the diversity of the soil culturable microrgansims in the Indian mid-Himalayas. *Applied Soil Ecology, 120*, 229–238.

Mahanta, D., Bhattacharyya, R., Sahoo, D. C., Tuti, M. D., Gopinath, K. A., Arunkumar, R., Mina, B. L., Pandey, B. M., Bisht, J. K., Srivastva, A. K., & Bhatt, J. C. (2015). Optimization of farmyard manure to substitute mineral fertilizer for sustainable productivity and higher carbon sequestration potential and profitability under gardenpea-frenchbean cropping system in the Indian Himalayas. *Journal of Plant Nutrition., 38*, 1709–1733.

Mandal, D., & Sharda, V. N. (2013). Appraisal of soil erosion risk in the eastern Himalayan region of India for soil conservation planning. *Land Degradation and Development, 24*, 430–437.

NAAS. (2010). *Degraded and Wastelands of India – Status of Spatial Distribution*. National Academy of Agricultural Sciences.

Narain, P., Singh, G., & Joshie, P. (1992, September 28–30). *Technological needs of vegetative land protection measures*. Proceedings of the 7th ISCO Conference, Sydney, Australia (pp. 638–643).

Narain, P., Singh, R. K., Sindhwal, N. S., & Joshie, P. (1998). Agroforestry for soil and water conservation in the western Himalayan Valley region of India 1. Runoff, soil and nutrient losses. *Agroforestry Systems, 39*(2), 175–189.

Palanisami, K., Suresh Kumar, D., & Chandrasekharan, B. (2002). *Watershed management: Issues and policies for 21st century*. Associated Publishing Company.

Pathak, P. S., Suresh, G., Khan, T. A., & Gupta, S. K. (2006). *Subabul in the rural economy of Andhra Pradesh—A case study*. Indian Grassland and Fodder Research Institute.

Prakash, V., Kumar, N., Bhattachariyya, R., Kumar, M., & Srivastva, A. K. (2007). Productivity, economics, energetics and soil properties of vegetables-based relay intercropping systems. *Indian Journal of Agronomy, 52*(4), 300–304.

Ranjan Bhattacharyya, R. B., Pandey, A. K., Gopinath, K. A., Mina, B. L., Bisht, J. K., & Bhatt, J. C. (2016). Fertilization and crop residue addition impacts on yield sustainability under a rainfed maize-wheat system in the Himalayas.

Rhoades, C. (1995). Seasonal pattern of nitrogen mineralization and soil moisture beneath *Faidherbia albida* (syn *Acacia albida*) in central Malawi. *Agroforestry Systems, 29*, 133–145.

Saha, S., Gopinath, K. A., Mina, B. L., Bhattacharyya, R., Kundu, S., & Gupta, H. S. (2010). Expression of soil chemical and biological behaviour on nutritional quality of aromatic rice under organic and conventional regimes. *Communications of Soil Science and Plant Analysis, 41*, 1816–1831.

Saha, S. R., Chaudhary, R. S., & Somasundaram, J. (2012). Soil health management under hill agro-ecosystem of north east India. *Applied and Environmental Soil Science, 2012*, Article 696174

Samra, J. S., Dhyani, B. L., & Sharma, A. R. (1999a). *Problems and prospectus of natural resource management in Indian Himalayas—A base paper*. Hill and Mountain Agro-ecosystem Directorate.

Samra, J. S., Sharda, V. N., & Sikka, A. K. (2002). *Water harvesting and recycling Indian experiences*. Central Soil & Water Conservation Research & Training Institute.

Samra, J. S., Vishwantham, M. K., & Sharma, A. R. (1999b). Biomass production of trees and grasses in a silvipasture systems on marginal lands of Doon Valley of north-west India. 2. *Performance of grass species. Agroforestry Systems, 46*, 197–212.

Sastry, G., Mathur, H. N., & Tejwani, K. G. (1981). *Landslide control in north-western outer Himalayas* (Bulletin no. R8/D6). Central Soil & Water Conservation Research & Training Institute.

Sathaye, J. A., & Ravindranath, N. H. (1998). Climate change mitigation in the energy and forestry sectors of developing countries. *Annual Review of Energy Environment, 23*, 387–437.

Schoreneberger, M., Bendrub, G., Gooijer, H., Soolanayakanahally, R., Sauer, T., Brandle, J., Zhan, X., & Current, D. (2012). Branching out: Agroforestry as a climate change mitigation and adaptation tool for agriculture. *Journal of Soil and Water Conservation, 67*(5), 128–136.

Sharda, V. N., & Dhyani, B. L. (2004, March 23–24). *Overview of land degradation assessment in India*. National Consultative Workshop on Assessment of Economic Cost of Land Degradation/Desertification in India, the Institute of Economic Growth, New Delhi.

Sharda, V. N., & Mandal, D. (2018). Prioritization and field validation of erosion risk areas for combating land degradation in north western Himalayas. *Catena, 164*, 71–78.

Sharda, V. N., Sikka, A. K., & Juyal, G. P. (2006). *Participatory integrated watershed management—A field manual*. Central Soil & Water Conservation Research & Training Institute.

Sharda, V. N., & Venkateswarlu, B. (2007, July 23–27.) *Crop diversification and alternate land use systems in watershed management*. Best-Bet Options for Integrated Watershed Management. Proceedings of the Comprehensive Assessment of Watershed Programs in India, Patancheru, Andhra Pradesh, India (pp. 111–128).

Sharma, N. K., Singh, R. J., Mandal, D., Kumar, A., Alam, N. M., & Keesstra, S. (2017). Increasing farmer's income and reducing soil erosion using inter-cropping in rainfed maize-wheat rotation of Himalaya, India. *Agriculture, Ecosystems & Environment, 247*, 43–53. https://doi.org/10.1016/j.agee.2017.06.026

Singh, G., Babu, R., Narain, P., Bhushan, L. S., & Abrol, I. P. (1992). Soil erosion rates in India. *Journal of Soil and Water Conservation, 47*, 97–99.

Singh, R. J., Ghosh, B. N., Sharma, N. K., Patra, S., Dadhwal, K. S., Meena, V. S., Deshwal, J. S., & Mishra, P. K. (2017). Effect of seven years of nutrient supplementation through organic and inorganic sources on productivity, soil and water conservation, and soil fertility changes of maize-wheat rotation in north-western Indian Himalayas. *Agriculture, Ecosystems & Environment, 249*, 177–186.

Siriri, D., Wilson, J., Coe, R., Tenywa, M. M., Bekunda, M. A., Ong, C. K., & Black, C. R. (2013). Trees improve water storage and reduce soil evaporation in agroforestry systems on bench terraces in SW Uganda. *Agroforestry Systems, 87*, 45–58.

Sonwal, D. K., Tripathi, A. K., & Kirekha, V. (1989). Zabo—An indigenous farming system of Nagaland. *Indian Journal of Hill Farming, 2*(1), 1–8.

Tuti, M. D., Mahanta, D., Bhattacharyya, R., Pandey, B. M., Bisht, J. K., & Bhatt, J. C. (2013a). Productivity, economics and energetics of pigeonpea (Cajanus cajan)-based cropping systems in mid-hills of northwest Himalaya. *Indian Journal of Agronomy, 58*(3), 303–308.

Tuti, M. D., Mahanta, D., Mina, B. L., Bhattacharyya, R., Bisht, J. K., & Bhatt, J. C. (2012). Performance of lentil (Lens culinaris) and toria (Brassica campestris) intercropping with wheat (Triticum aestivum) under rainfed conditions of north-west Himalaya. *Indian Journal of Agricultural Sciences, 82*(10), 841.

Tuti, M. D., Prakash, V., Pandey, B. M., Bhattacharyya, R., Mahanta, D., Bisht, J. K., Kumar, M., Mina, B. L., Kumar, N., Bhatt, J. C., & Srivastva, A. K. (2013b). Energy budgeting of colocasia-based cropping systems in the Indian sub-Himalayas. *Energy, 45*, 986–993.

Ved-Prakash, Bhattacharyya, R., Selvakumar, G., Kundu, S., & Gupta, H. S. (2007). Long-term effects of fertilization on some soil properties under rainfed soybean-wheat cropping in the Indian Himalayas. *Journal of Plant Nutrition and Soil Science, 170*, 1–10.

Velayutham, M., & Bhattacharya, T. (2000). Soil resources management. In J. S. P. Yadav & G. B. Singh (Eds.), *Natural resource management for agricultural production in India* (pp. 14–18). Indian Society of Soil Science.

Verma, A., Kaushal, R., Alam, N. M., Mehta, H., Chaturvedi, O. P., Mandal, D., Tomar, J. M. S., Rathore, A. C., & Singh, C. (2014). Predictive models for biomass and carbon stocks estimation in *Grewia optiva* on degraded lands in western Himalaya. *Agroforestry Sysytems, 88*, 895–905.

Vishwanatham, M. K., Samra, J. S., & Sharma, A. R. (1999). Biomass production of trees and grasses in a silvipasture system on marginal lands of Doon Valley of north-west India. 1. Performance of tree species. *Agroforestry Systems*, *46*, 181–196.

VPKAS Annual Report, 2010–2011, p. 55–57

Wani, S., Pathak, P., Sreedevi, T., Singh, H., & Singh, P. (2003). Efficient management of rainwater for increased crop productivity and groundwater recharge in Asia. In J. W. Kijne, R. Barker, & D. Molden (Eds.), *Water productivity in agriculture: Limits and opportunities for improvement* (pp. 199–215). CAB International.

Yadav, R. P., Prasad, R., Aggarwal, R. K., Agnihotri, Y., & Samra, J. S. (2005). Aonla (*Embilica officinalis*) based silvi-pastoral systems for soil and water conservation in degraded lands of Shivalik foothills. *Indian Journal of Soil Conservation*, *33*(3), 207–212.

Young, A. (1997). *Agroforestry for soil management* (2nd ed.). World Agroforestry (ICRAF).

Zuazo, V. H. D., & Pleguezuelo, C. R. R. (2008). Soil erosion and runoff prevention by plant covers: A review. *Agronomy for Sustainable Development*, *28*, 65–86.

3

Arid Eco-Region and Dryland Soil Health

Sumanta Kundu, Ch. Srinivasarao, V. K. Singh, and J. Naveen

Chapter Overview

Developing countries account for 72% of the planet's drylands. In India, dryland/rainfed agriculture accounts for around 51% of arable land and 44% of the land producing food, and farming on dryland thus plays a crucial role in the country's food security. Most farmers in India rely on rainfed agriculture. These areas are economically fragile and vulnerable to various types of environmental stresses. Moreover, the soil organic carbon content in these soils is too low for adequate crop production and soil fertility maintenance. The low productivity of soils and the lack of sustainability often have a negative impact on the development of livelihoods. Degraded soils with low moisture holding capacity, declining water tables, and multiple nutrient deficiencies lead to lower crop yields and further exacerbate soil degradation. Frequent drought in the form of dry spells, uneven rainfall distribution, soil crusting in the red soils and swelling and shrinking of the black soils, soil salinization, and low investment capacity of farmers lead to low productivity and yield. Therefore, a holistic approach to address edaphic, hydrological, and soil nutrient constraints is necessary to reduce the bottleneck. Increasing soil water storage, carbon sequestration, conservation agriculture, organic farming, mulching, and balanced and site-specific nutrient management, as well as application of soil amendments such as tank silt and other novel materials, namely biochar, polymers, zeolites, and bentonites, can improve water and nutrient storage and hence water and nutrient use efficiency. Improving water use efficiency through sprinkler or drip irrigation along with efficient nutrient management (i.e., foliar spraying of water-soluble fertilizers, nano fertilizers, combined use of chemical fertilizers, and locally available organic manures) can close the yield gaps in drylands.

Soil Health Series: Volume 4 Soil Health and Sustainability in India, First Edition.
Edited by Ranjan Bhattacharyya.

Introduction

About 80% of the world's farmland depends on rain-fed agriculture (Mahmood et al., 2020). India's National Rainfed Area Authority (NRAA) estimates that 40% of the total geographical area, 51% of the total cultivated area, and 42% of food production rely on rainfed agriculture. Almost three-quarters of the world's drylands are in developing countries. In India, most farmers depend on rainfed agriculture. For example, 80% of cereals, 50% of maize, 88% of soybean, and 65% of pigeon pea and chickpea are grown on dryland. The country's net cultivated area (141 million hectares [Mha]) has remained constant for several years, but the human population is growing. Overall, 40% of food is produced in dryland agriculture and two-thirds of livestock is fed (Panwar et al., 2018). India ranks first among countries practicing rainfed agriculture, both in terms of volume and value of produce. According to UN 2019 estimates, India will add nearly 273 million people between 2019 and 2050. By 2025, the population in India's drylands could reach 600 million (Singh et al., 2002). To meet the food needs of the growing population, it is essential to increase productivity without jeopardizing soil and water resources. In addition, people living below the poverty line live mainly in regions with high rainfall. The sustainable management of soil and water is the greatest challenge in increasing crop yields in arid regions. Soil erosion and degradation leading to frequent floods and droughts are the main reasons for low productivity in drylands. Poor soil conservation practices, lack of technology transfer, and low-yielding crop varieties, among other factors, further worsen the condition of soil fertility in rainfed areas.

Drylands are defined as lands where the rate of evaporation is greater than the total precipitation. Drylands cover 5.36 million km^2 (41% of our planet's surface) and are home to almost one-third of the world's population (Gaur & Squires, 2018). Major arid zones are found between 15° and 30° north and south latitude, including in North and South America, North Africa, the Sahel, Africa south of the equator, the Middle East, Asia, and the Pacific. Around 2.5 billion people live in dryland regions and depend on the ecosystem services provided by the drylands for their livelihoods (UN Environmental Management Group, 2011). The drylands are also home to 50% of the world's livestock and provide 44% of the world's food. In India, there are 38.7 Mha of dryland, of which 31.7 Mha are in hot drylands and 7 Mha in cold drylands. The hot, dry region is mainly located in the northwest of India (90.1%), with the remaining parts scattered throughout southern India. Despite the common characteristics of drought and high temperatures in drylands, wide regional variations exist in rainfall patterns, physiography, soils, the amount of available surface water and groundwater, and the extent of vegetation cover. Instead of the annual rainfall total, the decisive factor is how much rain falls during the growing season. The variability of rainfall is more important than the rainfall totals themselves. During the kharif season (Crop growing during the June to September monsoon) high temperatures, high winds, and low relative humidity exacerbate the negative effects of low and erratic rainfall. In this chapter, we have presented some technological solutions and policy measures to holistically address soil and nutrient-related constraints in arid and semi-arid drylands in India.

Challenges in Arid Regions

Climatic Stresses

Arid and semi-arid regions usually experience harsh environmental conditions, including low and sporadic precipitation, wind, intense solar radiation, and high evapotranspiration throughout each year (Rathore et al., 2019). In general, arid regions receive less than 200–250 mm of precipitation per year, with high evapotranspiration rates. Hyper-arid deserts receive less than 25 mm of precipitation per year and have no rainy season. Semi-arid regions receive more than 250 mm and less than 600 mm of precipitation per year. The main effects of heat and wind are an increase in evaporation that consequently makes droughts common. Hot air temperatures dry out soil quickly and increase the tendency of erosion by wind (Blanco-Canqui & Lal, 2009). Moist soils are not transported by the wind, as the soil particles are bound together by cohesive forces.

Fragile Land Resources

Soil is a component of terrestrial ecosystems, their productivity of which is influenced by several factors. Poor soil productivity and lack of sustainability are often detrimental to the development of livelihoods. Soil properties and climatic conditions are the main constraints in rainfed ecosystems and account for most causes of low soil productivity (Srinivasarao et al., 2013). Arid soils are typically sandy, loamy sandy to loamy with a weakly granular and subangular blocky structure with a light brown, yellowish brown, or grayish brown color. Arid ecosystems are dominated by Aridisols, Entisols, Mollisols, Alfisols, Vertisols, and other soils (Dregne, 1976; Noin & Clarke, 1997).

The dry soils in India are Inceptisols (95.8 Mha), Entisols (80.1 Mha), Alfisols (79.7 Mha), Vertisols (26.3 Mha), Aridisols (14.6 Mha), Oxisols (0.3 Mha), and miscellaneous soil types (23.1 Mha). Alluvial soils occur in the largest area, but the problems of dryland agriculture are not as acute in these soils. Soil problems are common in Vertisols and Alfisols, which are mainly found in northwest, central, and southern India. Vertisols are characterized by a high clay content (30%–60%). They are naturally rich in fertilizers but have problems with workability and develop subsurface sodicity in some areas. These soils are highly erodible and produce large amounts of runoff. The soils are low in nitrogen and often contain little available phosphorus. Alfisols are characterized by moderate (10%–20% clay) and low organic matter, low fertility especially in terms of nitrogen, phosphorus, sulfur, and calcium. Red soils have a low water retention capacity due to their shallow soil depth. These soils are prone to crusting, compaction of the subsoil, and severe soil erosion and produce large amounts of runoff water. Inceptisols and Entisols are characterized as deep soils with varying textures. These soils are generally deficient in available nitrogen and phosphorus but have a high agronomic potential for crop production.

Soils in arid areas usually have a high alkaline pH and accumulate high concentrations of potassium, sodium, and calcium, which are toxic to plants. This toxicity is because, in the absence of water, carbon combines with oxygen to form carbonates and then cations, which actively repel water. As a result, a gypsum crust or carbonates can form, which impedes the flow of water to the plant roots. The emergence of seedlings is hindered by the soil crust, which ultimately affects the plant population. The crusting of the soil surface also increases rainwater runoff and reduces the infiltration rate. Another problem that limits soil depth and the rooting zone is the presence of hard layers (pans) in the soil. The northwest and low rainfall regions of India are dominated by Aridisols. Low clay content, limited water retention capacity, and low soil fertility are the characteristics of these soils. The cultivation of crops is extremely difficult, so these soils are best suited to alternative land-use systems.

Nutrient Stress

Soils in arid and semi-arid regions areas show a low natural primary productivity and restricted fertility owing to the lack of water. The environmental extremes of hot and cold temperatures, in addition to the limited availability of water, are the primary factors that limit the fertility of the soil in arid and semi-arid conditions. These soils, with a few notable exceptions, have inherently poor fertility, which is measured by factors such as low availability of nitrogen and phosphorus, low water-holding capacity, high pH, low soil organic matter (varying from 0.1% to 3%), shallowness, stoniness, and other specific concerns (Husein et al., 2019). The poor fertility is because these places receive a limited amount of rainfall and are prone to high temperatures, both of which contribute to a lack of organic matter and nutritional inputs from external sources. These regions are quite extensive, covering somewhere between 30% and 40% of the world's terrestrial surface. In contrast to the temperate zone, where nutrient pollution is the main cause of soil environmental concerns, nutrient depletion and the loss of soil organic matter and its associated functions are the key issues in arid and semi-arid regions (Smaling, 1993). Under the conventional method of farming, there have been reports of nutrient depletion as well as loss (Agboola, 1993). Bush burning and an inadequate use of fertilizer have also contributed to nutrient depletion.

Another major issue affecting agricultural productivity is declining soil fertility and nutrient imbalance. Organic matter levels in intensively cropped areas have plummeted dramatically. In addition to widespread nitrogen deficiency, deficiencies in potassium, sulfur, and micronutrients are on the rise. Zinc deficiency has become the most severe, followed by sulfur and boron deficiency (Motsara, 2002). About 8–10 million metric tonnes of nutrients are leaching out of India's soils every year, contributing to a loss in fertility across 29.4 Mha of land (Lal, 2004). Another source of concern is poor nutrient use efficiency. Indian soils have low levels of organic matter; as a result, they have status ratings of 63% low, 26% medium, and 11% high. Approximately 80% of the soils tested had available phosphorus levels ranging from low to medium. In 50% of soil samples, the potassium status was low to medium. Similarly, zinc is the most deficient micronutrient, followed by iron, manganese, and copper (Table 3.1). Soil fertility issues have previously been primarily focused on irrigated agriculture, but recent research has revealed that drylands are not only thirsty, but also hungry (Srinivasarao & Vittal, 2007). Most rainfed regions have low levels of organic carbon and available nitrogen, and soils are deficient in multiple nutrients (Table 3.2).

Soil salinity is a widespread problem in arid and semiarid regions. Because of soil and water salinity, arid and semi-arid lowland agro-ecologies are marginal conditions for crop production (Qureshi et al., 2018).High daily temperatures and low annual rainfall have accelerated water evaporation rates, resulting in high concentrations of soluble salts accumulated in lowland areas (Sileshi et al., 2015). Due to the scarcity of surface water, hot arid zones must rely extensively on groundwater for irrigation. Because of the reliance on groundwater extraction, it is overutilized, resulting in a rapid decline in groundwater level. Soil degradation caused by a variety of factors reduces soil productivity and is a major issue on much of the land in arid and semiarid regions.

Table 3.1 Macro and Micronutrient Percentage Deficiencies in Some States of India.[a]

	Percentage deficiency of available macro and micronutrients						
Location	**N**	**P**	**K**	**Zn**	**Fe**	**Cu**	**Mn**
Madhya Pradesh	40	39	10	63	3	1	3
Uttar Pradesh	80	71	12	45	6	1	3
Maharashtra	67	86	8	—	—	—	—
Andhra Pradesh	62	57	9	51	2	1	2
Karnataka	29	31	7	78	39	5	19
Orissa	60	59	33	—	—	—	—
Tamil Nadu	75	24	12	53	15	3	8
India	63	42	13	—	—	—	—
Gujarat	—	—	—	24	8	4	4
Bihar	—	—	—	54	6	3	2

[a] *Sources:* Motsara (2002).

Fragile Water Resources

Rather than the availability of nutrients and the fertility of the soil, the scarcity of water is the most important limiting factor for crop production. The ability of the soil to supply crops with adequate moisture is the decisive factor for crop yields in arid and semi-arid regions. Water scarcity is a major obstacle that affects not only productivity but also the maintenance of soil fertility. Soil fertility in arid and semi-arid areas can be improved by increasing the soil's ability to retain moisture. Therefore, to maintain soil productivity, both water scarcity and soil sterility need to be addressed simultaneously. Canal irrigation in hot dry areas has led to considerable agricultural growth, but also to waterlogging and secondary salinization in some areas because of inadequate drainage, excessive irrigation, and seepage of water from the canals. Better availability and more effective uptake of nutrients require better water conservation and use. The use of vegetation barriers to intercept rainwater, the use of stones and green mulches to protect against evaporation, and the creation of ditches and ponds are all effective and low-cost strategies to improve soil moisture content and nutrient availability (Oweis et al., 2004). An arid moisture regime and a hyperthermal temperature regime are characteristic features of arid soils.

Table 3.2 Nutrient Deficiencies in Dryland Soils in Various States of India.

Area	Deficient nutrient	References
Varanasi	N, Zn, B	Srinivasarao and Vittal (2007)
Faizabad	N	
Phulbani	N, Ca, Mg, Zn, B	
Ranchi	Mg, B	
Rajkot	N, P, S, Zn, Fe, B	
Anantapur	N, K, Mg, Zn, B	
Indore	-	
Rewa	N, Zn	
Akola	N, P, S, Zn, B	
Kovilpatti	N, P	
Bellari	N, P, Zn, Fe	
Bijapur	N, Zn, Fe	
Jhansi	N	
Solapur	N, P, Zn	
Agra	N, K, Mg, Zn, B	
Hisar	N, Mg, B	
S.K. Nagar	N, K, S, Ca, Mg, Zn, B	
Bangalore	N, K, Ca, Mg, Zn, B	
Arjia	N, Mg, Zn, B	
Ballowal-Saunkri	N, K, S, Mg, Zn	
RakhDhiansar	N, K, Ca, Mg, Zn, B	
Andhra Pradesh (Nalgonda, Mahabubnagar, Kurnool, Ananthapur and Prakasam)	S, B, Zn	Sahrawat et al.(2007)
Karnataka (Kolar, Tumkur, Chitra Durga, Haveri, and Dharwad)	S, B, Zn	Sahrawat et al. (2007)
Rajasthan (Bundi, Dungarpur, and Udaipur)	S, B, Zn	Sahrawat et al. (2007)
Districts of MP (Vidisha and Dewas)	S, B, Zn	Sahrawat et al. (2007)
Gujarat (Junagadh)	S, B, Zn	Sahrawatet al. (2007)
Tamil Nadu (Tirunelveli, Salem, Kanchipuram, Vellore, and Karur)	S, B, Zn	Sahrawat et al. (2007)

Impact of Drought on Water Nutrient Relations

Drought reduced the percentage of nitrogen and P, suggesting that it hindered nutrient uptake rather than growth. Drought-induced decreases in phosphorus uptake correlated with decreases in the concentration and activity of phosphorus uptake proteins, but drought-induced decreases in nitrogen uptake correlated only weakly with the level of nitrogen uptake proteins. Nutrient uptake proteins per gram of root decreased despite the increase per gram of total protein, as total protein per gram decreased more. Thus, drought-induced decreases in nutrient concentration, particularly phosphorus content, were likely caused at least in part by a decrease in the concentration of nutrient uptake proteins in roots in both drought-sensitive and drought-tolerant species (Bista et al., 2018).

Impact of Drought on the Availability of Nutrients

Drought in agricultural areas, or the deficiency of moisture in the soil, directly affects crop productivity because a lack of water lessens the transportation of soil nutrients. Drought increases the likelihood of nutrient loss from the rooting zone through erosion (Gupta, 1993). Nutrients are transported to the roots by water, so reduced soil moisture decreases nutrient diffusion, especially over short distances. It also reduces the mass flow of water-soluble nutrients such as nitrate, sulfate, calcium, magnesium, and silicon over longer distances.

Land Degradation

Land degradation is a significant issue in arid and semiarid regions. About 10% of arid lands show signs of past land degradation. This degradation is sometimes known as "desertification," and it can occur owing to soil erosion, nutrient depletion, water scarcity, increased salinity, or disturbance of biological processes (UNEP 2007). According to estimates, only 1%–6% of people in arid regions live in arid areas, but a substantially higher percentage are at risk from future desertification (MEA 2005). In an arid climate, these soils are prone to desertification or soil degradation because of a lack of vegetation cover and a weak structure. Desertification and soil salinization go hand in hand (Amezketa, 2006). As in other ecosystems that have degraded, desertification is primarily caused by rapid wind and water erosion, fertilizer and soil organic carbon (SOC) depletion, a reduction in plant-available water reserves, a decline in biodiversity, and a disruption in the cycling of elements (carbon, nitrogen, phosphorus, sulfur, and others).

Increasing Population Pressure

Because of the region's high susceptibility to drought and famine, arid areas are home to some of the world's poorest people. Population distribution patterns exhibit regional and climatic variation in arid regions. More than 1,400 million people, or 42% of Asia's population, live in arid lands. Despite having a smaller population, Africa has nearly the same number of people living in arid regions (41%). South America's arid areas are home to 30% of its population, or approximately 87 million people. In all these areas, agricultural sustainability is threatened by population growth and the increased exploitation of natural resources. Population increases in arid countries owing to improved health conditions and other causes have put enormous strain on the natural resource base.

Low External Inputs

Rainfed crops use less production inputs than irrigated crops, such as fertilizers, extra irrigation, high-quality seeds, insecticides, and herbicides. As a result, crops produce a meager harvest. Despite evidence that rainfed soils are prone to nutrient deficiency, balanced use of essential inputs in rainfed crops is rare. There are wide differences in the extent of inputs used for irrigated areas compared with rainfed areas because of numerous uncertainties in crop production in rainfed agriculture. The majority of rainfed farms still engage in minimal external input or subsistence farming including livestock, especially small ruminants, in rural areas. In many remote areas of the country, approximately 30% of farmers engaged in rainfed agriculture do not use chemical fertilizers or pesticides (Venkateswarlu, 2008). As a result, the crop response ratio to applied fertilizer nutrients is rapidly declining (NPK).

Low-Yielding and Risky Agriculture

Rainfed farming accounts for most of the agricultural production in hot, arid regions, and yields are low. Inherently weak soil and soil deterioration caused by human activity both significantly limit system production. The percentage of rainfall that is available to plants on poorly managed land might be as low as 40%–50%. On severely damaged soil, as little as 5% of the total rainfall may be used productively. The typical rainfed crops, such as til (*Sesamum indicum*), moth bean (*Vigna aconitifolia*), mung bean (*Vigna radiata*), and pearl millet (*Pennisetum glaucum*), are sown as sole crops and as mixed crops in varying quantities. Leaving land fallow in alternate years is also a common practice, but it is on the decline. As a result, cropping intensity is less than 100%. Groundnut (*Arachis hypogaea*), cotton (*Gossypium* sp.), wheat (*Triticum aestivum*), Indian mustard (*Brassica juncea*), chickpea (*Cicer arietinum*), plantain (*Plantago ovata*), and cumin (*Cuminum cyminum*) are produced wherever water is available for irrigation. Indian plum (*Ziziphus mauritiana*), amla (*Phyllanthus emblica*), pomegranate, citrus, and date palm are among the fruit crops grown.

Vulnerability to Climate Change

Climate change is the most serious threat, affecting the environment, economy, and society in a variety of ways. Climate change, resulting from changes in precipitation (P) and potential evapotranspiration PET, would alter aridity (i.e. P/PET) and hence the area of arid lands. Observations show that the global drylands have been growing during the past 60 years (Feng & Fu, 2013). Climate change is anticipated to affect the spatiotemporal distribution, availability, and demand for water, and shifting temperatures and precipitation may drastically modify both water availability and agricultural water requirements. According to Goyal's (2004) projections, increases in temperature will raise overall ET demand by 14.8%. Groundwater supplies are also at risk from climate change, which will increase future water supply uncertainty and worsen water shortage for agricultural productivity. Land degradation (wind erosion, groundwater depletion, and soil salinization) is also expected to worsen if the frequency of droughts, extreme temperature, and rainfall events increases. Heat stress caused by climate change offers significant challenges to the animal husbandry business, affecting animal productivity and reproduction. Additionally, expected climate change, in conjunction with intensified agriculture, irrigation, industrialization, mining, tourism, and urbanization, could have a negative impact on the survival, abundance, and distribution of plant species, and hence plant biodiversity (Kumar, 2016). As a result, people's ability to make a living in the region is anticipated to become increasingly precarious because of climate change. An extension of the growing season may be advantageous in cold, arid places, but modifications to agricultural and farming practices may be necessary to adapt to climate change.

Yield Gap

Soil quality plays an important role in improving crop productivity. Erratic or uncertain rainfall on degraded soils are major constraints in rainfed regions (Maruthi Sankar et al., 2010). Soil and climatic factors are the main issues limiting high productivity levels of rainfed crops grown by resource-poor farmers. Thus, the yield gaps between potential, on-station farm yields and adopted crop management practices are high (Srinivasarao et al., 2010). Unless better water and nutrient management techniques are adopted, together with modern farming practices and the use of improved seed, improved yields are unlikely.

Rainfed crops are mainly grown on marginal and submarginal lands. These lands generally receive reduced applications of fertilizers and manures. Consequently, nutrient deficiencies in crops are the main reason for low crop yields. For example, in pulses (food legumes), the average on-farm yield is approximately one-third of the potential yields on research plots, and approximately half of the yield in demonstration plots. Consequently, rainfed crop land in India has a potential for increased yields. Response of different rainy and post–rainy season crops to fertilizer application under dryland conditions are depicted in Table 3.3. In addition, the adoption of recommended management practices (RMPs), synchronizing nutrient application with available moisture, such as precipitation, and regular weed control can improve crop yields (Table 3.4).

Table 3.3 Response of Rainy and Post–Rainy Season Crops to Fertilizer Application Under Dryland Conditions.[a]

Crop	Response range	
	kg grain kg^{-1} N	kg grain kg^{-1} P
Rainy season crops		
Sorghum	3.4–43.4	2.4–59.0
Pearl millet	2.1–24.8	1.7–14.3
Finger millet	5.0–42.4	6.4–38.9
Maize	4.1–67.6	6.8–80.0
Setaria	5.9–17.9	6.4–38.9
Cluster bean	—	6.7–10.7
Green gram	—	1.5–11.6
Black gram	—	1.8–6.7
Pigeon pea	—	3.1–8.3
Post–rainy season crops		
Chickpea	1.9–23.5	4.5–38.2
Peas	—	3.7–10.9
Lentil	—	3.2–6.3

[a] *Source:* Singh and Das (1990).

Table 3.4 Average Yield of Crops Under Improved Versus Common Practices.[a]

Region	Crop	Mean yield with improved practices (Mg ha^{-1})	Yield increase over the common practice (%)
Agra	Pearl millet	1.72	117.72
	Barley	1.89	71.82
Anand	Pearl millet	3.29	34.29
	Tobacco	0.73	40.38
Anantapur	Groundnut	1.71	78.13
Hyderabad	Sorghum	2.89	134.96
	Castor	1.84	68.81
Kovilpatti	Sorghum	1.31	627.78
Ranchi	Upland rice	3.11	122.14
	Barley	1.59	253.33
Solapur	Sorghum	2.48	367.92
	Safflower	1.56	77.27
Varanasi	Barley	1.39	82.89

[a] *Source:* AICRPDA annual reports, 2006, 2007, 2008, 2010 (n.d.).

Strategies for Soil Health Management

Soil Test–Based Balanced Fertilization

Most Indian soil are deficient in NPK and in many secondary nutrients such as sulfur, calcium, and magnesium, and micronutrients boron, zinc, copper, iron, and manganese, among others. Productivity could be improved by balancing the nutrient application to crops based on the nutrient requirement and the native nutrient-supplying capacity of soil and introducing specific targeted yields. This approach could also minimize nutrient losses and reduce the cost of cultivation. In Andhra Pradesh and Telangana, balanced nutrition was applied in eight districts with promising results. For example, in Warangal, balanced nutrition improved cotton yields significantly. In some fields, cotton yields reached 1.6 t ha^{-1} with balanced nutrition registering an increase in yield of 10%–25% over traditional farming practices. In addition, farmers also achieved 30%–40% yield improvements in other crops owingto balanced nutrition (Srinivasarao et al., 2010).

Need-Based Fertilizer Applications

Need-based fertilizer application is a system in which the amount and type of fertilizer applications to the soils and/or plants is based on soil testing. A leaf color chart (LCC; Ali et al., 2015) and SPAD chlorophyl meter (Varinderpal-Singh et al., 2012) are used. The need-based nitrogen management through an LCC or chlorophyll meter reduces nitrogen requirement of rice from 12.5% to 25%, with no loss in yield and higher nitrogen use efficiency.

Site-Specific Nutrient Management

Site-specific nutrient management (SSNM) often results in efficient utilization of applied nutrients and higher crop productivity besides environmental benefits. In rainfed regions, most of the soils are nitrogen deficient and therefore nitrogen application is essential depending on crop and soil conditions. Available phosphorus building up in Alfisol regions of southern India owing to continued diammonium phosphate application. Potassium application is concentrated on cash crops such as cotton, tobacco, and banana. Micronutrient application for rainfed crops is rather rare. Therefore, multiple nutrient deficiencies are emerging in these regions. Magnesium deficiency is emerging in Bt cotton growing in red soil regions of the country. Therefore, based on soil and crop analyses, SSNM strategies were developed in eight districts of Andhra Pradesh and Telangana. Several farm trails on SSNM showed higher yields and higher benefit–cost ratio over conventional recommendations. Bt cotton showed a significant response to SSNM in cotton-growing districts of

Table 3.5 Influence of Balanced Fertilizer Application on Maize Green Fodder Yield in Dryland.[a]

Treatments[b]	Green fodder yield (q ha^{-1})[c]	Dry fodder yield (q ha^{-1})	Percentage green fodder yield increased over control	Percentage green yield increased over RDF
No N control	418.52	83.23	—	—
As per recommended package of practices (50% N as basal, remaining 50% at 30 days after sowing)	462.18	121.13	10.44	—
150 kg N ha^{-1} (40% basal) + remaining based on SPAD meter critical value of 40	464.23	124.36	10.92	0.44
150 kg N ha^{-1} (40% basal) + remaining based on LCC 4	466.21	128.09	11.39	0.87
150 kg N ha^{-1} (40% basal) + remaining based on SPAD meter critical value of 50	467.12	130.07	11.61	1.07
150 kg N ha^{-1} (40% basal) + remaining based on LCC 5	470.01	135.02	12.30	1.69

[a] *Source:* Rashid et al. (2021).
[b] LCC, leaf color chart; RDF, recommended dose of fertilizer,.
[c] 1 quintal (q) = 100 kg.

Telangana (Adilabad, Khammam, Warangal, Nalgonda, and Mahaboobnagar). Yield and use efficiency of nutrients were significantly influenced by balanced nutrition, particularly with addition of phosphorus and potassium along with nitrogen in rain-fed crops. Split application of potassium also increased cotton yield. The Influence of balanced fertilizer application and SSNM on maize green fodder yield (Table 3.5) and other crops under rainfed condition is presented in Table 3.6.

Integrated Nutrient Management

Intensive farming that includes the use of chemical fertilizers has degraded soil fertility and crop yields since the post–Green Revolution period. The decline in soil health and in overall environmental conditions are the main constraints limiting sustainability in Indian agriculture. It is therefore important to consider the inclusion of organic fertilizers, such as manures, biofertilizers, and crop residues, in cultivation practices. The precision application of natural fertilizers will help ensure environmental security. An integrated nutrient management (INM)

Table 3.6 Influence of SSNM on Various Crops Under Rainfed Condition.[a]

Location	Climate	Crop	Treatment	Treatment details (kg ha^{-1})	Grain yield (kg ha^{-1})	% of yield improved	References
New Delhi	Semi-arid	Maize	SSNM	170:40:48 $N:P_2O_5:K_2O$	4,620	14.64	Kumar et al. (2014)
			RDF	150:60:40 $N:P_2O_5:K_2O$	4,030		
Agra	Semi-arid	Pearl millet	SSNM	$120N:60P_2O_5:40K_2O:20S:10Mn:2.5Zn$	3,440	1.33	Kumar and Singh (2019)
			RDF	$120N:60P_2O_5:40K_2O$	3,090		
Agra	Semi-arid	Wheat	SSNM	$150N:60P_2O_5:40K_2O:20S:10Mn:2.5Zn$	5,880	16.43	
			RDF	$150N:60P_2O_5:40K_2O$	5,050		
Bangalore	Semi-arid	Finger millet	SSNM	Targeted yield of 4.0t ha^{-1} (155 : 32 : 203 kg ha^{-1} NPK)	3,852	84.31	AICRPDA Annual reports
			RDF	100% NPK (50:40:25 kg ha^{-1})	2,090		
Arjia		Sorghum	SSNM	125% N + $ZnSO_4$at 25 kg ha^{-1}	2,610	10.45	
			RDF	100% NPK	2,363		

[a] RDF, recommended dose of fertilizer; SSNM, site-specific nutrient management.

approach is an effective way to improve productivity and poor soils. Recent studies on the effect of INM practices involving various treatment combinations of organic and inorganic nutrients under different cropping systems have demonstrated increased yields and net returns and benefit–cost ratio. Studies also suggest that INM could be an effective and innovative practice for climate-resilient agriculture that also offers ecosystem sustainability. Organic manures play a significant role in INM packaging by increasing SOC, which supports microbial activity in the soil; helping retain soil moisture longer, which imparts drought tolerance during dry spells; and reducing the leaching of plant nutrients. Advantages of INM practices has been reported in hundreds of research articles in the Indian subcontinent (Bijay-Singh & Ali, 2020).

Many organic amendments for soil are available to growers such as crop residue, weed biomass, green manures, compost, animal bedding material, seri-waste, and so forth. In addition, materials from other sources can be used, such as municipal biosolids, poultry manure, coir pith, biochar, and tank silt, among others. The basic concept of the INM model is to modify soil fertility and supplement plant nutrients using an available amendment at an optimum level to sustain crop productivity and maintain plant nutrient levels from available resources in an integrated approach (Srinivasarao et al., 2017). Adoption of new interventions has added new dimensions in the INM system. Improved farm mechanization, conservation agriculture system intensification technologies, various organic amendments, and a renewed focus on recycling existing natural organic nutrient flows have further broadened the concept of INM, making it more context-specific for local environmental conditions (Wu & Ma, 2015). There are a multitude of organic and inorganic nutrient source combinations adopted in INM model, as shown in Table 3.7. Furthermore, nutrient management has improved the rainwater use efficiency in many rainfed crops (Table 3.8). Furthermore, INM contributes to soil quality improvement by acting as a source of energy, organic carbon, and available nitrogen for soil microbes. This, in turn, enhances soil microbial activity and nutrient cycling, thus improving overall soil health. Key soil quality indicators and best management practices in different rainfed cropping systems are depicted in Table 3.9.

Application of Amendments and Other Novel Materials

Soil moisture is an important factor affecting the availability of nutrients. When choosing fertilizer/manure rates in rainfed areas, it is important to take soil moisture reserves into account because plant nutrient uptake is positively correlated with soil moisture availability, which in turn promotes biomass production and root biomass addition to the soil, both of which increase SOC levels. In rainfed areas, adopting soil moisture conservation techniques such as mulching and adding tank silt may be seen as acceptable solutions.

Table 3.7 Effective Integrated Nutrient Management (INM) Practices for Rainfed Crops Across the Country.[a]

Location/state	Crops/cropping system	INM package[b]
Jhansi	Sorghum+ Dolichos	NPK at 60:20:0 + *Rhizobium*
	Cluster bean	NPK at 15:60:0+ *Rhizobium*
Rajkot	Sorghum	NPK at 90:30:0+FYM at6 t ha^{-1}
	Pearl millet	NPK at 80:40:0+FYM at6 t ha^{-1}
	Groundnut	NPK at 12:25:0+FYM at6 t ha^{-1}
	Cotton	NPK at 40:0:0+FYM at6 t ha^{-1}
	Groundnut	NPK at 6:12:0+mulching of sunhemp in between rows + *Rhizobium* and PSB
Solapur	Sorghum	NPK at 50:0:0+subabul loppings 9–10 t ha^{-1}
Indore	Soybean	NPK at 20:13:0+FYM at6 t ha^{-1}
	Soybean–safflower system	NPK at 20:13:0+ FYM 6 t ha^{-1}
	Soybean–pigeon pea system (2:2)	NPK at 45:40:25
Arjia	Corn-pigeon pea	NPK at 50:30:0+50% N through organics
	Safflower and rapeseed-mustard	NPK at 30:15:0+follow legumes such as green gram/chickpea
	Horse gram	Application of 45 kg P_2O_5
	Ground nut	Rhizobium + PSB inoculation in groundnut.
	Maize + pigeon pea intercropping	Combination of 50% of inorganic and remaining of inorganic source to FYM + compost (60:30:0 NPK kg ha^{-1})
	Maize–black gram	
Agra	Barley	NPK at 60:30:0+FYM +*Azotobacter*
	Pearl millet	
Ranchi	Soybean	NPK at 20:80:40+ *Rhizobium*
	Groundnut	NPK at 25:50:20+ *Rhizobium*
	Pulses	NPK at 20:40:0+ *Rhizobium*
	Upland rice	NPK at 40:30:20
		NPK at 30:20:0
Dantiwada	Green gram	NPK at 0:20:0+ *Rhizobium*
Jodhpur	Pearl millet	NPK at 10:0:0+FYM at 10 t ha^{-1}
Hoshiarpur	Corn	NPK at 80:40:20+FYM
	Wheat	NPK at 80:40:0
	Chickpea	NPK at 15:40:0

(*Continued*)

Table 3.7 (Continued)

Location/state	Crops/cropping system	INM package[b]
Akola	Cotton+green gram	NPK at 25:25:0+FYM to meet 25 kg N ha^{-}
	Pigeon pea	Application of 5 t of FYM + 40 kg P_2O_5 + microbial culture at1.5 kg ha^{-1} to pigeonpea
Hyderabad	Sorghum	Application of 2 t *Gliricidia* loppings + 20 kg N through urea
		Application of 4 t FYM + 2t *Gliricidia* loppings
		Application of 4 t FYM + 20 kg N through urea
	Mung bean	Application of 2t FYM + 10 kg N through urea under conventional tillage
		Application of 2t FYM + 1t *Gliricidia*loppings under reduced tillage
Faizabad	Sorghum	Application of 15 kg N through compost + 20 kg N through inorganic fertilizer
Varanasi	Upland rice	NPK at 40:30:20
	Upland rice-lentil	100 % organic
Kovilpatti	Cotton+urdbean intercropping system	20 kg N through FYM + 20 kg N as urea
Anantapur	Groundnut	NPK at 20:40:40+*Rhizobium* at500 g ha^{-1}
Rewa	Soybean and Black gram	Seed treatment of *Rhizobium* bacteria and PSB in soybean and urd bean
	Chickpea	Use of bio-fertilizer, *Rhizobium* +PSB in chickpea
	Rice–wheat	
RakhDhiansar	Maize	NPK at 60:40:20 + $ZnSO_4$ at 20 kg ha^{-1}
Bangalore	Groundnut–finger millet	100% RDF +FYM at 10 t ha^{-1}
S.K. Nagar	Pearl millet–cluster bean–castor	50% RDN + 50% RDN FYM

[a] *Source:* AICRPDA Annual Reports, 2005–2016.
[b] FYM, farmyard manure; PSB, Phosphorus-Solubilizing Bacteria ; RDF, recommended dose of fertilizer; RDN, recommended dose of nitrogen.

Tank Silt Application

For decades, community tank systems have been an important component of rural life. Tank silt has a high-water retention capacity and is a rich source of nutrients. The potential of tank silt in supplying organic carbon and various nutrients was demonstrated by an analysis of many tanks in the Warangal, Anantapur, and Nalgonda districts of Andhra Pradesh and Telangana. Beneficial results are seen most clearly in soils with a light texture. Tank de-silting at Warangal revealed the

Table 3.8 Effect of Different Nutrient Management on Yield and Rainwater Use Efficiency in Different Dryland Cropping Systems.

Location	Cropping system	Treatment[a]	Yield increase over control (kg ha^{-1})	Rainwater use efficiency (kg $ha^{-1}mm^{-1}$)
Chianki	Maize	Raised bed sowing + polythene mulch	1,700	4.74
Chianki	Cereal–legume rotation	50% NPK (inorganic)+50% FYM	570	2.54
Faizabad	Pigeon pea+ sorghum intercrop system	Pigeon pea + sorghum	491	2.41
		PP:75% RDF + FYM 5 t ha^{-1}+ S at40 kg ha^{-1} + $ZnSO_4$ at 25 kg ha^{-1} + B at1.5 kg ha^{-1}		
Jagdalpur	Groundnut	100%RDF + lime at3 g ha^{-1} + $MgSO_4$at 15 kg ha^{-1} + FYM at5 t ha^{-1} in furrows	1,079	2.38
Jagdalpur	Direct seed drive-field pea system	100% NPK + FYM at5 t ha^{-1} + $ZnSO_4$ at25 kg ha^{-1} + lime at3 g ha^{-1}	1,735	2.88
Arjia	Maize	Kat 60 kg ha^{-1}	537	4.29
Arjia	Maize	Mgat 45 kg ha^{-1}	718	4.15
Arjia	Maize–black gram	25 kg N through FYM and 25 kg N through IF + P_2O_5 at 30 kg ha^{-1}	1,569	5.54
BallowalSaunkhri	Maize	K_2O at 40 kg ha^{-1}	516	4.85
Ballowal Saunkhri	Maize	$MgSO_4$ at 45 kg ha^{-1}	436	4.67
Ballowal Saunkhri	Maize	NPK (100% RDF) + FYM at10 t ha^{-1}	1785	6.79
Ballowal Saunkhri	Wheat	NPK (100% RDF) + FYM at10 t ha^{-1}	1058	24.0
Rakh Dhiansar	Maize	100% NPK + $ZnSO_4$at 20 kg ha^{-1}	1055	3.04
Rakh Dhiansar	Gobhisarson	FYM at10 tha^{-1}	579	2.84
Rakh Dhiansar	Wheat	Foliar spray of 0.5% K (KCl) + foliar spray of 0.5% N (urea)	1,369	22.0

[a] IF; FYM, farmyard manure; PP.

Table 3.9 Key Soil Quality Indicators and Best Management Practices in Different Rainfed Cropping Systems.

Location (Soil type)	Cropping systems	Key soil quality indicators[a]	Best management practices	References
Solapur, Maharashtra (Vertisols)	Post-monsoon sorghum	pH, EC, OC, NPK, Ca, Mg, S, Zn, Fe, Cu, Mn, B, DHA, MBC, LC, BD, MWD	The application of 25 kg nitrogen (N) ha^{-1} through crop residue (CR) + 25 kg N ha^{-1} (urea) showed the highest soil quality index (SQI) of 2.36, which was onpar with other treatments receiving farmyard manure (FYM) and crop residues along with urea.	Sharma et al. (2018)
Rakhdhiansar, Jammu and Kashmir (Inceptisols)	Maize	pH, EC, OC, NPK, Ca, Mg, S, Zn, Fe, Cu, Mn, B, DHA, MBC, LC, BD, MWD	The combination of low tillage practices with 100 % organic source of nutrients was quite superior in maintaining a relatively higher relative soil quality index(RSQI).	Sharma et al. (2020)
Kovilpatti, Tamil Nadu (Vertisols)	Sorghum + cowpea system	OC, available S, LC, BD	Application of 15 kg N (compost) + 20 kg N (inorganic) (2.40)	Sharma et al. (2019)
	Sole sorghum system	OC, available P, available S, available Zn, LC, BD	Application of 15 kg N (compost) + 20 kg N (inorganic) (3.69)	
	Sole cowpea system	EC, OC, available S, available Zn, MBC	1. Application of 25 kg N (compost) (2.27) 2. Application of 15 kg N (compost) + 20 kg N (inorganic) (2.17)	

Rakhdhiansar, Jammu and Kashmir Inceptisol	Maize–mustard	MBC, DHA, pH, EC, OC, available P, available S, available Zn	The conjunctive use of organic sources of N such as FYM and crop residues in combination with inorganic sources (at50:50 on the basis of N equivalents) proved very effective in improving the soil quality indicators and indices under maize–mustard crop rotation in rainfed hill Inceptisol soils.	Sharma et al. (2022)
Parbhani center, Maharashtra (Vertisols)	Sorghum + pigeon pea (4:2) system	MWD,LC,DHA, OC, available S	1. Application of FYM at5 t ha^{-1} (2.47) 2. Application of 25% RDF + FYM at2.5 t ha^{-1} (2.43) 3. Application of 25% RDF + Gliricidia at1.5 t ha^{-1} (2.31)	Sharma et al. (2021)
	Soybean + pigeon pea (4:2) system	LC, DHA, MWD, available S, pH	1. Application of FYM at5 t ha^{-1} (2.16) 2. Application of *Gliricidia* at3 t ha^{-1} (semi-dried) (2.00) 3. Application of 25% RDF + *Gliricidia* at1.5 t ha^{-1} (1.97)	
Rakh Dhiansar, Jammu and Kashmir Inceptisol	Maize–wheat cropping system	BD, available K,OC, available P available N, available Fe and Zn, MBC	The application of FYM at10 t ha^{-1} + 40 kg N ha^{-1} maintained significantly highest SQI of 5.66, which was onpar with the application of FYM at10 t ha^{-1} + 30 kg N ha^{-1} and green manuring with sunhemp + 20 kg N ha^{-1} both of which maintained SQI of 5.40	Chary et al. (2020)

[a] BD, bulk density; DHA, dehydrogenase assay; EC, electrical conductivity; LC, labile carbon; MBC, microbial biomass carbon; MWD, mean weight diameter; OC, organic carbon; FYM, Farm Yard Manure; RDF, Recommended dose of fertilizer; SQI, soil quality indices.

existence of all the vital nutrients required for plant growth in sufficient proportions. The recycling of tank silt could address nutritional deficiencies in many soils, particularly those containing zinc, boron, and sulfur. Moreover, it can increase the SOC content of the soil, enhancing soil health. Tank silt is a nutrient-rich substance, and its application at rates of 50, 100, 150, and 375 tractor loads ha^{-1} improved water content by 0.002, 0.007, 0.012, and 0.032 $g\,g^{-1}$ soil, respectively (Osman, 2008).

Vermicompost

Living organisms such as earthworms help to maintain nutrient flows in soil. They also minimize environmental degradation. Populations of earthworms decrease due to soil degradation, and such declines can be used as a indicator of soil health. Earthworms consume organic matter, reducing the volume by 40%–60%. An earthworm weighs about 0.5 to 0.6 g and can consume its body weight each day. It also produces casts equivalent to about 50% of the matter it consumes. Vermicasts are the main component of vermicompost. They contain nearly double the amounts of macro and micronutrients than regular compost.

Biochar

Biochar is a carbon-rich byproduct of biomass pyrolysis that forms under oxygen-depleted conditions. It is applied to soil to enhance its agronomic and ecological benefits (Diatta et al., 2020). Biochar may be especially beneficial in arid places since it can boost soil moisture (Novak et al., 2012) and improve soil carbonsequestration potential (Stewart et al., 2013), as well as mitigate the detrimental impacts of salinization (Akhtar et al., 2015). Biochar provides a favorable environment for microbes because of a porous structure and the presence of water and nutrients (Thies & Rilling, 2009). Biochar can improve plant growth by enhancing soil physical properties (bulk density, surface area, water-holding capacity, permeability; Sun & Lu, 2014) and soil chemical properties (high salinity, nutrient retention and availability, cation exchange capacity, and pH; Abel et al., 2013). Moreover, biochar enhances the biological characteristics of soil by fostering microbial communities in a favorable habitat and increasing their diversity (Abujabhah et al., 2016). Biochar's resistance to chemical and biological processes contributes to its long-term agronomic and environmental benefits (Fang et al., 2014). Biochar, like any other organic source, will decompose over time, although it will do so much more slowly than any other organic matter (O'Laughlin & McElligott, 2009). Although a portion of the carbon in biochar is labile and decomposes quickly, most of the carbon is in an extremely stable form owing to the presence of aromatic structures. Consequently, biochar takes hundreds to thousands of years to disintegrate in soil and even longer under marine conditions (Arif et al., 2020). The composition of stable and refractory forms of organic carbon after plant organic matter has undergone pyrolysis determines how well biochar will sequester carbon over the long run. A crop's response to biochar application may vary depending on soil type, char source, and other factors. The inherent fixed carbon in raw biomass that would otherwise decompose to greenhouse gases (GHGs) is trapped in soil for years, giving biochar the ability to mitigate climate change.

According to Lehmann et al. (2006), biochar production from agriculture and forestry wastes has a potential worldwide carbonsequestration of 0.16 Gt annually. Biochar could reduce non-CO_2GHG emissions from soil by restricting either stage of nitrification or by inhibiting denitrification. It could also lead to a reduction of N_2O and increase CH_4 uptake by soil and the long-term C sequestration in soil. It is predicted that India has an annual biochar production potential of 162 and 32.7 Mt from various crop and woody residues, respectively, with a total C sequestration potential of 95 Mt per year from the incorporation of this biochar into soil (Venkatesh et al., 2018).Studies over 9 years on the residual influence of different biochars in rainfed Alfisols growing maize revealed that biochar made from maize stalks provided better benefits than biochar prepared from castor, cotton, and pigeon pea stalks. It particularly improved the available nitrogen and phosphorus in soil. It also increased crop yield. Maize (DHM 117) grain yield varied from 0.26 to 3.14 t ha^{-1} under maize stalk biochar, 0.34 to 2.72 t ha^{-1} under castor stalk biochar, 0.33 to 2.97 t ha^{-1} under cotton stalk biochar, and 0.49–2.32 t ha^{-1} under pigeon pea stalk biochar, with different rates of combined organic and inorganic fertilizers (Venkatesh et al., 2018). In another study with pigeon pea in Alfisol, applications of 3–6 t ha^{-1} pigeon pea biochar used every year or alternate year produced higher pigeon pea seed yields comparable to cotton and castor biochar being used. Castor biochar applications had no significant effect on pigeon pea seed yield. Applications of 3–6 t ha^{-1} cotton biochar or pigeon pea biochar on alternate years produced higher yields compared with an application every year over 4 years. However, an application of biochar every year recorded higher organic carbon in soils compared with an application every other year. As expected, application of 6 t ha^{-1} biochar increased the organic carbon content in Alfisol significantly over the 3 t ha^{-1} application, irrespective of the type of biochar (Table 3.10; Venkatesh et al., 2018). The application of biochar resulted in an increase in soil organic carbon in various locations, as illustrated in Table 3.11.

Polyhalite

An emerging multi-nutrient fertilizer is polyhalite, which has the formula $K_2Ca_2Mg(SO_4)_4$, $2H_2O$ with a triclinic crystal system (Hudson Institute of Mineralogy, 2020). Crushing and granulating mined polyhalite without the use of chemicals yields a low-carbon-footprint fertilizer that can be used in organic farming. Anglo American

Table 3.10 Effect of Application of Different Biochars on Rainfed Pigeon Pea Yield and Organic C Content of Alfisol Soil.

Treatment[a]	4 years mean pigeon pea seed yield (kg ha^{-1})	Organic C content in soil after 4th year (g kg^{-1})
Control (no fertilizer and no biochar)	441.6	2.8
Recommended dose of fertilizers (RDF)	823.9	3.3
CASB at3 t ha^{-1} + RDF	1,445.4	6.4
CASB at6 t ha^{-1} + RDF	1,379.2	9.0
COSB at3 t ha^{-1} + RDF	1,388.2	6.4
COSB at6 t ha^{-1} + RDF	1,584.6	6.9
PPSB at3 t ha^{-1} + RDF	1,341.8	6.8
PPSB at6 t ha^{-1} + RDF	1,721.3	8.9

[a] CASB; Castor stalk biochar; COSB; Cotton stalk biochar; PPSB Pigeionpea stalk biochar
Venkatesh et al. (2018).

Table 3.11 Percent Difference in Soil Organic Carbon (SOC) with Biochar.

Location	Treatment	% SOC increased over the control	Reference
Egypt	Biochar rate 50 t ha^{-1}	49.51	El-Azeim and Haddad (2017)
	Biochar rate 100 t ha^{-1}	93.69	
	Biochar rate 150 t ha^{-1}	10.44	
India	Maize stalk biochar at 5.0 t ha^{-1}	33.33	Pandian et al. (2016)
	Cotton stalk biochar at 5.0 t ha^{-1}	30.56	
	Red gram stalk biochar at 5.0 t ha^{-1}	33.33	
China	Biochar applied at 50 t ha^{-1}	60.98	Yeboah et al. (2018)
India	75% recommended dose of fertilizers (RDF) + 5.0 t ha^{-1} Parthenium biochar	41.78	Choudhary et al. (2022)
	75% RDF+ 10.0 t ha^{-1} *Parthenium* biochar	51.84	
	75% RDF+ 5.0 t ha^{-1} *Lantana* biochar	46.23	
	75% RDF+ 10.0 t ha^{-1} *Lantana* biochar	54.55	
China	Biochar rate 7.5 t ha^{-1}	4.17	Peng et al. (2019)
	Biochar rate 15 t ha^{-1}	9.21	
	Biochar rate 30 t ha^{-1}	32.85	

Crop Nutrients, United Kingdom, is marketing this novel mineral fertilizer as POLY4, following deep-mine extraction in the United Kingdom. It contains calcium (17%), magnesium (6%), sulfur (19%), and potassium (14%), which are all essential plant nutrients. POLY4 also includes trace levels of many micronutrients such as boron, zinc, manganese, molybdenum, iron, and copper, which may greatly boost crop output and soil health all at once (Anglo American Crop Nutrients, 2020).In comparison with other fertilizers such as KCl and sulfate of potash, the POLY4 granular product's superior spreading qualities and prolonged nutrient release to the soil profile are particularly noteworthy (Vale, 2016).

Hydrogels

Hydrogels are a type of cross-linked polymer with a three-dimensional structure that expands when exposed to water and can store a large amount of liquid (Narjary et al., 2013). It performs as a super-absorbent polymer. In agriculture, hydrogel polymer compounds show great promise as a soil conditioner, increasing crop tolerance and growth in sandy soil (Ovalesha et al., 2017). Hydrogel polymers can absorb water up to 100 times their original weight in a short amount of time and then desorb the absorbed water under stress conditions. Polymers are essential in dryland agriculture because they can be used as structural elements to create a climate that is favorable to plant growth, while also boosting water use efficiency. Hydrogels are useful in agriculture because they hold water and prevent soil erosion, and slow-release polymer hydrogels act as carriers of nutrients in the soil. This is a key approach for improving fertilizers' efficiency, as it reduces nutrient losses related to leaching, lowers costs, and reduces pollution for the environment. Hydrogels can be used as a slow-release water source in soil because, upon hydration, they create an amorphous gelatinous mass that can absorb and release water over a long period of time. The hydrogel particles are also thought of as "miniature water reservoirs" in the soil, from which water will be released at the root's command via an osmotic pressure difference (Azzam, 1980). The use of hydrogels increased soil moisture volume in pigeon peas from 8.3% to 10%, demonstrating its effectiveness as a technique in arid and semi-arid settings (Gilbert et al., 2014).

Cover Crops

Cover crops are an essential component of sustainable agriculture. Using biological nitrogen fixation, these crops increase soil fertility without the use of chemical fertilizers (Srinivasarao et al., 2011). Using cover crops can reduce soil compaction, retain soil moisture, reduce energy use, and provide forage as a byproduct. Farmers can grow specific cover crops based on the overall requirements of the land as well as their desired needs and outcomes. Summer cover crops can be grown to fill gaps in crop rotations, help amend soils, suppress weeds, and avoid leaving the soil exposed. Winter cover crops provide ground cover over the winter and can help prevent erosion.

Cover crops can increase soil organic matter status by protecting soil layers from various types of erosion. They contribute to accumulation of soil biomass, particularly below the soil surface, and their roots create habitat for microorganisms. These contributions to soil biology help to support balanced nutrient management in soil. Another consideration is CO_2, which is a major GHG emitted by vehicles, power plants, and the combustion of fossil fuels, among other things. Sequestering atmospheric carbon in soils helps to mitigate GHG emissions. Cover crops are a viable option for soil carbon sequestration because they feed many soil organisms (bacteria, fungi, etc.), increasing soil carbon levels over time. The percent change in soil organic carbon (SOC) with cover crops is shown in Table 3.12 and the correlation between cover crops and enhanced carbon sequestration is quantified in Table 3.13.

Conservation Agriculture

Conservation agriculture promotes resource-saving agricultural crop production. Its goal is to produce acceptable profits from high and sustainable production levels while conserving the environment (FAO, 2007). Conservation agriculture lessens soil disturbance, which is essential to conserving minerals, limiting erosion, and preventing excess evaporation. Under conservation agriculture guidelines, topsoil is carefully managed so that an organic soil cover provides an environment for organisms to thrive. Conservation agriculture also encourages crop rotations or crop interactions. Crop rotation can be used for disease control by not allowing an environment for specific crop diseases to survive and spread. Rotating crops acts like a natural insecticide and herbicide against crop specific disease and pests (FAO, 2007).

Techniques to restore soil health include no-till or conservation tillage, leaving crop residue on the soil surface, and planting cover crops. These practices also promote an increase in soil organic matter. Long-term experiments conducted from 2005 to 2012 with sorghum-cowpea on Alfisol revealed that reducing tillage and an application of sorghum residue at 6 t ha^{-1} and 4 t ha^{-1} led to 21% and 16% higher sorghum yields, respectively, over control (No residue); whereas, the increase in the cowpea yield was 50% and 60%, respectively (Sharma et al., 2017; Table 3.14).In addition, the application of residues with reduced tillage had a positive effect on soil carbon fractions. The amount of the very labile fraction of carbon was significantly higher (3.96 g kg^{-1}) with the application of sorghum stover 6 t ha^{-1} in the 0–5 cm soil layer. Similar trends were observed in the case of labile and less labile organic carbon fractions. The lability index rose after increasing the application of residue. Smaller size soil aggregates (0.053 mm) contained a significantly higher amount of SOC compared with the large-sized fractions (2 mm). In 0.053-mm aggregate fractions, the SOC content was significantly higher (5.34 g kg^{-1}) after an application of sorghum stover at 6 t ha^{-1} and was 19%

Table 3.12 Percent Difference in Soil Organic Carbon (SOC) with Cover Crops.

Location	Climate	Cover crop	% SOC increased	Depth (cm)	References
China	Semi-arid	Huai bean–winter wheat	10.50	0–20	Zhang et al. (2019)
		Soybean–winter wheat	12		
		Mungbean–winter wheat	15.60		
USA	Semi-arid	Rye cover crop with no-tillage (R-NT)	11	0–15	Lewis et al. (2018)
		Mixed species cover crop with no-tillage (M-NT)	73		
USA	Semi-arid	No-till with rye cover crop (NT-rye)	77.33	0–10	DeLaune et al. (2019)
Spain	Semi-arid	Permanent cover crop of resident vegetation (RV)	140	0–5	Peregrina et al. (2012)
Italy	Semi-arid	High nitrogen supply legume cover crop	10.48	0–10	Mazzoncini et al. (2011)
Japan	Semi-arid	Winter cover cropping with hairy vetch in soya bean cultivation	10.87	0–2.5	Wulanningtyas et al. (2021)
		Winter cover cropping with rye in soya bean cultivation	15.22		
		Winter cover cropping with hairy vetch in soya bean cultivation	6.82	2.5–7.5	
		Winter cover cropping with rye in soya bean cultivation	9.76		

Table 3.13 Quantifying the Correlation Between Cover Cropping and Enhanced Carbon Sequestration.

Crop	C sequestration potential	References
Legume/non-legume mixes	$0.54\ t\ C\ ha^{-1} yr^{-1}$	Abdalla et al. (2019)
Legumes and non-legumes	$0.32\ t\ C\ ha^{-1} yr^{-1}$	Poeplau and Don (2015)
Legumes and grasses	$0.6–0.9\ t\ C\ ha^{-1} yr^{-1}$	Follett (2001)
Legumes and grasses	$0.3–0.7\ t\ C\ ha^{-1} yr^{-1}$	Lal et al. (1998)
Grassland	$0.8\ t\ C\ ha^{-1} yr^{-1}$	Bruce et al. (1999)
Winter cover crops	$0.2\ t\ C\ ha^{-1} yr^{-1}$	Sperow et al. (2003)
Legumes and grasses	$0.56\ t\ C\ ha^{-1} yr^{-1}$ (+15.5%)	Jian et al. (2020)
Legume/non-legume mixes	+8.5% increase of total soil carbon	McDaniel et al. (2014)

Table 3.14 Effect of Graded Levels of Residues and Minimum Tillage on Carbon Fractions in Alfisol (0–5 cm) Under a Sorghum-Cowpea System After 8 years of Experimentation.

	Soil carbon fractions ($g\ kg^{-1}$)					
Treatment	Organic C	Very labile C	Labile C	Less labile C	Lability index	OC in 0.053-mm aggregates
Control	5.77	1.75	0.62	0.39	1.19	4.50
Crop residue ($2\ t\ ha^{-1}$)	6.74	1.82	0.71	0.27	1.06	5.06
Crop residue ($4\ t\ ha^{-1}$)	6.85	3.80	0.89	0.30	1.97	5.19
Crop residue ($6\ t\ ha^{-1}$)	7.01	3.96	1.03	1.03	2.14	5.34
LSD ($P = 0.05$)	0.31	0.29	0.15	0.23	1.19	0.22

[a] *Source:* Sharma et al., 2017/Taylor & Francis Group.

higher over no residue control. Under a rainfed castor–pigeon pea sequence in Alfisols, conventional tillage recorded 24% higher yields over zero tillage, but conventional tillage and reduced tillage were on par with each other. Over time, the yield gap between the tillage treatments is narrowing. The zero tillage or reduced tillage with the retention of crop residue in different cropping systems not only produced similar or higher crop yields but also improved the long-term water retention of soils. They also reduced GHG emissions in comparison with conventional tillage practices.

Foliar Spraying

Split and foliar application of several nutrients leads to an improved nutrient use efficiency and contributes to ecosystem services. It also reduces the basal dose of nutrient application, and farmers can calculate the remaining doses based on crop conditions and other factors to achieve higher crop yields. Foliar nutrition of potassium also contributes to drought tolerance during inter-droughts or mid-season droughts. Spraying concentrations of different micronutrients are presented in Table 3.15.

Nano-fertilizers

Nanotechnology has greater potential in rainfed agriculture for efficient use of water and fertilizers by crop plants. Many techniques are available to synthesize different kinds of nanomaterials. Nano-fertilizers provide essential nutrients for plant growth with higher use efficiency and can deliver nutrients in a timely manner in the rhizosphere. There are numerous nano-products in different countries with immediate application to address the issue pertaining to low use efficiency of inputs such as fertilizers. Recent studies on nanotechnology are showing the importance of nanoparticles in increasing the bioavailability of nutrient elements in soil. An investigation was carried out with several commercially available nanoparticles of zinc oxide (<100 nm), copper oxide (<50 nm), and iron oxide (<50 nm) to study their stability in aqueous system and their assimilation by plant. The solution culture experiment with nanoparticles of ZnO (<100 nm) and soluble $ZnSO_4$, showed that ZnO nanoparticles had a stimulating effect, resulting in higher growth of maize at 0.28 mg kg^{-1} as compared with 0.50 mg kg^{-1} zinc as $ZnSO_4$. Yield increases associated with the application of nano-fertilizer in different crops are presented in Table 3.16.

Table 3.15 Spraying Concentrations of Different Micronutrients.

Micronutrient	Fertilizer	Spraying concentration (%)
Zn	$ZnSO_4$	0.5
B	Boric acid/Agri bor (H_3BO_3)	0.2
Mn	$MnSO_4$	0.2
Cu	$CuSO_4$	0.2
Fe	$FeSO_4$	1.0
Mo	$Na_2MoO_4/(NH_4)_6Mo_7O_{24}$	0.05

Table 3.16 Percent Yield Increased by the Application of Nano-fertilizer in Linseed Crop.[a]

Treatments[b]	Grain yield (kg ha^{-1})	Straw yield (kg ha^{-1})	% grain yield increased over the control	% grain yield increased over soil application of fertilizer
T_1, Control	850.32	1,823.14	—	—
T_2, 100% NPK (RDN through soil)	1,120.34	2,102.23	31.76	—
T_3, 100% NPK + 1 foliar application of 0.4% Nano-N at 40 DAS	1,205.55	2,295.39	41.78	7.61
T_4, 100% PK + 3 foliar applications of 0.4% Nano-N at 20, 40, and 60 DAS	1,322.22	2,512.74	55.50	18.02
T_5, 100% PK + 2 foliar sprayings of 0.4% Nano-N at 20 and 40 DAS	1,277.77	2,442.53	50.27	14.05
T_6, 75% N (RDN) and 100% PK +1 foliar application of 0.4% Nano-N at 40 DAS	1,155.55	2,249.13	35.90	3.14
T_7, 50% N (RDN) and 100% PK +2 foliar applications of 0.4% Nano-N at 20 and 40 DAS	1,440.33	2,894.92	69.39	28.56
T_8, 25% N (RDN) and 100% PK +3 sprayings of 0.4% Nano-N at 20, 40, and 60 DAS	1,398.96	2,742.46	64.52	24.87

[a] *Source:* Khule et al. (2023).
[b] DAS, days after sowing; RDN, recommended dose of nitrogen.

Fertigation

Water soluble fertilizers are used in sprinklers, drip irrigation systems, and foliar sprays and help improve crop yields and the quality of fruit and vegetable crops. These fertilizers have a low salt index, which avoids burning plant tissues; are 100% soluble in water; and are suitable for foliar application or fertigation. Water-soluble fertilizers are basically a combination of all essential nutrients with different ratios suited to the type of crop, quality of water, status of soil fertility, and climatic conditions (Malhotra, 2016). The nutrient composition of different water-soluble fertilizers is presented in Table 3.17. Various researchers have studied the effect of application of water-soluble fertilizers on increasing productivity. This research shows that much better nutrient use efficiency is observed with liquid and water-soluble fertilizers applied through fertigation in mango, litchi, grapevine, banana, kiwifruit, sweet cherry, citrus, apple, okra, garlic, squash, and bell pepper, among other crops.

Organic Farming

Organic farming promotes self-sufficiency in nitrogen. Mixed organic farms can recycle manure from livestock and compost crop residues to create natural fertilizers. Legume crops can deliver nitrogen in acceptable quantities especially on organic arable farms. Nitrous oxide emissions are directly related to the proportion of available mineral nitrogen in soils. High emission rates are observed immediately following the application of fertilizers but are extremely variable. Denitrification also increases in compacted soils. According to the Intergovernmental Panel on

Table 3.17 Fertilizer Control Order (FCO) Approved 100% Water-Soluble Fertilizers.[a]

	Nutrient composition (%)						
Product (grade)	**N**	**P**	**K**	**S**	**Ca**	**Mg**	**Zn**
Potassium nitrate (13-0-45)	13	0	45				
Mono potassium phosphate (0-52-34)	0	52	34				
Calcium nitrate	15.5				18.8		
Potassium magnesium sulfate			22	20		18	
Mono ammonium phosphate (12-61-0)	12	61	0				
Urea phosphate (17-44-0)	17	44	0				
Urea phosphate with sulfate of potash (18-18-18)	18	18	18	6.1			
NPK Zn (7.6-23.5-7.6-3.5)	17.6	23.5	7.6				3.5

[a] *Source:* Malhotra (2016).

Climate Change, 1.5% of applied nitrogen is released as nitrous oxide. The ban on mineral nitrogen and restricting livestock numbers per hectare in organic farming significantly reduces the concentration of mineral nitrogen in soils and thus reduces N_2O emissions. The following factors also help to lower nitrous oxide emissions:

- Rotating crops with the planting of green manure improves soil health and reduces nitrous oxide emissions.
- Organic soils are more aerated and have much lower mobile nitrogen concentrations.
- Organic farming practices, such as the use of green and animal manure, crop rotations with intercropping and cover cropping, and composting processes, may help to reduce soil erosion and help sequester carbon.

The results of field experiments conducted on Alfisol soil revealed that the seed yield of sunflower was 14% and 7% higher in the plots under INM than those under inorganic and organic management. However, plots under organic management gave slightly higher seed yields of green gram and pigeon pea. Plots organically managed recorded much higher soil organic carbon (Sammi Reddy et al., 2019). In comparison with INM and inorganic nutrient management, organic practices significantly reduced N_2O emissions in all three crops. However, organic management resulted in higher CO_2 emissions than INM and inorganic nutrient management (Sammi Reddy et al., 2019). Table 3.18 presents the package of organic nutrient management practices for various crops.

Natural Farming

Natural farming (NF), also known as zero budget natural farming (ZBNF), is a disruptive farm practice that aims to answer farmers' key concerns about rising production costs. It essentially envisions ecological or regenerative agriculture practices in which any type of chemical application to soil bio-systems is prohibited. It depends on soil biology rather than soil chemistry by encouraging multi-cropping, year-round soil cover, and the application of cow dung and urine to stimulate the microorganisms in the soil. However, the scientific community regards it as unscientific and unproven. Nevertheless, thousands of farmers across India are using NF practices. The supporters and practitioners are confident and upbeat about it, although many in the scientistic community have serious doubts about the claimed benefits. Hence, it is important to study the adoption of NF practices and their effect on crop choices, crop yield, and farm income, as well as the possibilities of scalability and sustainability for agriculture.

Table 3.18 Package of Organic Nutrient Management Practices of Various Crops.[a]

Location	Crop	Package of organic nutrient management practices[b]
Tamil Nadu	Cotton	Basal application of FYM 7.05 $t\,ha^{-1}$+ vermicompost 4.49 $t\,ha^{-1}$+ *Azospirillum* 2kg ha^{-1}+ phosphobacteria 2kg ha^{-1}+ *Pseudomonas* 2.5 kg ha^{-1} +*Trichoderma* 2.5 kg ha^{-1}. Top dressing of vermicompost 1 tha^{-1} at 45 DAS and Panchagavya 3% spray at 30, 60, and 90 DAS.
Maharashtra	Groundnut	Basal application of FYM 1,500 kg ha^{-1} + neem cake 160 kg ha^{-1}+ vermicompost 560 kg ha^{-1}. Top dressing of cow urine 50 lit ha^{-1} and Vermiwash 50 L ha^{-1} spraying at 30 and 60 DAS.
Madhya Pradesh	Soybean	Basal application of cow dung manure 5 $t\,ha^{-1}$ (0.95% N)
Himachal Pradesh	Maize (kharif)	Basal application of FYM 16 $t\,ha^{-1}$ + vermicompost 12 t+ rock phosphate 60 kg Top dressing of Cow urine 60 L in 600 L water ha^{-1} spraying at 30, 45, 60, and 90 DAS. Spraying of Panchagavaya 18 L in 600 L water ha^{-1} at 60 and 90 DAS Spraying of Vermiwash 60 L in 600 L water ha^{-1} at 60 and 90 DAS
Jharkhand	Lentil	Basal application of FYM 1.4 $t\,ha^{-1}$ + Karanj cake 0.2 $t\,ha^{-1}$ Top dressing of vermicompost 0.7 $t\,ha^{-1}$ at 25–30 DAS

[a] *Source:* Ravisankar et al., 2017, Scientific Package of Practices (PoPs) for organic production of crops in cropping systems ICAR-Network Project Organic Farming, ICAR-IIFSR.
[b] DAS, days after sowing; FYM, farmyard manure.

Zero Budget Natural Farming (ZBNF1) was promoted in the mid-1990s by agriculturist Sh. Subhash Palekar, who was awarded one of India's highest civilian awards, the Padma Shri, in 2016 for promoting these alternative farming practices (Khadse et al., 2017; Mishra, 2018; Niyogi, 2018; Economic Survey, 2019). It has led to widespread acceptance in many states, including Andhra Pradesh, Karnataka, Maharashtra, and Himachal Pradesh. It is thought that ZBNF1 can substantially reduce production costs by using intercropping, mulching and replacing chemical fertilizers and pesticides with home-grown products such as Jeevamritha, Beejamritha, and Neemastra.(Palekar, 2005). He claims that the process requires only one cow for every 30 acres of land. It is also intended to boost soil organic carbon without the need for large amounts of farmyard manure as in organic farming. The goal is the attainment of sustainable agriculture with a lower carbon footprint. The Economic Survey (2019) emphasized the importance of ZBNF as an alternative farming practice for improving farm income.

Sand Dune Stabilization Through Vegetation Cover

Shifting sand dunes are widespread in India's arid environments (150–400 mm per year), especially near human habitations. Sand causes significant problems for agriculture, but there are ways to combat the problem. These techniques include protecting against biotic interferences, the treatment of shifting sand dunes by installing barriers in parallel strips or in a "chessboard" design, using vegetation, or the afforestation of treated dunes through direct seeding and planting. *Ziziphus nummularia* and *Crotalaria burhia* are the two most widely used species for creating "brush-wood barriers" (micro windbreaks) (Kavia & Harsh, 1993). Both indigenous and nonnative species have proved successful in stabilizing sand dunes. These include trees, such as *Acacia senegal*, *Prosopis juliflora*, *Albizia lebbeck*, *Cordia rothii*, *Dalbergia sissoo* (in regions with mean annual rainfall of 250 mm), and *Ziziphus jujuba*; shrubs, such as *Calligonum polygonoides*, *Cassia auriculata*, *Ricinus communis*, and *Z.nummularia*; and grasses, such as *Lasiurus sindicus*, *Panicum turgidum*, and *Erianthus munja*. Nonnative species include *Eucalyptus oleosa* (Australia), *Acacia tortilis* (Israel), *Parkinsonia aculeata* and *Acacia victoriae* (Australia), *Acacia albida* (Middle East) were found to be very promising, particularly because they are frost resistant. Fast-growing species are also suitable for areas with low precipitation. They also provide fodder and fuel wood to the local people. Other species occurring on sand dunes are *Aerva pseudotomentosa*, *Leptadenia pyrotechnica*, *Citrullus colocynthis*,*L.sindicus*,and *Calotropis procera*, among others. Other suitable species for planting on sand dunes are *Colophospermum mopane* and *Prosopis cineraria* because these species are adapted to grow in areas where precipitation is less than 400 mm.

Future Research Needs

In India, operational issues of crop production and soil and environmental health have received little attention. To rapidly assess soil health and to identify degraded production systems, a database of important indicators must be generated for distinct soil types of rainfed agriculture, cropping systems, and management approaches under diverse agro-eco regions. There is a paucity of diagnostic tools/methods for measuring the soil's intrinsic fragility and/or recovery capacity under land-degrading production systems. Therefore, it is important to develop a diagnostic tool with simple, robust, and process-based indicators, both quantitative and qualitative, to determine why a given (management/cropping) system is beneficial or detrimental to soil health, as well as the causes behind these effects and the potential for a degraded system to recover (i.e., its resilience) following rehabilitation efforts. There is no quantifiable proof or information available to answer an essential research question: What level of soil organic carbon is required to sustain a desired range of soil functional characteristics in different climatic conditions? Because increasing overall SOC by a considerable amount is challenging, understanding and identifying certain fractions of SOC for which management has large impacts will be crucial in establishing effective management strategies.

Conclusion

Unpredictable climate, low levels of inherent fertility in the soil, fragile water and land resources, high population pressure, and poor resources make agriculture vulnerable and soil health maintenance difficult in arid and semi-arid regions. There have been many agricultural management changes in recent years. Despite these having positive soil health outcomes, increasing the adoption of practices that protect and improve soil will be needed in the future. Reduced tillage and crop residue retention on the fields play critical roles in improving soil health and water retention, as well as maintaining or increasing yields. Wider adoption of these practices will be encouraged by improved and available information. However, areas still exist where long-established practices work against these principles. Dryland soils will continue to be problematic, but the adoption of many of the new practices discussed in this chapter will help dryland areas become productive agricultural regions.

References

Abdalla, M., Hastings, A., Cheng, K., Yue, Q., Chadwick, D., Espenberg, M., & Smith, P. (2019). A critical review of the impacts of cover crops on nitrogen leaching, net greenhouse gas balance and crop productivity. *Global Change Biology*, *25*(8), 2530–2543.

Abel, S., Peters, A., Trinks, S., Schonsky, H., Facklam, M., & Wessolek, G. (2013). Impact of biochar and hydrochar addition on water retention and water repellency of sandy soil. *Geoderma*, *202*, 183–191.

Abujabhah, I. S., Bound, S. A., Doyle, R., & Bowman, J. P. (2016). Effects of biochar and compost amendments on soil physico-chemical properties and the total community within a temperate agricultural soil. *Applied Soil Ecology*, *98*, 243–253. https://doi.org/10.1016/j.apsoil.2015.10.021

Agboola, A. A. (1993). Farming systems in Nigeria. In E. A. Aiyelari, E. O. Lucas, M. O. Abattan, & O. A. Akinboade (Eds.), *Fundamentals of Agriculture* (pp. 14–32). Africa link Books.

AICRPDA. Annual Reports 2006, 2007, 2008, 2010. All India Coordinated Research Project for Dryland Agriculture,.

Akhtar, S. S., Andersen, M. N., & Liu, F. (2015). Biochar mitigates salinity stress in potato. *Journal of Agronomy and Crop Science*, *201*, 368–378. https://doi.org/10.1111/jac.12132

Ali, A. M., Thind, H. S., Sharma, S., & Singh, Y. (2015). Site-specific nitrogen management in dry direct-seeded rice using chlorophyll meter and leaf colour chart. *Pedosphere*, *25*(1), 72–81.

Amezketa, E. (2006). An integrated methodology for assessing soil salinization, a pre-condition for land desertification. *Journal of Arid Environments*, *67*(4), 594–606.

Anglo American Crop Nutrients.(2020).POLY4. https://www.poly4.com/site/assets/files/1225/poly4_powder_product_specification_sheet_-_2021_-_english-1.pdf

Arif, M., Jan, T., Riaz, M., Fahad, S., Adnan, M., Ali, K., & Rasul, F. (2020). Biochar; a remedy for climate change. In S. Fahad, M. Hasanuzzaman, M. Alam, H. Ullah, M. Saeed, I. A. Khan, & M. Adnan (Eds.), *Environment, climate, plant and vegetation growth* (pp. 151–171). Springer Nature.

Azzam, R. A. (1980). Agricultural polymers polyacrylamide preparation, application and prospects in soil conditioning. *Communications in Soil Science and Plant Analysis*, *11*(8), 767–834.

Bijay-Singh, & Ali, M. (2020). Advances in integrated plant nutrient management. In Z. Rengel (Ed.), *Achieving Sustainable Crop Nutrition*. Burleigh Dodds Science Publishing.

Bista, D. R., Heckathorn, S. A., Jayawardena, D. M., Mishra, S., & Boldt, J. K. (2018). Effects of drought on nutrient uptake and the levels of nutrient-uptake proteins in roots of drought-sensitive and-tolerant grasses. *Plants*, *7*(2), Article 28

Blanco-Canqui, H., & Lal, R. (2009). Crop residue removal impacts on soil productivity and environmental quality. *Critical Reviews in Plant Science*, *28*(3), 139–163.

Bruce, J. P., Frome, M., Haites, E., Janzen, H., Lal, R., & Paustian, K. (1999). Carbon sequestration in soils. *Journal of Soil and Water Conservation*, *54*(1), 382–389.

Chary, G. R., Sharma, K. L., Lal, M., Indoria, A. K., Singh, A. P., Abrol, V., & Gayatri, D. (2020). Impact of conjunctive nutrient management practices on soil quality indicators and soil quality indices under maize (*Zea mays*)-wheat (*Triticum aestivum*) cropping system in hill and mountainous Inceptisol soils of northern India. *Indian Journal of Dryland Agricultural Research and Development*, *35*(1), 27–35.

Choudhary, P., Prasad, M., Choudhary, M., Kumar, A., Mahawer, S. K., Srinivasan, R., & Kumar, S. (2022). Effect of invasive weed biochar mixed with inorganic fertilizer on soil physical and chemical properties in semi-arid region of India. *International Journal of Plant & Soil Science*, *34*(14), 92–98.

DeLaune, P. B., Mubvumba, P., Lewis, K. L., & Keeling, J. W. (2019). Rye cover crop impacts soil properties in a long-term cotton system. *Soil Science Society of America Journal*, *83*(5), 1451–1458.

Diatta, A. A., Fike, J. H., Battaglia, M. L., Galbraith, J. M., & Baig, M. B. (2020). Effects of biochar on soil fertility and crop productivity in arid regions: A review. *Arabian Journal of Geosciences*, *13*, 1–17.

Dregne, H. E. (1976). *Soils of arid regions*. Elsevier.

Economic Survey. (2019). *Economic survey 2018–19*. Department of Economic Affairs, Ministry of Finance, Government of India.

El-Azeim, M. A., & Haddad, S. A. (2017). Effect of biochar on sandy soil health under arid and semiarid conditions. In *Proceedings* of the The Sixth International Conference on Environmental Management, Engineering, Planning & Economics (CEMEPE) at Thessaloniki, Greece,. June 25-30, 2017. Vol 577, pp. 25-30.

Fang, Y., Singh, B., Singh, B., & Krull, E. (2014). Biochar carbon stability in four contrasting soils. *European Journal of Soil Science*, *65*, 60–71.

Feng, S., & Fu, Q. (2013). Expansion of global drylands under a warming climate. *Atmospheric Chemistry and Physics*, *13*(19), 10081–10094.

Follett, R. F. (2001). Soil management concepts and carbon sequestration in cropland soils. *Soil and Tillage Research*, *61*(1–2), 77–92.

Food and Agriculture Organization (FAO). (2007). *Agriculture and Consumer Protection Department*. Italy Available from. http://www.fao.org/ag/ca/

Gaur, M. K., & Squires, V. R. (2018). Geographic extent and characteristics of the world's arid zones and their peoples. In M. Gaur & V. Squires (Eds.), *Climate variability impacts on land use and livelihoods in drylands* (pp. 3–20). Springer.

Gilbert, C., Peter, S., Wilson, N., & Edward, M. (2014). Effect of hydrogels on soil moisture and growth of *Cajanus cajan* in semi-arid zone of Kongelai, West Pokot County. *Open Journal of Forestry*, *4*(1), 34–37.

Goyal, R. K. (2004). Sensitivity of evapotranspiration to global warming: a case study of arid zone of Rajasthan (India). *Agricultural Water Management*, *69*(1), 1–11.

Gupta, J. P. (1993). Wind erosion of soil in drought-prone areas. In A. Sen & A. Kar (Eds.), *Desertification and its control in the Thar Sahara and Sahel regions* (pp. 91–105). Scientific Publishers.

Hudson Institute of Mineralogy. (2020).Polyhalite. https://www.mindat.org/min-3259.html

Husein, H. H., Mousa, M., Sahwan, W., Bäumler, R., & Lucke, B. (2019). Spatial distribution of soil organic matter and soil organic carbon stocks in semi-arid area of northeastern Syria. *Natural Resources*, *10*(12), 415–432.

Jian, J., Du, X., Reiter, M. S., & Stewart, R. D. (2020). A meta-analysis of global cropland soil carbon changes due to cover cropping. *Soil Biology and Biochemistry*, *143*, Article 107735

Kavia, Z. D., & Harsh, L. N. (1993). Proven Technology of Sand Dune Stabilization—A Step to Combat Desertification. In *Proc. of National Seminar on Afforestation of Arid Lands*. Jodhpur: Institute of Arid Zone Forestry Research (IAZFR), Govt. of India, Jodhpur.

Khadse, A., Rosset, P. M., & Ferguson, B. G. (2017). Taking agro ecology to scale: the zero budget natural farming peasant movement in Karnataka, India. *The Journal of Peasant Studies*, *45*, 1–28.

Khule, Y. R., Waghmare, M. S., Shelake, M. S., Boradkar, S. G., & Chavan, A. L. (2023). Effect of foliar application of nano-N fertilizer on growth yield, and quality of linseed (*Linum usitatissimum* L.). *The Pharma Innovation Journal*, *12*(2), 3208–3211.

Kumar, M. (2016). Impact of climate change on crop yield and role of model for achieving food security. *Environmental Monitoring and Assessment*, *188*(8). http://rd.springer.com/journal/10661

Kumar, V., Singh, A. K., Jat, S. L., Parihar, C. M., Pooniya, V., Sharma, S., & Singh, B. (2014). Influence of site-specific nutrient management on growth and yield of maize (*Zea mays*) under conservation tillage. *Indian Journal of Agronomy*, *59*(4), 657–660.

Kumar, V., & Singh, V. (2019). Effect of site-specific nutrient management on yield and nutrient uptake in pearl millet (*Pennisetum glaucum*)-wheat (*Triticum aestivum*) cropping sequence. *Annals of Plant and Soil Research*, *21*(1), 1–6.

Lal, R. (2004). Soil carbon sequestration impacts on global climate change and food security. *Science*, *304*, 1623–1627.

Lal, R., Kimble, J., Follett, R. F., & Cole, C. V. (1998). *The Potential for U.S. Cropland to sequester carbon and mitigate the greenhouse effect*. Sleeping Bear Press.

Lehmann, J., Gaunt, J., & Rondon, M. (2006). Bio-char sequestration in terrestrial ecosystems—A review. *Mitigation and Adaptation Strategies for Global Change*, *11*, 403–427.

Lewis, K. L., Burke, J. A., Keeling, W. S., McCallister, D. M., DeLaune, P. B., & Keeling, J. W. (2018). Soil benefits and yield limitations of cover crop use in Texas High Plains cotton. *Agronomy Journal*, *110*(4), 1616–1623.

Mahmood, N., Arshad, M., Kaechele, H., Shahzad, M. F., Ullah, A., & Mueller, K. (2020). Fatalism, climate resiliency training and farmers' adaptation responses: Implications for sustainable rainfed-wheat production in Pakistan. *Sustainability*, *12*(4), Article 1650

Malhotra, S. K. (2016). Water soluble fertilizers in horticultural crops—An appraisal. *Indian Journal of Agricultural Sciences*, *86*(10), 1245–1256.

Maruthi Sankar, G. R., Sharma, K. L., Dhanapal, G. N., Shankar, M. A., Mishra, P. K., Venkateswarlu, B., & Kusuma Grace, J. (2010). Influence of soil and fertilizer nutrients on sustainability of rainfed finger millet yield and soil fertility in semi-arid Alfisols. *Communications in Soil Science and Plant Analysis, 42*(12), 1462–1483.

Mazzoncini, M., Sapkota, T. B., Barberi, P., Antichi, D., & Risaliti, R. (2011). Long-term effect of tillage, nitrogen fertilization and cover crops on soil organic carbon and total nitrogen content. *Soil and Tillage Research, 114*(2), 165–174.

McDaniel, M. D., Tiemann, L. K., & Grandy, A. S. (2014). Does agricultural crop diversity enhance soil microbial biomass and organic matter dynamics? A meta-analysis. *Ecological Applications, 24*(3), 560–570.

Millennium Ecosystem Assessment (MEA). (2005). *Ecosystems and Human Well-being: Synthesis.* Island Press.

Mishra S. (2018). *Zero budget natural farming: Are this and similar practices the answers?* [Working paper no. 70]. Nabakrushna Choudhury Centre for Development Studies.

Motsara, M. R. (2002). Available nitrogen, phosphorus and potassium status of Indian soils as depicted by soil fertilizer maps. *Fertilizer News, 47*(8), 15–21.

Narjary, B., Aggarwal, P., Kumar, S., & Meena, M. D. (2013). Significance of hydrogel and its applications in agriculture. *Indian Farming, 62*(10), 15–17.

Niyogi, D. G. (2018). Andhra farmers taste success with zero budget natural farming. *Down toEarth.* https://www.downtoearth.org.in/agriculture/andhra-farmers-taste-success-with-zero-budget-natural-farming-59445

Noin, D., & Clarke, J. I. (1997). Population and environment in arid regions of the world. In J. Clarke & D. Noin (Eds.), *Population and environment in arid regions* MAB/UNESCO (Vol. 19, pp. 1–18). The Parthenon Publishing Group.

Novak, J. M., Busscher, W. J., Watts, D. W., Amonette, J. E., Ippolito, J. A., Lima, I. M., Gaskin, J., Das, K., Steiner, C., & Ahmedna, M. (2012). Biochars impact on soil-moisture storage in an Ultisol and two Aridisols. *Soil Science, 177*, 310–320.

O'Laughlin, J., & McElligott, K. (2009). Biochar for environmental management: Science and technology. In J. Lehmann & S. M. Joseph (Eds.), *Forest Policy and Economics.* Earthscan.

Osman, M. (2008). Recycling of tank silt for improving soil and water productivity in rainfed areas. Lecture Notes for the Winter School on "Technological Advances in Conservation of Natural Resources in Rainfed Agriculture", Central Research Institute for Dryland Agriculture, Hyderabad

Ovalesha, M. A., Yadav, B., & Rai, P. K. (2017). Effects of polymer seed coating and seed treatment on plant growth, seed yield and quality of cowpea (*Vigna unguiculata*). *Journal of Pharmacognosy and Phytochemistry, 6*(4), 106–109.

Oweis, T., Hachum, A., & Bruggeman, A. (2004). *Indigenous Water Harvesting Systems in West Asia and North Africa* (pp. 173). Center for Agriculture Research in Dry Areas (ICARDA), vi+.

Palekar, S. (2005). *The philosophy of spiritual farming* (2nd ed.). Zero Budget Natural Farming Research, Development & Extension Movement.

Pandian, K., Subramaniayan, P., Gnasekaran, P., & Chitraputhirapillai, S. (2016). Effect of biochar amendment on soil physical, chemical and biological properties and groundnut yield in rainfed Alfisol of semi-arid tropics. *Archives of Agronomy and Soil Science, 62*(9), 1293–1310.

Panwar, A. S., Ravisankar, N., Shamim, M., & Prusty, A. K. (2018). Integrated farming systems: a viable option for doubling farm income of small and marginal farmers. *Bulletin of the Indian Society of Soil Science, 32*, 68–88.

Peng, X., Tong, X., Hao, L., & Wu, F. (2019). Applicability of biochar for limiting interrill erosion and organic carbon export of sloping cropland in a semi-arid area of China. *Agriculture, Ecosystems & Environment, 280*, 68–76.

Peregrina, F., Pérez-Álvarez, E. P., Colina, M., & García-Escudero, E. (2012). Cover crops and tillage influence soil organic matter and nitrogen availability in a semi-arid vineyard. *Archives of Agronomy and Soil Science, 58*(suppl. 1), SS95–SS102.

Poeplau, C., & Don, A. (2015). Carbon sequestration in agricultural soils via cultivation of cover crops–a meta-analysis. *Agriculture, Ecosystems & Environment, 200*, 33–41.

Qureshi, A. S., Ertebo, T., & Mehansiwala, M. (2018). Prospects of alternative copping systems for salt-affected soils in Ethiopia. *Journal of Soil Science and Environmental Management, 9*(7), 98–107.

Rashid, Z., Khuroo, N. S., Ahngar, T. A., Majid, S., Bashir, S., Nasseer, S., & Dar, Z. A. (2021). Precision nitrogen management for enhancing yield and quality of fodder maize. *International Journal of Plant & Soil Science, 33*(18), 236–241.

Rathore, V. S., Tanwar, S. P. S., Kumar, P., & Yadav, O. P. (2019). Integrated farming system: Key to sustainability in arid and semi-arid regions. *Indian Journal of Agricultural Sciences, 89*(2), 181–192.

Ravisankar, N., Panwar, A.S. Prasad, K., Kumar, V., & Bhaskar. S. (2017). *Organic Farming Crop Production Guide, Network Project on Organic Farming*, ICAR Indian Institute of Farming Systems Research, Modipuram, Meerut-250 110, Uttar Pradesh, India. p. 586.

Sahrawat, K. L., Wani, S. P., Rego, T. J., Pardhasaradhi, G., & Murthy, K. V. S. (2007). Widespread deficiencies of sulphur, boron and zinc in dryland soils of the Indian semi-arid tropics. *Current Science, 93*(10), 1428–1432.

Sammi Reddy, K., Prabhakar, M., Sharma, K. L., Rama Rao, C. A., Vanaja, M., Srinivasa Rao, M., Gopinath, K. A., Sarkar, B., Ramana, D. B. V., Subba Rao, A. V. M., SharathChandran, M. A., Pankaj, P. K., Prasad, J. V. N. S., Bhaskar, S. and Alagusundaram, K. (2019). *Research Highlights (2016–18)*. National Innovations in

Climate Resilient Agriculture (NICRA), ICAR – Central Research Institute for Dryland Agriculture, Hyderabad. P. 128.

Sharma, K., Chary, G. R., Reddy, K. S., Lal, M., Singh, A., Abrol, V., & Indoria, A. (2020). Effect of combined use of tillage and nutrient management practices on soil quality indicators and indices under maize (*Zea mays*) based cropping system in rainfed inceptisols. *Annals of Plant and Soil Research, 22*(4), 336–348.

Sharma, K., Reddy, K. S., Chary, G. R., Singh, V. K., Lal, M., Nath, K. G., & Gayatri, D. (2021). Effect of integrated nutrient management (INM) practices on soil quality indicators and indices under sorghum (*Sorghum bicolor* (L.) Moench)+ pigeon pea (*Cajanus cajan* (L.) Millsp.) (4:2) and soybean (*Glycine max* (L) Merr)+ pigeon pea (*Cajanus cajan* (L.) Millsp.) in rainfed Vertisols of Western India. *Annals of Plant and Soil Research, 23*(4), 379–389.

Sharma, K. L., Chary, G. R., Reddy, K. S., Lal, M., Indoria, A. K., Srinivas, K., & Sekhar, C. C. (2019). Effect of conjunctive use of inorganic and organic sources of nutrients on soil quality parameters, key indicators and soil quality indices (SQIs) under sorghum+ cowpea intercropping system in rainfed Vertisol soils of southern India. *Indian Journal of Dryland Agricultural Research and Development, 34*(2), 47–59.

Sharma, K. L., Lal, M., Chary, G. R., Reddy, K. S., Indoria, A. K., Srinivas, K., & Gayatri, D. L. A. (2022). Long term effects of conjunctive nutrient management practices on soil quality in maize (*Zea mays*)–mustard (*Brassica compestris* var. Sarson) rotation under rainfed hill Inceptisol soils. *Communications in Soil Science and Plant Analysis, 53*(4), 439–452.

Sharma, K. L., SrinivasaRao, C., Chandrika, D. S., Lal, M., Indoria, A. K., Reddy, K. S., & Srinivas, D. K. (2018). Effect of predominant integrated nutrient management practices on soil quality indicators and soil quality indices under post monsoon (rabi) sorghum (*Sorghum bicolor*) in rainfed black soils (Vertisols) of western India. *Communications in Soil Science and Plant Analysis, 49*(13), 1629–1637.

Sharma, K. L., Srinivasarao, C., Chandrika, D. S., Lal, M., Indoria, A. K., Thakur, P. B., & Usha Rani, K. (2017). Effect of graded levels of surface crop residue application under minimum tillage on carbon pools and carbon lability index in sorghum (*Sorghum bicolor*[L.] Moench)-cowpea (*Vigna unguiculata*) system in rainfed Alfisols. *Communications in Soil Science and Plant Analysis, 48*(21), 2506–2513.

Sileshi, A., Kibebew, K., &Amanuel, Z. (2015). *Temporal and spatial variations in of salt-affected soils using GIS and remote sensing at Dubti/Tendaho State Farm.* Doctoral dissertation,Haramaya University.

Singh, H.P., Sharma, K.D., & Subba Reddy, G. (2002). *Rainfed agriculture: Challenges of the 21st century—National perspectives.* Proceedings of the 89th Indian Science Congress—Part IV, Lucknow, India, pp. 3–4.

Singh, R. P., & Das, S. K. (1990). Nutrient management in drylands. In V. K. Virendra Kumar, G. C. Shotriya, & S. V. Kaore (Eds.), *Soil fertility and fertilizer use. Vol IV.*

Nutrient management and supply system for sustaining agriculture in 1990s (pp. 207–219). Indian Farmers Fertiliser Cooperative Limited.

Smaling, E. M. A. (1993). *An agro-ecological framework for integrated nutrient management, with special reference to Kenya*. Wageningen University and Research.

Sperow, M., Eve, M., & Paustian, K. (2003). Potential soil C sequestration on US agricultural soils. *Climatic Change, 57*, 319–339.

Srinivasarao, C., Gopinath, K. A., Venkatesh, G., Dubey, A. K., Wakudkar, H., Purakayastha, T. J., Pathak, H., Jha, P., Lakaria, B. L., Rajkhowa, D. J., Mandal, S., Jeyaraman, S., Venkateswarlu, B., & Sikka, A. K. (2013). *Use of biochar for soil health management and greenhouse gas mitigation in India: Potential and constraints*. Central Research Institute for Dryland Agriculture.

Srinivasarao, C., Prasad, J. V. N. S., Osman, M., Prabhakar, M., Kumara, B. H., & Singh, A. K. (2017). *Farm innovations in climate resilient agriculture*. Central Research Institute for Dryland Agriculture.

Srinivasarao, C., Venkateswarlu, B., Dixit, S., Kundu, S., & Gayatri Devi, K. (2011). *Livelihood Impacts of Soil Health Improvement in Backward and Tribal Districts of Andhra Pradesh*. Central Research Institute for Dryland Agriculture.

Srinivasarao, C., Venkateswarlu, B., Dixit, S., Wani, S. P., Sahrawat, K. L., Kundu, S., Gayatri Devi, K., Rajesh, C., & Pardhasaradhi, G. (2010). Productivity enhancement and improved livelihoods through participatory soil fertility management in tribal districts of Andhra Pradesh. *Indian Journal of Dryland Agriculture Research Development, 25*(2), 23–32.

Srinivasarao, C., & Vittal, K. P. R. (2007). Emerging nutrient deficiencies in different soil types under rainfed production systems of India. *Indian Journal of Fertilizers, 3*, 37–46.

Stewart, C. E., Zheng, J., Botte, J., & Cotrufo, M. F. (2013). Co-generated fast pyrolysis biochar mitigates green-house gas emissions and increases carbon sequestration in temperate soils. *GCB Bioenergy, 5*(2), 153–164.

Sun, F., & Lu, S. (2014). Biochars improve aggregate stability, water retention, and pore-space properties of clayey soil. *Journal of Plant Nutrition and Soil Science, 177*, 26–33.

Thies, E., & Rilling, M. (2009). Characteristics of biochar: Biological properties. In J. Lehmann & S. Joseph (Eds.), *Biochar for environmental management:Science and technology* (pp. 85–105). Earthscan.

UN Environment Programme (UNEP). (2007). *The global environment outlook 4: Environment for development*. UNEP.

UN Environmental Management Group. (2011). *Global drylands: A UN system-wide response*. United Nations Environment World Conservation Monitoring Centre.

Vale, F. (2016). *Calcium and magnesium movement in soil profile with polyhalite as potassium fertilizer for soybean crop*. Proceedings of FERTBIO, Goiana-Brazil. October -2016 16–20.

Varinderpal-Singh, V., Kaur, R., Singh, B., Singh, B. B., & Kaur, A. (2012). Precision nutrient management: A review. *Indian Journal of Fertilisers, 12*(11), 48–62.

Venkatesh, G., Gopinath, K. A., & Sammi Reddy, K. (2018). *Biochar production and its use in rainfed agriculture: Experiences from CRIDA*(CRIDA-NICRA Research Bulletin 02/2018). ICAR - Central Research Institute for Dryland Agriculture.

Venkateswarlu, B. (2008). Organic farming in rainfed agriculture: prospects and limitations. In B. Ventateswarlu, S. S. Balloli, & Y. S. Ramakrishna (Eds.), *Organic Farming in Rainfed Agriculture: Opportunities and Constraints* (pp. 7–11). Central Research Institute for Dryland Agriculture.

Wu, W., & Ma, B. (2015). Integrated nutrient management (INM) for sustaining crop productivity and reducing environmental impact: A review. *Science of Total Environment, 512–513*, 415–427.

Wulanningtyas, H. S., Gong, Y., Li, P., Sakagami, N., Nishiwaki, J., & Komatsuzaki, M. (2021). A cover crop and no-tillage system for enhancing soil health by increasing soil organic matter in soybean cultivation. *Soil and Tillage Research, 205*, Article 104749.

Yeboah, S., Lamptey, S., Cai, L., & Song, M. (2018). Short-term effects of biochar amendment on greenhouse gas emissions from rainfed agricultural soils of the semi–arid loess plateau region. *Agronomy, 8*(5), Article 74.

Zhang, D., Yao, P., Zhao, N., Cao, W., Zhang, S., Li, Y., & Gao, Y. (2019). Building up the soil carbon pool via the cultivation of green manure crops in the loess plateau of China. *Geoderma, 337*, 425–433.

4

Regional Soil Health Assessments: Central Semi-arid Region of India

Avijit Ghosh, Sukanya Misra, and Ranjan Bhattacharyya

Chapter Overview

The productivity potential in soils of semi-arid India is constrained by erosion, desertification, waterlogging and salinity. Soil health assessment and management will continue to play a prominent role in agricultural production systems of arid and semi-arid agroecosystems. The strategies like reduce soil disturbance (farmlands and rangelands), crop rotation practices (farmlands), cover cropping practices to promote plant diversity and living roots (farmlands), diversified production systems (farmlands and rangelands), adding soil organic amendments (farmlands and rangelands), integrate livestock into cropping systems (farmlands), and promoting diverse plant species with different rooting depths (rangelands) could improve the health of soil in semi-arid India. The chapter describes the health assessment tools and approaches used by various researchers in this region. These tools could be useful to prepare a Regional Soil Health Assessment system of the country.

Introduction

As the International Decade of Soils (2015–2024) approaches its end point, the significance of soils is being acknowledged, as evidenced by articles in the general public media (Velasquez–Manof, 2018) and worldwide climate modeling initiatives.

Soil Health Series: Volume 4 Soil Health and Sustainability in India, First Edition.
Edited by Ranjan Bhattacharyya.

Soil is among the many natural resources that sustain human civilization. Producing enough food to support an expanding human population is impossible without productive and healthy soil.

However, around the world, numerous areas have degraded soils and are no longer productive. During the past century, physical, chemical, and/or biological land degradation processes have resulted in serious consequences to soil resources globally (e.g., compaction, inorganic/organic contamination, and diminished microbial activity/diversity). Other soil-related problems include wasteland formation, nutrient imbalance, nutrient deficiency, and over-exploitation of soils. These problems mainly arise because of increased population, a rise in living standards, and urban encroachment of the agricultural fields.

Soils can only be restored to a productive state by using soil health management principles. According to worldwide trends, soil degradation results in a sustained reduction in productivity for 20% of cropland, 16% of forest land, 19% of grassland, and 27% of rangeland (UNCCD 2017). If not addressed, this diminished productivity can have significant effects on global food security.

While erosion can swiftly remove soil, the process of new soil formation takes a very long time. Depending on the soil-forming processes in a certain area, the time needed for an inch of soil to form may range from 100 to 10,000 years. Given the sluggish pace of its renewal, soil is a limited natural resource if it is not managed properly. In particular, an unhealthy soil is more susceptible to erosion losses from wind and water. Therefore, to ensure long-term soil productivity and food security, good soil development and maintenance are essential. The negative effects of degraded soils on the environment, the economy, and public health are additional reasons for concern about soil health. For instance, suspended sediments from dust storms have been shown to be dangerous for transportation safety and to pose substantial health risks to people with respiratory conditions. Additionally, sediments that are borne away by erosion may be deposited in undesirable locations, necessitating costly removal. Furthermore, owing to agri-chemicals adhering to the dust and soil, poorly managed soils can also result in the pollution of surface water and groundwater.

Beyond a focus on crop production and other overt benefits to humans, soil health is important for soil fertility. It is also part of the policy dimension of soil security. The multi-dimensionality of the soil health concept enables aligning soil management objectives with sustainability objectives and should account for a significant number of stakeholders, roles, and geographical and temporal scales. One of the most significant contributions of the soil health framework is the addition of the biological viewpoint to soil management, which is crucial for meeting the needs related to long-term sustainability of agricultural production. Expanding soil assessment and management to address issues with biodiversity, water quality, the climate, recreation, and the health of humans and the environment requires a biological approach as well.

The historical uneasiness with which scientists have viewed the concept of soil health is due to the challenges of defining soil health in a way that allows for a universal quantitative assessment that encompasses all of its ecosystem services, including human health. Reasons for the challenges include soil heterogeneity, the site-specific nature of soil management, and the varying ecosystem services that sometimes have conflicting or competing needs. Nevertheless, researchers, policymakers, and stakeholders have expressed widespread interest in the use of the soil health concept.

Central Semi-arid Region of India and its Geographical Landscapes

Semi-arid zones cover around 956 750 km^2 of India, chiefly located in the states of Rajasthan, Punjab, Haryana, Uttar Pradesh, Gujarat, Madhya Pradesh, Maharashtra, Karnataka, Andhra Pradesh, and Tamil Nadu.

Climate

In general, the mean maximum temperature exceeds 35°C in the south of the Vindhyan range and over 37°C in parts of Deccan in March. In April, temperatures of 38–43°C are common in Gujarat, southern Madhya Pradesh, and southwest Punjab. In May, the temperature commonly reaches 48°C and may even exceed 54°C, as recorded for Sri Ganganagar in Rajasthan. These high temperatures are associated with low relative humidity ranging between 30% and 50%. Mean annual rainfall within the semi-arid zone varies from 31.1 to 72.8 cm. In more northern latitudes, about 70–96% of the annual rainfall occurs during the southwest monsoon (June–September). In the southern localities, 7–63% of the annual precipitation results from the northeast monsoon during October to December. Variation in rainfall by as little as 10% is serious in areas of moderate precipitation, particularly when it is on the negative side. Therefore, any management plan for such areas has to recognize that the reliability of rainfall usually ranges from low to moderate within the semi-arid regions of India. Water balances show that none of the stations in the semi-arid zone have any water surplus, even on a monthly basis. Contributions to the groundwater reservoir from this region are thus normally absent. In certain years, however, when precipitation during the rainy season is much greater than normal, a local water surplus may occur for brief periods. These surpluses not only produce enormous surface runoff, resulting in inundation, but also contribute significantly to the groundwater resources.

Soils

The semi-arid regions support a variety of soils. Alluvial soils occupy extensive tracts. The new alluvium (riverine) occurs close to the banks of rivers, while the new alluvium (coastal) is formed along seacoasts. In both cases, the soil is sandy to loamy sand in texture. The old alluvial areas are no longer subjected to eluviation, because they occur in valleys above the level of inundation. Their texture varies from clay to sandy loam, and they have low lime, phosphorus, and humus content. The productivity of these soils is dependent on their texture, the amount of plant nutrients they contain, and their drainage conditions. An occasional feature of these soils is the presence of hard pans at a certain level in the soil profile. These hard pans are caused by the infiltration of silica or calcareous matter and the presence of calcareous nodules, known as Kankar. The pans impede penetration of plant roots and affect the percolation of water. Nevertheless, these soils can support luxuriant crops under irrigation.

Desert soils occur in parts of Rajasthan, southern Punjab, and adjacent arid lands. The sand with which the area is covered is partly derived from the disintegration of rocks. The soils thus contain a high percentage of soluble salts and varying amounts of calcium carbonate and are usually low in organic matter. The soils in depression areas contain a high percentage of salts; in addition, they are notably impervious and consequently have impeded drainage. The yellow color in soils is probably caused by a higher degree of dehydration, as found in Rajasthan and adjacent areas. Chestnut or gray-brown soils cover an area of 36 400 km^2 in Rajasthan. Saline and alkaline soils also occur in the semi-arid zone. Saline alluvium (riverine) occurs in drier tracts of the north where saline and alkaline efflorescence is very common. These once fertile tracts have now become impregnated with these salts as a result of poor drainage and a high-water table.

The semi-arid regions of peninsular India contain black, red, and lateritic soils; saline and alkaline soils; alluvial soil; red and yellow soils; and mixed red and black soils. Red soils constitute the largest soil group and cover the Archaean basement of peninsular India. These soils are light in texture and devoid of lime and free CO_3. Black soils originated from basaltic rocks of the lava plateau. These rich residual soils contain less clay than other soils and are highly retentive of moisture. Laterite and lateritic soils are poor in nitrogen, phosphorus, and potassium. Mountain and hill soils are formed in depressions within the mountains, in the valleys and basins and on the less steeply inclined slopes. In general, north-facing slopes have a thicker soil cover than the more precipitous southern faces. Most of the soils in the semi-arid region are comparatively dry. They are deficient in nitrogen and organic matter, but less deficient in phosphate, and potash deficiency is rare. Considerable use of proper commercial fertilizers is therefore essential because most of the soils are not suitable for sensitive crops.

Vegetation and Land Use

The major woodland types of the semi-arid region are subalpine and alpine forest, tropical dry deciduous forest, tropical thorn forest, tropical moist deciduous forest, littoral and swamp forest, and tropical dry evergreen forest. The grasslands of the semi-arid areas have mainly originated as a result of deforestation and intense browsing and grazing. By far the largest portion of the hot semi-arid area is covered by *Sehima–Dichanthium*, followed by *Dichanthium–Cenchrus–Lasiurus* and *Phragmites–Saccharum–Imperata*. The semi-arid regions exhibit a complex seasonal pattern of land use including arable cropping and grazing by different classes of livestock. Land not available for cultivation comprises such habitats as barren outcrops of rocky hills, sandy deserts, and alkaline and saline beds. The major crop regions within the semi-arid climate include cotton (*Gossypium* spp.), groundnut (*Arachis hypogea*), sugarcane (*Saccharum officinarum*), maize (*Zea mays*), jowar (*Sorghum vulgare*), wheat (*Triticum aestivum*), rice (*Oryza sativa*), millets (e.g., *Pennisetum*, *Eleusine*), pulses, and oilseeds. In semi-arid regions, 75–95% of the cultivation is rainfed, and since the rainfall is highly erratic besides being low, the frequency of crop failures owing to recurrent drought is high.

Soil Health Problems and Quality in the Central Semi-arid Region

Soils of the central semi-arid region face various types of problem, including erosion, desertification, waterlogging, and salinity. These issues are described in the following sections.

Soil Erosion—Water and Wind Erosion

Wind and water erosion are common types of soil erosion. Intensive agriculture, deforestation, roads, anthropogenic climate change, and urban sprawl are among the most significant human activities regarding their effect on stimulating erosion.

Soils of central semi-arid states, including Maharashtra, Rajasthan, Karnataka, and Madhya Pradesh, are degraded by water and wind erosion.

Desertification

Recent analysis reveals that the area under desertification in India is increasing and affects 29.3% of the country's total geographic area. About 24% of the area affected by desertification is in the states of Rajasthan, Maharashtra, Gujarat, Jammu and Kashmir, Karnataka, Jharkhand, Odisha, Madhya Pradesh, and Telangana (Figure 4.1).

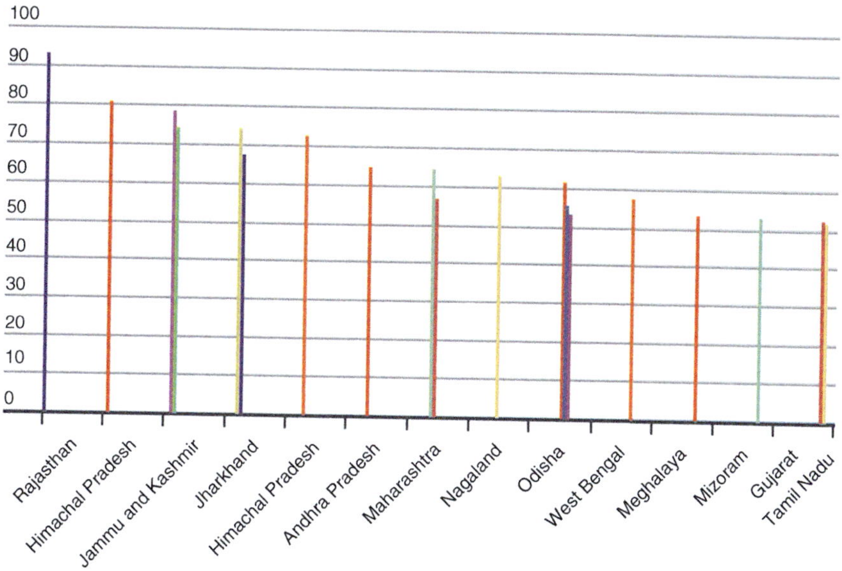

Figure 4.1 Number of districts in different Indian states with more than 50% desertification. *Source:* Government of India (2018) Desertification and land degradation atlas of selected districts of India.

Waterlogging

As per published reports, in large areas of Uttar Pradesh, Rajasthan, Gujarat Haryana, and few other states, coexisting problems of waterlogging and salinity are alarming. Waterlogging impedes the ability of soil to serve as an optimum medium for plant growth because it alters the soil's physical, chemical, electrochemical, and biological characteristics (Figure 4.2). Excessive water content in the soil due to waterlogging decreases O_2 diffusion capacity, leading to hypoxic or even anoxic environments that inhibit the activity of nitrifying communities, resulting in depleted soil nitrogen availability that will negatively affect nitrogen-dependent crop productivity.

Waterlogging also hinders the growth of plants by reducing the dispersal of oxygen through the pore spaces in the soil around the root zone. Under waterlogging conditions, oxygen is supplied to the root tip and to the rhizosphere through the formation of aerenchyma in roots through removal of some cells of the cortex. One function of the aerenchyma is to remove excess gases from the root and soil. The gas exchange between soil and atmosphere in a well-aerated soil is rapid enough to avert O_2 deficiency and prevent toxicity caused by excess CO_2 or other gases, such as ethylene and methane.

Figure 4.2 Effects of waterlogging on soil properties.

Soil management practices such as drainage, tillage, and traffic control can alter soil structure directly or indirectly. Many of these changes are relatively short term and reversible. Management-induced changes in the quantity and characteristics of soil can also lead to changes in soil structure that are much more persistent. Sustainable management practices must maintain the structure of soil over the long term in a state that is optimum for a range of processes related to crop production and environmental quality (Table 4.1).

Salinity

Salinity is one of the environmental factors that can most severely limit the productivity of crop plants. Most crop plants are sensitive to salinity caused by high concentrations of salts in the soil, and the area of land affected by it is constantly increasing. Central states such as Maharashtra, Rajasthan, Gujrat, and Uttar Pradesh are particularly affected by the problem of soil salinization.

Soil salinity is a particularly serious problem for agriculture under irrigation. The total area of affected soils continues to increase each year due to introduction of irrigation in new areas. In the hot and dry regions of the world, soils are frequently saline and have low agricultural potential. Most crops in these areas are grown under irrigation, and soil salinity is exacerbated by inadequate irrigation management. The poor management in turn leads to secondary salinization, which affects 20% of irrigated land worldwide.

Salts in the soil occur as ions (electrically charged forms of atoms or compounds). Ions are released from weathering minerals in the soil. They may also

Table 4.1 Summary of advantages and disadvantages of different soil and crop management practices.

Soil and crop management practices	Advantages[a]	Disadvantages
Surface drainage	Both installation and maintenance are simple and inexpensive	Open drains with less cropping area; needs periodic maintenance
Raised bed system	Improvements in soil structure	Efficiency depends on height of water table; poorer weed control in furrows; cost of modifying machinery; less cropping area
Pipe drains	Well-tested method for severe waterlogging	Needs outfall and periodic maintenance; cost of installation is high
Vertical drainage	Well-tested method for severe waterlogging	Maintenance and operational costs are higher than for horizontal pipe drainage systems
Mole drains	Well-tested method; less costly than other underground drainage	Needs periodic maintenance; will not maintain integrity in dispersive soils
Controlled traffic farming	Reduced soil compaction, erosion, tillage costs, water use, and nutrient losses	Variable results with different conditions, such as different crops, soil types, and tillage
Strategic deep tillage and subsoil manuring	Decreases soil strength resulting in deeper and denser rooting	Strategic deep tillage with no added amendment is often a short-term solution and is less effective in counteracting subsoil problems, such as acidity, sodicity, or subsoil salinity
Early sowing and vigorous crop	Use of existing soil water provides a buffer; avoids terminal waterlogging events	Minor benefit with severe waterlogging
Bio-drainage	Tried and tested at many locations with success	Needs proper plantation techniques, expertise, thinning, pruning, and harvesting
Nutrient application, particularly nitrogen	Improving plant growth and development	Appropriate methods, nutrient types, timing, and rate should be considered for large-scale application
Plant growth regulators	Promote stomatal conductance and photosynthetic capacity of waterlogged plants	Appropriate methods, timing, and rate should be considered for large-scale application; unproven in broad-scale agriculture

(Continued)

Table 4.1 (Continued)

Soil and crop management practices	Advantages[a]	Disadvantages
Use of anti-ethylene agents	Increase both photosynthesis and fruit retention; diminish crop loss induced by ethylene accumulation	Untested in broad-scale agriculture
Pretreatment with hydrogen peroxide	Protect crops from oxidative damage caused by waterlogging	Untested in broad-scale agriculture
Tolerant species and varieties	Cost effective for farmers	The introduction of waterlogging tolerance into existing plant varieties is time consuming and complex.

[a] Advantages are in addition to the reduction of waterlogging.

enter the soil through irrigation water or fertilizers or migrate upward in the soil from shallow groundwater. When precipitation is insufficient to leach ions from the soil profile, salts accumulate in the soil resulting salinity. All soils contain some water-soluble salts. Plants absorb essential nutrients in the form of soluble salts, but excessive accumulation strongly suppresses plant growth.

Salinization is recognized as one of the main threats to environmental resources and human health in many countries, affecting almost 1 billion hectares (ha) worldwide and representing about 7% of the earth's land mass. The total area affect is approximately 10 times the size of Venezuela and 20 times the size of France. In India, an estimated area of 7 million ha of land has saline soil. Most of this land occurs in Indo-Gangetic plains, including land within the states of Punjab, Haryana, Uttar Pradesh, Bihar, and some parts of Rajasthan. Arid tracts of Gujarat and Rajasthan and semi-arid tracts of Gujarat, Madhya Pradesh, Maharashtra, Karnataka, and Andhra Pradesh are also largely affected by saline lands.

Salinization can be restricted by leaching salt from the root zone and altering farm management practices. In addition, its impact can be reduced by the use of salt-tolerant plants. Irrigated agriculture can be sustained by better irrigation practices, such as the adoption of partial root zone drying methodology and the use of drip or micro-jet irrigation to optimize the use of water. Furthermore, the expansion of dryland salinity can be contained by reducing the amount of water that passes beyond the roots. This reduction can be achieved by re-introducing deep-rooted perennial plants that can still grow even in seasons that do not support annual crop plants. This may restore the balance between water use and water availability from rainfall and thus prevent rising water tables and the movement of salt to the soil surface. Farming systems can change to incorporate perennials in rotation with annual crops (phase farming), in mixed plantings (alley farming, intercropping), or in site-specific plantings (precision farming).

Although the use of these approaches for sustainable management can ameliorate yield reduction under salinity stress, implementation is often limited because of cost and availability of good water quality or water resources. Evolving efficient, low-cost, easily adaptable methods for abiotic stress management is a major challenge. Worldwide, extensive research is being carried out to develop strategies to cope with abiotic stresses through developing salt- and drought-tolerant plant varieties, shifting the crop calendars, and improving resource management practices, among other approaches.

Soil Quality Monitoring Systems in Central India

Vasu et al. (2016) assessed the soil quality index (SQI) using only surface soil parameters. However, this approach yields only partial information because subsurface features are inextricably tied to pedogenic processes that also affect crop yield.

Based on soil profile data from six identified soil series in a region of the semi-arid tropical Deccan plateau, two distinct SQIs were determined for the soil surface (0–15 cm) and a control section (0–100 cm) and then assessed with regard to crop production. Principal component analysis (PCA) and expert opinion approaches were applied to obtain a minimum data set. For SQI estimation, both additive and weighted index techniques were used and compared.

SQI measurements varied because PCA and expert opinion techniques yielded different outcomes. However, both approaches indicated that weighted SQIs were often more closely connected with crop yield than additive SQIs. Except for a few rare instances, the expert opinion-derived weighted SQI was equivalent for both the surface and the control section and consistent in terms of its link with crop production, showing that it performed better than PCA. This better performance is that PCA is essentially a data dimension reduction technique. In contrast, the expert opinion method is primarily developed by experts on cause-effect relationships of soil properties (such as hydraulic conductivity, $CaCO_3$, and exchangeable sodium percentage) that are influenced by regressive pedogenic processes in semi-arid tropical environments. Results indicated that a good association between soil functions and management objective may be established by considering both surface and control section soil characteristics.

In order to identify the most sensitive soil quality indicators, which may provide insight into the status of the soil, Jha and Mohaptra (2012) conducted a study on the ravinous landform of Ustifluvents in the Yamuna area. For the study, three alternative land-use systems—forest, farmland, and uncultivated ravinous wasteland—were chosen. The most common plant species in the semi-arid zones were chosen for assessing forest land usage, including *Jatropha curcas*, *Leucaena leucocephala*, *Acacia nilotica*, *Azadirachta indica*, and *Prosopis juliflora*.

As a baseline for comparison, an uncultivated ravinous wasteland close to agricultural and forestry land uses was chosen. To find sensitive indicators, the whole set of soil property data from various land-use systems was put through PCA. To eliminate the duplicate factors, correlation studies were conducted among the highly weighted variables. Sensitive indicators were first standardized to a 0–1 scale using various scoring procedures, then an index was created by applying an additive model to evaluate the relative state of deterioration associated with various land uses. According to factor analysis, the following sensitive indicators (minimal data set) under the ravinous Ustifluvents were listed, in descending order of significance: exchangeable sodium, available nitrogen, sand content, active carbon, available potassium, and available phosphorous. A decision-making tool for the creation of policies resulting in the reclamation of ravinous wastelands in India's semi-arid area might be the SQI created for the site. An enzyme-based treated soil quality index (T-SQI) was calculated by Ghosh et al. (2019) using activities of 11 soil enzymes, namely, dehydrogenase, cellulase, β-D-glucosidase, invertase, peroxidase, phenoloxidase, protease, urease, alkali phosphatase, fluorescein diacetate hydrolysis and aryl sulfatase.

Using geostatistical techniques, Mandal et al. (2011) examined the spatial variability of soil characteristics and soil quality at the watershed and district levels. In the southern Indian states of Andhra Pradesh and Nalgonda, soil samples were taken based on various land-use patterns. The SQIs were assessed using a framework based on a minimal data set, a linear scoring method, and additive indices. Using PCA, the minimal data sets were selected according to the importance of soil parameters to soil function as determined by professional opinion. Castor intercrop was placed first in the list of estimated SQIs for various land-use systems. A krigged map of SQIs was created for the watershed, Nalgonda, and Warangal districts (Figure 4.3). The results indicated that soil having higher SQIs were more productive and less prone to erosion.

The Ministry of Agriculture, Government of India, introduced the soil health card scheme (SHCS) on December 5, 2015, to help the agrarian community understand the soil health status of their agricultural fields and integrated nutrient management. The SHCS was motivated by the need to address the issue of managing soil health and in response to a United Nations resolution. The primary goal of the SHCS is to raise farmers' awareness of the need to manage soil health

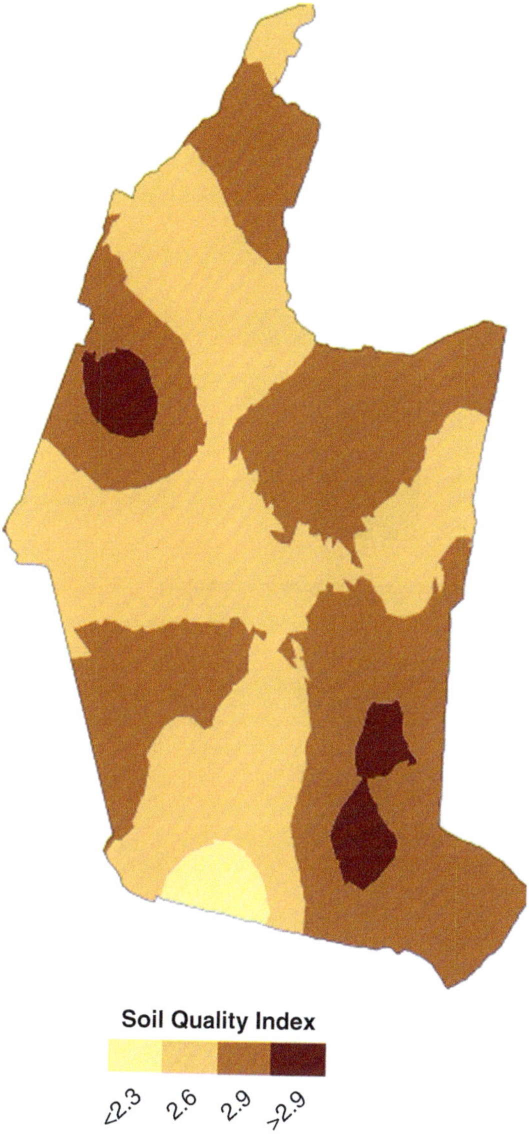

Figure 4.3 Thematic map for soil quality index (Andhra Pradesh).

through the prudent application of fertilizers, biofertilizers, and organic fertilizers. Farmers will also gain a better understanding of cropping patterns, cost-saving strategies, farm profitability, and sustainability. An example of a soil health card is shown in Figure 4.4.

General Recommendations	
1	Lime / Gypsum

Soil Test Results

District Agriculture Office, WEST GODAVARI , ANDHRA PRADESH

Soil Type: Black soil

Sr.No.	Parameter	Test Value	Unit	Rating	Normal Level
1	pH	6.39		Acidic	7, Neutral
2	EC	0.37	dS/m	Normal	0 - 2 dS/m
3	Organic Carbon (OC)	0.40	%	Low	0.51 - 0.75%
4	Available Nitrogen (N)	305.00	kg/ha	Medium	280 - 560 kg/ha
5	Available Phosphorus (P)	36.00	kg/ha	High	11-26kg/ha
6	Available Potassium (K)	69.00	kg/ha	Low	120-280kg/ha
7	Available Sulphur (S)	36.00	ppm	Sufficient	> 10 ppm
8	Available Zinc (Zn)	6.20	ppm	Sufficient	> 0.6 ppm
9	Available Boron (B)	2.21	ppm	Sufficient	> 0.5 ppm
10	Available Iron (Fe)	37.40	ppm	Sufficient	> 4.5 ppm
11	Available Manganese (Mn)	20.10	ppm	Sufficient	> 2.0 ppm
12	Available Copper (Cu)	5.90	ppm	Sufficient	> 0.2 ppm

Department of Agriculture, Cooperation & Farmers Welfare
Ministry of Agriculture and Farmers Welfare
Government of India
State Department of Agriculture(A.P.)

Soil Health Card

Soil Health Card Number - AP/2017-18/35499766/2

Validity - From: To:

Farmer's Details	
Farmer Name	**Jupalli aachutharamarao**
Father's/Husband Name	Venkatrao
Address	XXXXX
Mobile No.	9766-XXXX-XX
Gender	XXXXX
Category	XXXXX

Soil Sample Details	
Date of Sample Collection	02-05-2017
Survey No., Khasra No./ Dag No.	90/5,-
Farm Size	1.19 Acre Irrigated(Canal)

Figure 4.4 A soil health card issued to a farmer by the government of India.

Improvement of Soil Health in Central Semi-arid Regions of India

Soil health of central India can be improved by soil erosion control, water harvesting, legume intercropping, integrated nutrient management, organic manuring, reclamation of salt-affected soils, irrigation management for improving input use efficiency, reforestation, grassland and horticulture development, vegetative barriers and use of mulching and diversified cropping, and promotion of agroforestry and conservation agriculture, among other approaches. For example, Ghosh et al. (2018) reported that carbon accumulation and sequestration in integrated nutrient management were 0.74 and 0.22 Mg C ha^{-1} yr^{-1}, respectively, in the 0- to 90-cm soil layer, with >50% of the accumulated carbon being in deep soil layers (30–90 cm). Baradwal et al. (2022) estimated that the *Lucaenea*, *Hardwickia*, and *Azadirachta* improved nutrient supply capacity by 2.5, 2.2, and 1.6 times, respectively, and soil functionality by 9.3, 5.3, and 5.1 times, respectively, over fallow land in the surface soil layer. A similar trend was observed in lower layers. Kumar et al. (2019) estimated that the sustainable yield index for fruit and pasture in moisture-conserved plots was approximately (a) 1.85 and 1.68 and (b) 1.33 and 1.25 times higher than control for *Emblica*- and *Psidium*-based agroforestry systems, respectively. Soil moisture conservation could enhance profit by ~30% and 20% in those horti-pasture systems, respectively, over control. Soil moisture conservation can potentially recycle approximately (a) 83%, 77%, and 71% and (b) 62%, 122%, and 81% higher nitrogen, phosphorus, and potassium than control in those systems, at surface and subsurface soil, respectively.

As emphasized in previous sections, the purpose of a soil health assessment is to manage the soil for sustainable production and to achieve sustainable productivity of the soil. Figure 4.5 illustrates several strategies that can be used. These strategies include the following:

- Reduce soil disturbance (farmlands and rangelands)
- Crop rotation practices (farmlands)
- Cover cropping practices to promote plant diversity and living roots (farmlands)
- Diversify production systems (farmlands and rangelands)
- Add soil organic amendments (farmlands and rangelands)
- Integrate livestock into cropping systems (farmlands)
- Promote diverse plant species with different rooting depths (rangelands)
- Sustainable animal grazing practices (rangelands)
- Be patient for the results and do not expect positive change overnight (farmlands and rangelands)

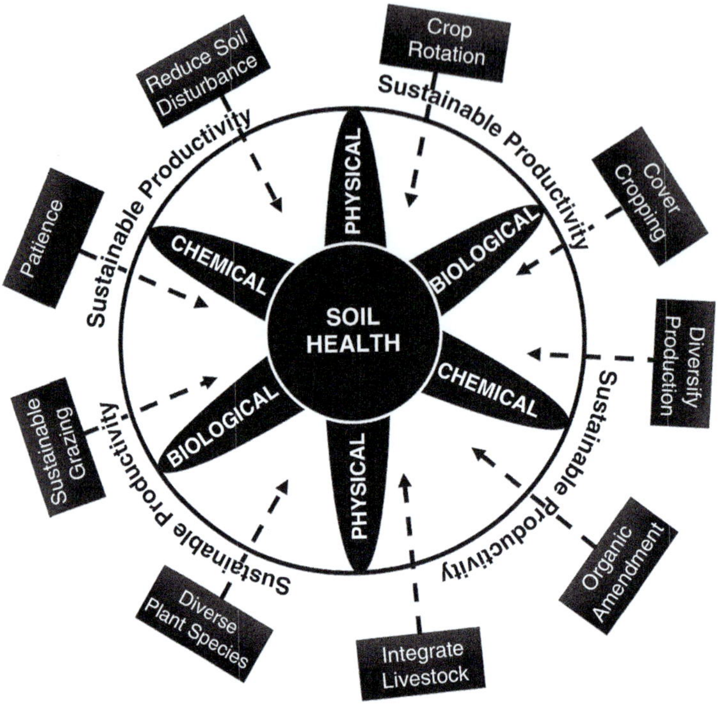

Figure 4.5 Management strategies that lead to soil health maintenance and improvement.

Conclusions

Soil health assessment and management will continue to play a prominent role in agricultural production systems of arid and semi-arid agroecosystems. Healthy soil is more resilient against fluctuations in growing conditions. Cover cropping practices, organic soil matter, sustainable grazing, and diversified food systems could be effective soil management strategy for rehabilitation. With yearly weather uncertainties, the resiliency of the soil system needs to be enhanced to cope with these variations. Building and improving the soil health of central India will ensure continued productivity, enhance farmers' incomes, and promote food security. Building and maintaining healthy soil is not an easy task in the arid and semi-arid Southwest, especially in the face of droughts and other production challenges. With the resolve of the farming and ranching community, it will be possible to adopt soil health practices that will maintain and improve soil productivity.

References

Baradwal, H., Ghosh, A., Kumar, A., Singh, P. D., Sannagoudar, M. S., Ahamad, S., Jha, P., Singh, A. K., Bhattacharyya, R., Manna, M. C., & Kumar, S. (2022). Ecological restoration of degraded lands with alternate land use systems improves soil functionality in semiarid tropical India. *Land Degradation & Development*, *33*(7), 1076–1087.

Ghosh, A., Bhattacharyya, R., Meena, M. C., Dwivedi, B. S., Singh, G., Agnihotri, R., & Sharma, C. (2018). Long-term fertilization effects on soil organic carbon sequestration in an Inceptisol. *Soil and Tillage Research*, *177*, 134–144.

Ghosh, A., Kumar, S., Manna, M. C., Singh, A. K., Sharma, P., Sarkar, A., Saha, M., Bhattacharyya, R., Misra, S., Biswas, S. S., & Biswas, D. R. (2019). Long-term in situ moisture conservation in horti-pasture system improves biological health of degraded land. *Journal of environmental management*, *248*, 109339.

Jha, P., & Mohapatra, K. P. (2012). Developing a soil quality index for ravinous Ustifluvents of the semi-arid region of India. *Journal of the Indian Society of Soil Science*, *60*(2), 116–124.

Kumar, S., Singh, A. K., Singh, R., Ghosh, A., Chaudhary, M., Shukla, A. K., Kumar, S., Singh, H. V., Ahmed, A., & Kumar, R. V. (2019). Degraded land restoration ecological way through horti-pasture systems and soil moisture conservation to sustain productive economic viability. *Land Degradation & Development*, *30*(12), 1516–1529.

Mandal, U. K., Kausalya, R., Sharma, K. L., Satyam, B., Venkanna, K., Bhanu, M. U., Mandal, M., Masane, R. N., Narsimlu, B., Rao, K. V., & Srinivasarao, C. (2011). Assessing soil quality under different land-use systems in a watershed of semi-arid tropics in India using GIS. *Soil Science Society of America*.

UNCCD. 2017. *Global land outlook*. 1st ed. Bonn: UN Convention to Combat Desertification.

Vasu, D., Singh, S. K., Ray, S. K., Duraisami, V. P., Tiwary, P., Chandran, P., Nimkar, A. M., & Anantwar, S. G. (2016). Soil quality index (SQI) as a tool to evaluate crop productivity in semi-arid Deccan plateau, India. *Geoderma*, *282*, 70–79.

Velasquez-Manoff, M., 2018. Can dirt save the Earth. *The New York Times Magazine*, *4*(18), p.18.

5

Regional Soil Health Assessments: Eastern Sub-humid Soil Health in India

Dibyendu Chatterjee, Rajendiran Selladurai, Mohammad Shahid, Dibyendu Sarkar, and Saikat Ranjan Das

Chapter Overview

Soil health degradation is a major challenge that leads to ecosystem imbalance and reduced productivity. Soil health assessment is essential for eastern sub-humid India with its diverse soil types, varied climatic zones, and wide variety of crops. It is important to have a systematic approach to determine and evaluate appropriate soil parameters as indicators for accurate assessment of soil quality. Assessing soil health involves observing the physical, chemical and biological properties of the soils and their interactions. Indicators are used to measure changes in these properties and various methods such as field observations and data synthesis are used for their assessment. To maintain and improve the health of agricultural soils in the eastern sub-humid region, it is important to follow five basic principles of soil health management. These principles include protecting the soil by covering it, minimizing disturbance, promoting plant diversity, maintaining living roots, and integrating livestock farming. The government supports various programs aimed at improving soil health, increasing productivity, reducing the cost of agricultural production, increasing farmers' income and minimizing the environmental impact of nutrient loss to ensure food security in this region.

Soil Health Series: Volume 4 Soil Health and Sustainability in India, First Edition.
Edited by Ranjan Bhattacharyya.

Introduction

Soil health deterioration is the major problem with modern-day intensive cultivation practices. It results in ecosystem imbalance and declines in productivity. Maintaining soil quality poses a substantial challenge for sustaining production and ecosystems in emerging economies owing to intensive land use and rapid economic growth (Doran et al., 1996). Periodic soil health assessment is invaluable for understanding the impact of various management interventions on a particular production or ecological system. Further, such assessments will help identify management practices that are suitable for a particular ecosystem or region. Eastern sub-humid India consists of West Bengal, Bihar, Odisha, Jharkhand, and Andaman and Nicobar Islands. The states of Bihar and West Bengal are situated on the Indo-Gangetic plains, Jharkhand is on the Chota Nagpur Plateau, Odisha is on the Eastern Ghats and the Deccan Plateau, and Andaman and Nicobar Islands are at the junction of the Bay of Bengal and the Andaman Sea. The Eastern India region is in a humid sub-tropical zone with a monsoon period during July–October, mild winters from November to February, and hot summers from March to June. In each state, there are different agro-climatic zones based on the effects of landform, topography, climate, soil, and crop adaptability. The soils of the region vary mainly because of differences in parent material and climatic conditions. In West Bengal, alluvial soil, mountain soil, saline soil, and red soil are the four major soil groups; in Bihar, Piedmont swamp soil, terai soil, and gangetic alluvium are the three major soil groups; in Jharkhand, red soil, sandy soil, black soil, laterite soil, and red micaceous soil are the five major soil groups; in Odisha, red soil, mixed red and yellow soil, black soil, laterite soil, deltaic alluvial soil, coastal alluvial and saline soil, brown forest soil, and mixed red and black soil are the eight major soil groups; and in Andaman and Nicobar Islands, alluvial soil, black soil, laterite soil, and coastal saline soil are the four major soil groups. Major crops grown in this region are rice, wheat, maize, pulses, oil seeds, wheat, barley, potatoes and other vegetables, fruits, jute, and coconut, among others. Paddy, areca nut, and coconut are the major crops of Andaman and Nicobar Islands. Vegetables, pulses, root crops, oilseeds, orange, mango, banana, sapota, pineapple, and papaya are grown on hilly lands. Pepper, clove, nutmeg, and cinnamon are grown in multi-tier cropping system. Rubber, red oil, palm, noni, and cashew are grown on a limited scale on these islands.

Soil quality is assessed by various means and methods. In India (Kundu et al., 2012; Mahajan et al., 2020; Singh, 2007; Vasu et al., 2016) and throughout the world (Andrews et al., 2004; Bünemann et al., 2018), different approaches have been adopted for soil quality assessment. These approaches range from the simple approach of visible evaluation to complex statistical procedures (Andrews et al., 2004). Identification of a minimum data set of soil quality indicators that largely influence and represent the ecosystem function is very important for accurate soil quality assessment. Moreover, scoring soil quality indicators is an essential criterion for getting a better and accurate picture of soil quality of the region. The soil quality index (SQI) is widely agreed to be the best tool for assessing soil quality decline (Wang & Gong, 1998). However, there is no procedure for selecting, evaluating, and indicating soil quality under all possible field conditions. Several studies have used different methods to index soil quality (Andrews et al., 2004; Masto et al., 2008). The purpose of assessing soil quality is to identify major constraints or variability of soil quality indicators that largely affect the production and other ecological functions in the region. This information will be useful in recommendations for suitable management measures to overcome the constraints and improve soil quality and ecosystem sustainability. Soil quality assessment focuses on dynamic soil properties and processes and on the sustainability of soil management systems. Soil degradation status and change in land use patterns following intervention through smallholder management are additionally included in soil quality assessment (Lal & Stewart, 1994). The successful use of agricultural resources to meet evolving end-user requirements, while preserving or improving the natural resource base and preventing environmental deterioration, represents the foundation of sustainable agriculture. Natural, environmentally sound, low-input, organic, and alternative agricultural systems are basically clean production systems compared with conventional fertilizer-based systems. The former systems are all included in sustainable agriculture. They place a strong emphasis on the long-term viability of the soil resource and soil health, which sustain our food production alongside the other critical resources of light, air, and water (Hatfield & Karlen, 1993).

Soil Health Assessment Suitable for Eastern Sub-humid Region

Soil Quality Indicators

Soil quality indicators are used to assess soil function, as direct measurement of soil function is challenging. As a measure of the soil's quality, soil characteristics that are sensitive to changes through anthropogenic or environmental factors can be utilized (Schloter et al., 2003). Soil quality parameters can be qualitative or quantitative in nature. Ideal soil quality indicators should align with ecosystem processes; encompass physical, chemical, and biological aspects of soil; be easily accessible to various users; be sensitive to climate and management practices; be part of existing databases; and be understandable and interpretable (Doran & Parkin, 1997). As a result, they are used to create indexes based on the links between various soil indicators. A systematic approach to determining and evaluating soil parameters that are suitable for serving as soil quality indicators are necessary for accurate and reliable soil quality evaluation (Granatstein & Bezdicek, 1992). It has already been established that a single soil characteristic may not be a reliable indicator of soil quality. An integrated soil quality indicator based on a mixture of soil parameters may be more accurate than a single soil property (Dick, 1994; Elliott, 1994).

Categories of Indicators

Soil quality indicators can be divided into three main categories: physical, chemical, and biological (Karlen et al., 2001). Chemical indicators provide information on soil solution and exchange site equilibrium, plant health, nutrient requirements, and levels of available soil pollutants for plant uptake (e.g., soil nitrate, electrical conductivity [EC], pH). Physical indicators relate to the hydrologic properties of the soil, such as water entry and retention, which affect plant availability and nutrient accessibility through their impact on rooting volume, aeration, and erosion status. Physical characteristics include soil structure, available water capacity, soil infiltration, bulk density, aggregate stability, and the presence of features such as macropores, micropores, and soil crusts. Biological indicators provide information about the organisms involved in the soil food web, responsible for organic matter decomposition and nutrient cycling. These indicators encompass soil enzymes, respiration, potentially mineralizable nitrogen, earthworms, particulate organic matter, and total organic carbon.

Selection of Indicators

Selection of correct soil quality indicators is necessary for assessing SQI. It depends on various factors such as the ecosystem function one looks for and scale of soil quality measurement (farm, village, block, taluk, district, state, region, country level), among others. A soil quality indicator should be very simple, measurable, closely related to ecosystem function, and influenced by management intervention. Although there is no clear approach or agreement on how to choose the indicators, they should be chosen based on the climate, soil type, hydrology, and other characteristics of the area under study. Many soil quality indicators have been suggested, but only a few have been extensively studied and validated (Ghaemi et al., 2014; Obade & Lal, 2016b).

Soil Quality Indicators for Eastern India

Eastern India has diversified soil systems, climatic conditions, and cropping patterns. Therefore, identification of suitable soil quality indicators for different production systems or regions is essential. In eastern coastal soils under different land use systems, Srinivasan et al. (2021) used soil organic carbon

(SOC), total nitrogen, pH, EC (Electrical conductivity), available nutrients, cation exchange capacity, particle size, soil aggregation, available water capacity, porosity, penetration resistance, biological activity, and invertebrates. Similarly, Rai et al. (2009) assessed the topsoil quality of Jharia Coalfield, Jharkhand, India using pH, EC, moisture content, organic matter, and available nitrogen and phosphorus. In agricultural systems of alluvial soils of Koch Bihar District, West Bengal, India, soil pH, soil EC, SOC content, nitrogen content, potassium content, phosphorus content, soil texture, the groundwater level in the winter season, cropping intensity, and drainage frequency were used to assess the soil quality (Das & Paul, 2022). In Bansloi river basin (area, 1,860 km^2) of eastern India, 12 soil quality indicators (nitrogen, phosphorus, potassium, pH, EC, organic carbon, iron, copper, manganese, zinc, boron, and sulfur) were used, but nitrogen, phosphorus, potassium, iron, and manganese had significant control toward soil quality (Paul et al., 2020). In the Indo-Gangetic Plains of eastern India, Bazaya et al. (2009) employed bulk density, organic carbon, soil moisture content, infiltration rate, soil temperature, pH, and EC, in addition to the population of bacteria, fungi, *Azotobacter*, and actinomycetes. Similarly, in alluvial soils of eastern India, available phosphorus, SOC, available potassium, phosphorus solubilizing microorganisms, available nitrogen, microbial biomass carbon, soil respiration, and enzymatic activity were used as minimum data set (MDS; Singh et al., 2017). Although several studies have used different soil quality indicators as MDS, all these indicators are tested and validated under different conditions.

Tools/Methods of Assessing Soil Health

Soil health assessment involves evaluating the ability of soil to perform all its functions, including crop output and water quality, which cannot accurately indicate the overall health of the soil on their own. Soil's physical, chemical, and biological attributes interact together to give it this functionality; hence, capacity must be indirectly measured by monitoring changes in its properties or ecosystem attributes called indicators. These quantifiable properties resulting from functional relationships can be assessed through field observations, remote sensing surveys, or data synthesis. To evaluate soil health effectively regarding various land use practices and management techniques, a comprehensive qualitative or quantitative index that defines observed indicators is needed (Dick, 1994).

Attempts have been made to construct quantitative and qualitative markers to describe soil health, despite the variety of features and the disagreement over common soil qualities. If an indicator does not incorporate evaluated parameters into a numerical index that ranks soils along a continuous scale, it is characterized as nonquantitative. Whereas, quantitative indicators produce a numerical index, often with the greatest value being attributed to the soils having the best quality (Singer & Ewing, 2000).

Numerous attempts have been made to create indices that measure soil health and quality (Andrews et al., 2004; Bünemann et al., 2018; Obade & Lal, 2016a; Lilburne et al., 2004; Shahid et al., 2013). According to Ewing and Singer (2012), it is crucial to generate a set of biotic and abiotic indicators to create a workable soil health index and to take advantage of the interactions between the different indicators to give a comprehensive picture.

Analytical Approaches to Soil Health Assessment

Significant research has gone into developing a quantitative assessment system that combines the most important indicators of soil health and developing a health index of soils that can quantify the effectiveness of soil management strategies. Several integrative scoring curve methods have been proposed, including the Soil Management Assessment Framework (Andrews et al., 2004; Veum et al., 2017), Comprehensive Assessment of Soil Health (Moebius-Clune et al., 2017), Soil Health Assessment Protocol and Evaluation (Nunes et al., 2021), Soil Management Index (Williams et al., 2020), Soil Health Calculator (Jian et al., 2020), Soil Conditioning Index (SQI, 2003), and Agro-Ecosystem Performance Assessment Tool (Wienhold et al., 2006).

A soil health assessment technique created for one agricultural system or geographic area may not be relevant to others. To accurately reflect the characteristics and reveal the problems that underlie soil health, multiple indicators must be measured when assessing soil health. Natural soil performs several important environmental tasks, including plant support, nutrient recycling, and organic residue breakdown, which are directly controlled by the physical and chemical environments (Kibblewhite et al., 2008). When evaluating soil health, the three types of soil attributes should be given comparable or equal weight. In the past, soil physical and chemical features were given more weight than biological processes because they are often easier to measure and interpret (Lehmann et al., 2020).

As a result of the development of the soil health system, numerous biological indicators have been introduced recently, including soil microbial biomass, soil enzymes, soil protein index, short-term carbon mineralization (soil respiration rate), microbial activity, fatty acid methyl ester profile, phospholipid fatty acid analysis, biodiversity, abundance of earthworms, presence of pathogens and parasites, and DNA sequencing for soil microbial community structure (Allen et al., 2011; Lehmann et al., 2020; Norris et al., 2020; USDA-NRCS, 2019). For instance, Nunes et al. (2021) chose seven biological markers to build a soil health assessment system (SOC, active C, microbial biomass C, microbial biomass N, respiration rate, beta-glucosidase activity, and protein index). An extensive analysis of soil biological indices for assessing soil health was undertaken by Paz-Ferreiro and Fu (2016).

Visual Assessment Approaches to Soil Health

Key quantitative markers of soil health are closely correlated with several visual indicators, including color, structure, aggregation, texture, porosity, moisture levels, and earthworm casts. Soil research and soil advisory services use the same standard field diagnostic criteria inside predetermined frameworks and validate them on greater scales, just as indigenous people have done in the past.

Although visual soil health assessment has the advantage of being simple to interpret, it cannot evaluate the state of ecosystem services that are influenced by biological and chemical soil processes (Ball et al., 2017). The combination of both would be beneficial because visual soil assessment provides different information than laboratory procedures (Emmet-Booth et al., 2016; Pulido Moncada et al., 2014). Ultimately, it is believed that the rising use of visual soil assessment is significant in yield gap analyses and land management programs (McKenzie et al., 2015).

Soil Health Card Methods

A soil health card (SHC) is used to evaluate the current state of soil health and make suggestions on the application of fertilizer and manure. It seeks to prevent fertilizer overuse by offering crop-specific advice for nutrient administration. Through the balanced use of nutrients determined by a soil test in conjunction with organic manures, it encourages integrated nutrient management (INM). The program seeks to cover all farmers completely.

Owing to the lack of local soil testing laboratories, soil testing is a laborious process for farmers. Without the right information or training, collecting soil samples may not yield reliable findings. Twelve significant soil characteristics, including phosphorus, nitrogen, potassium, pH, EC, organic carbon, sulfur, zinc, boron, iron, manganese, and copper, were examined using SHC. Six crops are recommended using a uniform sample strategy, using a grid of 2.5 ha for irrigated crops and 10 ha for rainfed crops (Dey, 2016). Each farmer receives a soil health card every 3 years. The SHC Scheme's entire operations are carried out using a single, universal, web-based, generic piece of software (https://www.soilhealth.dac.gov.in).

Impact of Soil Degradation on Soil Health in the Eastern Sub-humid Region

Maintaining soil health is crucial for sustainability because it allows the soil to perform its fundamental activities. When multiple threats happen simultaneously, their effects are compounded. They can cause soil degradation

Table 5.1 State-wise extent of various kinds of land degradation in eastern and northeastern states of India.

State	Water erosion	Wind erosion	Water logging	Salinity/ alkalinity	Soil acidity	Complex problem	% of degraded area to TGA[a]
				Mha			
Northeastern Hill Region	12.4	0	0.4	0	5.6	2.0	33.0–89.2
Assam	0.7	0	0	0	0.6	0.9	28.2
Tripura	0.1	0	0.2	0	0.2	0.1	59.9
Bihar + Jharkhand	3.0	0	2.0	0.2	1.0	0	36.1
West Bengal	1.2	0	0.7	0.2	0.6	0.1	31.0
Odisha	5.0	0	0.7	0	0.3	0.1	39.3

Data source: Bhattacharyya et al. (2015); NBSS&LUP-ICAR (2004) on 1:250,000 scale.
[a] TGA, total ground area.

and a reduction in the soil's ability to perform its tasks if they are left unchecked. Loss of one or more of the functions of the soil and potential water contamination are both possible effects of contaminated soil. Table 5.1 shows the estimated level of land degradation in India according to the National Bureau of Soil Survey and Land Use Planning (NBSS & LUP) and the Indian Council of Agricultural Research (ICAR).

Contaminants, both natural and human-related, accumulate in soil and are released into the atmosphere or water. The irreversibility threshold for storage and buffering capacity is exceeded by some pollutants. Thus, early warning systems are crucial to preventing environmental harm and public health problems. Soil degradation leads to a decline in soil health, causing negative impacts on productivity, air and water quality, and human and animal health (Manna & Sharma, 2009). Currently available research suggests that the soil may be deteriorating due to a variety of anthropogenic activities. Physical factors and chemical factors are two main categories for abiotic factors that affect soil health or soil degradation. Soil response, salinity, acidity, and alkalinity changes in the soil nutrients status, and physical characteristics such as texture, structure, water, and temperature all have an impact on the soil health and quality. These components are essential for the production of the soil. Within the soil system, all the chemical and physical characteristics of the soil are clearly interrelated, which makes it difficult to classify the individual components.

Erosion, a drop in organic matter, local and widespread contamination, compaction, acidification, salinization, a loss of biodiversity, and pollution are some of the threats to soil quality (European Commission, 2002). Around 25% of the world's land area has already been degraded. Soil acidity followed by water erosion is the most serious land degradation problem in eastern and northeastern India, resulting in loss of productivity, topsoil, nutrients, and terrain deformation. Degradation of the soil resource means destruction of a valuable asset for today's farmers as well as future generations of farmers. Farmland sustainability, soil fertility maintenance, and contamination prevention should all receive special attention. Since soil health serves as the foundation for supporting life in numerous ways, it is extremely important.

Living organisms, humus (the by-product of decomposing organic matter), and organic material (plant roots, leaves, and feces) make up the organic matter in soil. For the soil to be fertile, organic matter is crucial for vital soil processes. It enhances the soil's potential for binding and buffering, which reduces water pollution. Through the sequestration of CO_2 into the SOC pool, carbon in soil contributes significantly to the global carbon cycle. Lower bulk densities are associated with soils that contain more organic matter. Soil is home to a wide variety of living things. The physical and biochemical qualities required for soil fertility are maintained by soil bacteria, fungi, protozoa, and other tiny creatures. Microorganisms further degrade organic matter after it has been broken down by larger organisms such as insects, worms, snails, and small arthropods. In addition to acting as a nutrient source, foreign diseases are suppressed and pollutants are broken down into harmless components by soil organisms.

Healthy soil smells like fresh earth, has residues at different decomposition stages on the soil surface and in the topsoil, and has more earthworms, casts, and many holes. Plant roots and other soil organisms, such as worms, create macropores in the soil through their activities. The biological activity of the soil is influenced by its physical and chemical properties. Living and working in the pores are soil microbes and plant roots. Roots cannot develop as well in soil that has lost its porosity, and many species struggle to survive.

Principles and Practices of Management of Soil Health for Eastern Sub-humid Region

Inappropriate land use, crop husbandry, and animal grazing cause deterioration of soil health and loss of soil functioning. Good soil management practices are necessary to maintain and improve the health of agricultural soil.

Principles of Soil Health Management

To maintain and promote soil health, USDA-NRCS has defined five basic principles for soil and land management (USDA-NRCS, 2016):

Principle 1—Soil armor: to protect as much soil as possible with natural vegetation, agricultural waste, compost, or synthetic covers. Better soil structure, reduced soil temperature fluctuations, weed control in fields, and better habitats for soil microorganisms are all benefits of soil cover.

Principle 2—Minimize soil disturbance: to minimize mechanical (e.g., tillage), chemical (e.g., pesticide use), and biological (e.g., overgrazing) soil disturbance. Conservation tillage, integrated nutrient and pest management, and rotational grazing are effective management techniques for minimizing soil disturbance.

Principle 3—Plant diversity: to grow a variety of crops on farms to prevent disease and pest infestation and maintain soil food web health.

Principle 4—Continuous living plants/roots: to maintain a living root throughout the year to enhance soil biodiversity, achieve high microbial populations, and prevent soil erosion.

Principle 5—Integrate livestock management: to improve animal welfare, reduce pesticide use, promote nutrient cycling, and reduce nutrient export from agriculture, it is beneficial to incorporate livestock grazing into crop residue, cover crop, and weed management.

Technological Options for Soil Health Management

To conserve or improve soil health, the ideal soil management techniques include suitable cropping systems, utilization organic amendments, conservation tillage, crop fertilization, engineering soil conservation structures, and appropriate land use (Troeh et al., 2004; Lal, 2016). These procedures include the five fundamental concepts of soil health management and have proven to be highly efficient in enhancing several soil functions: (a) transformations of carbon, (b) nutrient cycling, (c) soil structure, (d) pest control, and (e) disease control. Important indices of soil health are responsible for these actions.

Suitable Land Use

The most important factor affecting the health of the soil is suitable land use. Depending on their potential to sustainably grow crops and pastures plant without degrading the soil over an extended period, the lands are divided into eight groups. Class I lands have no restrictions on producing intensive crops, but Class II lands have mild restrictions that limit the types of crops that may be grown or necessitate modest conservation measures. The land capacity classes III–V in the mountainous eastern states of India include steep slopes, which have significant restrictions on crop production. These areas may effectively conserve their soil by using conservation practices such contour terraces, benches, half-moon terraces, grassed waterways, and water harvesting ponds (Baishya & Sarkar, 2015; Baruah et al., 2014). The implementation of need-based agricultural planning on a priority basis based on soil health, drainage features, crop intensity, and degree of erosion threats may be assisted further by the development of watershed programs for landowners. Class VI–VIII properties are often utilized for forestry, wildlife, and recreational activities since they are unsuitable for farming. The order of predominant land use in the eastern Himalayan region regarding soil health based on SQI is natural forestland (0.93) > grassland (0.87) > shifting cultivated land (0.60) > cultivated lowland (0.57) > plantation land (0.54) > cultivated upland terrace (Singh et al., 2014). In terms of managing soil health, agroforestry systems surpassed shifting cultivation, with improvements in SOC (44.8%), dispersion ratio (52.9%), mean weight diameter (29.4%), and in situ soil moisture conservation (20.6%), while reducing soil loss (99.3%) and erosion (45.9%) in the northeastern hill region's conditions of high rainfall, moderate to steep slopes, and shallow soil depth (Saha et al., 2012).

To reduce the impact of shade and biochemical interactions on the cultivation of agricultural crops, sufficient tree density in the case of an agri-horticulture system is 400 trees ha^{-1}, while it is 200 plants ha^{-1} in the case of an agri-silviculture system (Bhatt, 2003; Umashankar, 2005). Multipurpose trees have been crucial in the region's efforts to improve the health of the soil (Saha & Mishra 2007; Saha et al., 2012). According to reports, *Michelia oblonga* and *Pinus kesiya* are the best tree species to use as bio-ameliorants on these soils because their constant leaf litter and root exudates greatly enhance the physical properties and SOC. Farmers of the northeastern region have been regenerating shifting cultivated lands by practicing agro-forestry and tree farming (Nongkynrih et al., 2018; Pant et al., 2018; Ritse et al., 2020). For example, people in Nagaland have adopted the growing of alder trees (*Alnus nepalensis*) in shifting cultivated areas (Baruah et al., 2014; Cairns et al., 2010). Additionally, Sarkar et al. (2015) revealed that fallow soil under shifting cultivation functions as a carbon sink throughout the growth of forest plants, highlighting the essential need to enact appropriate land management policies to preserve organic carbon in soil and prevent further soil degradation. They believed that in order to sustain soil organic carbon stock under shifting agriculture in northeast India, a fallow interval of at least 10 years was necessary.

Owing to their poor water holding capacity, light-textured soils (such sand and sandy loam) should not be used for the rice cultivation. Poor land use causes soil erosion, and ultimately, the degradation of soil health. In the eastern sub-humid region, poor drainage and flood water management are concerns because they cause water stagnation in some rice-growing alluvial soils for about 6 months in a year (Bhattacharyya et al., 2015). This poor physical condition of the soil allows only one lowland submerged rice cultivation with a very low yield potential. The limited functionality of the soil is the result of various land management practices, including repeated cultivation with extensive tillage, monocropping, crop residue removal, excessive use of inorganic fertilizers, and inadequate use of organic amendments (Bandyopadhyay et al., 2016; Kumar et al., 2019a; Mandal, 2005; Singh et al., 2022).

Crop Rotation and Diversity

Crop rotation or diversity, the alternating or simultaneous cultivation of two or more different crops on the same plot, is a necessary part of a good cropping system. Crop rotation and diversity are most effective in controlling diseases, pests, and weeds on cropland because they break the cycles of their host plants. Plant species differ in terms of their root structures, nutrient requirements, root exudates, and nutrient cycling ability. Because of this, crop rotation and diversification increase crop yields and enhance C sequestration, soil fertility, soil structure, erosion resistance, biodiversity, and biological activities, and thus the overall soil health (Ansari et al., 2021; Das et al., 2018; Singh et al., 2022). Rice is the dominant crop and about 30% area remains fallow after wet season rice cultivation in the northeastern hill and eastern plains (NAAS, 2013; Subbarao et al., 2001). After the harvest of rainy season (*Kharif*) rice, agro-climatic conditions of rice fallow lands in many places of eastern India are favorable for raising short duration pulses or oil seed crops with the judicious use of residual soil moisture. Short-duration and high-yielding rice varieties that leave the field in September or October make this approach more efficient. Compared with chickpea, lentil grows better in low-lying areas with high soil moisture. Rice–lentil cultivation is widespread in the lowlands of eastern Uttar Pradesh, Bihar, West Bengal, and Jharkhand. In the Char region of Bihar, the Mahananda region of Odisha, eastern Uttar Pradesh, the Brahmaputra Valley of Tripura, Assam, Manipur, and Meghalaya, groundnut cultivation in rice fallows can also be a viable option. Together with black gram and mung, mustard offers significant potential for zero-tillage cultivation in eastern Uttar Pradesh (NAAS, 2013).

Specific studies showed that systematic intensification of rice-fallows with the inclusion of winter crops, such as safflower, lentil, and chickpea, after rice harvest with conservation tillage would increase system productivity and energy use efficiency with reduction in greenhouse gas emission (Kumar et al., 2020a). In addition, diversification of rice-growing areas with rice–mustard–jute system under integrated soil–crop management would improve soil health by minimizing water and nutrient loss (Singh et al., 2022). Crop rotation and diversification with leguminous crops (e.g., black gram, green gram, groundnut, and soybean) improved soil health over the existing cereal-based cropping in the eastern Himalayas (Ansari et al., 2021; Baruah et al., 2014; Das et al., 2018). Strip cropping with cereals and erosion-resistant legumes can store rainwater and slow runoff. Growing groundnut as an intercrop with maize is a useful practice for soil conservation to improve soil health in the northeastern region (Baishya & Sarkar, 2015).

Cover Crops

Cover crops, which are cultivated to cover the surface of the soil, are useful in improving biological, chemical, and physical soil properties. These crops are critically important to control soil erosion and loss nutrients due to leaching and surface runoff, maintain soil microbial diversity and activity, and improve soil health. In the middle altitudes of northeast India, perennial tropical grasses, such as broom grass (*Thysanolaena maxima*), signal grsss (*Brachiaria rosenesis*), setaria (*Setaria sphacelata*), napier grass (*Pennisetum purpureum*), guinea grass (*Panicum maximum*), and perennial groundnut (*Arachis pintoi*), and the legumes Gliricidia (*Gliricidia sepium)* and lucerne (*Medicago sativa)* grow well. These grasses and legumes provide forage for livestock (Gupta, 2004) and produce high amount of biomass, which contributes to increase SOC, suppresses weeds and pests, improves soil structure, reduces soil compaction, and improves soil health (Baruah et al., 2014; Das et al., 2016; Ghosh et al., 2009; Sarkar et al., 2018a).

Conservation Tillage

Conservation tillage is defined as tillage or a combination of tillage and cultivation methods that leave at least 30% of the soil surface covered by crop residues. Conservation tillage techniques include no-tillage, ridge tillage, strip tillage, and mulch tillage. The main variant of no-tillage is direct drilling of seeds along with basal dose of fertilizers (also termed as zero-tillage), while other variants are reduced tillage and minimum tillage. Conservation tillage provides benefits such as soil conservation, reduction in tillage cost and crop establishment time, reduction in soil erosion and nonpoint pollution, and increased storage of SOC and improvement of soil health. Conservation tillage with judicious plant nutrient management and residue recycling enhanced soil health by improving carbon and nitrogen sequestration in soils under double rice cropping systems in the northeastern region (Yadav et al., 2017, 2019). Conversion of conventional to conservation tillage improved soil quality at mid-altitudes of northeastern India by improving biological, chemical, and physical properties (Das et al., 2018). In eastern sub-humid India, no-till tillage with incorporation of residues under the jute–rice–lentil system has also been found to be helpful in maintaining crop yield and improving soil health (Saha et al., 2022).

Use of Organic Amendments

The main factor affecting soil health is its organic matter content. Application of organic residues or organic amendments such as farmyard manure (FYM), compost, solid animal manures, biosolid, green manure, agro-industrial wastes, food processing refuse, and biochar are useful in increasing SOC and improving soil health. These organic amendments contain substantial amount of plant nutrients (e.g., nitrogen, phosphorus, potassium, calcium, magnesium, sulfur, iron, manganese, zinc, copper, and boron) along with organic carbon (Batabyal et al., 2016; Sarkar et al., 2018b). Efficient use of bio-wastes (crop residues, weed biomass, forest litter, and animal dung) has great potential to improve the soil health and crop productivity in northeastern India (Kumar et al., 2020b). Waste composting allows for the potential organic amendment of the large resources of crop residues and animal manure (Chatterjee et al., 2016, 2021).

Long-term application of organic amendments at sufficient rates, either alone or in combination with inorganic fertilizers for crop production, would improve soil health. A balanced supply of nutrients through integrated use of inorganic and organic sources could maintain or enhance soil health/quality owing to improvement in biological, chemical, and physical properties under rice–rice system (Majhi et al., 2021; Mandal, 2005), rice–wheat system (Ghosh et al., 2012; Kumari et al., 2017), rice–wheat–jute system (Manna et al., 2005; Mazumdar et al., 2018), jute–rice–lentil system (Mazumdar et al., 2022), and soybean–wheat system (Chakraborty et al., 2021) and under vegetable production (Batabyal et al., 2016, 2017; Tamang et al., 2017) in the eastern plains of India. Similarly, integration of FYM with inorganic fertilizers improved soil quality by increasing SOC, microbial biomass carbon, and plant available nutrients under maize–mustard cropping system (Das et al., 2010; Sarkar et al., 2018a), rice–rice system (Gogoi et al., 2021), and vegetable-based systems (Patel et al., 2015) in acidic soils of northeast India.

Crop-Livestock Integration and Controlled Grazing

In India, most agriculture is based on crop and livestock production. Livestock and agriculture complement each other by sharing benefits. Animal waste and agricultural by-products are used to maintain the animal component, while animal dung is used to fertilize crops. Animal dung improves soil health by increasing the SOC and reducing the need for inorganic fertilizers (Kochiwad et al., 2017; Nayak et al., 2020). However, overgrazing can prevent the realization of these benefits (e.g., cover crop biomass and limited crop residues are overloaded with livestock and grazing). Excessive trampling by livestock, which leads to soil compaction, soil structure degradation, and reduced infiltration, can accelerate soil health degradation. Overgrazing can be avoided by rotating grazing areas. In rotational grazing, animals are allowed to graze on a selected area of land at a given time and then to move on to another area of land to allow for restoration of forage in the previous area. In the case of open grazing, farmers let their animals graze in the village fields after they have harvested *Kharif* rice and farmers can rarely grow a second crop after rice (Saud & Bezbaruah, 2020). Under this circumstance the village community can come forward to limit open grazing to a part of the village land and let farmers grow a second crop on the remaining land. But such understanding lies in community initiatives, and extension workers can encourage villagers for this solution to facilitate controlled grazing.

Management of Specific Soil Problems

Soil Acidity Acidic soils cover about 21.2 million hectares (Mha; ~81% of the geographic area) of the northeastern hill region, and most of these soils (16.2 Mha) are extremely acidic (pH < 5.5; Sharma et al., 2006). Similarly, in Bihar, West Bengal, Jharkhand, and Odisha, about ~40,000, 420,000, 735,000, and 200,000 ha, respectively, are covered with acidic soils (Maji et al., 2008). Crop productivity on these soils is mainly affected by the adverse effects of aluminum, deficiency of zinc and phosphorus, and low biological activity (Patiram, 2007). Management of soil acidity and improvement of crop productivity on these soils are important for enhancing food security and soil health. The preferred management technique for improving acidic soils is liming. This method raises the pH of acidic soils, increasing the availability of nutrients to plants and reducing the toxicity of iron and aluminum (Bhat et al., 2007; Sharma & Sarkar, 2005). It has proven useful in this area to apply a lower dose of lime (i.e., one-tenth to one-fifth of the required lime) in furrows at the time of crop sowing. To gain acceptance among farmers, liming must be enabled by the availability of affordable and efficient lime supplies that meet the required standards. Commercial lime (calcite) is a useful liming material, but it can be expensive. However, in West Bengal, inexpensive basic slag and paper mill waste are available as substitute lime materials (about 0.3 million tonnes [Mt]). Large deposits of limestone available in Arunachal Pradesh (140 Mt), Assam (159 Mt), Manipur (46 Mt), Meghalaya (1,400 Mt), and Nagaland (357 Mt) can be exploited as agricultural lime through minimum processing at an affordable cost (Baishya & Sarkar, 2015).

Salt-Affected Soils Sodic soils cover about 106,000 ha in Bhojpur and Patna districts, whereas saline soils cover 47,000 ha in Vaishali and Muzaffapur districts of Bihar (Sharma et al., 2004). Saline soils have low osmotic potential, which often causes water stress in plants. Reduced availability of phosphorus, potassium, or essential ions due to high calcium concentrations, reduced uptake of potassium and calcium due to high sodium concentrations (in sodic soils), and reduced uptake of nitrate-nitrogen due to excess sulfate and chloride in the soil are all additional effects of salinity on plants. However, loss of soil physical health may have a greater impact on plant development in sodic soils than chemical health of soil.

Management of salt-affected soils requires tillage, irrigation, and leaching. The most effective method of removing salinity from the crop root zone can be done by applying good-quality irrigation water with low EC and cultivating tolerant crops. Drainage of water from the soil profile is also important for controlling salinity. Good irrigation management options for saline soils include: (a) increasing the amount of water applied to the soil to increase water infiltration, and (b) reusing the runoff water for irrigation. Rice is the potential crop for reclamation of sodic soil in this region. Leaching soluble salts from the root zone and replacing exchangeable sodium from the soil exchange complex are two steps in improving sodic soils. Soluble calcium salts, such as phosphogypsum or gypsum or acidifiers such as sulfuric acid, sulfur, aluminum sulfate, and pyrite are used as amendments to improve soil conditions. However, phosphogypsum or gypsum are readily available and are the most cost-effective soil amendments. Agricultural grade phosphogypsum or gypsum should be added to the soil at least 15–30 days before sowing a *Kharif* crop. To promote reaction and any leaching, the treated field must remain submerged in good-quality water. Preferably, rice followed by wheat should be grown in the treated plots during *Kharif* and *Rabi* (winter) seasons, respectively, in sodic soil of the region.

Soil Health Policy and its Implementation for the Sub-humid Region

Most of the nutrients essential to plants are relatively lower use efficiency in India. One of the various causes is the lack of awareness of the inherent fertility of the soil. A necessity for the era of modern technology, distribution and capacity building through human resource development, and infrastructure construction or improvement is adequate government and policy support. The Department of Agriculture, Cooperation and Farmers' Welfare within the Ministry of Agriculture and Farmers' Welfare has always given high priority to soil health management. The Government of India has implemented several centrally funded schemes under various 5-year plans in order to improve their effectiveness, ensure higher productivity, reduce the cost of agricultural production, maximize farmers' income, and minimize the environmental impact of nutrient losses. These programs focus on improvement of soil health and judicious use of crop nutrients and soil amendments.

Government Schemes for Promoting Soil Health in Eastern Sub-humid Region

Soil Health Management

Soil health management objectives include promoting crop-specific and site-specific sustainable soil health management, establishing and integrating soil fertility mapping with macro- and micronutrient management, using judicious fertilizer application, reducing soil degradation and erosion, and adopting organic farming practices. Support will be provided for a series of improved practice packages based on land use and soil characteristics created using thematic databases and maps generated using geographical information systems and derived from intensive scientific field studies. In addition, this component will support the reclamation of problematic soils (alkaline, saline, and acidic). The establishment of soil testing laboratories is one of the features of soil health management; they are often found at the block or district level. Soil testing programs at the village level are established with the aim of providing job opportunities to rural youth and expediting soil testing. Table 5.2 describes the different soil health schemes in the sub-humid region of India.

Soil Health Card

The benefits of all potential sources of biological, organic, and inorganic components are maximized in an integrated manner by using SHCs to maintain plant nutrient supply and soil fertility status at optimal levels to maintain target performance (Das et al., 2022). The main objective of the launch is to help farmers understand soil health and INM. The first step to ensure the long-term profitability of farmers is to maintain soil health and fertility. As part of this plan, the use of fertilizers at the proper dose is also critical. To determine the status of SOC and facilitate regular assessment of soil carbon improvement, farmers receiving benefits under the program will be required to prepare SHCs. Application of crop- and location-specific INM techniques will be used to increase agricultural yield and to maintain and improve soil health. The map contains important information about the nutrient status of the soil and suggestions for the dose of the various nutrients needed. It also advises the farmer on the fertilizers to use and the amount of each, as well as the soil amendments to make, in order to obtain the best yields. The impact of SHC on farming community are manifold and include the following:

- According to the 2017 impact analysis from MANAGE, farmers who received SHC use less fertilizer, which has a positive impact (Reddy, 2017).
- Savings of 8%–10% were achieved by applying fertilizers and micronutrients according to SHC guidelines (Singh et al., 2020).
- The implementation of SHC recommendations increased the overall crop yield by 5%–6% (Reddy, 2019).

Table 5.2 Different Soil Health Schemes in the Sub-humid Region of India.

Scheme name	Target state	Details	Component of the scheme
Bhoochetana soil health scheme (Chander et al., 2020)	Odisha	• The scheme has been proposed as a beneficial scheme that is expected to benefit nearly 4.35 million farmers in the state. Over the life of the scheme, 30 districts in the state will enjoy the benefits of the scheme. • Under the 3-year agreement, 6.1 Mha of agricultural land in 30 districts of Odisha will be reached through a science-led development approach. • The program is also authorized to provide assistance in the form of soil health information.	International Crops Research Institute for the Semi-Arid Tropics, Government of Odisha
New soil health card (SHC; Singh et al., 2018)	Bihar	• The new SHC has been written in the regional language, with additional images and symbols to help farmers understand and interpret the content, and hopefully, help them implement the necessary recommendations.	Cereal Systems Initiative for South Asia, Bihar Agricultural University
Attracting and Retaining Youth in Agriculture (Singh et al., 2019)	Jharkhand	• The main objective of the scheme is to encourage young farmers in rural areas to use fertilizers according to the soil health card. • The agriculture and allied sector recorded an impressive average annual growth of over 5% between 2011 and 2016.	Government of Jharkhand
Bharatiya Prakritik Krishi Paddhati (Pathak et al., 2020)	Chhattisgarh	• The scheme has a total budget of 4645.69 crore for the period of 6 years (2019–2025). • Financial support of 12,200 rupees ha^{-1} for 3 years will be provided for capacity building, cluster formation and continuous support from certification, trained personnel and residue analysis.	National Mission on Sustainable Agriculture (NMSA)
Jaivik Kheti Mission (Khurana & Kumar, 2020)	Chhattisgarh	• Application of the principles of natural agriculture, such as jivamrit, ghana jivamrit, beejamrit, and botanical pesticides, among others. • The area of Jaivik Kheti Mission in Chhattisgarh is 80 ha.	Government of Chhattisgarh

- Rice farmers increased the use of potassium by 20%, while they reduced the use of urea by 9% and of diammonium phosphate by 7%. This is a positive indication that fertilizer use is becoming more balanced (Fishman et al., 2016).
- Expenditures per acre were found to decrease by 4%–10%.
- Net yields increased by almost 22% for soybean, 11% for rice, and 16% for cotton and wheat after SHC adoption (Bhaskar et al., 2021).
- Most importantly, the use of FYM increased with SHC implementation, especially for rice, soybeans, and cotton. Sixty percent of farmers use commercial FYM (Bhaskar et al., 2021).

SHCs were distributed in the sub-humid region. In Phase I of the SHC program (years 2015–2017), 107.4 million cards were issued to farmers, and in Phase II (years 2017–2019), 124.2 million cards were issued (https://www.soilhealth.dac.gov.in). The Krishi Vigyan Kendras also provided SHCs (456,000) to farmers. With 247–544 samples per 1,000 ha, Kerala was found to have the largest sample collection among the states, followed by Bihar and Uttar Pradesh (Reddy, 2019). Most eastern states, including Jharkhand, Chhattisgarh, and Odisha, collected between 27 and 156 samples per 1,000 ha. Himachal Pradesh and Bihar, on the other hand, fell in the red zone with the lowest samples per 1,000 farmers. West Bengal and Uttar Pradesh changed from light green to yellow zone with lower samples collected per 1000 cultivators. With 687–1,001 SHCs per 1,000 ha, Bihar led the eastern Indian states, followed by Uttar Pradesh, West Bengal, and Orissa with 355–686 SHCs per 1,000 ha (http://social.niti.gov.in/health-index) (Table 5.3).

Paramparagat Krishi Vikas Yojana

Organic farming uses low-cost, environmentally friendly technologies to produce agricultural products without residues of chemicals and pesticides. The "Paramparagat Krishi Vikas Yojana" (PKVY) is an expanded part of National Mission on Sustainable Agriculture's major soil health management initiative, which emphasizes the widespread adoption of organic farming across the country. The PKVY was launched and implemented through a cluster strategy that institutionalized farmers and used the Participatory Guarantee System (PGS) for quality assurance.

PKVY is designed to improve soil health through a combination of conventional knowledge and state-of-the-art technical practices to achieve the desired outcome of better soil fertility and health in the long term. By adopting a cluster-based strategy to improve agricultural production, quality assurance, input production, value addition, and product marketing facilitation through innovative methods, PKVY is also addressing the long-standing issue of empowering farmers and improving their incomes. According to the PGS standards, all the requirements that farmers must meet are followed. The program therefore promotes the nation's conversion to organic farming and the improvement of soil health through sustainable agricultural practices. This, in turn, will meet the population's increasing demand for sustainable food.

Table 5.3 Status of the Soil Health Card (SHC) Scheme in Eastern India.

States	No. of soil samples collected	No. of soil samples tested	No. of SHCs printed	No. of SHCs distributed
Cycle I (2015–2017)				
Uttar Pradesh	4,770,399	4,770,399	16,991,346	16,991,346
Bihar	1,308,778	1 308,778	6,469,650	6,469,650
West Bengal	1,300,349	1 300,349	5,040,510	5,040,510
Jharkhand	115,302	115,302	637,507	637,507
Odisha	668,635	668,635	2,374,233	2,374,233
Chhattisgarh	703,691	703,691	3,890,709	3,890,709
Cycle II (2017–2018)				
Uttar Pradesh	4,799,725	3,822,417	13,045,920	13,045,920
Bihar	1,266,473	587,001	2,135,795	2057,670
West Bengal	876,000	542,000	2,102,000	1,370,000
Jharkhand	71,740	60,845	193,185	193,185
Odisha	474,231	395,039	1,026,725	965,222
Chhattisgarh	880,890	759,933	3,814,635	3,719,054

Data source: http://social.niti.gov.in/health-index.

About 1.493 million farmers have benefited from PKVY and a total of 597,000 ha of land has been covered for 29,859 clusters in the form of physical coverage. In addition, around 600,000 certifications have been issued by more than 300 regional councils under the PGS-India certification program. According to the ministry's physical performance measures, the program has achieved 170% of its overall targets from 2015–2016 to 2017–2018 (Amarender, 2017).

Mobile Apps for Improvement of Soil Health

For the benefit of farmers and stakeholders, mobile applications in agriculture have been developed following the introduction of cell phones. Services that began with random messages have evolved into multimodal, multimedia, and mobile-agriculture applications for smartphones. With the help of these applications, farmers now have a stronger bargaining position, and the communication and information gap between them and extension personnel is closed (Saravanan & Bhattacharjee, 2014).

Clearly, communication and information technologies can facilitate farmers' ability to obtain and share information. Among these technologies, the use of cell phones has increased, changing the way agriculture communicates. The use of cell phones has given rise to new applications and programs (Table 5.4). In agriculture, they include access to soil health monitoring, weather data, market data, and other services.

Table 5.4 Different mobile apps related to soil health improvement.

App	Features	Developed by	Reference
Kisan Suvidha Mobile app	• Kisan Suvidha is an integrated mobile app designed to support farmers by delivering relevant information of soil health. • The current languages in which the information is available are Odia, Gujarati, Tamil, Marathi, Hindi, and English.	Ministry of Agriculture and Farmers Welfare, Government of India	https://mkisan.gov.in/Home/Downloads [visited on November 30, 2024]
Soil Health Card Mobile App	• Farmers who have registered can receive soil health information through the Soil Health Card program. • This will also help address soil nutrient deficiencies noted in soil health cards.	Department of Agriculture, Co-operation and Farmers Welfare under Ministry of Agriculture and Farmers Welfare, Government of India	https://play.google.com/store/apps/details?id=com.nic.soilhealthcard&hl=en [visited on November 30, 2024]

Policy Gap

In southern and northwestern states of India such as Tamil Nadu, Punjab, Andhra Pradesh, and Karnataka, more farmers (92.4%) were aware of the soil health and various soil health programs/schemes of the government than in the eastern states such as Uttar Pradesh (eastern), Bihar, West Bengal, Jharkhand, and Chhattisgarh (55.1%) (Kumar et al., 2019b). A different policy plan is required to increase awareness in the eastern sun-humid region. Regular soil sampling, analysis, printing and dissemination of SHCs were essential for successful implementation of the system. However, this is related to the equipment of soil testing laboratories and the expertise of laboratory staff and soil sample collectors. In several states in the eastern sub-humid zone, agricultural officials still used manual instruments to locate farmers' fields and collect soil samples (Reddy, 2019). To improve the accuracy of soil sampling, GPS devices need to be deployed. Therefore, a larger budget is required for the distribution of tablets with the GPS feature. Although some areas in the western and southern states had better results, the eastern states lacked the tools needed to collect and analyze soil samples. Most farmers in eastern India initially tried to implement SHC recommendations on small farms, but they did not fully understand them. The importance of education in interpreting the cards cannot be underestimated. Since the main goal of this program is to change the way inputs are used, farmers—the end users of the program—should receive information in a language they understand and from a reliable source (Sinha, 2017).

Implementation Requirement

Although the magnitude of farmers' profit varies by crop grown, with greater benefits for crops with high fertilizer use such as rice and cotton and lesser benefits for crops with low fertilizer use, such as soybeans and wheat, the soil health scheme/program has greatly benefited farmers (Reddy 2017). The following lessons may be useful for replication in the sub-humid region of India:

- The SHC should be printed with the soil health index, which shows how healthy the soil is overall. Depending on the index, soils can be classified for easy implementation of government programs.
- Observations show that farmers have little confidence in soil health advice. Therefore, to increase farmer confidence, nongovernmental organizations and governments need to maintain model farms to demonstrate the benefits.
- The application of soil health should be flexible enough to accommodate local variables such as water quality, grid size, and soil physical characteristics such as depth and slope.
- The state should improve its capacity for trained workers, GPS-enabled tablets for soil sampling, and computerized soil analysis laboratories.

- Unlike commercial companies that have a motive to suggest higher rates of fertilizer application to make more money from fertilizer sales, farmer cooperatives or farmer-owned production companies are better able to advise farmers based on soil health recommendations without a conflict of interest (Weber & McCann, 2015). Incentives can also be provided to farmer-owned cooperatives or farmer-produced farms to deliver recommended fertilizers to their doorsteps at discounted prices.
- Funding and resources should be focused on smallholder farmers and rain-fed agriculture, where adoption rates are lower but benefits are expected to be higher.

Conclusion

The eastern sub-humid region is known for its diverse agricultural practices and a wide range of crops are grown. In this region, a variety of crops are cultivated to meet the agricultural needs, including rice, maize, wheat, pulses, oilseeds, vegetables, sugarcane, jute, and tobacco, among others. However, rice-based cropping systems such as lowland rice-fallow system, rice-wheat rotation, rice-maize rotation, and upland rice system are dominant. Several good agricultural practices, such as crop rotation, application of organic matter, conservation tillage, suitable nutrient and water management, erosion control measures, and so forth are recommended based on crop and location. Implementing good agricultural practices is crucial for maintaining and improving soil health in this region.

To ensure food security, proper exploitation of natural resources, including soil, is required to achieve the potential yield of rice and other crops grown in the eastern sub-humid region of India. Therefore, soil health is one of the most important parameters to ensure food security. The main causes of low awareness and low confidence in soil health recommendations in eastern India were poor communication approach, lack of proper extension machinery, and discrepancies between common and recommended practices. A proper extension system that will organize meetings, information visits, and training for the farming community of the eastern sub-humid region will improve the situation.

References

Allen, D. E., Singh, B. P., & Dalal, R. C. (2011). Soil health indicators under climate change: A review of current knowledge. In B. P. Singh, A. L. Cowie, & K. Y. Chan (Eds.), *Soil health and climate change* (pp. 25–45). Springer.

Amarender, R. A. (2017). *Impact study of Paramparagat Krishi Vikas Yojana*. National Institute of Agricultural Extension Management.

Andrews, S. S., Karlen, D. L., & Cambardella, C. A. (2004). The soil management assessment framework: A quantitative soil quality evaluation method. *Soil Science Society of America Journal*, *68*(6), 1945–1962.

Ansari, M. A., Saha, S., Das, A., Lal, R., Das, B., Choudhury, B. U., Roy, S. S., Sharma, S. K., Singh, I. M., Meitei, C. B., Changloi, K. L., Singh, L. S., Singh, N. A., Saraswat, P. K., Ramakrishna, Y., Singh, D., Hazarika, S., Punitha, P., Sandhu, S. K., & Prakash, N. (2021). Energy and carbon budgeting of traditional land use change with groundnut based cropping system for environmental quality, resilient soil health and farmers income in eastern Indian Himalayas. *Journal of Environmental Management*, *293*, Article 112892

Baishya, L. K., & Sarkar, D. (2015). Land–soil resources of north eastern region of India: constraints and management options. In D. RajKhowa, A. Das, S. V. Ngachan, A. K. Sikka, & M. Lyngdoh (Eds.), *Integrated soil and water resource management for livelihood and environmental security* (pp. 20–33). ICAR Research Complex for NEH Region.

Ball, B. C., Guimarães, R. M. L., Cloy, J. M., Hargreaves, P. R., Shepherd, T. G., & McKenzie, B. M. (2017). Visual soil evaluation: A summary of some applications and potential developments for agriculture. *Soil and Tillage Research*, *173*, 114–124.

Bandyopadhyay, P. K., Singh, K. C., Mondal, K., Nath, R., Ghosh, P. K., Kumar, N., Basu, P. S., & Singh, S. S. (2016). Effects of stubble length of rice in mitigating soil moisture stress and on yield of lentil (*Lens culinaris* Medik) in rice-lentil relay crop. *Agricultural Water Management*, *173*, 91–102.

Baruah, U., Bandyopadhyay, S., & Reza, S. K. (2014). Land use planning and strategic measures in north eastern region of India. *Agropedology*, *24*(02), 292–303.

Batabyal, K., Mandal, B., Sarkar, D., & Murmu, S. (2017). Assessment of nutrient management technologies for eggplant production under subtropical conditions: a comprehensive approach. *Experimental Agriculture*, *53*(4), 588–608.

Batabyal, K., Mandal, B., Sarkar, D., Murmu, S., Tamang, A., Das, I., Hazra, G. C., & Chattopadhyay, P. S. (2016). Comprehensive assessment of nutrient management technologies for cauliflower production under subtropical conditions. *European Journal of Agronomy*, *79*, 1–13.

Bazaya, B. R., Sen, A., & Srivastava, V. K. (2009). Planting methods and nitrogen effects on crop yield and soil quality under direct seeded rice in the indo-Gangetic plains of eastern India. *Soil and Tillage Research*, *105*(1), 27–32.

Bhaskar, S., Ravisankar, N., Dey, P., Raghavendra, K. J., & Panwar, A. S. (2021). Impact and refinement of soil health card-based nutrient management in major cropping systems in India. *Indian Journal of Fertilisers*, *17*(11), 1108–1122.

Bhat, B. P. (2003). Agroforestry for sustainable mountain development in N.E.H. region. In M. S. S. Rawat (Ed.), *Central Himalaya environment and development: Potentials, actions and challenges* (pp. 206–223). Transmedia Publisher.

Bhat, J. A., Mandal, B., & Hazra, G. C. (2007). Basic slag as a liming material to ameliorate soil acidity in Alfisols of sub-tropical India. *American-Eurasian Journal of Agricultural & Environmental Sciences*, *2*(4), 321–327.

Bhattacharyya, R., Ghosh, B. N., Mishra, P. K., Mandal, B., Rao, C. S., Sarkar, D., Das, K., Anil, K. S., Lalitha, M., Hati, K. M., & Franzluebbers, A. J. (2015). Soil degradation in India: Challenges and potential solutions. *Sustainability*, *7*(4), 3528–3570.

Bünemann, E. K., Bongiorno, G., Bai, Z., Creamer, R. E., De Deyn, G., De Goede, R., Fleskens, L., Geissen, V., Kuyper, T. W., Mader, P., Pulleman, M., Sukkel, W., Groenigen, J. W. V., & Brussaard, L. (2018). Soil quality—A critical review. *Soil Biology and Biochemistry*, *120*, 105–125.

Cairns, M., Keitzar, S., & Yaden, T. A. (2010). Shifting forests in Northeast India: Management of alnus nepalensis as an improved fallow in Nagaland. In *Voices from the Forest*. Routledge.

Chakraborty, R., Sharma, V. K., Das, D., Biswas, D. R., Mahapatra, P., Shahi, D. K., Barman, M., Chobhe, K. A., & Anil, A. S. (2021). Fertility status in wheat (*Triticum aestivum*) under long-term nutrient management practices in an Alfisol. *Indian Journal of Agricultural Sciences*, *91*(8), 1190–1193.

Chander, G., Dixit, S., & Sawargaonkar, G. (2020). Bhoochetana: Reviving soils for agriculture. In *Mapping the Nutrient Status of Odisha's Soils* (pp. 1–6). ICRISAT. https://oar.icrisat.org/11938/

Chatterjee, D., Dutta, S. K., Kikon, Z. J., Kuotsu, R., Sarkar, D., Satapathy, B. S., & Deka, B. C. (2021). Recycling of agricultural wastes to vermicomposts: characterization and application for clean and quality production of green bell pepper (*Capsicum annuum* L.). *Journal of Cleaner Production*, *315*, Article 128115

Chatterjee, D., Kuotsu, R., Kikon, Z. J., Sarkar, D., Ao, M., Ray, S. K., Bera, T., & Deka, B. C. (2016). Characterization of vermicomposts prepared from agricultural solid wastes in north Eastern Hill region of Nagaland, India. *Proceedings of the National Academy of Sciences, India Section B: Biological Sciences*, *86*(4), 823–833.

Das, A., Lyngdoh, D., Ghosh, P. K., Lal, R., Layek, J., & Ramkrushna, G. I. (2018). Tillage and cropping sequence effect on physico-chemical and biological properties of soil in eastern Himalayas, India. *Soil and Tillage Research*, *180*, 182–193.

Das, A., Patel, D. P., Lal, R., Kumar, M., Ramkrushna, G. I., Layek, J., Buragohain, J., Ngachan, S. V., Ghosh, P. K., Choudhury, B. U., Mohapatra, K. P., & Shivakumar, B. G. (2016). Impact of fodder grasses and organic amendments on productivity and soil and crop quality in a subtropical region of eastern Himalayas, India. *Agriculture, Ecosystems and Environment*, *216*, 274–282.

Das, A., Patel, D. P., Munda, G. C., & Ghosh, P. K. (2010). Effect of organic and inorganic sources of nutrients on yield, nutrient uptake and soil fertility of maize (*Zea mays*)-mustard (*Brassica campestris*) cropping system. *Indian Journal of Agricultural Sciences*, *80*(1), 85–88.

Das, B. S., Wani, S. P., Benbi, D. K., Muddu, S., Bhattacharyya, T., Mandal, B., Santra, P., Chakraborty, D., Bhattacharyya, R., Basak, N., & Reddy, N. N. (2022). Soil health and its relationship with food security and human health to meet the sustainable development goals in India. *Soil Security, 8*, Article 100071

Das, S., & Paul, S. (2022). An assessment of soil quality and agricultural production status in the alluvial soil region: A case study in Koch Bihar District, West Bengal. *India. Current World Environment, 17*(1). http://dx.doi.org/10.12944/CWE.17.1.24

Dey, P. (2016). Soil health management. In J. C. Katyal, S. K. Chaudhari, B. S. Dwivedi, D. R. Biswas, R. K. Rattan, & K. Majumda (Eds.), *Soil health: Concept, status and monitoring, Bulletin of the Indian Society of Soil Science* (Vol. *30*, pp. 79–97).

Dick, R. P. (1994). Soil enzyme activities as indicators of soil quality. In J. W. Doran, D. C. Coleman, D. F. Bezdicek, & B. A. Stewart (Eds.), *Defining soil quality for a sustainable environment* (Vol. 35, pp. 107–124). Soil Science Society of America.

Doran, J. W., & Parkin, T. B. (1997). Quantitative indicators of soil quality: A minimum data set. *Methods for Assessing Soil Quality, 49*, 25–37.

Doran, J. W., Sarrantonio, M., & Liebig, M. (1996). Soil health and sustainability. In D. L. Sparks (Ed.), *Advances in agronomy* (Vol. 56, pp. 1–54). Academic Press.

Elliott, E. T. (1994). The potential use of soil biotic activity as an indicator of productivity, sustainability, and pollution. In *Soil biota: Management in sustainable farming systems* (pp. 250–256).

Emmet-Booth, J. P., Forristal, P. D., Fenton, O., Ball, B. C., & Holden, N. M. (2016). A review of visual soil evaluation techniques for soil structure. *Soil Use and Management, 32*(4), 623–634.

European Commission. (2002). Communication of 19 April 2002 from the Commission to the Council, the European Parliament, the Economic and Social Committee and the Committee of the Regions: Towards a Thematic Strategy for Soil Protection [COM (2002) 179 Final]. European Commission.

Ewing, S. A., & Singer, M. J. (2012). Soil quality. In P. M. Huang, Y. Li, & M. E. Sumner (Eds.), *Handbook of soil science: Resources, management and environment impacts* (pp. 1–26). Taylor & Francis Group.

Fishman, R., Kishore, A., Rothler, Y., Ward, P. S., Jha, S., & Singh, R. K. P. (2016). *Can information help reduce imbalanced application of fertilizers in India? Experimental evidence from Bihar* (IFPRI discussion paper 1517). International Food Policy Research Institute.

Ghaemi, M., Astaraei, A., Emami, H., Mahalati, M., & Sanaeinejad, S. (2014). Determining soil indicators for soil sustainability assessment using principal component analysis of astan quds-east of Mashhad Iran. *Journal of Soil Science and Plant Nutrition, 14*(4), 987–1004.

Ghosh, P. K., Saha, R., Gupta, J. J., Ramesh, T., Das, A., Lama, T. D., Munda, G. C., Bordoloi, J. S., Verma, M. R., & Ngachan, S. V. (2009). Long-term effect of pastures on soil quality in acid soil of north–east India. *Australian Journal of Soil Research, 47*(4), 372–379.

Ghosh, S., Wilson, B. R., Ghoshal, S. K., Senapati, N., & Mandal, B. (2012). Organic amendments influence soil quality and carbon sequestration in the indo-gangetic plains of India. *Agriculture, Ecosystems and Environment, 156*, 134–141.

Gogoi, B., Borah, N., Baishya, A., Nath, D. J., Dutta, S., Das, R., Bhattacharyya, D., Sharma, K. K., Valente, D., & Petrosillo, I. (2021). Enhancing soil ecosystem services through sustainable integrated nutrient management in double rice-cropping system of north-east India. *Ecological Indicators, 132*, Article 108262

Granatstein, D., & Bezdicek, D. F. (1992). The need for a soil quality index: Local and regional perspectives. *American Journal of Alternative Agriculture, 7*(1–2), 12–16.

Gupta, J. J. (2004). Production and use of broom grass in north east India. *Journal of Hill Research, 17*, 29–30.

Hatfield, J. L., & Karlen, D. L. (1993). *Sustainable agriculture systems*. CRC Press Inc.

Jian, J., Du, X., & Stewart, R. D. (2020). Quantifying cover crop effects on soil health and productivity. *Data Brief, 29*, Article 105376

Karlen, D. L., Andrews, S. S., & Doran, J. W. (2001). Soil quality: Current concepts and applications. *Advances in Agronomy, 74*, 1–40.

Khurana, A., & Kumar, V. (2020). *State of organic and natural farming: Challenges and possibilities*. Centre for Science and Environment. https://www.cseindia.org/state-of-organic-and-natural-farming-in-india-10346

Kibblewhite, M. G., Ritz, K., & Swift, M. J. (2008). Soil health in agricultural systems. *Philosophical Transactions of the Royal Society B: Biological Sciences, 363*(1492), 685–701.

Kochiwad, S. A., Meena, L. R., Kumar, D., Kumar, S., Meena, L. K., Singh, S. P., & Singh, K. (2017). Livestock based integrated farming system for livelihood improvements in small and marginal farmers. *South Asian Journal of Food Technology and Environment, 3*(1), 526–532.

Kumar, M., Rajkhowa, D. J., Mahanta, K., Verma, B. C., Choudhury, B. U., Rao, K. K., Saurabh, K., & Rakshit, R. (2020b). Biowaste utilisation for improving soil health and crop productivity in north eastern India. *Research Biotica, 2*(2), 44–49.

Kumar, R., Mahajan, N. C., Naresh, R. K., Dhaliwal, S. S., Kumar, A., Chandra, M. S., Pandey, A. K., Zaidi, S. F. A., Kumar, S., & Gupta, S. K. (2019a). Agro-technological options for scaling up crop productivity, soil health and water footprint in Rice based cropping system in sub-humid (Purvanchal) region of Uttar Pradesh, India: A review. *International Journal of Current Microbiology and Applied Sciences, 8*(7), 2679–2700.

Kumar, R., Mishra, J. S., Rao, K. K., Mondal, S., Hazra, K. K., Singh Choudhary, J., Hans, H., & Bhatt, B. P. (2020a). Crop rotation and tillage management options for sustainable intensification of rice-fallow agro-ecosystem in eastern India. *Scientific Reports, 10*, Article 11146

Kumar, R., Mishra, J. S., Upadhyay, P. K., & Hans, H. (2019b). Rice fallows in the eastern India: problems and prospects. *Indian Journal of Agricultural Sciences, 89*(4), 567–577.

Kumari, R., Kumar, S., Kumar, R., Das, A., Kumari, R., Choudhary, C. D., & Sharma, R. P. (2017). Effect of long-term integrated nutrient management on crop yield, nutrition and soil fertility under rice-wheat system. *Journal of Applied and Natural Science*, *9*(3), 1801–1807.

Kundu, S., Vassanda Coumar, M., Saha, J. K., Rajendiran, S., Hati, K. M., Biswas, A. K., Reddy, K. S., & Subba Rao, A. (2012). Assessing soil health of vertisol of AESR10.1 using selected physical, chemical and biological attributes of soils. *Journal of the Indian Society of Soil Science*, *60*(4), 281–287.

Lal, R. (2016). Soil health and carbon management. *Food and Energy Security*, *5*(4), 212–222.

Lal, R., & Stewart, B. A. (1994). *Soil processes and water quality*. Lewis Publishers.

Lehmann, J., Bossio, D. A., Kogel-Knabner, I., & Rillig, M. C. (2020). The concept and future prospects of soil health. *Nature Reviews Earth & Environment*, *1*(10), 544–553.

Lilburne, L., Sparling, G., & Schipper, L. (2004). Soil quality monitoring in New Zealand: Development of an interpretative framework. *Agriculture, Ecosystems & Environment*, *104*(3), 535–544.

Mahajan, G., Das, B., Morajkar, S., Desai, A., Murgaokar, D., Kulkarni, R., Sale, R., & Patel, K. (2020). Soil quality assessment of coastal salt-affected acid soils of India. *Environment Science and Pollution Research*, *27*(21), 26221–26238.

Majhi, P., Rout, K. K., Nanda, G., & Singh, M. (2021). Long term effects of fertilizer and manure application on productivity, sustainability and soil properties in a rice-rice system on Inceptisols of eastern India. *Communications in Soil Science and Plant Analysis*, *52*(14), 1631–1644.

Maji, A. K., Obi Reddy, G. P., & Meshram, S. (2008). *Acid soil map of India; Annual report 2008*. NBSS&LUP.

Mandal, B. (2005). Assessment and improvement of soil quality and resilience for rainfed production system. Completion report, National Agricultural Technology Project, Indian Council of Agricultural Research.

Manna, M. C., & Sharma, K. L. (2009). Soil and water quality. In N. N. Goswami, R. K. Rattan, G. Dev, G. Narayanaswamy, D. K. Das, D. K. Pal, & D. L. N. Rao (Eds.), *Fundamentals of soil science* (pp. 642–668). Indian Society of Soil Science.

Manna, M. C., Swarup, A., Wanjari, R. H., Ravankar, H. N., Mishra, B., Saha, M. N., Singh, Y. V., Sahi, D. K., & Sarap, P. A. (2005). Long-term effect of fertilizer and manure application on soil organic carbon storage, soil quality and yield sustainability under sub-humid and semi-arid tropical India. *Field Crops Research*, *93*, 264–280.

Masto, R. E., Chhonkar, P. K., Singh, D., & Patra, A. K. (2008). Alternative soil quality indices for evaluating the effect of intensive cropping, fertilisation and manuring for 31 years in the semi-arid soils of India. *Environmental Monitoring and Assessment*, *136*(1), 419–435.

Mazumdar, S. P., Kundu, D. K., Saha, A. R., Majumdar, B., Saha, R., Singh, A. K., Barman, D., Ghosh, D., Dey, R., Behera, M. S., Mitra, S., & Bhattacharyya, P. (2018). Carbon and nutrient dynamics under long-term nutrient management in tropical rice-wheat-jute system. *Archives of Agronomy and Soil Science*, *64*(11), 1595–1607.

Mazumdar, S. P., Saha, A. R., Majumdar, B., Kumar, M., Alam, N. M., Saha, R., Dey, P., & Kar, G. (2022). Impact of balanced fertilization on carbon and nutrient dynamics under long-term jute-rice-lentil cropping system in alluvial soils of eastern India. *Communications in Soil Science and Plant Analysis*, *53*(13), 1574–1591.

McKenzie, M. C., Moncada, M. A. P., & Ball, B. C. (2015). Reductions of yield gaps and improvement of ecological function through local-to-global applications of visual soil assessment. In B. C. Ball & L. J. Munkholm (Eds.), *Visual soil evaluation: Realizing potential crop production with minimum environmental impact* (pp. 31–48). CABI.

Moebius-Clune, B. N., Moebiue-Clune, D. J., Gugino, B. K., Idowu, O., Schindelbeck, R., Rostow, A., van Es, H., Thies, J., Shyler, H., McBride, M., Kurtz, K. S. M., Wolfe, D. W., & Abaw, G. S. (2017). *Comprehensive assessment of soil health* (3rd ed., pp. 19–101). Cornell University.

NAAS. (2013). *Improving productivity of rice fallows (Policy paper 64)*. National Academy of Agricultural Sciences.

National Bureau of Soil Survey & Land Use Planning (NBSS & LUP). (2004). *Soil map (1,1 million scale)*. NBSS & LUP.

Nayak, P. K., Nayak, A. K., Kumar, A., Kumar, U., Panda, B. B., Satapathy, B. S., Poonam, A., Mohapatra, S. D., Tripathi, R., Shahid, M., Chatterjee, D., Panneerselvam, P., Mohanty, S., Das, S. K., & Pathak, H. (2020). *Rice based integrated farming systems in eastern India: A viable technology for productivity and ecological security* (NRRI research bulletin No. 24). ICAR-National Rice Research Institute.

Nongkynrih, J. M., Pohshna, C., & Sarma, K. K. (2018). Dynamics of shifting cultivation in relation to slope and elevation in parts of Nagaland, India. *Current Science*, *114*(5), 1094.

Norris, C. E., Bean, G. C., Cappellazzi, S. B., Cope, M., Greub, K., Liptzin, D., Rieke, E., Tracy, P., Morgan, C., & Honeycutt, C. W. (2020). Introducing the North American project to evaluate soil health. *Agronomy Journal*, *112*(4), 3195–3215.

Nunes, M. R., Veum, K. S., Parker, P. A., Holan, S., Karlen, D., Amsili, J., van Es, H., Wills, S., Seybold, C., & Moorman, T. (2021). The soil health assessment protocol and evaluation applied to soil organic carbon. *Soil Science Society of America Journal*, *85*(4), 1196–1213.

Obade, V. D. P., & Lal, R. (2016a). A standardized soil quality index for diverse field conditions. *Science of the Total Environment*, *541*, 424–434.

Obade, V. D. P., & Lal, R. (2016b). Towards a standard technique for soil quality assessment. *Geoderma*, *265*, 96–102.

Pant, R. M., Tiwari, B. K., & Choudhury, D. (2018). *Report of Working Group III. Shifting cultivation: Towards a transformational approach*. NITI Aayog.

Patel, D., Das, A., Kumar, M., Munda, G., Ngachan, S. V., Ramkrushna, G. I., Layek, J., Pongla, N., Buragohain, J., & Somireddy, U. (2015). Continuous application of organic amendments enhances soil health, produce quality and system productivity of vegetable-based cropping systems in subtropical eastern Himalayas. *Experimental Agriculture, 51*(1), 85–106.

Pathak, H., Pal, S., & Mohapatra, T. (2020). *Mahatma Gandhi's vision of agriculture: achievements of ICAR*. Indian Council of Agricultural Research.

Patiram. (2007). Management and future research strategies for enhancing productivity of crops on the acid soils. *Journal of the Indian Society of Soil Science, 55*, 411–420.

Paul, G. C., Saha, S., & Ghosh, K. G. (2020). Assessing the soil quality of Bansloi river basin, eastern India using soil-quality indices (SQIs) and random Forest machine learning technique. *Ecological Indicators, 118*, Article 106804

Paz-Ferreiro, J., & Fu, S. (2016). Biological indices for soil quality evaluation: perspectives and limitations. *Land Degradation & Development, 27*(1), 14–25.

Pulido Moncada, M., Gabriels, D., & Cornelis, W. M. (2014). Data-driven analysis of soil quality indicators using limited data. *Geoderma, 235*, 271–278.

Rai, A. K., Paul, B., & Singh, G. (2009). Assessment of top soil quality in the vicinity of subsided area in the south eastern part of Jharia coalfield, Jharkhand, India. *Report and Opinion, 1*(6), 18–23.

Reddy, A. A. (2017). *Impact study of soil health card scheme*. National Institute of Agricultural Extension Management (MANAGE).

Reddy, A. A. (2019). The soil health card scheme in India: Lessons learned and challenges for replication in other developing countries. *Journal of Natural Resources Policy Research, 9*(2), 124–156.

Ritse, V., Basumatary, H., Kulnu, A. S., Dutta, G., Phukan, M. M., & Hazarika, N. (2020). Monitoring land use land cover changes in the eastern Himalayan landscape of Nagaland, northeast India. *Environmental Monitoring and Assessment, 192*, Article 711

Saha, R., Chaudhary, R. S., & Somasundaram, J. (2012). Soil health management under hill agroecosystem of north East India. *Applied and Environmental Soil Science, 2012*, Article 696174

Saha, R., & Mishra, V. K. (2007). Long-term effect of various land use systems on physical properties of silty clay loam soil of NE hills. *Journal of the Indian Society of Soil Science, 55*(2), 112–118.

Saha, R., Paswan, A., Mazumdar, S. P., Barman, D., Majumdar, B., Behera, M. S., & Saha, A. R. (2022). Improvement in soil quality through tillage and residue management in jute (*Corchorus* spp.) based cropping systems of Indo-Gangetic plains. *Carbon Management, 13*(1), 205–215.

Saravanan, R., & Bhattacharjee, S. (2014). Mobile phone applications for agricultural extension in India. In S. Raj (Ed.), *Mobile phones for agricultural extension: Worldwide mAgri innovations and promise for future* (pp. 1–75). New India Publishing Agency.

Sarkar, D., Baishya, L. K., Meitei, C. B., Naorem, G. C., Thokchom, R. C., Singh, J., Bhuvaneswari, S., Batabyal, K., Das, R., Padhan, D., Prakash, N., & Rahman, F. H. (2018a). Can sustainability of maize-mustard cropping system be achieved through balanced nutrient management? *Field Crops Research*, *225*, 9–21.

Sarkar, D., Meitei, C. B., Baishya, L. K., Das, A., Ghosh, S., Chongloi, K. L., & Rajkhowa, D. J. (2015). Potential of fallow chronosequence in shifting cultivation to conserve soil organic carbon in Northeast India. *Catena*, *135*, 321–327.

Sarkar, D., Meitei, C. B., Das, A., Ghosh, P. K., & Mandal, B. (2018b). Changes in soil organic carbon pools in a long-term trial with perennial fodder crops in acid soils of north-east India. *Grass and Forage Science*, *73*(2), 473–481.

Saud, J., & Bezbaruah, M. P. (2020). Open grazing and cropping intensity in the Brahmaputra Valley. *Agricultural Economics Research Review*, *33*(1), 109–118.

Schloter, M., Dilly, O., & Munch, J. C. (2003). Indicators for evaluating soil quality. *Agriculture, Ecosystems and Environment*, *98*(1–3), 255–262.

Shahid, M., Nayak, A. K., Shukla, A. K., Tripathi, R., Kumar, A., Mohanty, S., Bhattacharyya, P., Raja, R., & Panda, B. B. (2013). Long-term effects of fertilizer and manure applications on soil quality and yields in a sub-humid tropical rice-rice system. *Soil Use and Management*, *29*(3), 322–332.

Sharma, P. D., Baruah, T. C., Maji, A. K., & Patiram. (2006). Management of acid soils in NEH Region [Technical bulletin]. Natural Resource Management Division (ICAR).

Sharma, P. D., & Sarkar, A. K. (2005). *Managing acid soils for enhancing productivity* [Technical bulletin]. Natural Resource Management Division, Indian Council of Agricultural Research.

Sharma, R. C., Rao, B. R. M., & Saxena, R. K. (2004). *Salt affected soils in India—Current assessment*. Advances in Sodic Land Reclamation; Proceeding of the International Conference on Sustainable Management of Sodic Lands, Lucknow, India.

Singer, M. J., & Ewing, S. (2000). Soil quality. In M. E. Sumner (Ed.), *Handbook of soil science* (pp. G271–G298). CRC Press.

Singh, A. K. (2007). Evaluation of soil quality under integrated nutrient management. *Journal of the Indian Society of Soil Science*, *55*(1), 58–61.

Singh, A. K., Bordoloi, L. J., Kumar, M., Hazarika, S., & Parmar, B. (2014). Land use impact on soil quality in eastern Himalayan region of India. *Environmental Monitoring and Assessment*, *186*(4), 2013–2024.

Singh, A. K., Ghorai, A. K., & Kar, G. (2022). Diversification of rice growing areas in eastern India with integrated soil–crop system management for GHGs mitigation and higher productivity. *Carbon Management, 13*(1), 105–116.

Singh, A. K., Singh, R., Adhiguru, A., Chandre Gowda M J, Thimmappa, K., & Hanji, M. B. (2019). ARYA Report 2019.

Singh, S. R., Kundu, D. K., Dey, P., & Mahapatra, B. S. (2017). Identification of minimum data set under balanced fertilization for sustainable rice production and maintaining soil quality in alluvial soils of eastern India. *Communications in Soil Science and Plant Analysis, 48*(18), 2170–2192.

Singh, U., Singh, A., & Pandey, A. K. (2020). Soil health card: A scheme for sustainable agriculture. In *International Web Conference Soil Health Management for Sustainable Crop Productivity Bihar Agriculture University, Sabour, Bhagalpur–813 210 (Bihar) India* (pp. 3–8).

Singh, V., Ganguly, S., & Dakshinamurthy, V. (2018). *Evaluation of India's Soil Health Card from users' perspectives* (Vol. 12). International Food Policy Research Institute (IFPRI).

Sinha, R. K. (2017). *Impact of soil health scheme on production. Productivity and Soil Health in Bihar.* Agro-Economic Research Centre for Bihar and Jharkhand, TM Bhagalpur University.

SQI. (2003). *Interpreting soil conditioning index: A tool for measuring soil organic matter trends* (Soil quality-agronomy technical note no. 16). Soil Quality Institute.

Srinivasan, R., Gangopadhyay, S. K., Reza, S. K., Hegde, R., & Singh, S. K. (2021). Assessment of soil quality in different land-use systems of eastern coastal plains of Odisha, India. *Climate Change and Environmental Sustainability, 9*(1), 53–63.

Subbarao, G. V., Kumar, R., Kumar, J., Johansen, C., Deb, U. K., Ahmed, I., Krishna, R., Venkataratnam, L., Hebbar, K. R., Sai, M. V. S. R., & Harris, D. (2001). *Spatial distribution and quantification of rice-fallows in South Asia—Potential for legumes.* International Crops Research Institute for the Semi-Arid Tropics.

Tamang, A., Das, I., Batabyal, K., Sarkar, D., Murmu, S., Mandal, B., Hazra, G. C., & Bhattacharyya, R. (2017). Assessment of nutrient management technologies for broccoli to improve productivity and quality and soil resources in the subtropics. *International Journal of Vegetable Science, 23*(2), 102–124.

Troeh, F. R., Donahue, R. L., & Hobbs, J. A. (2004). *Soil and water conservation for productivity and environmental production.* Pearson Education.

Umashankar (2005). Indigenous agroforestry tree species for conservation and rural livelihood. In B. P. Bhatt & K. M. Bujarbaruah (Eds.), *Agroforestry in north east India: Opportunities and challenges* (pp. 149–174). ICAR Research Complex for NEH Region.

USDA-NRCS. (2016). *Soil health: Principle 1 of 5—Soil armor.* Natural Resources Conservation Service. https://www.nrcs.usda.gov/sites/default/files/2022-09/Soil_Health_Principle_1_of_5_Final.pdf

USDA-NRCS. (2019). *Recommended soil health indicators and associated laboratory procedures* (Soil health technical note no. 450-03). Natural Resources Conservation Service.

Vasu, D., Singh, S. K., Ray, S. K., Duraisami, V. P., Tiwary, P., Chandran, P., Nimkar, A. M., & Anantwar, S. G. (2016). Soil quality index (SQI) as a tool to evaluate crop productivity in semi-arid Deccan plateau, India. *Geoderma*, *282*, 70–79.

Veum, K. S., Sudduth, K. A., Kremer, R. J., & Kitchen, N. R. (2017). Sensor data fusion for soil health assessment. *Geoderma*, *305*, 53–61.

Wang, X. J., & Gong, Z. T. (1998). Assessment and analysis of soil quality changes after eleven years of reclamation in subtropical China. *Geoderma*, *81*(3–4), 339–355.

Weber, C., & McCann, L. (2015). Adoption of nitrogen-efficient technologies by US corn farmers. *Journal of Environmental Quality*, *44*(2), 391–401.

Wienhold, B. J., Pikul, J. L., Liebig, M. A., Mikha, M. M., Varvel, G. E., Doran, J. W., & Andrews, S. S. (2006). Cropping system effects on soil quality in the Great Plains: synthesis from a regional project. *Renewable Agriculture and Food Systems*, *21*(1), 49–59.

Williams, H., Colombi, T., & Keller, T. (2020). The influence of soil management on soil health: an on-farm study in southern Sweden. *Geoderma*, *360*, Article 114010

Yadav, G. S., Datta, R., Imran Pathan, S., Lal, R., Meena, R. S., Babu, S., Das, A., Bhowmik, S. N., Datta, M., Saha, P., & Mishra, P. K. (2017). Effects of conservation tillage and nutrient management practices on soil fertility and productivity of rice (*Oryza sativa* L.)–rice system in north eastern region of India. *Sustainability*, *9*(10), Article 1816

Yadav, G. S., Lal, R., Meena, R. S., Babu, S., Das, A., Bhowmik, S. N., Datta, M., Layak, J., & Saha, P. (2019). Conservation tillage and nutrient management effects on productivity and soil carbon sequestration under double cropping of rice in north eastern region of India. *Ecological Indicators*, *105*, 303–315.

6

Coastal Region Soil Health

Problems and Perspectives from India

Uttam Kumar Mandal, V. Ramesh, Sudipa Mal, T. D. Lama, S. K. Sarangi, Sourav Mullick, Amit Ghosh, Dibyendu Bikas Nayak, and Dhiman Burman

Chapter Overview

The Indian coastline is approximately 7,517 km long, spread across nine coastal states, two island groups, and two union territories. It covers an area of 10.78 million hectares, of which 3.094 million hectares are salt affected. Owing to salinity buildup in the soil and water, the scarcity of high-quality irrigation water during the dry season, and flooding during the wet season from excessive monsoon rains, farming in coastal regions is difficult. Farmers' livelihoods are further affected by the impacts of climate change on agriculture. A variety of soil characteristics play a role in the assessment of soil quality in coastal lands. These characteristics include inherent electrolyte concentrations; pH; relative distribution of various ions such as Na^+, Mg^{2+}, Ca^{2+}, K^+, SO_4^{2-}, Cl^-, and HCO_3^- in soil solution; soil aggregation-dispersion; hydraulic conductivity; metabolic activities; and the abundance of beneficial soil microbial populations. Electrolyte concentrations can change during the *kharif* (monsoon) and *rabi* (winter) seasons, owing to the presence of brackish water that can damage the physical structure of soil, disrupt the balance of soil nutrients, decrease the availability of nutrients, increase the osmotic effect on plant roots, and limit the proliferation of soil microbiomes. Accordingly, regular assessment of key soil quality indicators is essential for

Soil Health Series: Volume 4 Soil Health and Sustainability in India, First Edition.
Edited by Ranjan Bhattacharyya.

understanding the health of the soil. Determining the critical values (optimum and threshold) of these indicators may be useful in identifying soil-related problems and recommending potential solutions for achieving profitable crop yields and improving the sustainable use of the region's water and soil resources. Recommendations for restoring soil quality and improving the security of the coastal farming community include land modification, surface or underground drainage, surface water harvesting for *rabi* crops, calcium supplementation for acid neutralization, adoption of sustainable agricultural practices, and integrated nutrient management by introducing organic amendments such as manure, green manure, vermicompost, and well-managed city waste compost.

Introduction

Coastal regions comprise the coastline, the surrounding terrain that is frequently submerged in seawater, and coastal seas over the continental shelf to a depth of 50 m. The coastal and marine environment can extend up to 100 km inland (Mehvar et al., 2018). Coastal regions house one-third of the world's population and are twice as densely populated as inland places, although they represent only 4% of the planet's total land area and 11% of its ocean area (Barbier, 2013; Mehvar et al., 2018).

In India, 7,517 km of coastline is spread across nine states, two island groups, and two union territories. The country's coastal region contains 77 towns and cities, including the megacities Mumbai, Kolkata, and Chennai, and it is home to almost 250 million people. Economic activities, including ports, power plants, fishing, tourism, and mining, are all supported by the coastline and have an impact on the coastal ecosystem.

Coastal environments, especially mangroves and coral reefs, are among the richest repositories of biodiversity and natural resources. The human population living along the coast benefits from the ecosystems in many ways. For example, the ecosystems provide a variety of coastal resources that support people's livelihoods, and they also shield the population from pollution, storm surges, tsunamis, erosion, and cyclones.

The west coast and the east coast are the two geomorphic settings of the Indian coast. The west coast is generally exposed, with strong surf and rocky headlands and marked by backwaters, hillocks, and mud flats. The east coast, which includes West Bengal, Odisha, Andhra Pradesh, and Tamil Nadu states, is characterized by relatively low-lying beaches, lagoons, alluvial plains, deltas, and marshes. Most of its land area is made up of flat deltas formed by large rivers, with a sizable portion covered in mangrove vegetation. The east coast is also subject to the powerful wave actions of the Bay of Bengal, and it is more vulnerable to natural disasters than the west coast. Owing to the unique climatic, edaphic, and geomorphologic factors of the east coast, its ecosystem is extremely susceptible to degradation (Maji et al., 2020).

In addition to a variety of water- and soil-related constraints that limit the yield of food grain crops, anthropogenic activities (human activity) pose a serious threat to the fragile natural state of the entire shoreline. Its degradation and the ensuing loss of biodiversity persist at an unprecedented rate, which compromises the resilience and functioning of coastal ecosystems and endangers their capacity to provide services to both current and future generations. These risks are expected to worsen as a result of global population growth and rising consumption demands.

Major Issues and Degradation in the Coastal Zone of India

Numerous resources are found in the coastal zone, including fertile land, fishing, mangrove forests, marine and terrestrial biodiversity, and scenic beauty marine resources. The coastal zone is also the meeting point of flora and fauna of fresh and saline water. It encompasses salt fields and mineral resources, such as quartz, zircon, uranium, and others, and provides easy transportation and sailing facilities. These characteristics and features have facilitated development of port, industry, and tourism growth.

At the same time, the activities associated with ship-breaking companies, tourism, unplanned building, overfishing, and cutting of hills have accelerated the devastation of the ecosystem. The continuous dangers to the coasts are exacerbated by the uncontrolled removal of sand, gravel, and pebble deposits from beaches and underwater coastal slopes. More than 34 species of tropical rainforest plants, including *Podocarpus neriifolius* and *Entada phaseoloides*, are facing extinction in the coastal forests. Numerous animals, including pythons, wild boars, cats, bears, porcupines, and anteaters, are rapidly disappearing from coastal regions. In addition, although they were the dominant marine and coastal species prior to 1980, red crabs, jellyfish, sharks, and dolphins are becoming less common. Depletion of coastal resources results in major socio-economic and cultural issues, including the weakening of the social fabric, marginalization, unemployment, and erosion-related property degradation. It also frequently creates conflict amongst users.

Furthermore, owing to a rise in relative sea level and an increase in cyclonic activity, the Sundarbans, the largest contiguous mangrove ecosystem in the world located in coastal West Bengal, India and Bangladesh, face serious erosion and drowning risks. Ghoramara Island is partially flooded, Lohachara Island and

New Moore Island/South Talpatti Island have already vanished beneath the waves. The mangrove forests of this area support coastal livelihoods, fulfill significant socioeconomic and environmental roles, and offer vital ecosystem services. These functions of mangroves are due to their dense root systems, which capture large amounts of sediment from both maritime and upland origins and hold it in place. Change detection analyses have clearly indicated natural internal mangrove dynamics with small-scale land loss and gain by erosion and accretion of sediments within the tidal channels, even though the overall mangrove area has not changed considerably over the past few millennia.

Mangrove forests are generally very resilient to natural disasters such as tsunamis and tropical storms, but Islam (2014) reported that mangrove ecosystems take about 25 years to recover after storm damage. Following a sequence of recent severe events, with several cyclones in a row—Sidr in 2007, Nargis in 2008, Aila in 2009, Bulbul in 2019, Amphan in 2020, and Yaas in 2021—and the Asian tsunami of 2004, concern has arisen that the mangroves' regeneration ability has been weakened. In particular, in the Bay of Bengal along the Sundarbans, the occurrence of cyclones has increased by 26% between 1881 and 2001.

The resilience of mangroves is also strained by human-related activities. The wastewater pollution from big cities and industries and the creation of upstream embankments to divert irrigation water represent the two human activities with the most detrimental impacts on the mangrove forest. A particularly notable example is the 1975 construction of the Farakka Barrage, which significantly changed the regional hydrological balance, contributed to variations in freshwater distribution, and enhanced coastal soil salinization (Mandal et al., 2021a).

Aquaculture represents another source of detrimental impacts. Because the practice provided stable economic profits for decades, numerous rice fields in coastal Andhra Pradesh and West Bengal were converted into brackish water fish farms. Saline aquaculture occupies around 200,000 hectares (ha) in Andhra Pradesh's coastal areas alone. Seawater is mostly used to raise valuable prawns (Singh, 2009), and aquaculture is the primary source of income for almost every progressive and large farmer in the area, as well as for small and marginal farmers. Creeks have been used to draw brackish water from the sea, and in many cases, seepage of this water into cultivated land has led to salinization of neighboring agricultural fields, reduced crop yields, and widespread land degradation.

The extreme salinity of the pond soils prevents farmers from returning to rice farming. Owing to a deadly virus that was introduced into ponds by the addition of water drained from canals and creeks, farmers have been giving up on raising prawns. There is no treatment for infected prawns, and growers have reported that an entire harvest is lost within a few days of prawns being exposed to the virus. Even if aquaculture declines though, the serious environmental problems, declines in the water and soil quality, and associated social problems will linger.

Coastal regions are also threatened because they are heavily inhabited. Coastal cities are a popular destination for migration, and the increased population leads to urban growth. The development associated with this growth, such as roads, houses, and industries, ultimately increases pollution of the soil, water, and air. Dramatic population growth in coastal cities also results in a concentration of people and businesses that are producing waste that remains mostly untreated prior to its release into the water and soil. Furthermore, over-extraction of groundwater in the geologically young sedimentary coastal areas can result in subsidence, which can intensify the relative rise in sea level.

In summary, littoralization, or the increase of population and development along the coast, is altering the morpho-dynamics of the coast and putting a strain on the sustainability of soil, agriculture, and coastal ecosystems (Gopalakrishnan et al., 2019).

Coastal Soils Resources

India's coastal regions have diverse soil types. Soil characteristics can range from alluvial to lateritic, have coarse sand to clay, be non-saline to very saline, be alkaline to highly acidic, have well-drained or poorly drained conditions, have low to high levels of organic matter, and be toxic or deficient in nutrients (Velayutham et al., 1999). The salt-affected soils cover a large area on the east and west coasts and comprise subgroups that are low-lying and have saline, sodic, acid sulfate, marshy, or waterlogged conditions. A thorough description of the various coastal soil types is provided in Table 6.1. In India, coastal saline soils have varying salinity levels throughout the year, with the electrical conductivity of a saturated soil extract (EC_e) ranging from $0.5\,dS\,m^{-1}$ during monsoon season to $50\,dS\,m^{-1}$ during summer. Except for a few isolated areas along the south and west coast, soils devoid of soluble cations often have an abundance of soluble salts, namely NaCl and Na_2SO_4. According to Subba Rao et al. (2011), the sequence is $Na > Mg > Ca > K$, with bicarbonate in traces and chloride as the main anion. Table 6.2 lists the salient physical and chemical characteristics of the soils on India's east coast.

Three main processes affect the soil issues that occur in coastal regions: subsurface intrusion of saline water, which occurs continuously; cyclones, tsunamis, and other natural disasters; and sedimentation, or the deposition of terrigenous sediment load, which occurs where rivers meet the coast (Mandal et al., 2022). In places that receive a lot of precipitation, excess soluble salts can be leached from soils within two to three years. In areas with high levels of humidity, however, it may take several years for the soils to return to their native physico-chemical condition, which is necessary for healthy crop growth.

Table 6.1 Characteristics of Soils and AESRs of Coastal Area of India by State[a].

State	AESR number	AESR	Length of growing period (days)	Soil/textures	OC (%)	Dominant soils	Area (km^2)
West Bengal	15.1	Bengal Basin north Bihar plains; hot moist sub-humid	210–240	Silty clay loam to silty clay, deep loamy to clayey alluvium	0.2–0.5	Vertic Haplaquepts	14,152
	18.5	Gangetic delta; hot moist sub-humid to humid	240–270	Silty loam to clay loam, deep loamy to clayey coastal and deltaic alluvium	0.3–0.8	Typic Fluvaquents	
Odisha	18.4	Utkal plains and east Godavari delta; hot dry sub-humid	180–210	Sand to silty clay loam, deep loamy to clayey coastal and deltaic alluvium	0.1–0.2	Aertic Tropaquepts	7,900
	18.5	Gangetic delta; hot moist sub-humid to humid	240–270	Clay, deep loamy to clayey coastal and deltaic alluvium	Tr-0.2	Vertic Tropaquepts	
Andhra Pradesh	18.3	Andhra plains; hot dry sub-humid	150–180	Silty clay to clay, deep clayey coastal and deltaic alluvium	0.9–1.1	—	35,500
	18.4	Utkal plains east Godavari delta; hot dry sub-humid	180–210	Loam to clay, deep loamy to clayey coastal and deltaic alluvium	0.9–1.2	Chromic Haplusterts	
Tamil Nadu	18.1	South Tamil Nadu plains; hot dry semiarid	90–120	Sand and sandy loam, deep loamy to clayey alkaline coastal and deltaic alluvium	0.1	Psammentic Paleustalfs	7,424
	18.2	North Tamil Nadu plains; hot moist semiarid	120–150	Clay, cracking deep clayey and cracking coastal and deltaic alluvium	0.6–1.0	Psammentic Paleustalfs	
Kerala	19.3	Kerala coastal plains; hot humid to per-humid	240–270	Clay to loam, deep, clayey to loamy, acidic, coastal alluvium	0.3–2.6	Typic Sulfaquents	7,719

Karnataka	19.3	Karnataka coastal plains; hot humid to per-humid	240–270	Sand to sandy loam, deep, clayey to loamy, acidic, coastal alluvium	0.07–1.2	Aquic Ustifluvents	7,424
Maharashtra	19.1	Konkan coast; hot humid	210–240	Loam to clay, medium to deep loamy to clayey mixed red and black soils	0.3–0.6	Udic Paleustalfs	10,000
	19.3	Konkan coast plains; hot humid to per-humid	240–270	Clay to loam, deep, clayey to loamy, acidic, coastal alluvium	0.4–0.7	Udic Paleustalfs	
Goa	19.1	North Sahyadris and Konkan coast; hot humid	210–240	Clay to silty clay loam, medium to deep mixed red and black soils	0.3–0.6	Fluventic Ustropepts	220
	19.3	Coastal plains; hot humid to per-humid	240–270	Clay to loam, deep, clayey to loamy, acidic, coastal alluvium	0.2–0.6	Fluventic Ustropepts	
Gujarat	2.2	Kachchh peninsula (Great Rann of Kachchh); hot hyper-arid	<60	Saline marsh of the Rann of Kachchh, salts, light textured and high EC, preponderance of chloride and sulphate, low available water-holding capacity	0.1	Typic Natrargids	17,465
	2.4	South Kachchh and North Kathiawar peninsula; hot arid	60–90	Loam, deep loamy saline and alkali soils, low available water-holding capacity	0.3–0.5	Typic Calciorthids	
	5.3	Coastal Kathiawar peninsula; hot moist semi-arid	120–130	Medium to deep black soils of deltaic and coastal semi-arid region, Loam to clay, deep loamy coastal and deltaic alluvium	0.1–0.2	Fluventic Ustochrepts, Typic Haplusterts	
Andaman and Nicobar	20.1	Coasts of Islands; hot per-humid	>300	Loam to clay, shallow to deep loamy coastal alluvium			
Lakshadweep	20.2	Coastal Lakshadweep and group of Islands; hot humid	240–270	Sand, shallow to medium deep sandy and littoral soils	0.1–0.8	Typic Ustipsamments	26
Pondicherry and Karaikal	18.4	Coast Pondicherry and Karaikal; hot dry sub-humid	180–210	Loam to clay, deep loamy to clay coastal and deltaic alluvium		Typic Ustorthents	3

[a] AESR, agro-ecological subregion; OC, organic carbon.
Note. Velayutham et al. (1999).

Table 6.2 Physical and Chemical Properties of Soils of East Coast of India[a].

State	AESR number	pH (1:2.5, w/v)	EC (dS m^{-1})	OC (%)	CEC (cmol (P^{+}) kg^{-1})	BS (%)
West Bengal	15.1; 18.5	6.3–7.4; 6.5–7.6	1.5–9.8; 4.1–35.0	0.15–0.52; 0.3–0.78	19.4–22.7; 6.6–10.6	70–80; 74–82
Odisha	18.4; 18.5	6.7–7.4; 4.5–6.3	—; —	0.10–0.22; Tr.-0.20	12.8–13.4; 33.8–55.5	75–85; 49–69
Andhra Pradesh	18.3	8.0–8.4	4.6–27.0	0.9–1.1	2–6	—
Tamil Nadu	18.1	5.1–6.1	—	0.06–0.09	4.0–6.6	89–100

[a] AESR, agro-ecological subregion; BS, base saturation; CEC, cation exchange capacity; EC, electrical conductivity; OC, organic carbon.

Soil salinity hampers crop production in the coastal ecosystem to such an extent that the term "coastal saline soil" has become almost synonymous, although not correctly, with the entire coastal ecosystem. The development of saline and alkali soils due to impeded drainage, inundation, and seawater intrusion has left large areas of Kari (Khar), Pokkali, and Kole lands unsuitable for the profitable cultivation of key crops (Yadav, 2004). Nonetheless, the coastal region's soil salinity varies throughout time. Owing to the upward capillary migration of saline groundwater, the salinity of the soil gradually increases after the rainy season until the start of the following monsoon season. Because of the leaching and washing of salts by rainwater, the majority of lands become non-saline during the rainy season. Development of salinity reduces the variety of crops that can be grown, lowers the quality and quantity of production, and necessitates the use of beneficial, economical, and sustainable management techniques (Dhanushkodi and Subrahmaniyan, 2012).

Coastal soils can also present certain location-specific issues, such as iron toxicity in Odisha's soils, highly permeable sandy soils in some areas of Gujarat, and highly leached, poor fertility lateritic soils in Maharashtra, Karnataka, and Kerala. Crop failures due to acidification and salinization are common in the acid sulfate and tidal marshy areas of Kerala, West Bengal, and Andaman and Nicobar Islands (Table 6.3 and Figure 6.1). The low-lying regions of Kerala, the West Bengal Sundarbans, and the Andaman and Nicobar Islands are home to acid sulfate soils that exhibit unique characteristics of high acidity. These soils are typically characterized by a hazardous level of soluble iron and aluminum. These soils are rich in organic matter but very low in available phosphorus due to drainage of soils rich in pyrites (FeS_2), which when oxidized, form sulfuric acid in the presence of excess SO_4^{2-} ions. Because of the significant leaching that takes place during the

Table 6.3 Physico-chemical Characteristics of Typical Coastal Saline Soils of West Bengal, Odisha, and Kerala[a].

	Saline soils		Acid sulfate soils	
Physico-chemical characteristics	**West Bengal (Kamalpur) Typic Endoaquepts**	**Odisha (Balimunda) Vertic Fluvaquents**	**West Bengal (Nirdeshkhali)**	**Kerala (Arikalm, Calicut)**
pH (1:2)	6.5–8.0	6.1–7.9	4.0–5.9	3.5–4.8
EC_e ($dS\,m^{-1}$)	7.0–10.5	9.3–19.7	2.1–5.8	8.4–43.6
OC (%)	0.26–0.78	0.65–0.92	0.6–2.2	2.4–4.8
CEC (cmol (p^+) kg^{-1})	19.7–21.6	25.7–28.9	19.4–23.4	14.7–69.2
ESP (%)	10.9–15.2	4.7–19.2	11.6–20.6	5.8–20.5
SAR	10.5–12.1	12.9–31.7	3.9–5.4	—

[a] CEC, cation exchange capacity; EC_e, electrical conductivity of a saturated soil extract; ESP, exchangeable sodium percentage; OC, organic carbon; SAR, sodium adsorption ratio.

Figure 6.1 Typical soil profile from West Bengal coast. (a) Soil profile from West Bengal coast; and (b) Jarosite mottles (hue 2.5Y or Yellower and chroma 6 or more) found in lower horizons.

dunes' development, transportation, and deposition, coastal sand dunes are lacking in plant nutrients. The following sections describe the soil resources in the nine coastal states.

West Bengal

The delta region of the Ganges River occurs in the coastal tracts of North and South 24-Parganas districts of West Bengal, constituting approximately 820,000 ha and 220 km of coastline with wide variability in climatic, topographical, and edaphic conditions (Mitran et al., 2014). After Gujarat (described in a later section), the West Bengal coast has the second highest saline soil, at 508,343 ha, and accounts for around 38% of Indian coastal saline soils (Mandal et al., 2023). The soils range in texture from silt loam to clay, and they are brownish, deep to very deep. These soils have a low to high organic carbon content (0.08%–1.45%), a very strongly acidic to moderately alkaline (pH 3.9–8.8) soil reaction, slight to moderate salinity (EC_e of 0.57–14.3 dS m^{-1}), a low cation exchange capacity (CEC; 6.9–19.2 cmol (p+) kg^{-1}), and a moderate to high base saturation (52%–90%). In all soils, sodium is the most prevalent exchangeable cation, followed by magnesium, calcium, and potassium. The primary use of these lands in *kharif* is paddy, which is followed by *boro* (rice) paddy in the summer. In the *rabi* and summer months, vegetables and oilseeds are cultivated with the aid of tube well irrigation or rainfall collected in ponds.

The soils have been classified as Inceptisols. Soils are categorized as Fluventic Endoaquepts, Typic Halaquepts, Typic Endoaquepts, and Typic Haplustepts based on their features (Bandyopadhyay et al., 2003; Srinivasan et al., 2019).

Odisha

The state of Odisha (formerly Orissa) boasts a 485-km-long coastline, known as coastal Odisha or the Utkal Plains. The Eastern Ghats in the west, the Bay of Bengal in the east, Andhra Pradesh in the south, and the lower Ganges plain in the north define the region's borders. This region has the largest lake in the nation, Chilika Lake; the historic Kalinga kingdom; beaches such as Chandipur and Gopalpur; and the Bhitarkanika mangrove habitat. The coast is depositional, formed by the Mahanadi, Brahmani, and Baitarani delta. Around 400,000 ha of the state is covered by coastal soils, which are found in a narrow band, 476.4 km in length and 2–15 km in width, along the seacoast (Chaudhary et al., 2008). The soils include alluvial soils with high levels of total soluble salts (EC_e, 4 dS m^{-1}). Because of creeks and estuarial incursion of brackish tidal water from the sea, the littoral deposits contain salt. The districts of Balasore, Bhadrak, Jagatsinghpur, Kendrapara, Puri, Khurda, and Ganjam are home to about 254,000 ha of saline soils. Together with sodium and magnesium, soluble salts of chloride and sulfate are abundant in saline soils. Soils of lacustrine sediments of Chilika Lake are also affected by salts due to flooding of brackish lake water in the districts of Puri, Khurda, and Ganjam. Subsoil salt builds up during the monsoon because of the high groundwater table in low-lying areas. The majority of these soils have a columnar structure, and their texture ranges from clay to clay loam. These soils have an EC_e of 10–40 dS m^{-1} and a pH range of 6.0–8.0 during summer. The percentage of exchangeable sodium ranges from 18% to 27%. The soils include low to medium levels of phosphate and rich in potassium, and nitrogen. The amounts of sulfate, boron, molybdenum, and chloride in the saline soils are adequate. Crop failure is typically caused by (a) plasmolysis of roots and seeds that are germinating, (b) young seedlings dying; (c) reduced uptake of potassium, calcium, and magnesium because of excess sodium; (d) toxicity from boron; and (e) damage from hydrogen sulfide. The application of lime, which substitutes some of the exchangeable Na^+ or H^+ with Ca^{2+}, can restore coastal saline soils that have a high exchangeable sodium content but are acidic in response. Utilizing green manure crops, recycling straw, and applying organic manures all help to improve soil structure by lowering salinity and alkalinity levels (Sahu & Mishra, 2005).

Andhra Pradesh

Extending between the Eastern Ghats and the Bay of Bengal, India's second-longest coastline after that of Gujarat, is in the state of Andhra Pradesh and measures 975 km. The state's soil composition is made up of about 5% deltaic and coastal alluvial soils, and there are three different kinds of soils: coastal, saline, and saline alkali.

The coastal soils, which fall under the Entisol order, are coarse-textured, sandy subsoil. Coastal soils of the state are mostly deep (1.8–5.0 m or more). The shallow water table causes the soils to be quickly permeable, neutral in reaction (pH 6.5–7.5), and low in CEC because of very low clay concentrations. A significant portion of this state is made up of saline soils, which have salt concentrations that range from 0.2% to more than 4.0% and with an exchangeable sodium percentage (ESP) of less than 15. The pH seldom rises above 8.5 and is typically around 7.0. Saline alkali soils can be found along the banks of streams and in certain hinterland areas. While the salt level has been found to range from 0.3% to 1.5% or even above, the pH ranges from 8.0 to 10.0. The water table is usually shallow and shows not only high salt content but also the presence of alkali carbonates.

Tamil Nadu

Tamil Nadu's coastline stretches 907 km along the Coromandel Coast. The boundaries are the Eastern Ghats to the west, the Kaveri Delta to the south, Andhra Pradesh to the north, and the Bay of Bengal to the east. The coastline includes 14 coastal districts: Thiruvallur, Chennai, Chengalpattu, Villupuram, Cuddalore, Mayiladuthurai, Nagapattinam, Tiruvarur, Thanjavur, Pudukottai, Ramanathapuram, Thoothukudi, Tirunelveli, and Kanyakumari. About 7% of the state's soil is made up of coastal soils, and there are three different kinds of soils: laterite, coastal alluvium, and coastal saline. Only the Coromandel Coast has saline soils. They contain large proportions of sodium, potassium, and magnesium, and consequently, the soils are infertile and do not support vegetative growth. The soils also lack calcium and nitrogen and have higher salt content because of inadequate drainage in arid conditions. The soils have a sandy to loamy texture.

Soil salinity is exacerbated by sea water intrusion in deltas. In areas with intensive agriculture supported by irrigation, saline soil is formed due to capillarity action that draws salts to the soil surface. Sand is also deposited by rivers, creating alluvial soils that can be found in the district's coastal plains and river valley areas. This type of soil is typically found in the districts of Thanjavur, Tiruvarur, Nagapattinam, Villupuram, Cuddalore, Tirunelveli, and Kanyakumari. In a few interior locations, minor amounts of alluvial soils occur along the river valleys. The rear adjoining the coast is characterized by laterite capping. Along the entire Kanyakumari sea coast, lateral deposits or bay deposits of sand, zircon, rutile, ilmenite, and garnet are fairly widespread.

Kerala

The Malabar Coast includes Kerala's 590-km coastline. It begins from the western coast of Karnataka, runs along the length of Kerala, and ends at Kanyakumari in Tamil Nadu. Within Kerala, nine districts are along the coastline: Thiruvananthapuram, Kollam, Alappuzha, Ernakulam, Thrissur, Malappuram, Kozhikode, Kannur, and Kasaragod. There are four different kinds of soils: coastal alluvium soil, acid saline soil (Kuttanad Alluvium), brown hydromorphic soil, and hydromorphic saline soils. Coastal alluvium soil is the result of recent marine deposits of fluvial sediment. Low water-holding capacity, high sand content, and somewhat acidic reaction ($pH < 6.5$) are its main characteristics. Its texture ranges from sandy loam to loamy sand. Kuttanad alluvium, or acid saline soil, is mostly found in the Kuttanad region. Because Kuttanad is mostly below sea level, the Kuttanad Alluvium is further separated into three groups based on morphological and physiochemical characteristics: (a) soil from reclaimed regions (Kayal soil; high clay content), (b) black soil with high organic content (Kari soil; see Figure 6.2), and (c) soil from Karappadam (high silt content).

The majority of brown hydromorphic soil is found in wetlands, which lack phosphate and lime, but are fairly rich in other minerals and organic materials.

Figure 6.2 Profile picture of Kari soil (Kuttanad Kari soil, Acid saline soils): Typic Sulfohemists Courtesy: ICAR-NBSS & LUP Archives.

The soil has a texture that ranges from sandy loam to clay and has a deep brownish color. It was created by the movement and deposition of earth from nearby hill slopes and has an acidic reaction. In addition to having hydromorphic features, such as hard pans, gray horizons, and organic matter deposition, it has poor drainage. Hydromorphic saline soils are found near the coastal tracts of the state. The direct cause of their salinity is due to the intrusion of tidal and backwater into these regions. These soils range in texture from sandy to clayey, with a brownish color. They support agricultural crops such as Pokkali and Kaipad rice that can withstand salt. The ultra-acid soils in the inundated areas of Kuttanad, Kole, Pokkali, and Kaipad lands have a pH of less than 3.5. The primary source of these soils is marine sediments rich in sulfur. Owing to seawater intrusion, the salinity of the soils primarily affects the rice environment in coastal areas. This issue is seasonal, becoming worse in the summer when tide flows are high.

Karnataka

The 300-km-long Kanara coastline region of Karnataka is bordered by the Arabian Sea to the west, the Kerala Plains in the south, the Konkan in the north, and the Western Ghats in the east. Three districts fall within the region: Uttara Kannada,

Dakshina Kannada, and Udupi. The Kali Nadi, Gangavalli, Bedti, Tadri, Sharavati, and Netravati rivers drain the coast; they have carved up small valleys with sharp grades and run mostly westward.

Southern states have coastal alluvial soils, which are created by the deposition of sediments and alluvium brought by rivers and waves of the sea. The soil is acidic and has a high content of organic nutrients, with low content of bases and low CEC. The red soils along the rest of the coast are unproductive and frequently sandy and gravelly in texture.

Goa

Goa, the smallest state in India, has a 131-km coastline, which is bordered by the Arabian Sea to the west and the states of Maharashtra and Karnataka to the north and east. Goa's soil is reddish in color and rich in ferric-aluminum oxides. Most of the inland and riverbank soil is loamy and alluvial. The soil reaction of the salty coastal soil is acidic ($pH < 6.57$). According to Bandyopadhyay et al. (2003), the intrusion of saltwater raises the salinity levels of the soil, resulting in the formation of both saline (high pH) and saline acid (low pH) soils, which are characteristic of the coastal saline soils. The soil $EC_{1:2}$ (1:2 soil-water suspension) vary from $0.09\,dS\,m^{-1}$ at the lowest to $6.29\,dS\,m^{-1}$ at the highest.

Maharashtra

The Konkan coast consists of 720 km of coastline in the state of Maharashtra. It borders the Arabian Sea to the west, the Daman Ganga River to the north, the Gangavalli River to the south, and the Western Ghats to the east, and it includes the districts of Palghar, Thane, Raigad, Ratnagiri, and Sindhudurg. These districts' coastal saline soils result from the regular flooding of arable land by creek or sea water during high tides. Locally, these soils are referred to as khajan or khar soils.

The state's two agroclimatic zones contain coastal saline soils. While the southern part of Raigad and the entirety of Ratnagiri and Sindhudurg districts are covered by the very high rainfall lateritic soil (VRL) zone, Thane district and the northern part of Raigad district are covered by very high rainfall non-lateritic soil (VRNL) zone. This soil contains high accumulations of basalt.

The coastline region is nearly level topographically, with an elevation of less than 10 m above sea level. The principal rivers run towards the Arabian Sea from the Sahyadri hills and are influenced by tidal water. While the soils in the VRL zone have a coarse texture and moderate conductivity, the soils in the VRNL zone are characterized by a fine texture and extremely low hydraulic conductivity. In contrast to the moderately acidic to strongly acidic pH found in the VRL zone, the soil pH varies from slightly acidic to slightly alkaline in the VRNL zone. Because of the shallow saline water table, restricted drainage, and relatively little rainfall, the average ECe of the soils in the VRNL zone is much higher ($25.4\,dS\,m^{-1}$).

The sodium adsorption ratio (SAR) has an average value of 28.1. Sodium, magnesium, and calcium concentrations are comparatively high, and there is a limited range for potassium. In these soils, sodium chloride predominates. For cations, the order is $Na^+ > Mg^{2+} > Ca^{2+} > K^+$, while for anions, it is $Cl^- > SO_4^{2-} > HCO_3^- > CO_3^{2-}$.

Gujarat

Gujarat has the longest coastline in India. The Arabian Sea envelops this shoreline, which spans 1,600 km and is located in the Kathiawar portion of the state. Deep black and coastal alluvial soils are the most common soil types in south Gujarat. Central Gujarat has medium black and calcareous medium black soils, and the Bhal region and Saurashtra peninsula contain shallow to medium-depth clay soil with high concentrations of salt. The region's soils have become increasingly salty due to salinity ingress, subsurface seawater intrusion in areas with high groundwater tables, and excessive groundwater extraction for extremely intensive cropping techniques.

Industrialization along the Vapi to Mehsana belt and the increasing release of effluents have also affected natural resources, particularly the groundwater, soils, and natural vegetation. According to a study by ICAR-Central Soil Salinity Research Institute, Regional Research Station (ICAR-CSSRI, RRS), Bharuch, heavy metal contamination has had a negative effect on 31 of the 40 villages along the Amla Khadi, a tributary of the Narmada River (Chinchmalatpure, 2018). Across an area of $590.7\,km^2$, 22, 24, 15, 18, and 11 villages have been affected by cadmium, cobalt, chromium, nickel, and manganese, respectively. A significant level of groundwater contamination was discovered, with contaminants exceeding the threshold levels for cadmium (0.01 parts per million [ppm]), cobalt (0.05 ppm), chromium (0.10 ppm), nickel (0.20 ppm), zinc (2.0 ppm), and copper (0.2 ppm). The contaminants included cadmium (0–1.81 ppm), cobalt (0–0.12 ppm), chromium (0–0.23 ppm), manganese (0–0.88 ppm), nickel (0.03–0.35 ppm), zinc (0.01–2.92 ppm), and copper (0–0.25 ppm).

The Valsad and Surat districts' coastal salt-affected soils have a medium to heavy texture and a grayish to dark brown color. They have extremely little to no permeability. At all depths, the pH is alkaline and tends to rise with depth. The coastal soils of the Bharuch district are heavy and have a clay texture at all depths. The pH ranges from 8.6 to 9.5 and the $EC_{1:2}$ from 0.22 to $8.1\,dS\,m^{-1}$; the latter tends to increase with depth. The ESP shows an increasing pattern with depth, ranging from roughly 5 to about 18.

In the Bharuch district of coastal Gujarat, the Bara tract area is primarily composed of vertisols and their accompanying fine-textured, deep to very-deep soils, with smectite being the predominant clay mineral. These soils have demonstrated a high potential for shrink-swell, as evidenced by the formation of fissures that are 100 cm deep and 4–6 cm wide. Although the soils have a large water-holding capacity, their permeability is slow to very sluggish and the drainability is inadequate. There is 0.4%–0.6% organic carbon content, low to medium concentration of available phosphorus, and high to very high content of available potassium.

Despite the soluble salt concentration in the surface layer being low, the Bara tract soils have a substantial concentration of soluble salts in their sub-soils (Table 6.4). Different levels of salinization are apparent in the soils in the Bara tract. The Saurashtra region's coastal salt-affected soils have a clay texture and a high lime concentration. There is minimal fluctuation in pH with depth, ranging from 7.7 to 9.0. The ECe also varies widely from 2 dS m^{-1} to as high as 20 dS m^{-1}, with a tendency for the salinity to decrease with depth. Additionally, the ESP levels are typically high, with values of about 50 being typical. The physiography of the Bhal area's soils causes a wide range of salinity/sodicity issues. The salinity reportedly varies from 0.79 to 282.5 dS m^{-1}, whereas the range of ESP is 11.3–30.0. These soils are essentially calcareous in nature, with very low permeability, and range from clay to clay loam in texture. The top layer of these soils in the Rann of Kachchh has a blocky structure and a texture of clay or silty clay, and it becomes lighter with depth. The pH is between 7.8 and 8.4, and there is no consistent pattern in the salt's behavior with depth.

Table 6.4 Physico-chemical Characteristics of Typical Soil Profile (Typic Haplusterts) of the Bara Tract (Village Samni, Taluka Amod District, Bharuch)[a].

Horizon	Depth (m)	ECe (dS m^{-1})	pH	$CaCO_3$ (%)	OC (g kg^{-1})	Sand (%)	Silt (%)	Clay (%)	ESP	CEC (cmol (p+) kg^{-1})
Ap	0.00–0.20	2.8	8.2	5.7	5.5	17.8	30.4	51.8	9.8	44.3
Bw1	0.20–0.44	2.7	8.0	5.7	3.7	22.0	25.8	52.2	10.5	45.4
Bw2	0.44–0.73	10.9	7.5	6.4	2.5	14.2	26.9	58.9	11.3	49.9
Bss	0.73–0.96	10.9	7.5	6.3	1.4	11.4	27.2	61.4	10.4	50.1
BC	0.96–1.25	10.5	7.7	6.4	2.3	21.8	26.8	51.4	5.7	44.4
C	1.25–1.50	11.4	7.7	17.7	0.1	37.2	18.0	44.8	4.7	36.2

[a] CEC, cation exchange capacity; ECe, electrical conductivity of a saturated soil extract; ESP, exchangeable sodium percentage; OC, organic carbon.

Note. Chinchmalatpure et al., 2008

The Bara tract's farmers use saline tubewell water or canal water for irrigated crops. Although the quantity of soluble salts in the surface layer of the Bara tract soils is modest, the subsoil has a notable concentration. When cotton crops were irrigated with saline tube well water over saline Vertisols, salt buildup was seen in the surface layer. When soil profiles were watered with saline groundwater, secondary salinization was shown to develop; however, soil salinity decreased when soils were irrigated with fresh canal water. The availability of high-quality canal water allowed for an improvement in crop yield in the study area; however, the excessive use of canal water for irrigation resulted in an increase in the ESP of soils and the formation of sub-soil sodicity in the lower horizon, resulting in the start of the pedogenic process, or sodification.

Andaman and Nicobar

The Andaman and Nicobar Islands are situated in the Bay of Bengal, 1,200 km to the east of the Indian mainland. They are part of the hot, humid to slightly humid Islands ecoregion, which is known for its abundance of biodiversity. The low-lying coastal region experiences salinity, acidity, and waterlogging. Acid sulfate soils can be found in coastal locations that are vulnerable to tidal floods. Red loamy soils ranging from medium to deep, generated from coastal alluvium, are found along the shore. They qualify for the great groups of Hapludalfs, Dystropepts, Eutropepts and Sulfaquents (along the coast). The soils have low to medium available water-holding capacity (Ganeshamurthy et al., 2002). Land management techniques such as drainage and bunding allowed salts to leach more easily and have restored agriculture to foothills. In addition to improving drainage, broad beds and furrows assisted in collecting rain water. They also resulted in higher cropping intensity and greater water and fish productivity. Depth of submergence and salinity were correlated with the rainfall and soil moisture (Velmurugan et al., 2015).

Lakshadweep

Scattered tiny coral islands in the Arabian Sea, roughly 400 km from the mainland (southern tip of the Indian Peninsula), make up the Lakshadweep islands. The parent material of the soil was obtained from physical weathering of the organogenic calcium carbonate, as the islands are composed entirely of coral limestone. The majority of the light-textured soils are sand or loamy sand, with sporadic sandy loam. Soils are underlain by impervious coral limestone. The soils of the islands are alkaline in reaction with surface soil pH ranging from 7.9 to 9.1. The subsurface horizons' pH rises with depth.

According to Krishnan et al. (2004), the $EC_{1:2.5}$ of the soils decreased with depth and ranged from 0.23 to 1.0 $dS\,m^{-1}$ at the surface. With an average organic carbon content in the surface horizon of 2.5% (0.77%–4.78%), the soils are rich in organic matter. Surface soils had CECs ranging from 1.0 to 17 cmol (p+) kg^{-1}. Calcium predominates in the exchange complex, with magnesium, sodium, and potassium following in decreasing order.

The islands' soils are poor in water retention capacity, highly drained, and porous. Because the soils are permeable, precipitation swiftly seeps into the coral limestone substratum underneath. The calcareous alkaline soils have low levels of nitrogen, potassium, iron, copper, and zinc nutrients. The island lacks significant surface freshwater bodies, such as streams (Krishnan et al., 2004).

Water Resources in Coastal India

The coastal ecosystem occurs in the vicinity of creeks, rivers, deltas, low-lying lands, and estuaries, and it is characterized by high salinity of groundwater and a paucity of good-quality water in the post-monsoon season. The water table's depth varies greatly as well. While the majority of Gujarat experiences very little rainfall, areas in the east and southwest receive a large amount. Areas such as the Sundarbans experience continuous heavy waterlogging of land during the monsoon season due to an overabundance of rainwater. Rice is the main crop, and the entire region is essentially mono cropped. Monsoon rainfall in these locations ranges from 1,000 to 1,800 mm and there is essentially no other source of irrigation for the crop. In addition to the risk of increasing the salt of the soil and groundwater, the scarcity of high-quality irrigation water is a constant barrier to planting a second post-monsoon crop. The nature and extent of the problems vary from region to region, but in most places the coastal ecosystem presents a unique problem of water management. India's water resources, particularly in the coastal regions, may be severely affected by the changing climate.

Assessment of Soil Quality for Coastal Soil Resources

The ability of the soil to function, or "its fitness for use," is known as soil quality. In the assessment of soil quality, conventional processes are typically followed, utilizing affordable, site-specific indicators that farmers can easily use. Because they are less expensive, morphological and visual indicators are occasionally also recommended for the evaluation of soil quality. However, the identification of indicators in coastal soil presents a variety of challenges owing to the ambiguity surrounding the hazards posed by both natural disasters and seawater intrusion into agricultural land.

The main issues with parameterization include large temporal and geographical variation, the soil ecosystem being governed primarily by physical features and climatic conditions, the compromised biogeochemical cycling of vital nutrients, and the intricate link between nutrient displacement and salt-loading. According to Mandal et al. (2022), indicators used to evaluate coastal soils should be location specific. The complexity of soil quality necessitates the use of a suite of indicators for creating a thorough, low-cost assessment. The following guidelines can be used to choose indicators: (a) inherent indicators, (b) dynamic indicators, (c) a logical sieve rather than a straitjacket approach, (d) cross-functional/all-encompassing (aggregative/blackbox) type, (e) early warning type, (f) surrogacy (i.e., not evident due to lack of database values), (g) ability to meet unique needs, (h) indicators of climate change, and (i) farmer friendly.

About 40% of the total master indicators used to determine the quality of coastal soil were pH and EC. Although biological characteristics are a key component of assessing the quality of soil, they are not specified for screening when it comes to coastal soil. Furthermore, in a coastal ecosystem, the loadings of salt of calcium and magnesium impart physico-chemical recalcitrance to soil organic carbon. This can be achieved through the formation of chelates with the metal ions and soil mineral matrix or by the compartmentalization of the SOC within the aggregates, which keeps it out of the reach of microbes (Deb et al., 2016, 2020). This compartmentalization could be the cause of the reduced biodiversity of the soil beneath the coastal habitat. A notable example of the uneven richness of microorganisms is the link between EC and soil microbial biomass carbon (Tripathi et al., 2006, 2007).

Rehabilitation of Coastal Arable Land

Soil and Water Management

During the *kharif* season, proper flood control measures in the low-lying coastal areas must be combined with adequate drainage to remove water congestion. Protective embankments with a 3:1 slope on the river side and a 2:1 slope on the land side, with 1 m of free board above the high tide level, should be built to shield agricultural land from tidal inundation. Tidal inundation can be prevented by planting mangrove trees as wind barriers and brick-pitching earthen embankments. Waterlogged land drains more quickly when one-way sluice gates are installed on riverbanks or in any other suitable site to let surplus water flow from the land to the river during low tides (Maji et al., 2020).

Leaching of soluble salts from crop root zones by high-quality irrigation water during the *rabi*/summer season is one of the key management practices for coastal saline soils. Leaching requirement is mostly determined by the type of irrigation water used, the technique used, the soil's texture, the crop's maximum tolerance to salinity, and so forth. The leaching requirement is 0.90 $cm\,cm^{-1}$ of soil for sensitive crops and 0.60–0.75 $cm\,cm^{-1}$ for tolerant crops in order to leach 70% of the salts from the 0- to 30-cm soil layer. SALTMOD, DRAINMOD-S, and SAHYSMOD are a few of the agro-hydro salinity models that have been developed to desalinate salty soil through drainage under specific circumstances. The SALTMOD model was also used in Andhra Pradesh's coastal clay soils, where a subsurface drainage system was set up with multiple drain spacings. According to the study, the model could be confidently utilized to assess different subsurface drainage system drain spacing and provide realistic reclamation period prediction.

The following strategies have been proposed for the management of acid sulfate soils, both potential and existing. Leaching after drying and aeration can remove pyrite and acidity from the soil. Limiting or stopping pyrite oxidation and inactivating pre-existing acidity can be achieved by keeping the water table high, either with or without further liming and phosphorus fertilization. When there is waterlogging, the microbial breakdown of organic waste reduces the acidity of the soil and can quickly raise pH levels to almost neutral.

This strategy works very well for growing rice. Having access to fresh water for irrigation is essential. Rice can be grown in acid sulfate soils that are less toxic and have developed acidity in deeper layers; these soils can be made better with good agronomic practices. The productivity of these extremely acid soils has increased considerably as a result of extensive irrigation and improved water management. In India, it has been discovered that applying lime, rock phosphate, super phosphate, and green manuring with sesbania to the coastal acid sulfate soils of the Sundarbans improves the soil's characteristics and promotes rice development. It was discovered that using powdered oyster shell, which is abundant in calcium and found in coastal areas, was a cost-effective substitute for other soil-improving agents.

Integrated Water Management

Deeper freshwater aquifers in coastal regions do not contribute to the buildup of salt in the crop root zone. However, the formation of salinity in soil is caused by a salt-rich water table that is present at a relatively short depth—typically no more than 2 m below the soil's surface. Depending on the ratio between the rate of salt recharge by upward rise to the rate of salt flux by leaching, the net salt loading in the root zone will either be positive (salinity will increase) or negative (desalinization will occur). The relative salt accumulation will thus be treated as positive during the dry season and negative during the wet season owing to waterlogging caused by high rainfall. This cycle will repeat each year.

Given the susceptibility of the coastal plains to seawater intrusion and its detrimental effects on the soil and plant growth, it may be best to avoid using groundwater for irrigation in coastal locations. Thus, more surface storage of rainwater in excavated ponds, furrows, ditches, community canals, and so forth should be used to produce irrigation sources. Therefore, in order to preserve the stability of the ecosystem, water management in the coastal plains should primarily focus on increasing the number of fresh surface water sources and managing them appropriately, with minimal reliance on subsurface sources.

Rainwater harvesting is becoming more and more popular throughout the coastal region, especially in the Sundarbans and Odisha. It involves the on-farm collection and storage of excess rainwater during the monsoon and recycles it for irrigation of crops during the dry season (deficit water period). The goal is to introduce multi-cropping in areas that are primarily mono-cropped. A few land shape models for water harvesting have been created by the ICAR-CSSRI in an effort to improve the productivity of the region's low-lying degraded land, address protracted waterlogging during the monsoon season, and increase water availability during the *rabi* season. A variety of land-forming methods, including bunding, the broad bed and furrow system, and the cultivation of rice and fish in the Andaman and Nicobar Islands, were shown to enhance crop diversity by permitting multiple cropping and to raise total farm yield (Ravisankar et al., 2008). Through the application of various land-shaping techniques such as farm pond, deep furrow and high ridge, paddy-cum-fish, broad bed and furrow, three-tier system, paired bed system, and drainage improvement network, about 370 ha of land in underprivileged areas in Sundarbans and Andaman and Nicobar Islands have been converted from mono-cropped to multi-cropped with integrated crop and fish cultivation under the GEF-funded National Agricultural Innovation Project (CSSRI, NAIP, 2014). Different land situations, such as high land, medium land, and low (original) land, were formed in low-lying and degraded farmers' fields using land-shaping techniques. These circumstances also included rainwater collection structures, such as farm ponds, furrows, and trenches. The issue of drainage congestion during the *kharif* season was lessened by raising land and constructing water-harvesting facilities. These measures allowed for the growth of high-value crops such as vegetables during any time of year and allowed for the early sowing of *rabi* crops.

Land-shaping approaches resulted in decreased salinity buildup in the soil of various land conditions, particularly medium land and highland/ridges/dikes in the land-shaped area, and increased fertility status and biological activity in the soil (Table 6.5). Techniques for shaping land were an appealing and financially feasible option for the area affected by coastal salt (Burman et al., 2015).

Subsurface drainage for reclamation of saline waterlogged soils: In saline, wet environments, the sub-surface drainage (SSD) method was created to

Table 6.5 Effect of Land-Shaping Systems on Soil Properties[a].

Parameters	ECe winter	ECe summer	OC	Av N	Av P	Av K	Bulk density	MBC
	dS m^{-1}	dS m^{-1}	%		kg ha^{-1}		Mg m^{-3}	µg g^{-1}
Rice-monocrop	3.82aA	9.77aB	0.762a	122.8a	9.30a	514.6ac	1.502ab	187.6a
Rice-rice	3.28bA	5.33bB	0.913b	124.1a	18.10b	474.5a	1.474a	288.2bc
Farm pond system- lowland	2.1cA	2.64cB	0.855b	114.3a	27.49c	394.5b	1.589b	243.8b
Farm pond system- medium land	2.47cA	2.62cA	0.865b	129.5a	36.93d	479.2a	1.542b	279.1b
Farm pond system- upland	2.57cA	2.82cA	0.926b	194.2b	49.78e	519.3ac	1.562b	342.2cd
Deep furrow and high ridge system-lowland	2.89cbA	3.59cB	0.793ac	169.3b	24.72c	642.2d	1.584b	291.1bc
Deep furrow and high ridge system- ridge	3.12bA	3.55cB	0.822c	234.8c	46.56e	499.8ac	1.512ab	376.3d
Paddy cum fish system-lowland	2.46cA	2.74cA	0.810c	178.3b	39.22de	442.0ab	1.601b	225.6ab
Paddy cum fish system-dikes	2.88cbA	3.92cB	0.847bc	198.2b	46.12e	549.1c	1.541b	325.9c

[a] Av N, available nitrogen; Av P, available phosphorus; Av K, available potassium; ECe, electrical conductivity of saturation paste extract; MBC, microbial biomass carbon; OC, organic carbon. Values with the same lowercase within land uses are not significantly different at $P < 0.05$; values with the same uppercase letters within summer and winter season are not significantly different at $P < 0.05$.

Note. Mandal et al., 2019/with permision of Elsevier.

lower the water table. To remove excess water and soluble salts from the affected region, perforated corrugated/smooth PVC pipes are physically inserted at a calculated spacing and depth using laser-controlled trencher machines or by hand in an SSD system. To stop clogging, a synthetic or gravel filter is placed over these pipes. In a few coastal areas, including Karnataka, Gujarat, Maharashtra, and Andhra Pradesh, the success of a few manually laid pilot SSDs proved beneficial (Sharma et al., 2016). Significant gains have been made in soil health, crop yields, cropping intensity, and overall environmental quality on the reclaimed soils (Sharma & Singh, 2015). Again, adoption of SSD in landlocked places is frequently impeded by issues related to the safe disposal of saline drainage water. A number of suggestions have been made to resolve these challenges, including evaporation ponds, bio-drainage, the simultaneous use of fresh and saltwater for irrigation, and the adoption of salt-resistant cultivars.

Bio-drainage to fix waterlogging and secondary salinity: Large tracts of agricultural land have become unusable due to waterlogging and the ensuing secondary salinization in many of the nation's canal-irrigated areas. Water seepage from canals and faulty water management practices on farms create shallow water table conditions and significantly increase root zone salinity.

Traditionally, subsurface drainage systems—which are costly and challenging to maintain and pose challenges related to the disposal of saline drainage effluents—are installed to reclaim waterlogged saline areas. Due to these operational challenges, there is interest in alternate workable options, including bio-drainage, to maximize the productivity of water-logged soils.

The term "bio-drainage" describes how fast transpiration by perennial trees pumps out surplus soil water and dissolved salts using bioenergy. Comparable to energy-driven water pumps, this technology has been shown to stop saline accumulation in canal channels when the right trees (such as bamboo, eucalyptus, and poplar) are planted early to avoid salinization and water-logging. Recently, research has been done on the combination of bio-drainage and appropriate land improvements to make productive use of the waterlogged salt-affected soils (Sharma et al., 2016).

Agronomic research on salt-affected Vertisols: Gujarat's coastal regions are home to Vertisols, or "black soil." Even when salt and exchangeable sodium concentrations are low, soils suffer from excessive clay content. Furthermore, the extent of salinization is expanding quickly due to the advent of massive irrigation projects. When salinization of these lands occurs, restoring their productivity is far harder than it is for the alluvial sandy loam soils of the Indo-Gangetic Plains. Farmers in Gujarat have devised and accepted the cultivation of Salvadora, which can withstand salinities of up to $50\,dS\,m^{-1}$, to recover extremely salinized soils ($>40\,dS\,m^{-1}$), and dill, which can withstand salinities of $4–6\,dS\,m^{-1}$ (Sharma & Chaudhari, 2012) either irrigated with salty groundwater or with the remaining soil moisture (Chinchmalatpure et al., 2015).

Innovative practices to restore soil quality: Mulching limits the salinization of surface soil. It increases the infiltration capacity, while decreasing the rate of subsurface water and salt evaporation and capillary rise. Thus, in saline croplands along the shore, mulching stops land degradation. Zero tillage technology allows for the efficient conservation of carryover soil moisture. Zero tillage techniques allow for the early seeding of potatoes during the post-rainy season. It conserves soil moisture in the soil profile, which in turn reduces the crop's need for irrigation water (Sarangi et al., 2021).

In many tropical coastal and deltaic locations, growing rice is the primary land use. Under unfavorable conditions of fluctuating water levels and extremely salinized soil, only rice can be produced. Puddling or high tillage during rice farming has been found to cause soil disturbance, which accelerates the rate at which SOC decomposes and transfers carbon from the soil to the atmosphere. Therefore, reducing disturbance through the use of reduced tillage techniques, in combination with residue management, is a highly effective way to increase soil carbon (Mandal et al., 2021b). Adding crop residues or organic amendments, such as farmyard manure, green manure, compost, vermicompost, farm trash, and municipal garbage without disturbing the soil are simple methods for increasing the amount of carbon in soils.

In Tamil Nadu's coastal soils, the addition of organic waste materials, such as press mud at $12\,t\,ha^{-1}$, has been proven to lower EC and improve soil quality by increasing SOC and accessible nutrients. While conservation tillage is becoming more popular in India's Indo-Gangetic plains, its use in the fragile coastal agro-ecosystem is minimal. In a rice-based cropping system, reduce tillage combined with residue management would be helpful in improving the soil health in coastal areas (Mandal et al., 2021).

Researchable Issues

If well managed, the coastal ecology is a treasure that provides secure livelihoods for a large number of people worldwide. However, because of changes in land use patterns, the coastal environment is becoming increasingly exposed to the pressures of human activity. The ability of the coastal environment to sequester blue carbon stocks and halt degradation is enormous. It is necessary to engage in management and conservation efforts that enhance the soil's quality. To achieve this goal, it may be useful to create a spatial and temporal map of the coastal regions that highlights important soil-related issues specific to individual locations, together with their master indicators. An attempt might be made to pinpoint a small number of straightforward, reliable, affordable, and highly productive indicators for evaluating the quality of the soils under various coastal regions across the globe. Understanding the coastal region's hydrological cycle is a fundamental component of coastal system rehabilitation. A detailed database on the hydrological cycles may be generated for planning different activities in different coastal regions. A set of site-specific soil management practices needs to be developed to check degradation and improve agronomic performance of coastal soils.

References

Bandyopadhyay, B. K., Maji, B., Sen, H. S., & Tyagi, N. K. (2003). *Coastal soils of West Bengal—Their nature, distribution and characteristics*. Bulletin No. 1/2003. Central Soil Salinity Research Institute, Regional Research Station.

Barbier, E. B. (2013). Valuing ecosystem services for coastal wetland protection and restoration: Progress and challenges. *Resources, 2*, 213–230.

Burman, D., Mandal, S., Bandopadhyay, B. K., Maji, B., Sharma, D. K., Mahanta, K. K., Sarangi, S. K., Mandal, U. K., Patra, P., De, S., Patra, S., Mandal, B., Maitra, N. J., Ghoshal, T. K., & Velmurugan, A. (2015). Unlocking production potential of degraded coastal land through innovative land management practices: A synthesis. *Journal of Soil Salinity and Water Quality, 7*, 12–18.

Chaudhary, D. R., Ghosh, A. R. U. P., & Boricha, G. N. (2008). Characterization and classification of coastal saline soils of Paradip, Orissa. *Agropedology, 18*(2), 129–133.

Chinchmalatpure, A. R. (2018). Assessment and management of natural resources of coastal Gujarat towards enhancing productivity and ensuring food security. *Journal of Indian Society of Coastal Agricultural Research, 36*(2), 27–39.

Chinchmalatpure, A. R, Ali, S., Kulshrestha, N., Singh, R. K., Bundela, D. S., Kumar, P., & Sharma, D. K. (Eds.). (2015). *Intellectual property management and commercialization of ICAR-CSSRI technologies for management of salt-affected and waterlogged soils of India*. ICAR-Central Soil Salinity Research Institute.

Chinchmalatpure, A. R., Khandelwal, M. K., & Rao, G. G. (2008). Characterization and classification of salt affected soils of Samni farm, Bharuch district, Gujarat. *Agropedology, 18*, 71–75.

CSSRI, NAIP. (2014). *Final report of NAIP sub-project on strategies for sustainable management of degraded coastal land and water for enhancing livelihood security of farming communities*. ICAR-Central Soil Salinity Research Institute, Regional Research Station.

Deb, S., Mandal, B., & Bhadoria, P. B. S. (2020). Influence of sea water ingression on carbon sequestration in soils under coastal agro-ecosystems of eastern India. *Agriculture Research, 9*, 622–630.

Deb, S., Mandal, B., Bhadoria, P. B. S., Schulz, E., Ghosh, S., & Debnath, M. K. (2016). Microbial biomass and activity in relation to accessibility of organic carbon in saline soils of coastal agro-ecosystem. *Proceedings of the National Academy of*

Sciences India. Section B, Biological Sciences, *88*(2), 633–643. https://doi.org/10.1007/s40011-016-0801-4

Dhanushkodi, V., & Subrahmaniyan, K. (2012). Soil management to increase rice yield in salt affected coastal soil. *International Journal of Research in Chemistry and Environment*, *2*(4), 1–5.

Ganeshamurthy, A. N., Dinesh, R., Ravisankar, N., Nair, A. K., & Ahlawat, S. P. S. (2002). *Land resources of Andaman and Nicobar islands*. Central Agricultural Research Institute.

Gopalakrishnan, T., Hasan, K. M., Haque, A. T. M. S., Jayasinghe, S. L., & Kumar, L. (2019). Sustainability of coastal agriculture under climate change. *Sustainability*, *11*, Article 7200.

Islam, M. T. (2014). Vegetation changes of Sundarbans based on Landsat imagery analysis between 1975 and 2006. *Landscape Environment*, *8*, 1–9.

Krishnan, P., Nair, K. M., Naidu, L. G. K., Srinivas, S., Koyal, A., Nasre, R. A., Ramesh, M., & Gajbhiye, K. S. (2004). Land, soil and land use of Lakshadweep coral islands. *Journal of the Indian Society of Soil Science*, *52*(3), 226–231.

Maji, B., Sarangi, S. K., Mandal, U. K., Burman, D., & Balasubramaniam, P. (2020). Climate change and management options for sustainable soil health and crop production: Eastern coast of India. *Advanced Agricultural Research & Technology Journal, IV*, *1*, 28–41.

Mandal, B., Dash, B., Jaison, M., Das, A., Ghosh, S., Mukherjee, S., Dey, A., Sahu, B., Singh, P., Deb, P., Choudhury, A., Basak, N., & Shovik, D. (2022). Soil quality assessment for coastal agroecosystem—Problems and perspectives. In T. D. Lama, D. Burman, U. K. Mandal, S. K. Sarangi, & H. S. Sen (Eds.), *Transforming coastal zone for sustainable food and income security* (pp. 607–620). Springer International Publishing AG.

Mandal, U. K., Bhardwaj, A. K., Lama, T. D., Nayak, D. B., Samui, A., Burman, D., Mahanta, K. K., Sarangi, S. K., Mandal, S., & Raut, S. (2021b). Net ecosystem exchange of carbon, greenhouse gases, and energy budget in coastal lowland double cropped rice ecology. *Soil & Tillage Research*, *212*, Article 105076

Mandal, U. K., Burman, D., Bhardwaj, A. K., Nayak, D. B., Samui, A., Mullick, S., Mahanta, K. K., Lama, T. D., Maji, B., Mandal, S., Raut, S., & Sarangi, S. K. (2019). Waterlogging and coastal salinity management through land shaping and cropping intensification in climatically vulnerable Indian Sundarbans. *Agricultural Water Management*, *216*, 12–26.

Mandal, U. K., Mandal, S., Lama, T. D., & Burman, D. (2021a). Innovative approaches for agriculture in coastal fragile ecosystem in India. In P. K. Ghosh, P. Kumar, D. Chakraborty, D. Mandal, & P. N. Sivalingam (Eds.), *Innovations in agriculture for a self-reliant India* (pp. 421–464). New India Publishing Agency.

Mandal, U. K., Nayak, D. B., Ghosh, A., Bhardwaj, A. K., Lama, T. D., Mahajan, G. R., Das, B., Nagaraja, M. S., Vittal, B. K., Rani, P. P., Mal, S., Samui, A.,

Mahanta, K. K., Mandal, S., Raut, S., & Burman, D. (2023). Delineation of saline soils in coastal India using satellite remote sensing. *Current Science, 125*(12), 1339–1353.

Mehvar, S., Filatova, T., Dastgheib, A., de, van Steveninck, E., R., & Ranasinghe, R. (2018). Quantifying economic value of coastal ecosystem services: A review. *Journal of Marine Science and Engineering, 6*, Article 5. http://dx.doi.org/10.3390/jmse6010005

Mitran, T., Mani, P. K., Basak, N., Mandal, B., & Mukhopadhyay, S. K. (2014). Soil fertility constraint assessment using spatial nutrient map at three selected villages of coastal Sundarbans. *Journal of Soil Salinity and Water Quality, 6*(1), 1–8.

Ravisankar, N., Ambast, S. K., & Srivastava, R. C. (2008). *Crop diversification through broad bed and furrow system in coastal regions.* Central Agricultural Research Institute.

Sahu, G. C., & Mishra, A. (2005). Soil of Orissa and its management. *Orissa Review, 11*(4), 56–60.

Sarangi, S.K., Maji, B., Sharma, P.C., Digar, S., Mahanta, K.K., Burman, D., Mandal, U.K., Mandal, S., & Mainuddin, M. 2021. Potato (*Solanum tuberosum* L.) cultivation by zero tillage and paddy straw mulching in the saline soils of the Ganges Delta. *Potato Research, 64* (2), 277–305.

Sharma, D. K., & Chaudhari, S. K. (2012). Agronomic research in salt affected soils of India: An overview. *Indian Journal of Agronomy, 57*(3rd IAC Special Issue), 175–185.

Sharma, D. K., & Singh, A. (2015). Salinity research in India: Achievements, challenges and future prospects. *Water and Energy International, 58*, 35–45.

Sharma D. K., Singh, A., & Sharma, P. C. (2016). Role of ICAR-CSSRI in sustainable management of salt-affected soils-Achievements, current trends and future perspectives. Fourth International Agronomy Congress, November 22–26, p. 91–103.

Singh, G. (2009). Salinity-related desertification and management strategies: Indian experience. *Land Degradation and Development, 20*, 367–385.

Srinivasan, R., Singh, S. K., Nayak, D. C., Hegde, R., & Ramesh, M. (2019). Estimation of soil loss by USLE model using remote sensing and GIS Techniques-A case study of Coastal Odisha, India. *Eurasian Journal of Soil Science, 8*(4), 321–328.

Subba Rao, A., Biswas, A. K., & Maji, B. (2011). Improved soil management practices for enhancing crop production and reducing soil degradation in coastal regions on India. *Journal of the Indian Society of Coastal Agricultural Research, 29*(1), 10–18.

Tripathi, S., Chakraborty, A., Chakrabarti, K., & Bandyopadhyay, B. K. (2007). Enzyme activities and microbial biomass in coastal soils of India. *Soil Biology and Biochemistry, 39*, 2840–2848.

Tripathi, S., Kumari, S., Chakraborty, A., Gupta, A., Chakrabarti, K., & Bandyopadhyay, B. K. (2006). Microbial biomass and its activities in salt-affected coastal soils. *Biology and Fertility of Soils*, *42*, 273–327.

Velayutham, M., Sarkar, D., Reddy, R. S., Natarajan, A., Krishan, P., Shiva Prasad, C. R., Challa, O., Harindranath, C. S., Shyampura, R. L., Sharma, J. P., & Bhattacharyya, T. (1999). Soil resource and their potentialities in coastal areas of India. *Journal of the Indian Society of Coastal Agricultural Research*, *17*, 29–47.

Velmurugan, A., Swarnam, T. P., & Lal, R. (2015). Effect of land shaping on soil properties and crop yield in tsunami inundated coastal soils of Southern Andaman Island. *Agriculture, Ecosystems and Environment*, *206*, 1–9.

Yadav, J. S. P. (2004). Coastal agricultural research-present status and future perspective with special reference to value addition. *Journal of the Indian Society of Coastal Agricultural Research*, *22*(1&2), 1–8.

7

Soil Health and Its Quality in Southern Peninsular India

K. S. Anil Kumar, M. Lalitha, K. M. Nair, Rajendra Hegde, and B. S. Dwivedi

Chapter Overview

Soil health and quality are critical factors influencing agricultural practices in southern peninsular India, particularly in the humid tropics encompassing states like Kerala, Tamil Nadu, Karnataka, and Goa. The highly weathered laterite soils of this region, characterized by low cation exchange capacity and high phosphorus-fixing ability, often face challenges such as acidification. A comprehensive soil fertility assessment in Kerala revealed that about 90% of soils are affected by acidity, significantly impacting crop productivity. Additionally, the degradation of soil quality due to the extensive use of chemical fertilizers has led to both nutrient imbalances and deficiencies, emphasizing the need for sustainable agricultural practices to improve soil conditions and maintain productivity. To address these soil health issues, proactive measures such as establishing soil quality monitoring sites have been initiated, particularly in rubber and coffee growing areas. These monitoring efforts aim to evaluate soil properties and inform management practices that enhance soil fertility and resilience. The findings underscore the importance of integrating agronomic practices that not only improve soil quality but also support ecosystem functions, ensuring the long-term sustainability of agricultural systems in the region. Effective soil management is crucial for optimizing crop yields, enhancing nutrient cycling, and maintaining overall environmental health.

Soil Health Series: Volume 4 Soil Health and Sustainability in India, First Edition.
Edited by Ranjan Bhattacharyya.

Introduction

Healthy and productive soil is essential for feeding the growing world population; therefore, soil health management is crucial for all agricultural systems. Soil quality or health is an approach for linking soil functions and management goals. Soil quality is the capacity of a specific type of soil to function, within natural and managed limits, to support plant and animal growth, improve water and air quality, and enhance human health and living conditions (Karlen et al., 1997). This concept has led to nutrient critical limits and definitions for optimal soil health and sustainability. Soil health is expressed as a dynamic balance between plants and animals and their agroecological regions that enables the metabolic processes of plants to thrive without any impedance, stress, or disruption from the animals. The concept of a perfectly healthy soil is difficult to articulate, but the impaired functioning of the soil is easily identifiable and quantifiable as limitations. Hence, Carter et al. (1997) noted that soil health could be viewed as the status of the soil at a specific moment, reflecting the dynamic soil properties that change in the short term due to management factors, whereas soil quality can be seen as the soil's suitability for a specific purpose over a long period, reflecting intrinsic or static soil quality. The Natural Resources Conservation Service emphasizes the four key principles of managing healthy soil: maximizing living roots, soil cover, and biodiversity and minimizing soil disturbance.

Our Perspective on Soil Health and Quality

Healthy soil not only aids in crop production but also plays a crucial role in five essential functions (Table 7.1): regulating water, sustaining plant and animal life, filtering and buffering potential pollutants, cycling nutrients, and providing physical stability and support. Given that healthy soils are essential for sustainable agriculture, the connection between soil and human health is significant, and effective soil management strategies are thus necessary. However, nutrient imbalances are a common issue in intensive agriculture, leading to deficiencies in secondary nutrients and micronutrients, a lower water table, reduced water quality, and overall deterioration of soil health (Department of Agriculture & Cooperation, Government of India [DAC, GOI], 2013).

In response to such problems, the government of India introduced the "National Mission for Sustainable Agriculture" scheme to promote natural resource conservation through comprehensive soil health management and improved water utilization practices, aiming to enhance agricultural productivity, sustainability, and climate resilience (DAC, GOI, 2014). Similarly, developed countries have also prioritized soil health, as exemplified by the U.S. Department of Agriculture's

Table 7.1 Five Essential Functions of Soil and Their Importance.

Key essential functions	Importance
Regulating water	Soil helps to control water flow and purification.
Sustaining plant and animal life	The diversity and productivity of the living things on earth depend on the soil.
Filtering and buffering potential pollutants	Minerals and microbes in soil are responsible for filtering, buffering, degrading, immobilizing, and detoxifying organic and inorganic material, including industrial and municipal by-products and atmospheric deposits.
Cycling nutrients	Soil facilitates storing, transforming, and cycling the major, secondary, micro, and other nutrients in the soil.
Providing physical stability and support	Soil structure supports root growth and human structures.

(USDA's) initiative "Unlock the Secrets in the Soil," which highlights the crucial role of healthy soil in sustaining productivity and promoting a healthy life.

Soil quality and soil health are primarily assessed by evaluating critical soil functions within the ecosystem that are or are not being adequately performed, or by identifying the ecosystem services for which specific soil resources could be utilized. For evaluating the dynamic and inherent characteristics of soil, assessment tools, test kits, minimum data sets, and weightage scores have been established for many physical, chemical, and biological quality indicators. Establishing a universal set of indicators for soil quality is a challenging task, however, as the indicators vary across different agro-ecological regions. The key characteristics associated with healthy tropical soil are provided in Table 7.2.

With this background, this chapter focuses on soil quality and health appraisal of southern peninsular India, emphasizing both inherent and dynamic soil quality parameters.

Southern Peninsular India and Its Geographical Settings

The Indian subcontinent covers a variety of geological assemblages from Archaean to recent. The evolutionary history of peninsular India is complex, as reflected in its tectonic structures, topographic features, and climatic zones. In southern peninsular India, for instance, the Ghats and coastlines of the west and east are distinct and intriguing. Rivers originating not far from the west coast travel

Table 7.2 Characteristics of Soil Health and Quality in Tropical Regions[a].

Characteristic	Description
Good soil tilth	Soil tilth refers to the overall physical characteristics of the soil in the context of its suitability for crop production.
Sufficient depth	Sufficient depth refers to the extent of the soil profile to which roots can grow and function. A soil with a shallow depth because of past erosion is more susceptible to extreme fluctuations in the weather, thus predisposing a crop to drought or flooding stress.
Good soil drainage	Even after heavy rain, healthy soil drains more rapidly than unhealthy soil due to good soil texture and structure, gravel content, and optimum bulk density. Healthy soil also retains adequate water for plant uptake.
Organic matter	Soil organic matter (SOM) is vital for maintaining the vast population of soil microbes, which are key to soil fertility. SOM serves as the primary source of energy for microorganisms. SOM also affects soil physical properties, including moisture and nutrient retention characteristics. Due to rapid mineralization under favorable climatic conditions, tropical soils are often poor in organic matter. Maintenance of adequate SOM levels, especially in the labile carbon pool, is necessary for enhancing microbial activities and sustaining soil health.
Soil reaction (pH) and electrical conductivity (EC)	The availability of plant nutrients is greatly affected by soil pH. Most plant nutrients are available in the pH range of 6.5 to 7.3. Soil pH also influences microbial activity. Soil acidity management through liming is essential for maintaining soil fertility. EC measures the type and quantity of salts in the soil and is an essential indicator of soil health. It affects crop yields, suitability, plant nutrient availability, and the activity of soil microorganisms. Soils with EC less than 2 dS m^{-1} are considered non-saline and are not harmful to most crops and microbial processes.
Sufficient but not excessive supply of nutrients	An adequate and accessible supply of nutrients is necessary for optimal plant growth and balanced nutrient cycling within the system. Excess nutrients can lead to leaching, potential groundwater pollution, high nutrient runoff and greenhouse gas losses, and toxicity to plants and microbial communities.
Large population of beneficial organisms	Soil microbes are essential to the functioning of the soil. They help nutrient cycling, decomposition of organic matter, maintenance of soil structure, and biological suppression of plant pests, among other functions. Healthy soil has a high and diverse population of beneficial organisms to carry out these functions and thus help maintain a healthy soil status.

[a] Source: Venugopal, V. K et al., 2019/Department of Agriculture Development and Farmer's Welfare.

thousands of kilometers to the east coast. The unique geomorphic evolution, drainage patterns, basin development, river systems, and coastal deltas, combined with the region's climatic diversity, have resulted in diverse soil and vegetation types. For example, the soils in Kerala are situated in the humid tropical zone with high levels of rainfall and temperatures that are conducive to laterization, an intense form of rock weathering, and are predominantly acidic. The inherent soil quality in the rest of the southern peninsular region likewise differs according to geology and climate. The response to agronomic management depends on soil health and its inherent soil quality. Southern peninsular India is divided into humid and semi-arid regions based on the prevailing climate despite geology, land use, and soil.

The major soil orders in southern peninsular India include Inceptisols, Vertisols, and Alfisols. The major soil suborders are Haplustepts, Haplusterts, Paleustalfs, Haplustalfs, Ustorthents, Rhodustalfs, Haplocambids, Haplohumults, Kandihumults, Palehumults, and Kandiustults (Figure 7.1).

Productivity Constraints

The soil constraints for crop production primarily depend on prevailing climatic situations, landform occurrence, land use, and management practices. The related factors include slope tendency; rainfall probability, intensity, and distribution; and length of growing period.

Generally, the significant constraints of Inceptisols for crop production include effective root depth; the limited volume of root anchorage, water, and nutrient holding capacity; and soil erosion, particularly in soils on sloping landforms. The soil in alluvial or low-lying areas has drainage limitations, salinity/sodicity, waterlogging, and restricted land use. The presence of sulfuric properties, duripan, calcic and petro calcic horizons, and fragipan influence the physical and chemical constraints for crop growth.

Vertisols have constraints due to imperfect to poor drainage caused by high shrinking swelling clays. The soil workability is limited when the soil is at field capacity. The poor leaching environment accumulates salt in the soil system, causing subsurface salinity and sodicity. The soils are rich in basic cations. In some soils, the high pH and presence of $CaCO_3$ restrict the availability of phosphorus and iron. In addition, the sodic vertisols have poor hydraulic conductivity.

Soil Health and Quality in the Humid Tropics

The humid tropics of southern India encompass Kerala and parts of Tamil Nadu, Karnataka, and Goa. These areas experience high rainfall and temperatures. These conditions are ideal for the laterization process, which involves the depletion

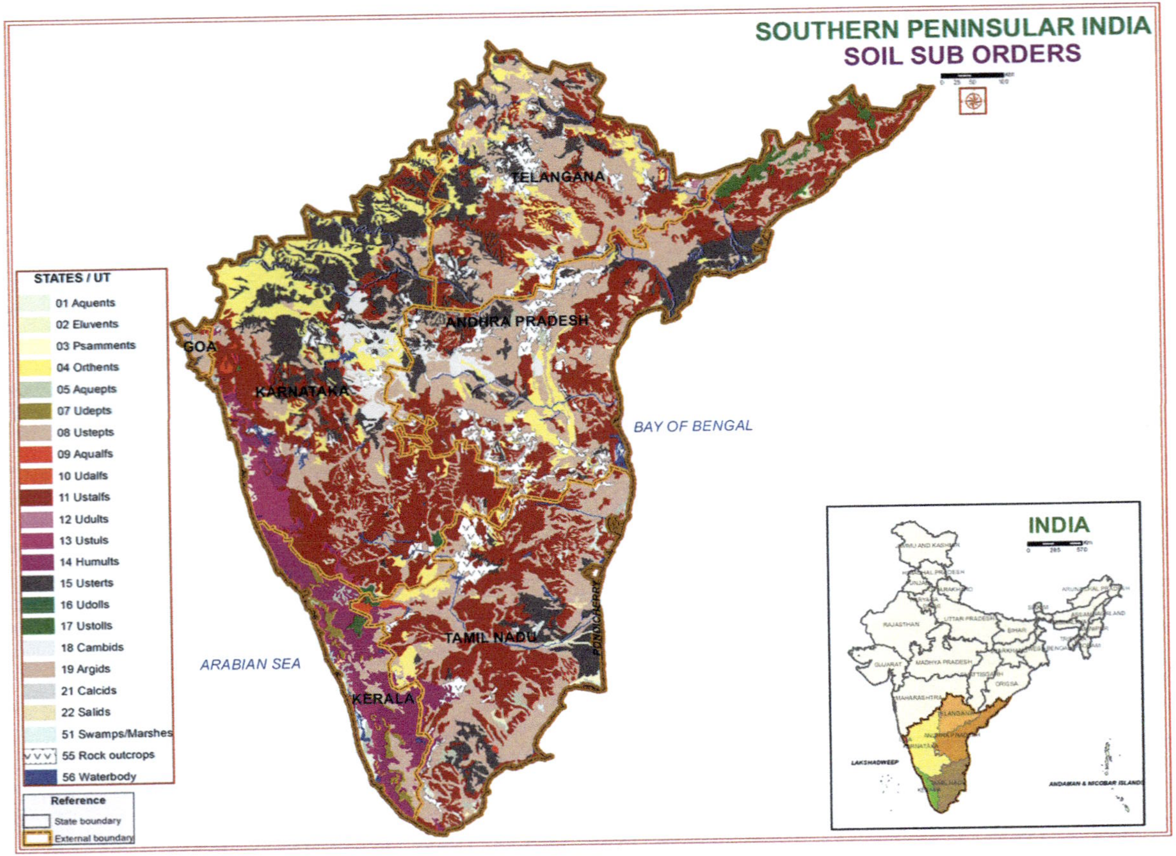

Figure 7.1 Major soil suborders of southern peninsular India.

of soluble substances, such as alkaline and silica-rich components from soils and rocks, and enrichment with insoluble substances, such as hydrated iron and aluminum oxides. Consequently, the soils are predominantly highly weathered, acidic, gravelly, kaolinitic, and lateritic, with low cation exchange capacity (CEC) and water holding capacity, as well as high phosphorus fixing capacity. They are naturally deficient in soil organic carbon (SOC), bases, and other nutrients, but they nonetheless respond well to agronomic practices (Venugopal et al., 2019). To enhance the sustainability of land management decisions, Nair et al. (2016) and Anil Kumar et al. (2016) established soil quality monitoring sites in coffee and rubber cultivation areas of southern India.

Kerala Soil Fertility Characteristics

After observing the land-use changes and the decreases in crop production and productivity, the state planning board of Kerala initiated a project for the systematic evaluation of soil fertility and soil-related constraints to crop production in the state. Approximately two lakhs (200,000) surface samples were gathered from all regions of the state and examined for all nutrient properties (Figure 7.2). This comprehensive soil fertility assessment revealed the types, extent, and severity of soil-related constraints to crop production across the entire state. The discoveries included the following:

- Soil acidification has reached a critical level, negatively affecting the productivity of most crop plants. Ninety percent of the soils are affected by acidity, with 50% being strongly to extremely acidic. Immediate action is required to ensure the regular application of liming materials to neutralize the soils and enhance their productivity.
- Uncontrolled use of phosphate fertilizers over an extended period has led to high phosphorus levels in Kerala's soils. The excessive nutrient levels in the soil hinder the uptake of other essential nutrients by plants. The application rates of this nutrient could be reduced by at least half.
- There is widespread deficiency of secondary nutrients, calcium, and magnesium. Addressing soil acidity through liming will help alleviate calcium deficiency. Regular application of $MgSO_4$ at $80\,kg\,ha^{-1}$ is crucial to maintain sufficient magnesium levels in the soil.
- Among the examined micronutrients, boron deficiency stands out as significant and extensive, necessitating immediate action. The deficiency of boron could be rectified by applying borax at a rate of $10\,kg\,ha^{-1}$.

The study highlighted that the situation is severe, extensive, and debilitating, while also showing that soil fertility-related constraints were limited to four significant issues, with simple solutions. Taking affirmative action to address

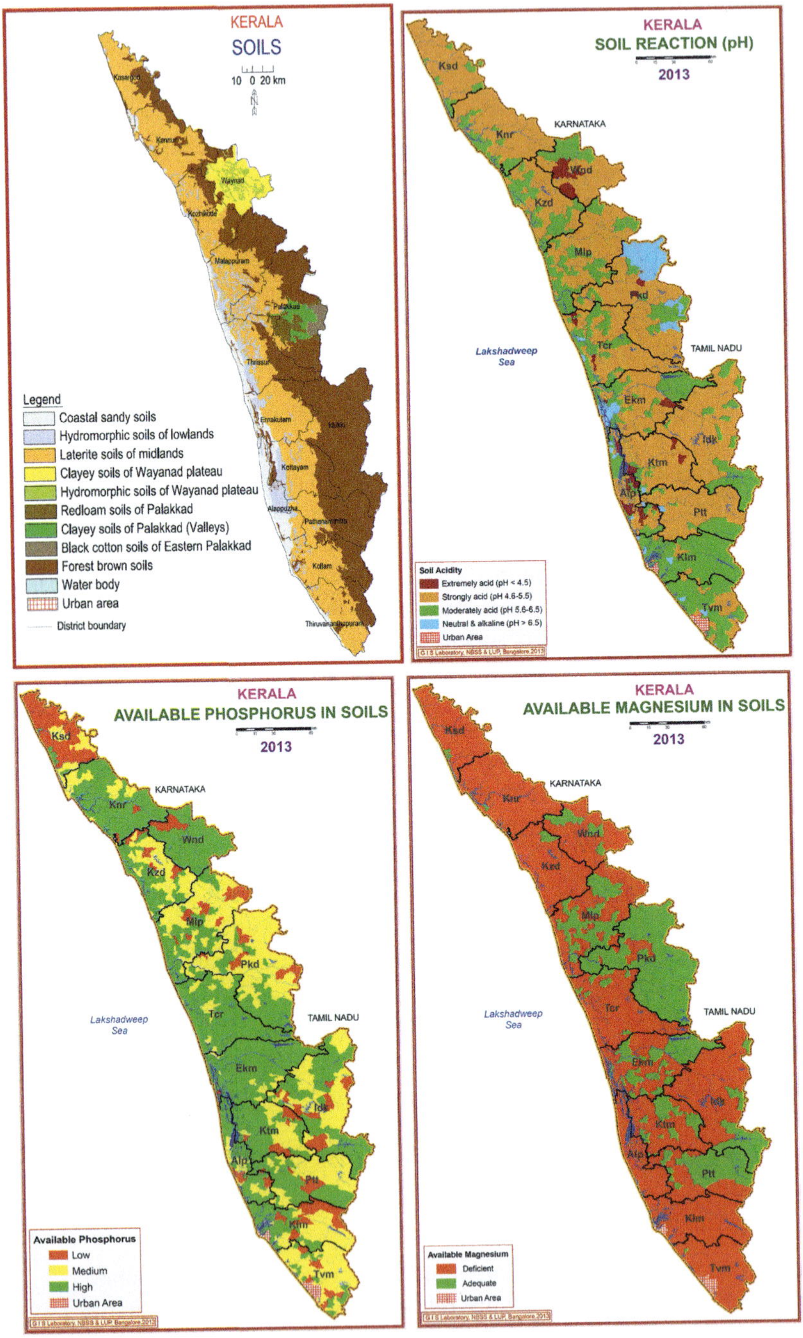

Figure 7.2 Soil fertility status map of Kerala.

these constraints could greatly enhance the productivity and profitability of all crop production systems in the state. Immediate attention is required for the reorganization of soil testing services, incorporating secondary and micro-nutrients in soil test programs, and conducting focused extension activities to raise awareness among farmers about soil-related issues (Rajasekharan et al., 2013).

Soil Quality Monitoring Systems

Soil quality monitoring systems involve developing monitoring sites at which soil qualities are assessed at regular intervals. The systems not only reveal the basic properties of soils but also provide information on changes in these properties.

Soil Quality Monitoring Sites Within Traditional Coffee-Growing Areas of Southern India

Managed land-use systems, such as those related to coffee production, involve significant anthropogenic activity and inputs and outputs of energy and materials. The external inputs are mainly soil conditioners to make the medium more suited for coffee and inputs of plant nutrients to enhance economic productivity. The output from coffee plantations is primarily the coffee beans. The partial loss of biodiversity, external inputs, the output of products, and anthropogenic activities are bound to influence plantation systems and the immediate environment.

Monitoring sites were also established at a few forest sites adjoining the coffee plantations to assess the changes in soil qualities due to the conversion of forests to plantations. The locations of the monitoring sites are based on climate and soil variability in coffee-growing tracts of three states. Sixty soil-monitoring sites have been established for coffee and forest sites together (46 for coffee, 14 for forest; Figure 7.3). The distribution of sites in these states (Kerala, Tamil Nadu, and Karnataka) roughly followed the proportion of plantation areas in the states. In addition to surface soil qualities being assessed in coffee-growing areas, soil profiles were studied at each monitoring site.

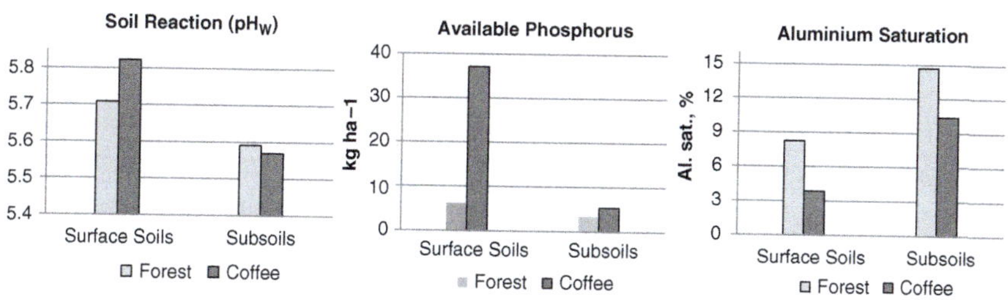

Figure 7.3 Conversion of forests to coffee: consequences for soil pH, phosphorus, and Aluminium saturation.

Soil acidity declined in the surface soils of coffee plantations compared with forest soils (Figure 7.4). This increase in pH is likely due to the regular application of lime to coffee soils. Lime applied to soils seldom moves down due to the insolubility of calcium and magnesium carbonates. Therefore, the pH of subsoils does not change significantly.

Available phosphorous in surface soil ranged from 15 to 52 $kg\,ha^{-1}$ in coffee-growing soils, as assessed through analysis of samples from soil quality monitoring sites (Figure 7.3). Phosphorous accumulates in coffee-growing soils because of regular inputs of phosphatic fertilizers; however, high levels are confined to the surface soils, with negligible movement into subsoil layers.

Overall, there was no significant decline in soil qualities because of the conversion of forests to coffee plantations. Favorable changes were observed for soil reaction and nutrient availability for higher plants and soil-inhabiting organisms. The significant build-up of phosphorous was the only cause for concern, and this concern could translate to an opportunity to reduce the inputs of costly phosphatic fertilizers (Nair et al., 2016).

Soil Quality Monitoring Sites Within Traditional Rubber-Growing Areas of Southern India

When the rubber plantations were established on the West Coast of southern India at the beginning of 1990, the climate and soil qualities of the area were conducive to satisfactory productivity. Soil qualities play a significant role in the sustained productivity of any crop, particularly in long-duration plantation production systems. Hence, 121 soil quality monitoring sites have been established for rubber-growing areas, including forest sites near rubber plantation sites (Figure 7.4).

Soil data sets from 70 rubber soil quality monitoring sites (where plantations were established by forest clearing) and 21 forest soil quality monitoring sites were analyzed to evaluate changes in soil quality following the conversion of forests to the managed system of rubber plantations. The distribution of sites in these states roughly reflected the proportion of plantation areas within the respective agro-ecological units in the states (Anil Kumar et al., 2016).

The highly weathered soils of rubber-growing areas are often described as laterite soils. These soils are developed through the laterization process and often contain plinthite or laterite fragments. They are naturally acidic, base depleted, and dominated by low-activity clay minerals of kaolinite type and iron and aluminium oxides. Although the general term "laterite soils" distinguishes rubber-growing highly weathered soils from other kinds of soils, it fails to reflect the finer variability of soil properties within the large spatial extent of rubber-growing areas (Figure 7.5). To evaluate the soil variability across the rubber-growing areas of three states, we classified the soils according to the international system of soil

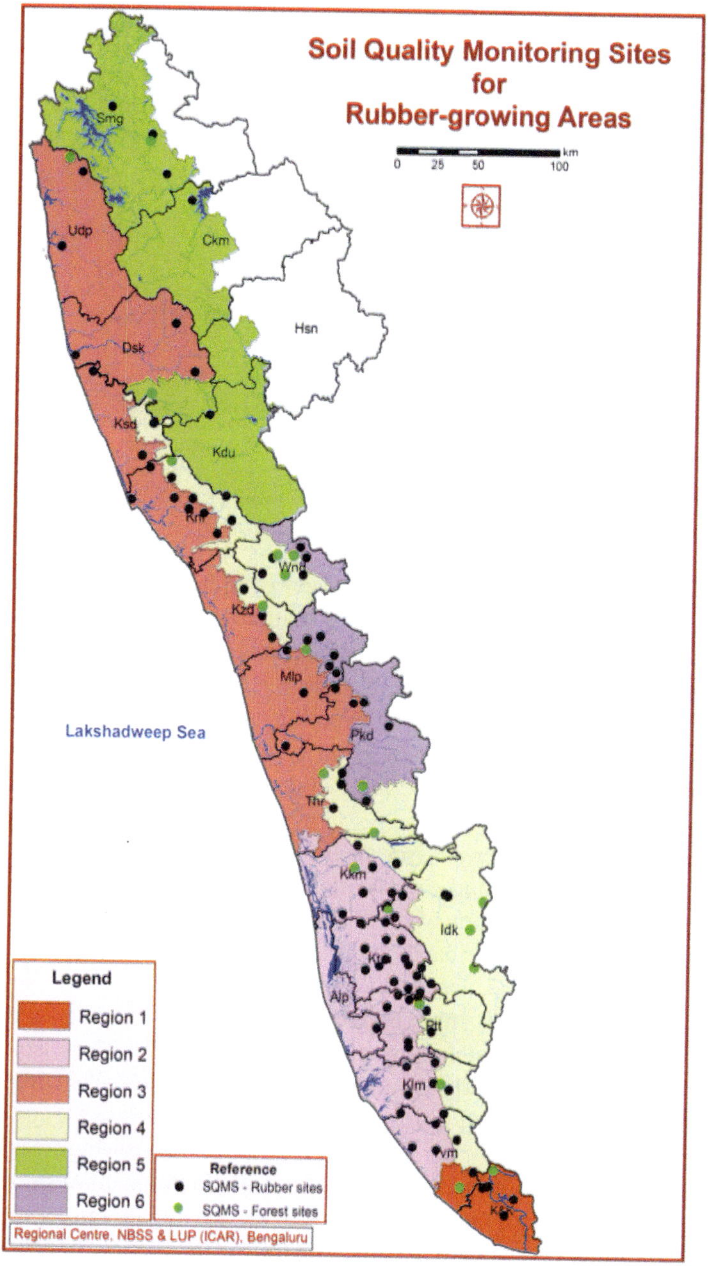

Figure 7.4 Soil quality monitoring sites (SQMS) established in rubber-growing areas of southern India.

Figure 7.5 Major soils identified in rubber-growing areas of southern India.

classification, Soil Taxonomy of the USDA. *Hevea* evolved in hot, humid tropical climates (native to tropical Brazil) in soils with low natural fertility. This history explains their adaptation to the hot, humid tropical climate of the west coast of South India and satisfactory productivity there. However, the low inherent fertility of highly weathered tropical soils, mono-crop production system, and removal of the large quantity of biomass at the cessation of the productive period of the trees (around 35 years) necessitate periodic evaluation and monitoring of soil quality.

Systematic study of soils of rubber-growing areas for their morphological, physical, and chemical characteristics and classification according to the USDA soil taxonomy resulted in the identification of six distinct taxa of soils at the subgroup level: Ustic Kandihumults, Ustic Kanhaplohumults, Typic Plinthohumults, Ustic Palehumults, Ustic Haplohumults, and Pachic Paleustolls. The low-activity clay soils of Kandihumults, spatially interspersed with Kanhaplohumults and Plinthohumults, dominate the rubber-growing areas. The other three soil types, Ustic Palehumults, Ustic Haplohumults, and Pachic Paleustolls, cover only limited areas (Anil Kumar et al., 2016).

Ustic Kandihumults are the most extensive type of soil in rubber-growing areas, Kanniyakumari to Malanad in Karnataka, the coast to the foothills of Western Ghats, and to a limited extent, in highland plateaus. Ustic Kanhaplohumults are also low-base, high organic carbon (OC)-containing soils with a kandic horizon (low activity clay) and ustic soil moisture regime. The significant differences are the root limiting layer within 150 cm of mineral soil surface and the decline in clay from the maximum reached in the kandic horizon by as much as size (relative) within the solum. Typic Plinthohumults, interspersed with Kandihumults and Kanhaplohumults, in rubber-growing areas of coastal uplands and midlands, are characterized by argillic or kandic horizons. These soils are base depleted and rich in organic matter, with plinthite forming a continuous layer within 150 cm from the mineral soil surface. Plinthite is an iron-rich, humus-poor mixture of quartz, clay, and other minerals. It typically appears as dark red redox concentrations of sesquioxides with platy or reticulate patterns. Plinthite undergoes irreversible changes to ironstone or irregular aggregates when exposed to wetting and drying (Anil Kumar et al., 2016). The presence of plinthite in the lowlands of the lateritic region also aids in resisting landslips during heavy rainfall, as it consists of a hardened mix of iron-bearing minerals with a honeycomb structure and voids filled with kaolinite, holding up against supersaturation until reaching the critical limit (Lalitha et al., 2021c).

The OC content of rubber-growing soils is similar to that of the forest owing to near-zero tillage in the plantations. The external inputs of nitrogenous fertilizers coupled with the negation of liming has led to the acidification of rubber-growing soils to unacceptable levels. The very strong acidity has in turn led to very high levels of aluminium saturation in the exchange complex of the soils and depletion of secondary plant nutrients calcium and magnesium, especially in the plantation areas of Central Kerala (Figure 7.6). As part of fertilizer inputs, the cation potassium is not deficient. The rubber-growing soils do not have any significant deficiency of micronutrients, with the exception of zinc.

The evaluation of the consequences of converting forests to rubber plantations revealed a mix of positive and negative impacts on the environment and soils. On the positive side are the slightly higher content of OC, the total stock of OC in soils in rubber plantations relative to forest soils, and rubber plantations and natural forest ecosystems having equally good protection of soils from erosion. On the negative side are the loss of biodiversity, attenuation of soil dryness, soil acidification, and depletion of basic cations from soils in rubber plantations. Strongly acid soils depleted of basic cations are stressed environments for the macro- and microflora

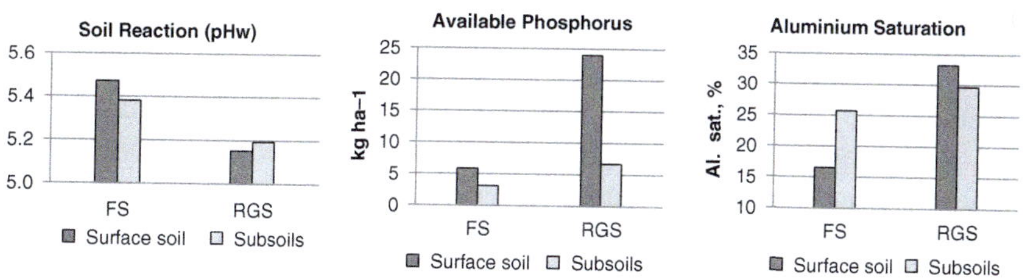

Figure 7.6 Conversion of forests to rubber plantations: consequences for soil pH, phosphorus, and aluminium Saturation.

in soil, and such soils constrain biological activity, which is essential for the fundamental processes of rapid mineralization and turnover of organic matter in soils of humid tropics.

Soil Health and Quality in the Semi-arid Tropics

The changes in soil health and quality, particularly in semi-arid tropics, are more prominent due to the rate of physical and chemical weathering triggered by the high temperature and precipitation. The rate of change is minimal for inherent soil quality parameters, whereas the dynamic properties change according to soil management practices. The soil quality parameters or indicators are sensitive to soil functions. Vasu et al. (2016b) assessed the soil quality indicators of the southern semi-arid Deccan plateau and identified the five minimum data sets, such as silt, clay, base saturation, sodium absorption ratio, and OC. The study used surface and profile sample parameters to arrive at the minimum data set since the soil functions are controlled by pedogenic processes operating in the soil control section. Similarly, Mandal et al. (2011) stressed the importance of resource management amid declining productivity in rainfed and irrigated agriculture of semi-arid tropical regions of India. The study in the Nalgonda district of Andhra Pradesh prioritized soil CEC, exchangeable sodium percentage, plant diethylenetriamine-pentaacetic acid (DTPA)-extractable zinc, phosphorus, water, and dehydrogenase activity as indicators for assessing the soil quality.

Vijaya Kumar et al. (2013) reported that the OC was as low as $<5\,g\,kg^{-1}$ in 72% of the samples, and dehydrogenase activity ranged from 4.50 to 31.90 μg of triphenylformazan per gram of soil per day from salt-affected soils of Andhra Pradesh. The low value of OC might be attributed to the high decomposition rate and mineralization rate under a semi-arid climate (Lalitha & Kumar 2016) and or the solubilization of OC under an alkaline environment. The low substrate content is the most likely reason for low dehydrogenase activity in the soil. The available phosphorus and iron content was also found to be low in salt-affected soil due to the precipitation of Fe^{+} and P by calcium carbonate.

In another study, Mandal et al. (2011) found that nitrogen, phosphorus, potassium, DTPA-extractable zinc, CEC, OC, microbial biomass carbon, carbon mineralization, total nitrogen, available water capacity, bulk density, clay content, pH, and exchangeable sodium percentage are crucial indicators for soil nutrient cycling, water availability, resilience, and saline-sodic characteristics (Table 7.3). The research also indicated that soils with a higher soil quality index were more productive and exhibited greater resilience. Maintaining a soil organic stock of $8.6\,Mg\,ha^{-1}$ per 15 cm was deemed essential for preserving soil quality, with an annual addition of around $2.2\,Mg\,ha^{-1}$ organic matter required

Table 7.3 Minimum Data Sets Identified for Assessing Soil Quality in Andhra Pradesh, India[a].

Alfisols	Inceptisols	Vertisols and Vertic Inceptisols
Soil moisture at 15 bar tension, available N, exchangeable sodium percentage, C-mineralization, available P, DTPA-extractable Zn, bulk density	Soil moisture at 1/3 bar tension, pH, mineralizable N, available P, C-mineralization, available N, and bulk density	Soil moisture at 15 bar tension, CEC, organic C, exchangeable sodium percentage, mean weight diameter of soil aggregates, mineralizable N, and C-mineralization

[a] *Source:* Mandal et al. (2011).

to achieve and sustain this SOC stock (Mandal et al., 2017). In particular, the degradation of Alfisols in semi-arid tropics is attributed to physical, chemical, and biological constraints. A study by Sharma et al. (2008) at ICRISAT's long-term experiment on nutrients and tillage under a sorghum-mung bean cropping system highlighted that oxidizable nitrogen, DTPA-extractable zinc, copper, microbial biomass carbon, soil aggregates, and hydraulic conductivity are key indicators of soil quality in Alfisols. Similarly, the fertility capability classification of Alfisols in Kangayam grasslands, Tamil Nadu, for surface (0–15 cm) and subsoils (15–50 cm) identified significant soil constraints such as dry soil moisture, low CEC, basic reaction, alkalinity, and gravel content (Lalitha et al., 2018b).

High evapotranspiration also aids the development of subsoil sodicity in the semi-arid tropics of southern India. Studies revealed that prevailing high aridity (102.5) in northern parts of Tamil Nadu abets the precipitation of calcium carbonate and magnesium carbonate from calcium and magnesium ions present in the soil solutions, which results in the sodium ion dominating the exchange complex with concomitant development of subsoil sodicity. It is prominent in Vertisols and associated Vertic intergrades with high clay content (Lalitha et al., 2018a; Kalaiselvi et al., 2020). The development of subsoil sodicity also impairs the soil hydraulic conductivity of soils of heavy texture in the southern Deccan plateau (Vasu et al., 2018).

The soil resource inventory of Goa indicates that the dominant soils of the state are brown soils on hills and ridges (28.9%), followed by deep alluvial clay soils of the lowlands (22.5%). Although the state has sufficient forest cover, most areas are subject to water erosion due to gradient slope, and lateritic soil accelerates soil erosion and prevents percolation of water into the subsoil. Analysis of data for soil erosion revealed that the state has severe (20–40 $t\,ha^{-1}\,yr^{-1}$) and very severe (40–80 $t\,ha^{-1}\,yr^{-1}$) potential soil loss in 26.9% and 22% of its area, respectively, followed by slight potential soil loss (5–10 $t\,ha^{-1}\,yr^{-1}$) elsewhere (Obi Reddy et al., 2013).

Multinutrient Deficiencies

In addition to single-nutrient deficiency, multinutrient deficiency is more common in semi-arid tropical and rainfed regions. Rainfed cultivation in the semi-arid tropics of India entirely depends on the onset of the monsoon. Uncertain and uneven distribution of rainfall, irrigation water scarcity, and poor-quality irrigation water (wherever it is found sufficient) hinder cropping patterns and associated crop management and possibly also undermine soil health. Hence, agronomic management is meager, and nutrient management is particularly difficult. The chances of crop failure are high under uncertain climates, and even if a crop thrives with minimum rainfall, productivity may still be low. The low productivity in rainfed regions is believed to have less nutrient mining in comparison with irrigated regions with high cropping intensity. Thus, the application of macro- and micronutrients in rainfed regions usually ranges from low to nil categories.

A survey conducted by ICRISAT on soil fertility assessment in different states under semi-arid tropics revealed multinutrient deficiencies of sulfur, boron, and zinc in the following districts of Andhra Pradesh: Nalgonda (86%, 93%, and 73%), Mahabubnagar (73%, 94%, and 62%), Kurnool (88%, 83%, and 94%), Ananthapur (95%, 100%, and 86%), and Prakasam (94%, 70%, and 88%). Likewise, samples were highly deficient in sulfur, boron, and zinc in the districts of Tirunelveli (97%, 97%, and 72%), Salem (60%, 100%, and 45%), Kanchipuram (70%, 85%, and 55%), Vellore (55%, 90%, and 65%), and Karur (50%, 65%, 60%) in Tamil Nadu. Sulfur, boron, and zinc deficiencies were also reported to be high in Kolar (87%, 90%, and 64%), Tumkur (93%, 96%, and 88%), Chitradurga (82%, 75%, and 93%), Haveri (81%, 635, and 79%), and Dharwad (83%, 54%, and 34%) districts of Karnataka, respectively. The extent and severity of widespread multinutrient deficiencies in different rainfed regions of semi-arid tropics highlight that rainfed regions are not only thirsty but also hungry (Sahrawat et al., 2007; Figure 7.7).

Food grain demand can be met by expanding cultivated lands or improving cultivation practices. However, both approaches require consistent monitoring of soil fertility to prevent nutrient deficiencies, a key factor limiting crop yield. This monitoring is essential for farmers and policymakers to manage soil nutrients effectively. Soil fertility in rainfed areas of arid and semi-arid regions has become more complex due to widespread nutrient deficiencies exacerbated by climate change (Shukla et al., 2019). The watershed management program, implemented in Karnataka, assessed soil fertility in rainfed areas of the region. Around 30 000 geo-referenced surface soil samples at grid intervals were collected for soil fertility

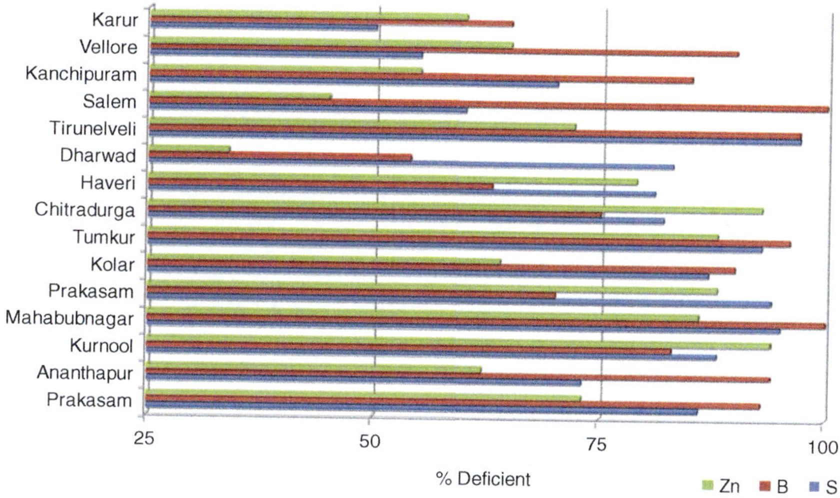

Figure 7.7 Multinutrient deficiency in soils of semi-arid tropics of southern peninsular India (Sahrawat et al., 2007).

evaluation. The finding pinpointed widespread deficiencies of soil available zinc (80%), phosphorus (60%), boron (60%), sulfur (50%), and iron (40%). Deficiencies of available phosphorus, sulfur, boron, iron, and zinc were evident in areas where fertilizer application was insufficient to meet crop requirements, and where little attention was paid to micronutrients and land reclamation. The SOC strongly influences the availability of copper and manganese, while soil pH strongly affects the availability of available phosphorus, potassium, iron, and zinc. Therefore, while nutrient concentrations may vary based on the parent material, their availability to plants is primarily influenced by soil properties such as pH, organic matter, texture, moisture, and clay mineralogy. Soil electrical conductivity also plays a significant role in the availability of sulfur and boron. Thus, it is crucial to apply secondary micronutrients, along with major nutrients according to crop requirements and nutrient availability, to enhance nutrient use efficiency and factor productivity in rainfed regions (Lalitha et al., 2021a; Figure 7.8).

A recent study by Vasu et al. (2017) in the semi-arid tropics of Mahabubnager district of Telangana state assessed both active factors (soil chemical properties) and passive factors (topographic variables) on soil health through large-scale spatial variability mapping. The semivariogram analysis revealed that the weak relationship between soil nutrients (iron, copper, manganese, zinc, and boron) and topography and the spatial variance of soil nutrients that occurred might be due to management practices. Results also revealed that inevitable deficiencies in multiple nutrients, particularly zinc and boron, were invariably present.

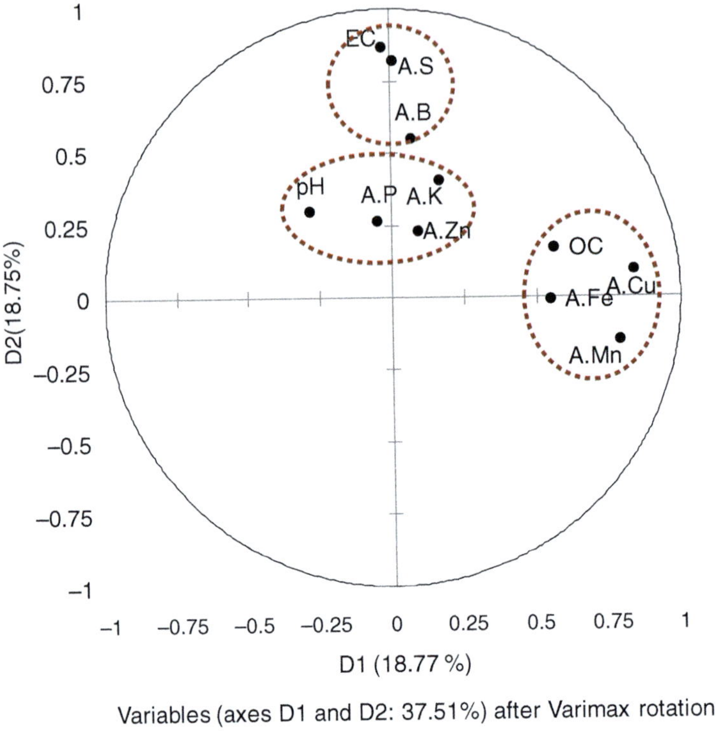

Figure 7.8 Factor analysis representing the relationships between soil nutrient parameters.

Irrigation water containing a high amount of sodium increases the soil pH and thereby affects nutrient availability, particularly micronutrients such as manganese and zinc in cotton-growing soils of Thimmajipet, Telangana (Vasu et al., 2016a). The same was reported in Cauvery command areas of Tamil Nadu, where the high sodium in irrigation water was found to impair the soil hydraulic properties by saturating the exchange surface with sodium cations (Lalitha et al., 2021b).

Red Ferruginous Soils of Southern India

Soils termed red, brown, yellow, laterite, lateritic, ferralitic, and latosols are collectively referred to as red ferruginous soils. They are mainly 2.5 YR to 7.5 YR color and have a poor base status, a highly to moderately acidic condition, and a stable structure influenced by iron oxides (Pal et al., 2000). These soils are distributed throughout southern India across varying climates and land uses from agriculture, forestry, and horticulture, including production of spices and cash crops. They are mainly classified under Alfisols, Ultisols, Mollisols, and Inceptisols in southern peninsular India. Except for Inceptisols, these soils have a clay-enriched B-horizon and are generally slightly acidic. Although they are in an acidic situation, Alfisols and Mollisols maintain sufficient base saturation because of underlying parent materials such as limestone and calc-gneiss constantly supply basic cations to the soil system. The presence of kaolinite alongside smectite in red ferruginous soils of India under slight to alkaline conditions of present-day soils does not favor the transformation of smectite to kaolinite, indicating such transformation is the result of high rainfall and more significant temperature fluctuations in the earlier geological period (Pal et al., 2014). The intermittent product of dioctahedral smectite's stability was ephemeral in humid tropical climates, as indicated by its transformation into kaolinite. Similarly, trioctahedral smectite is formed from sand and silt biotite inherited from earlier humid climates (Chandran et al., 2006). The formation of pedogenic carbonate and the concomitant development of subsoil sodicity indicates the prevailing semi-arid tropical climate, which may cause natural chemical degradation and impair drainage in the long run (Srivastava et al., 2001).

Soil acidity in humid tropics develops on soil colloids mainly through isomorphous substitution or the formation of polymers with soil organic matter. Studies on soils of Kerala, Goa, parts of Tamil Nadu, and Karnataka indicated that the total acidity is higher than KCl extractable acidity. The low pH and associated low KCl extractable acidity (<1 c mol (+) kg^{-1}) suggest the chances of gibbsite or presence of other amorphous minerals. However, the presence of gibbsite in a highly acidic humid climate is obliquity (Chandran et al., 2006). Thus, the low KCl extractable acidity might be due to the fixation of Al^{3+} by clay minerals other than kaolinite, which is not easily extractable by a KCl solution. The carbon sequestration potential is relatively high in Alfisols, Ultisols, and Mollisols in humid climates (1%–5%) compared with semi-arid climates (<1%). The larger surface area of clay minerals is the fundamental soil substrate quality in sequestering carbon under varying land uses and climates (Velayutham et al., 2000).

Factors Affecting Soil Quality

The soils in the rainfed region are on the verge of degradation for numerous reasons. The primary causes of deteriorated soil quality and declines in crop productivity are the following:

- Soil erosion that washes away all the topsoil rich in nutrients and the organic matter associated with soil aggregates
- Intensive tillage that disrupts soil aggregates, which speeds the decomposition rate of crop residues, aggravates the SOC oxidation aided by high temperature, and reduces microbial biodiversity by disturbing habitats

- Nutrient removal that is not adequately offset by fertilizer addition, resulting in widespread multinutrient deficiencies
- Low nutrient use efficiency due to imbalanced nutrition of soil caused by inappropriate addition of fertilizer with negligible application of organic manures
- Indiscriminate use of agricultural chemicals for controlling weeds, pests, and diseases, which eradicates soil microbial load
- Mono cropping without legumes in crop rotation, affecting nutrient cycling
- Less attention towards management of soil quality constraints such as acidity, salinity, sodicity, drainage, an so forth, in surface and subsurface soil
- Lack of knowledge about soil water conservation measures to improve water use efficiency and arrest the erosion in rainfed areas
- Soil pollution due to disposal of urban and industrial waste and the resultant detrimental effect on soil health
- Mining of nutrient-rich topsoil for nonagricultural purposes

Conclusion

Soil health is crucial for environmental, human, and societal well-being. Farmers play a vital role in maintaining soil health by effectively managing plant nutrients to sustain production, minimize nutrient loss, and ensure high-quality products for human and animal consumption. Therefore, soil management is interconnected with biodiversity, environmental integrity, and human and animal well-being. It is essential to establish specific thresholds for soil quality and management in each agro-ecological region as part of land use planning to support soil productivity and environmental health.

References

Anil Kumar, K. S., Nair, K. M., Ramesh Kumar, S. C., Srinivas, S., Ramamurthy, V., Maske, S. P., Jessy, M. D., Hegde, R., Jacob, J., & Singh, S. K. (2016). *Soil quality monitoring sites (SQMS) for traditional rubber-growing areas of South India* (NBSS Publication No. 1096). ICAR-National Bureau of Soil Survey and Land Use Planning.

Carter, M. R., Gregorich, E. G., Anderson, D. W., Doran, J. W., Janzen, H. H., & Pierce, F. J. (1997). Concepts of soil quality and their significance. In E. G. Gregorich & M. R. Carter (Eds.), *Developments in Soil Science* (pp. 1–19). Elsevier.

Chandran, P., Ray, S. K., Bhattacharyya, T., Sen, T. K., Sarkar, D., & Pal, D. K. (2006). *Rationale for mineralogy class of ferruginous soils of India* [Abstract]. 15th Annual Convention and National Symposium on "Clay Research in Relation to Agriculture, Environment and Forestry" of the Clay Minerals Society of India. BCKVV, Mohanpur, West Bengal.

Department of Agriculture & Cooperation, Government of India (DAC, GOI). (2013). *State of Indian agriculture 2012–13.* Ministry of Agriculture, Government of India.

Department of Agriculture & Cooperation, Government of India (DAC, GOI). (2014). *Operational guidelines for national mission for sustainable agriculture (NMSA).* Ministry of Agriculture, Department of Agriculture & Cooperation, Government of India.

Kalaiselvi, B., Lalitha, M., Dharumarajan, S., Anil Kumar, K. S., Srinivasan, R., Hegde, R., Niranjana, K. V., & Chandran, P. (2020). Evidence of sub-soil sodicity in Tamil Nadu uplands and their characterization. *Journal of Soil Salinity and Water Quality, 12*(2), 155–162.

Karlen, D. L., Mausbach, M. J., Doran, J. W., Cline, R. G., Harris, R. F., & Schuman, G. E. (1997). Soil quality: A concept, definition, and framework for evaluation (a guest editorial). *Soil Science Society of America Journal, 61*, 4–10.

Lalitha, M., Anil Kumar, K. S., Nair, K. M., Dharumarajan, S., Koyal, A., Khandal, S., Kaliraj, S., & Hegde, R. (2021c). Evaluating pedogenesis and soil Atterberg limits for inducing landslides in the Western Ghats, Idukki District of Kerala, South India. *Natural Hazards, 106*, 487–507. https://doi.org/10.1007/s11069-020-04472-0

Lalitha, M., Dharumarajan, S., Hegde, R., & Singh, S. K. (2018b). Application of fertility capability classification in Kangayam grasslands representing dry semi-arid climate of Tamil Nadu. *Range Management and Agroforestry, 39*(1), 1–7.

Lalitha, M., Dharumarajan, S., Kalaiselvi, B., Shivanand, K., Koyal, A., Kaliraj, S., & Hegde, R. (2021b). Hydrochemical characterization and groundwater quality in Cauvery deltaic fluvial plains of Southern India. *Environmental Science and Pollution Research, 28*(1), 44861–44876. https://doi.org/10.1007/s11356-021-13467-8

Lalitha, M., Dharumarajan, S., Natarajan, A., Thayalan, S., Naidu, L. G. K., Anil Kumar, K. S., Vasundhara, R., Nair, K. M., Hegde, R., & Singh, S. K. (2018a). Assessment of soil sodicity and its relationship with aridity and other soil properties under semi-arid environment. *Journal of Soil Salinity and Water Quality, 10*(2), 149–156.

Lalitha, M., Hegde, R., Dharumarajan, S., & Koyal, A. (2021a). Soil fertility evaluation in rainfed regions of different agro-climatic zones of Karnataka, India. *Agricultural Research, 11*, 215–228. https://doi.org/10.1007/s40003-021-00561-zS

Lalitha, M., & Kumar, P. (2016). Soil carbon fractions influenced by temperature sensitivity and land use management. *Agroforestry Systems, 90*, 961–964. https://doi.org/10.1007/s10457-015-9876-9

Mandal, U. K., Kausalya, R., Sharma, K. L., Satyam, B., Venkanna, K., Bhanu, M. U., Mandal, M., Masane, R. N., Narsimlu, B., Rao, K. V., Srinivasarao, C., Korwar, G. R., & Venkateswarlu, B. (2011). Assessing soil quality in a semiarid tropical watershed using a geographic information system. *Soil Science Society of America Journal, 75*(3), 1144–1160.

Mandal, U. K., Sharma, K. L., Venkanna, K., Pushpanjali, P., Adake, R. V., Masane, R. N., Prasad, J. V. N. S., Venkatesh, G., & Srinivasa Rao, C. (2017). Sustaining soil quality, resilience, and critical carbon level under different cropping systems in semi-arid tropical Alfisol soils. *Current Science, 112*(9), 1882–1895.

Nair, K. M., Anil Kumar, K. S., Srinivas, S., Nagaraj, J. S., D'Souza, M. V., Raghuramulu, Y., Hegde, R., & Singh, S. K. (2016). *Soil quality monitoring sites (SQMS) for traditional coffee-growing areas of India* (NBSS Publ. No. 1095). ICAR-National Bureau of Soil Survey and Land Use Planning.

Obi Reddy, GP., Kurothe, R.S., Sena, D.R., Harindranath, C.S., Naidu, LGK., Dipak Sarkar and V.N. Sharda. (2013). Soil erosion in Goa. NBSS Publ. 155. NBSS&LUP (ICAR), Nagpur, p34.

Pal, D. K., Bhattacharyya, T., Deshpande, S. B., Sarma, V. A. K., & Velayutham, M. (2000). *Significance of minerals in soil environment of India* (NBSS Review Series 1). NBSS & LUP.

Pal, D. K., Wani, S. P., Sahrawat, K. L., & Srivastava, P. (2014). Red ferruginous soils of tropical Indian environments: A review of the pedogenic processes and its implications for edaphology. *Catena, 121*, 260–278.

Rajasekharan, P., Nair, K. M., Rajasree, G., Sureshkumar, P., & Narayanan Kutty, M. C. (2013). *Soil fertility assessment and information management for enhancing crop productivity in Kerala*. Kerala State Planning Board.

Sahrawat, K. L., Wani, S. P., Rego, T. J., Pardhasaradhi, G., & Murthy, K. V. S. (2007). Widespread deficiencies of sulphur, boron, and zinc in dryland soils of the Indian semi-arid tropics. *Current Science, 93*(10), 1428–1432.

Sharma, L., Grace, J. K., Mandal, U. K., Gajbhiye, P. N., Srinivas, K., Korwar, G. R., Bindu, V. H., Ramesh, V., Ramachandran, K., & Yadav, S. K. (2008). Evaluation of long-term soil management practices using key indicators and soil quality indices in a semi-arid tropical Alfisol. *Australian Journal of Soil Research, 46*, 368–377.

Shukla, A. K., Behera, S. K., Satyanarayana, T., & Majumdar, K. (2019). Importance of micronutrients in Indian agriculture. *Better Crops, 11*(1), 6–10.

Srivastava, P., Chandran, P., Ray, S. K., & Bhattacharyya, T. (2001). Evidence of chemical degradation in tropical ferruginous soils of southern India. *Clay Research, 20*, 31–41.

Vasu, D., Singh, S. K., Duraisami, V. P., Jangir, A., & Butte, P. S. (2016a). Fertility status of cotton (*Gossypium hirsutum*) growing soils in Thimmajipet Mandal, Mahabubnagar District, Telangana. *Andhra Pradesh Journal of Agricultural Science, 2*(1), 10–15.

Vasu, D., Singh, S. K., Ray, S. K., Duraisami, V. P., Tiwary, P., Chandran, P., Nimkar, A. M., & Anantwar, S. G. (2016b). Soil quality index (SQI) as a tool to evaluate crop productivity in semi-arid Deccan plateau, India. *Geoderma, 282*, 70–79.

Vasu, D., Singh, S. K., Sahu, N., Pramod Tiwary, P., Chandran, V. P., Duraisami, V., Ramamurthy, M., & Lalitha, B. K. (2017). Assessment of spatial variability of soil properties using geospatial techniques for farm level nutrient management. *Soil and Tillage Research*, *169*, 25–34. https://doi.org/10.1016/j.still.2017.01.006

Vasu, D., Tiwary, P., Chandran, P., Singh, S. K., Ray, S. K., Butte, P., & Parhad, V. (2018). A conceptual model of natural land degradation based on regressive pedogenesis in semi-arid tropical environments. *Land Degradation and Development*, *29*(8), 2554–2567.

Velayutham, M., Pal, D. K., & Bhattacharyya, T. (2000). Organic carbon stock in soils of India. In R. Lal, J. M. Kimble, & B. A. Stewart (Eds.), *Global climate change and tropical ecosystems* (pp. 71–96). Lewis Publishers.

Venugopal, V. K., Nair, K. M., Rajasekharan, P., & Sasidharan Nair, A. N. (2019). *Soil health management for sustainable crop production in Kerala*. Department of Agriculture Development and Farmer's Welfare.

Vijaya Kumar, M., Lakshmi, G. V., & Madhuvani, P. (2013). Appraisal of soil fertility status in salt-affected soils of Ongole Division, Prakasam District, Andhra Pradesh. *Journal of the Indian Society of Soil Science*, *61*(4), 333–340.

8

Soil Health in Indian Vertisols

Pramod Jha, K. M. Hati, N. K. Lenka, B. L. Lakaria, M. V. Coumar, B. P. Meena, J. K. Thakur, A. K. Biswas, and A. K. Patra

Chapter Overview

Vertisols occupy approximately 26.6 million hectares, making up about 8.1% of India's total geographical area. These soils are primarily located in the central and southern regions, as well as parts of the western plateau. These soils are characterized by dark clay that undergoes significant expansion and contraction and are typically found in semi-arid to sub-humid climates. They support a variety of crops under both irrigated and rainfed conditions. Rainfed crops commonly grown include cotton, sorghum, maize, chickpea, pearl millet, pigeon pea, groundnut, soybean, and wheat, while rice and sugarcane are the main irrigated crops. The pH of Vertisols ranges from 6.0 to 8.0, potentially increasing to between 8.0 and 9.5 in areas with higher exchageable sodium percentage. Sustainable crop production in salt-affected Vertisols are restricted due to surface leaching of accumulated salts and poor water retention capacity. Vertisols account for nearly 13% of India's total carbon stock and hold significant potential for carbon sequestration. In peninsular India, these soils typically contain less than 1% organic carbon and total nitrogen levels rarely exceed 0.1%. These soils are generally enriched in calcium phosphates, available phosphorus, and both organic and total phosphorus. In older Vertisols derived from basalt, are rich in aluminum phosphates, iron phosphates, and occluded phosphorus, with a notable correlation to smectite content and crack volume. Potassium is abundant in Vertisols, but the swelling process could lead to potassium loss through cracks. Additionally, low hydraulic conductivity and infiltration rates result in significant potassium loss via surface runoff. With 30%–80%

Soil Health Series: Volume 4 Soil Health and Sustainability in India, First Edition.
Edited by Ranjan Bhattacharyya.

clay, mainly smectitic, Vertisols become extremely hard when dry and very sticky when wet, complicating agricultural practices. Crop yields have stagnated or declined due to low water and nutrient use efficiency, poor infiltration, high run-off, soil erosion, and nutrient depletion from intensive cultivation. Poor drainage, both on the surface and subsurface, is a major factor contributing to low crop yields. In semi-arid tropical areas, these soils face significant challenges related to soil quality. Research suggests that integrated nutrient management (INM), legume-based crop rotations, recycling crop residues and implementing efficient rainwater harvesting techniques are effective strategies for enhancing soil health and yield sustainability in Indian Vertisols. Summer plowing and deep tillage, increase soil water storage through better infiltration and reduce crop water stress during early growth stages. Retaining crop residues while minimizing tillage has been shown to significantly improve soil organic carbon content, aggregation, and saturated hydraulic conductivity compared to conventional tillage methods.

Introduction

Vertisols are primarily found on weathered Deccan basalt under various climatic zones of the Indian Peninsula. These include humid tropic, sub-humid moist, sub-humid dry, semi-arid moist, semi-arid dry, and arid areas. This distribution shows how basaltic parent materials can lead to the formation of similar soils across different climatic zone (Pal et al., 2009). The depth of Vertisols depends on the type of underlying rock; moderately shallow (50–75 cm) to moderately deep (75–100 cm) on gneisses and schists, but deeper (100–150 cm) to very deep (>150 cm) on basalt. Vertisols cover an area of 26.6 million ha (Mha), about 8.1% of India's geographical area, and are predominantly found in the central and southern regions and in parts of the western plateau (Pal et al., 2010). Moreover, Vertisols and soils with Vertic properties extend over roughly 72.9 Mha, constituting nearly 23% of India's total land area. These soils are present in various zones across the country including Uttar Pradesh in the north; Madhya Pradesh, Maharashtra, and Chhattisgarh in the center; Karnataka, Andhra Pradesh, Telangana, and Tamil Nadu in the south; and Bihar and Odisha in the east. Maharashtra constitutes 34.4% of India's Vertisols, while Madhya Pradesh has 30.1%, where soybean, maize-wheat, and chickpea are common crops (Ahmad, 1996; Murthy et al., 1982). Other states with notable Vertisol coverage include Andhra Pradesh (13.4%), Karnataka (8.0%), Gujarat (7.0%), Tamil Nadu (4.0%), Rajasthan (1.5%), and Uttar Pradesh (1.6%) (Mandal et al., 2013; Murthy et al., 1982). They are the third most widely distributed soil order in India after Inceptisols and Alfisols.

Vertisols are rich in dark clay that expand and contract significantly and are found in regions with semi-arid to sub-humid climates. These soils are also known as black cotton soils, black earths, dark clays, grumusols, or regurs in older soil classification systems (Dudal, 1965). They are characterized by their wedge-shaped soil aggregates or slickensides, high clay content of over 30%, and the ability to form deep cracks that open and close with change in soil moisture content. Vertisols typically have an ustic moisture regime, leading to their classification as Chromusterts with a hyperthermic or isohyperthermic temperature regime. They are generally deep with wide cracks during dry season, gilgai microrelief with slickensides, and good structural aggregates. Smectitic clays are the dominant clay mineral present in this soil. The prevalence of smectite clay minerals with extensive swell-shrink properties prohibits horizon differentiation. Vertisols are characterized by higher cation exchange capacity (CEC) and water holding capacity, which plays a major role in nutrient and water retention for sustainable crop production. During wet season (rainy season), Vertisols are very difficult to cultivate and manageable only under a narrow workable soil moisture condition; hence, cropping in Vertisols are mostly concentrated during rabi season (winter/post-rainy season) and some remains fallow during rainy season. Vertisols and associated vertic intergrades of soils possess similar characteristics and thus necessitate similar land management for various agriculture and allied uses.

In Vertisols, agricultural crops are extensively grown under irrigated and rainfed conditions. Vertisols are used for cultivated during two major seasons, the *kharif* season (July–September) and *rabi* season (October–December). Under rainfed conditions, crops such as cotton, sorghum, maize, chickpea, pearl millet, pigeon pea, groundnut, soybean, and wheat are cultivated mainly in Vertisols. Rice and sugarcane are the predominant crops cultivated in Vertisols under irrigated conditions. Cotton- and soybean-based systems are followed mainly in the sub-humid (moist) to semi-arid (dry) bioclimatic zones under the Vertisols in the states of Maharashtra and Madhya Pradesh. Cotton-based systems are the most predominant cropping system followed in these soils. In some areas, groundnut, an oilseed crop, is also planted alongside cotton. Following cotton, soybean-based cropping systems are the second most prevalent agricultural practice in Vertisols. Soybean-wheat and soybean-chickpea are mainly planted by the farmers in Madhya Pradesh. Pigeon pea alone or intercropped with soybean is also an important cropping practice followed in central Indian Vertisols (Mandal et al., 2013). Paddy and sorghum are the main cereals grown by the farmers in cereal-based cropping systems. In areas with paddy-based agriculture, paddy-wheat and paddy-paddy rotations are prevalent in sub-humid, low-lying regions. In semi-arid areas, the most widespread cropping system involves sorghum, often paired with legumes such as pigeon pea, black gram, green gram, or chickpea, or alternated biennially with cotton. This system is mainly practiced in the states of Karnataka, Andhra Pradesh, and Maharashtra.

Status of Land Degradation in Vertisols of India

Fifty years after the Green Revolution, India faces multiple issues in sustaining agricultural production and managing soil health. These include the loss of soil organic carbon (SOC) due to residue removal or burning, decreased productivity, nutrient depletion, pollution of soil and water, rising costs of farm inputs, inefficient resource management, reduced farm incomes, and impact of climate change (Paroda, 2018). With 60% of its geographical area used for agriculture, much of this land suffers from various forms of degradation (National Academy of Agricultural Science, 2010). An estimated 36.5% of India's total land area (120.4 million hectares [Mha]) impacted by degradation, including 82.6 Mha from water erosion, 12 Mha from wind erosion, 24.8 Mha from chemical degradation, and 1 Mha from physical degradation. Sharda et al. (2010) calculated that soil erosion in rainfed areas leads to a loss of 13.4 million tonnes (Mt) in crop production for major cereals, oilseeds, and pulses, which translates to an economic loss of Rs. 162.8 billion. Additionally, Dhruvanarayana and Ram Babu (1983) estimated that about 5,000 tonnes (t) of topsoil are eroded annually, with roughly ~1,600 Mt (30%) being permanently lost to the sea.

In India, out of the 3.77 Mha of sodic soil, nearly 1.77 Mha is in the irrigated/rainfed Vertisols in Karnataka, Madhya Pradesh, Maharashtra, Gujarat, Andhra Pradesh, Tamil Nadu, and Rajasthan states (NRSA and Associates, 1996). An area of 55,000 ha of sodic Vertisols occurs in Surat, Bharuch, Navsari, and Valsad districts of Gujarat state. Patel et al. (2000) reported that nearly 40% of soils in the Ukai-Kakrapar command area are affected by sodicity problem. Due to perennial water availability in the command area, farmers cultivate paddy and sugarcane, and as a result of water logging and soil salinity/sodicity problem, the crop productivity is declining (Rana & Raman, 1999). Lingade et al. (2008) reported that nearly 5.6 Mha of sub-soil sodicity of Vertisols occurs in the basaltic terrain of Maharashtra state. Vertisols are generally calcareous in nature due to their characteristic high clay content, high base saturation, and low SOC. Normally, the pH of Vertisol ranges from 6.0 to 8.0 and in areas where the exchangeable sodium percentage (ESP) is >15, the soil pH may go even higher (8.0–9.5). Constrains in surface leaching of accumulated salt and poor available water holding capacity are the major limitations for sustainable crop production in salt-affected Vertisols. The extent of loss in crop productivity and soil fertility owing to soil degradation depends mainly on soil type. For example, significantly higher reduction in soil productivity is observed in black soil than in alluvial soil under moderate water-erosion-related soil degradation. Similarly, the extent of soil productivity loss under ESP value at 15 was reported as 25% in Vertisols versus 10% in alluvial soil.

Soil Carbon Dynamics in Vertisols of Central India

In rainfed agro-ecosystems in India, soil degradation due to SOC depletion is the major reason for poor soil quality and declining crop productivity. Management practices, which enhance soil carbon sequestration in terrestrial ecosystems, makes individual ecosystem components more balanced and sustainable. Vertisols contribute nearly 13% of total carbon stock in India and offer a greater scope for carbon sequestration (Kundu et al., 2001). Since the Vertisols are inherently high in fertility, soil productivity and sustainability can be achieved under proper management conditions without declining soil quality.

In a study examining different land uses in Vertisols, the total SOC content ranged between 2.2 and 24.8 g kg^{-1}. Forest and horticulture lands had the highest organic carbon (OC) levels due to greater turnover of biomass for over 25–50 years and minimal soil disturbance compared with agricultural and agri-horticultural systems (Lakaria et al., 2012). Significantly higher OC was found in fields receiving organic additions under agriculture land use system. In another study, the carbon management index indicated that the use of organic farming practices is 1.31-fold better than no fertilizer and no manure application in the Vertisols of Central India. Similarly, the plantation crops such as mango, eucalyptus, and *mahua* and *tendu* have 4.02, 3.28, and 3.88 times better impact over no manure and no fertilizer application (Lakaria et al., 2012).

The Fourier-transform infrared spectra of humic acids and fulvic acids under control and organic and inorganic fertilizer application studied in a Vertisol of Central India revealed differences in absorption intensity of several absorption bands. The bands were indicative of amines, amide, aliphatic C-H stretching, carboxylic groups, aromatic bonds, and C=C bonds. In comparison with the control, the spectra of humic acids under inorganic fertilizers application revealed that the decomposition rate of organic matter was faster (Lakaria et al., 2017). The minimum quantity of carbon required to maintain the existing SOC levels without any increase or decrease is 554 kg C ha^{-1} (Lakaria et al., 2012).

In Vertisols of central India (Jabalpur), the carbon pool (% of total C) in the acid non-hydrolyzable fraction of soil was found to be 52% in untreated soil, 65% in soil treated with NPK fertilizer, and 66% in soil treated with both NPK fertilizer and farmyard manure (FYM). The slow pool's carbon concentration decreased by 18% with NPK alone compared with untreated soil. The extra carbon input with NPK and NPK + FYM treatments contributed more to the acid non-hydrolyzable fraction. The combined application of NPK + FYM raised the carbon levels in the active, slow, and acid non-hydrolyzable fractions by 99%, 22%, and 33%, respectively, compared with NPK alone (Jha et al., 2014).

Nitrogen, Phosphorus, and Potassium Status of Indian Vertisols

The Vertisols found in peninsular India contain less than 1% OC, and their total nitrogen content rarely goes above 0.1% (Dudal, 1965; Virmani et al., 1982; Tandon & Kanwar, 1984). When continuously cropped, they are mostly deficient in nitrogen and phosphorus. Crop responses to nitrogen were notable, with fertilizer band placement proving more effective than broadcast application. The fertility of central India's Vertisols varies widely in the semi-arid tropics, with nitrogen and phosphorus being the most critical and commonly deficient nutrients. Variations in nitrogen and phosphorus levels across the semi-arid tropics reflect differences in soil origin and past crop cultivation. Nitrogen fractionation studies showed that increasing nutrient levels from 50% to 150% and adding FYM led to an increase in all soil fractions, suggesting that balanced nutrition is key to maintaining higher available soil nitrogen, thereby supporting sustained productivity. Phosphorus fractionation studies found that calcium-bound phosphorus (Ca-P) is predominant in Vertisols (Jabalpur, Coimbatore, Akola), and the absence of phosphorus led to a reduction in these fractions, indicating that balanced phosphorus fractions are crucial for providing plants with phosphorus.

Vertisols of semi-arid tropics are generally characterized by moderate to high pH in the context of high temperature and frequently scant rainfall in the early rainy season, which is favorable for nitrogen application losses in ammoniacal form (El-Swaify et al., 1985). Information on nitrogen dynamics in high clay soil is sparse. Dalal (1977) reported that NH_4-N in fixed form is considerably high in swell-shrink soil, and several researchers have reported that tropical soils are generally enriched with large amounts of fixed ammonium; however, information is limited on this form of nitrogen, especially in the semi-arid tropics (Burford & Sahrawat, 1989). In benchmark sites of Kasireddipalli at the International Crops Research Institute for Semi-Arid Tropics (ICRISAT) Center and Patancheru and Barsi in Maharashtra state in semi-arid tropics, fixed NH_4-N distribution has a share of 22%–59% in the former and 16%–31% in the latter in the total soil nitrogen, and its percentage was higher in the sub-soils (Sahrawat, 1995).

Vertisols of India are inherently low in nitrogen concentrations and the response of crops to nitrogen application vary by climatic zone. The practice of sequentially planting crops without adequately addressing their nutritional needs has led to reduced soil fertility, particularly in terms of nitrogen content (Ghosh et al., 2003). Jha et al. (2014) demonstrated that in Jabalpur soils, the use of balanced fertilizers (NPK) along with an integrated nutrient management (INM) approach (NPK + FYM) resulted in an increased total nitrogen level. Specifically, the combination of NPK and FYM boosted the total N levels by 28% compared with untreated soil at Jabalpur. Similar results were also reported by Meshram et al. (2018), who found that application of 100% NPK + FYM at 5 $Mg\,ha^{-1}$ significantly improved soil health and crop productivity and was economically feasible for soybean-safflower cropping sequence grown in Vertisols. While working on black soils of Akola, Katkar et al. (2012) observed that the total nitrogen content of soil varied significantly with long-term application of chemical fertilization and manuring. The significantly highest soil total nitrogen (0.059%) were recorded in treatment where 10 $t\,ha^{-1}$ of FYM was applied along with chemical fertilizer for 23 years. Adoption of a no-till system along with residue retention helps improve soil available and microbial biomass nitrogen (Sonune et al., 2021). Patra et al. (1996) observed that nitrogen depletion in deep black soils was significantly affected by the soil pH level and moisture content. They found nitrogen losses to be 29.5%, 33.5%, and 33% in wet soils, and 37%, 42%, and 40.5% in moist dry soils at pH levels of 7.7, 8.2, and 9.3, respectively. These findings indicate that surface application of urea to Vertisols can lead to considerable nitrogen losses.

Subba Rao et al. (2015) reported that soils of India had either low or medium available phosphorus in analyses of the phosphorus availability status of Indian soils during the period 1969–2002. Motsara (2002) reported that 42%, 38%, and 20% districts of the country fell in low, medium, and high phosphorus status categories, respectively. Singh et al. (2003) observed that younger Vertisols are higher in Ca-P, available phosphorus, and organic and total phosphorus; whereas, older Vertisols originating from basalt parent material are rich in Al-P, Fe-P, and occluded phosphorus and had a significant relationship with smectite content and crack volume of the soils. Murthy et al. (2002) suggested that parent material apparently does not have much influence in modifying soils' ability to differentially enrich the added phosphorus into different phosphorus fractions. According to him, despite low phosphorus status of Vertisols, response to phosphatic fertilizer is small (ICRISAT, 1981, 1988).

Vertisols have abundant soil potassium. The dynamics of potassium ions in soils and clay minerals are complex. The mobility of potassium within the soil and the dynamic reactions that exist between the different forms of potassium are dependent on the texture, structure, CEC, moisture saturation percentage, and temperature of the soil. In Vertisols, which are characterized by high CEC, nitrates leach out more easily than calcium. In addition, potassium leaches out more easily than calcium because calcium binds more strongly to soil particles than potassium. However, clay and organic matter in soil can hold onto potassium ions in a form that plants can use, making it less likely to leach out in silty or clayey soil. The prolonged swelling process in Vertisols can cause potassium to be lost through cracks that go deeper than the roots of plants. Furthermore, due to low hydraulic conductivity and low infiltration rate in Vertisols, a considerable quantity of potassium is lost through surface runoff. Vertisols develop deep cracks of sometimes more than 2 m deep and more than 5 cm wide. When these soils are inundated with heavy rain, a perceptible percentage of rainwater initially moves into the soil through the cracks, by-passing the soil matrix. This infiltrating water can carry potassium fertilizer present on the surface deep into the soil, where it is lost from the rhizophere and inaccessible to the plants. Paltineanu (2001) found that in Vertisols, the movement of nutrients was affected by the presence and shape of cracks. He suggested that because the swelling process in these soils takes a long time, there is an ongoing risk of fertilizers being washed away through cracks that extend below the depth of plant roots. But as the soil saturates and swells with time, this movement of potassium through bypass flows stops.

Physical Attributes of Indian Vertisols

Swelling and shrinking are the key pedogenic processes in Vertisols and are responsible for the genesis, distinctive behavior, and characteristic properties of this soil. Repeated swelling–shrinking affects the spatial (both vertical and lateral) distribution of physical and chemical properties of the soil. Owing to the presence of a high percentage of smectitic clays, these soils become very hard when dry and develop extreme stickiness when wet. Consequently, these soils are suitable for cultivation only within a narrow soil-moisture range, which is a major crop production constraint (Mandal et al., 2013). Measurement of soil properties, particularly the physical properties in Vertisols, entails operational challenges due to swelling and shrinking, changing pore-geometry, deep crack formation, and hardening and stickiness of the soil during the seasonal wetting and drying cycles.

Vertisols contain 30%–80% clay, primarily smectitic, which expands and contracts significantly. This leads to unique features like gilgai micro-relief, deep and wide cracks, and slickensides. Vertisols formed from calcareous clay have high levels of $CaCO_3$, which can increase with depth. The pH of these soils is usually between 7.8 and 8.7 but can go up to 9.4 in sodic soils. While Vertisols can hold a lot of moisture (150–250 mm m^{-1}), not all of it is available to plants because the smectitic clays hold moisture very tightly. These soils also have low permeability, making them prone to waterlogging during the rainy season, which is a significant challenge in regions with Vertisols. These soils have low permeability, which makes these soils susceptible to water congestion. Prolonged periods of inundation or water logging during the rainy season are major constraints for the Vertisols region during the rainy season.

Constraints of Vertisols to Crop Production

In Vertisols, crop yields have been either stagnating or declining over the years owing to low water and nutrient water use efficiency, poor water infiltration, high runoff and soil erosion as a result of unpredictable rainfall pattern, and nutrient mining due inappropriate intensive cultivation practices (Ahmad, 1996). Imbalanced application of chemical fertilizer coupled with low use of organic resources lead to reduced level of soil organic matter in Vertisols of central India. This has in turn resulted in a decline in soil quality of Indian Vertisols. Unless good soil management practices are adopted, yield reduction will continue and sustainable crop production will be difficult.

Indian Vertisols are in general poor in soil organic matter and plant nutrient content mainly because of a lack of crop residue addition, poor nutrient inputs, and the use of intensive tillage practices. It is generally difficult to increase the SOC content in these soils by using traditional soil management methods because they have reached a balance due to the high temperature and rapid decomposition of SOC (Jha et al., 2012). The only ways to boost and store more SOC are by regularly adding organic materials such as FYM, compost, or crop residues (Hati et al., 2007), or by changing from traditional plowing to no-till farming or other conservation tillage methods (Lal & Kimble, 1997). Besides low fertility, Vertisols also struggle with limited moisture and poor drainage (Murthy, 1988). They are very susceptible to erosion, especially where plant cover is thin and water flows unchecked through channels. If not managed well, these soils can degrade even further (Sharma et al., 2011). The main constraints of Indian Vertisols towards crop production are described in the following sections.

Poor Workability

The high amount of clay and the predominance of montmorillonitic clays in Vertisols give them unique properties that make farming difficult. They develop cracks when dry, and plowing can create large, hard clumps. When they are wet, they become sticky and hard to work with because machinery cannot easily move on them. This leaves farmers with only a short period for planting crops during the rainy season. Hence, some Vertisols remain fallow even during the rainy season.

Poor Drainage and Water Erosion

One of the main reasons for low crop yields from Vertisols is poor drainage, both on the surface and within the soil. When it first starts to rain, especially if the soil is dry, water can seep in quickly because of the cracks and loose structure. However, once the soil is saturated and starts to swell, the water transmission property becomes affected, which causes problems with water moving through the soil. The flat landscape and slow absorption rates in Vertisols lead to water pooling on the surface when they are completely wet. Therefore, if Vertisols are not properly managed with appropriate land treatment and crops during the rainy season, they can lose a lot of soil and water, mainly through erosion.

Soil Fertility Loss and Residue Burning

Vertisols found in India are generally low in soil organic matter, nitrogen, phosphorus, sulfur, and zinc. Low and imbalanced use of chemical fertilizers cause micro-nutrient deficiency in the soil. In addition, the recycling of nutrients from plant residues is low because crop residues are often taken away from the fields or burned. When deep soils are watered, they are very susceptible to becoming salty or sodic, especially below the surface. The tendency of soil to retain $CaCO_3$ can also make it harder for plants to get nutrients.

Biological Health and Soil Microbial Biodiversity in Indian Vertisols

Microbial and biochemical properties of soil constitute important soil quality indicators as they are involved in nutrient cycling and decomposition of soil organic matter, and carbon storage determines the availability of major and micro nutrients. Soil biological properties are considered to be sensitive indicators of soil quality and need to be assessed under intensive tillage systems and land use changes (Mirzavand et al., 2020). However, soil biological properties are not uniform, either spatially or temporally, and they differ with land use, management practices adopted, soil type, crop grown, weather conditions, and ecosystems. Vertisols are mainly found in the central and western parts of India, extending through Madhya Pradesh, Maharashtra, and parts of Gujarat, where weather remains semi-arid with relatively higher soil temperature during most of the year. The cultivated land area is mainly occupied by crops such as soybean, wheat, cotton, sugarcane, maize, and mustard, among others.

Nutrient management practices affect soil biological activity by regulating sources of carbon and energy for proliferation of different groups of soil flora and fauna. Combinations of organic inputs from various sources such as poultry manure, vermicompost, cattle manure, Panchagavya, and so forth are comparable with or even better (higher by 5%–13%) than the recommended dose of fertilizers for soybean and also significantly improve the available nutrient status of soil. According to Aher et al. (2019), soil enzyme activities, specifically dehydrogenase and alkaline phosphatase, increased by 62%–72% and 27%–35%, respectively, when organic nutrients were used instead of the recommended dose of chemical fertilizers. The number of soil microbes such as bacteria, fungi, and actinomycetes and the amount of microbial biomass carbon were higher in soils treated with organic manure (Aher et al., 2018).

Bhat et al. (2017) also showed that the stage of crop growth affects microbial activity. They found differences in soil enzyme activities (dehydrogenase, β-glucosidase, acid and alkaline phosphatase), microbial respiration, substrate-induced respiration, and microbial biomass carbon between organic and conventional farming systems under a soybean-wheat crop rotation in Vertisols. Under organically managed treatment. The highest biological activity was recorded at the reproductive stage in soybean and the panicle initiation stage in wheat. The effectiveness of organic manures and biofertilizers on productivity, grain quality, soil fertility, and microbial activity was studied by Behera (2009) in Vertisols of central India. There was an improvement in soil fertility, including both chemical and biological qualities, from the starting point when FYM and poultry manure were used. The amount of microbial biomass carbon and the activities of enzymes such as phosphatase and dehydrogenase went up with the addition of these manures.

The burning of crop residues in the field significantly influences microbial community dynamics. Burning temporarily elevates the surface soil temperature to a level that can kill the native microbial flora and also tends to oxidize the stored soil carbon, which may have negative repercussions in the long run if burning is practiced continuously (Thakur et al., 2021).

Similarly, tillage also modulates the microbial diversity and activity by influencing soil aeration, carbon storage, moisture conservation and soil temperature. The surface soil (0–15 cm) microbial biomass carbon was found to be greater in reduced tillage practice than conventional tillage in rainfed Vertisols of central India (Somasundaram et al., 2019). The levels of crop residue under no tillage significantly influence the population of soil microbes, particularly fungi and actinobacteria. Kumawat et al. (2022) reported the highest fungal density (81.4×10^3 CFU g^{-1} soil) in zero tillage with 90% residue retention of preceding crop levels as compared with conventional tillage without residue retention (5.7×103 CFU g^{-1} soil).

In long-term fertilizer experiments, the highest numbers of microbes, as well as higher microbial biomass carbon and nitrogen, were found in soil treated with a balanced mix of inorganic fertilizers and FYM (100% NPK + FYM). In contrast, using only nitrogen fertilizer (100% N) or no fertilizer at all had a negative impact on soil microbes and nitrogen fixation (Bairwa et al., 2021). Therefore, careful use of both organic and mineral fertilizers together seems to be beneficial over time. Mohanty et al. (2022) studied soil processes mediated by microbes and reported the potential nitrification rate was 0.88, 1.86, 1.278, and 1.493 μg NO_3^- produced per gram of soil per day in fallow, control, 100% NPK, and 100% NPK + FYM, respectively. These results indicate better mineralization of nitrogen in the presence of added organic matter in a long-term fertilizer experiment on a soybean-wheat cropping system.

Soil Quality Concept and Soil Health Assessment in Vertisols

The concept of soil quality and health, which deals with the inherent and dynamic nature of soil, is new and still evolving. However, recognition of the importance and role of maintaining soil health and optimum soil environment for sustained agricultural production dates back to the 1980s and 1990s (Goswami, 1987, 1995; Goswami & Rattan, 1992). Das and Dutta (2000) highlighted the indicators of soil quality function in land resource management that is sustainable. The idea of soil quality in relation to sustainability and its applicability to Indian agriculture were reviewed and critically examined by Goswami (2005).

A quantitative technique for establishing a connection between a management objective and the physical, chemical, and biological characteristics of soil is the soil quality index (SQI) approach. This approach combines information on key soil attributes, which supports decision-making for sustainable soil management. With the evolution in soil quality research, the definition of soil quality has also undergone modifications. Nonetheless, the vital roles that soils play are the main focus of research on soil quality. Karlen et al. (1997) defined soil quality as "the capacity of soil to function," which is the most basic meaning of the term. A few more definitions also draw attention to the important topic of soil functions or services. In several industrialized countries, indicators for evaluating the quality of soil have been identified for main soil types, eco-regions, and management approaches. These indicators have threshold values for certain systems (Herrick et al., 2002; Sparling and Schipper 2002). In India, several studies have attempted and successfully screened key indicators for soil quality assessment for a few selected cropping systems and soil types.

The normal procedure of soil quality assessment is to start by selecting a set of critical soil properties that are considered indicators of soil quality. In this assessment procedure, the indicators should be easy to measure. The selection of a minimum soil data set (MDS) is either carried out through statistical data reduction procedures such as principal component analysis (PCA) or through the opinion of subject matter experts.

Using a PCA approach, Kundu et al. (2012) identified key soil indicators from a field study in two districts (Sehore and Vidisha) in the Vertisols of agro-ecological sub-region (AESR) 10.1. A stratified random sampling procedure was used for soil sample collection from the study region, which was dominated by a soybean-wheat cropping system. Based on the study and the statistical analysis, eight critical indicators for the MDS were selected, including total OC, non-exchangeable potassium, total zinc, total manganese, bulk density, $CaCO_3$ content, available sulfur, and alkaline phosphatase activity. The computed SQI values varied from

0.894 to 1.809 and explained 57% of the production potential of the study region. Kundu et al. (2013) identified the key indicators as water retention at 0.3 bar, CEC, mineralizable carbon and nitrogen, available nitrogen and zinc, mean weight diameter through PCA for yield sustainability in Vertisols of Warangal and Nalgonda districts (AESR 7.2). Similarly, the key indicators for yield sustainability in Vertisols of Sehore and Vidisha districts of AESR 10.1 are total OC, non-exchangeable potassium, total zinc, total manganese, bulk density, $CaCO_3$, available sulfur, and alkaline phosphatase activity for soybean-wheat/chickpea and rice-wheat cropping system. Another study, conducted in the semi-arid tropical black lands of Central India, indicated that OC, microbial biomass carbon, available Zn and mean weight diameter were emerged as the key indicators that contribute towards soil quality under soybean system (Sharma et al., 2016). Sharma et al. (2011) identified OC, pH, electrical conductivity, available potassium, exchangeable magnesium, microbial biomass carbon, and dehydrogenase activity as major soil quality indicators for semi-arid tropical rainfed Vertisols, Akola, Maharashtra state under cotton + green gram system.

In a field study from the Mahbubnagar district of Telangana in the semi-arid Deccan plateau region of India, Vasu et al. (2016) employed a PCA approach to select the critical soil indicators. The study district was characterized by semi-arid dry climate with ustic soil moisture regime and predominated by cotton crop and clay soil. The percentage of clay in the study region varied from 20% to 37% on the surface and from 5% to 52% in the control section. Pedogenic $CaCO_3$ content was the major reason for high pH in parts of the study region. The mean SOC content on the surface soil was 0.65%, and CEC varied from 3.9 to 32.7 cmol (p+) kg^{-1} on the surface and from 2.59 to 40.20 cmol (p+) kg^{-1} in the control section. The PCA showed sodium absorption ratio, clay content, base saturation, SOC, and silt content to be the critical variables for soil quality assessment. Wide variation in soil quality was due to soil heterogeneity and soil degradation caused by subsoil sodicity. As compared with PCA, computation of SQI by the expert opinion method, using clay, soil hydraulic conductivity, $CaCO_3$, SOC, CEC, and ESP as indicators, produced better results.

Using the PCA approach, Lenka et al. (2022a) identified the critical variables for AESR 6.1, which is dominated by Vertisols, under hot, dry semi-arid eco-subregion and constituting parts of southwestern Maharashtra and northern Karnataka. Grid sampling was followed for soil sampling covering the entire AESR, which encompasses about 7.6 Mha, constituting 2.3% total geographical area of India. It is commonly considered as a drought-prone water scarcity area of central Peninsular India. With regard to cropping, the area is dominated by sugarcane along with cotton, green gram/sorghum, wheat, and maize. The northern part of the studied AESR is dominated by cotton-based system, while sugarcane is prominent in the southern part. In the two districts of Karnataka, rice is the dominant crop. Based on the component coefficients and Pearson's correlation, eight soil parameters, including pH, clay content, active pool of SOC, available water capacity, soil bulk density, available zinc, available boron, and available phosphorus, were included in the MDS for developing SQIs for the soils of AESR 6.1 dominated by cotton- and sugarcane-based cropping systems. SQI and relative SQI for the area varied from a minimum to maximum of 0.48 to 0.75 and 64% to 100%, respectively.

Using a quantitative approach based on soil functions, Lenka et al. (2022b) prepared a spatial distribution map of SQI for Hoshangabad district of Madhya Pradesh. The study district was predominantly covered under a rice-wheat cropping system with deep clay soils. The results showed that the SQI varied from 0.40 to 0.60, with 71% of area of the district having SQI values of more than 0.50. The spatial distribution map of critical variables showed 36% samples as being low in oxidizable SOC, whereas 25% of the samples had high SOC (>1.0%). Further, 99% samples were low in available nitrogen, but in contrast, there was a phosphorus saturation in the soil profile. In the surface soil, 89% of samples were high in available phosphorus.

Microbial activity and microbial mediated soil processes of agricultural importance are also significantly inhibited in soils contaminated with heavy metals. Soil enzymatic activities and microbial biomass carbon are significantly reduced in soil contaminated with copper. Microbial activity of soil under copper stress (500 $\mu g\,Cu\,g^{-1}$ soil) was reduced, as shown by decreases of 74.7% in dehydrogenase activity, 59.9% in microbial biomass carbon, 48.2% in alkaline phosphatase activity, and 15.1% in acid phosphatase activity (Coumar et al., 2017).

Management Interventions and Strategies for Improvement of Soil Health

Vertisols, particularly in rainfed semi-arid tropical environments, face significant challenges due to soil quality factors. These challenges often lead to poor crop yields. However, there are effective management practices that can enhance the quality and productivity of these black soils, such as planting legumes as intercrops or in rotation with main crops, integrating organic and inorganic nutrient sources, using balanced fertilization, recycling crop residues, and harvesting rainwater. These practices not only enhance soil fertility but also have a positive impact on physical and biological properties.

A soil and nutrient management practice plays major roles in identifying key soil quality indicators that influence soil health and promote sustainability. The soil organic matter is important for soil quality related to many physical, chemical, and biological aspects of soil (Smith et al., 2000; Fortuna et al., 2003). Depletion of organic matter and poor capacity of the soil to supply nutrients have been considered a major factor for yield decline under intensive cropping systems (Dawe et al., 2000). The best management practices have been screened out of a few studied systems in Vertisols using soil quality assessment tool (Mandal et al., 2005; Sharma et al., 2005; Chaudhury et al., 2005; Masto et al., 2007). Under soybean-wheat cropping system of Vertisols of Sehore and Bhopal districts of Madhya Pradesh, long-term application of organic manure through organic farming or INM have yielded significant increases in SOC and soil quality as compared with inorganic nutrient management (Chaudhary et al., 2022). The highest SQI value was recorded under organic nutrient management (0.61), followed by integrated (0.52) and inorganic (0.34) nutrient management, indicating that manure-based nutrient management practices improve soil's physical, chemical, and biological properties and also sustain crop productivity. Under rainfed conditions on Vertisols in central India, conjoint application of chemical fertilizer and organic manure was essential for improving SOC sequestration in a soybean-safflower sequence (Sharma et al., 2011; Srinivasarao et al., 2012). Coumar et al. (2020) reported that among the cropping system in Vertisols of Sehore and Vidisha districts of AESR 10.1, SOC pool values and SQI are relatively higher under legume-based cropping systems (soybean-wheat and soybean-chickpea) than a cereal-cereal cropping system (rice-wheat). They concluded that growing soybean-wheat and soybean-chickpea rotations in Vertisols helps to sequester carbon into the soil better than a rice-wheat system. Another study also confirms that adoption of a cropping system based on cereal-legume (rice-chickpea) performed better in improving key soil quality indicators, soil quality, and yield sustainability than cereal-based (rice-wheat) cropping system in Vertisols of the central plains in the Chhattisgarh region. Similarly, Mohanty et al. (2007a, 2007b) reported that direct-seeded rice with residue addition and conventional tillage have better crop productivity, soil health, and sustainability in Vertisols of Madhya Pradesh, India, under a rice-wheat system. The time for sustainable productivity was increased from 6 to 15 years for direct-seeded rice, 5–11 years with low-intensity puddling, and 1–8 years for high-intensity puddling under residue addition/retention with conventional tillage system. Sharma et al. (2016) concluded from a 6-year experiment that the SQI values were reasonably good in low-tillage practice with compost addition and hand weeding as compared with other management practices under a soybean system in Vertisol.

In short, based on several research findings, INM and legume-based cropping rotation are better strategies for improving soil health and yield sustainability in Vertisols of India. The quality and productivity of black soils can be improved by adopting suitable management practices, such as the inclusion of legumes as intercrops or in rotation with the main crops, integrated use of organic and inorganic sources of nutrients, balanced fertilization, residue recycling, and rainwater harvesting and its efficient recycling (Burford & Sahrawat, 1989). These practices may not only help in improving the soil fertility and organic-matter status of soil but also help in improving physical and biological parameters. Management interventions for mitigation of land degradation, improvement in soil health, and maintenance of sustainable high production in Vertisols are listed in the following sections.

Rainwater Harvesting and Its Efficient Recycling

Most of the areas under Vertisols do not receive adequate or well-distributed rainfall. About 75%–85% of the total annual rainfall is received during the four monsoonal months of the year. Harvesting of the excess rainfall received during the monsoon and its use during the rain deficit or lean period is imperative for growing the winter season crops. Rainwater can be collected in situ or ex situ, but in order to maximize production, it must be used carefully and with site-specific methods. Maximizing production per unit of irrigation water (pumped or transported) or rainwater (harvested or retained in situ) in a sustainable manner is the goal of every effective water management system. Some of the management techniques are elucidated in the following sections.

Land Surface Management and Crop Diversification

A graded broad bed and furrow (BBF) system of land configuration improves surface drainage and allows better water infiltration. The BBF system allows safe drainage of excess water through grassed waterways into a water harvesting pond. However, it is less effective for shallower Vertic soils, as it can lead to runoff if not properly aligned with a graded slope. The BBF system considerably reduces the runoff and soil losses in Vertisols compared with the traditional flat-on-grade (FOG) system. A 4-year study on a Vertisol in central India showed that the BBF system reduced the runoff volume by 24% to 32% and the soil loss by 31% to 55% over the years compared with the FOG system. In a year with little rainfall, the BBF system held onto more soil water in the profile than FOG, contributing to a 13%–18% higher soybean yield than FOG throughout the later stages of crop growth. Similarly, maize grain yield in the BBF system was 12%–16% greater than that under FOG.

Sub-soiling

A low rate of infiltration is a major problem for Vertisols, which induces accelerated surface runoff of the fertile surface soil and poor water storage in the profile during the initial phase of the rainy season. Improved tillage practices such as summer plowing and deep tillage (subsoiling with chisel plough) can improve soil water storage through greater infiltration and minimize crop water stress during the initial growth stages. Research on a Vertisol in central India compared three tillage treatments: conventional tillage, conventional tillage with alternate-year sub-soiling, and conventional tillage with annual sub-soiling. The study found that the basic infiltration rate and soil water storage at a 90-cm depth were higher in the annual subsoiling treatment than in conventional tillage. Additionally, annual subsoiling improved soybean-wheat system productivity (Mohanty et al., 2007a, 2007b).

Summer Cultivation

In the post-rainy season, plowing the land immediately after crop harvest, when the soil is still slightly moist, prevents weeds from producing seeds, reducing their survival and multiplication. Crop residues left in the field after harvest often serve as shelter for insects during the off-season, leading to increased populations during cropping. Stubble from crops such as sorghum, maize, and paddy provides shelter for stem borers. Additionally, many weeds act as alternate hosts for pests and diseases during the off-season. Summer cultivation exposes hibernating insect eggs and pupae to intense solar heat, effectively eliminating them (Patil & Jawaregowda, 2000). Moreover, it also enhances soil fertility, fills cracks, and pulverizes the soil, thereby improving moisture retention.

Conservation Agriculture

Conservation agriculture, based on principles of continuous soil cover, minimal soil disturbance, and crop rotations, holds great potential for sustainable agriculture in Vertisols. It addresses key challenges such as increasing productivity while safeguarding natural resources and the environment. Under conventional cultivation, bare and pulverized top soils are susceptible to accelerated soil loss through runoff, especially on rolling slopes. Additionally, conventional tillage disrupts soil aggregates, exposing SOC trapped within them to rapid microbial degradation. This results in a net loss of OC from the arable ecosystem. The breakdown of aggregates makes the carbon accessible to microbes for decomposition, and exposure to oxygen in the air leads to faster aerobic decomposition. Conservation practices minimize surface soil disturbances during tillage, ensuring that the soil remains covered with anchored crop residues or cover crops, particularly during heavy rainfall. This approach reduces the negative impact of agriculture on the natural ecosystem. Furthermore, conservation tillage provides a sustainable option for managing crop residues, especially in regions where burning residues contribute to air pollution and the loss of valuable nutrients from the ecosystem. While working on conservation tillage in Vertisols of central India, Hati et al. (2015) reported that retention of crop residues, coupled with minimum tillage, resulted in significant improvement in SOC content, aggregation, and saturated hydraulic conductivity of the soil compared with the conventional tillage system. The no-tillage system accumulated a higher amount of OC near the surface soil layer. The study showed that no tillage to soybean and wheat with balanced doses of fertilizer can be a viable alternative to conventional tillage for sustainable production, with concomitant improvement in soil health and sequestration of carbon in Vertisols of central India. Somasundaram et al. (2018) also reported that conservation agriculture practices enhanced soil aggregation, aggregate stability, and SOC content in Vertisols of central India. These practices positively affected aggregate-associated carbon for large macroaggregates in the top 0–15 cm of the soil layer. Additionally, conservation agriculture not only improved soil aggregation and carbon retention but also led to increased crop yields over time (Somasundaram et al., 2018). Conservation agriculture has helped to improve soil health and crop productivity in rainfed Vertisols of central India.

Soil organic matter loss, high wind and soil erosion, low infiltration rate, and destruction of soil structure are harmful effects of high-tillage practices under traditional agriculture and ultimately affect crop production and soil health. However, it is well established that the adoption of reduced tillage with residue incorporation is beneficial for improving soil health and crop productivity (Meena et al., 2016; Somasundaram et al., 2018). Furthermore, cropland has a great potential to sequester carbon, particularly when crop biomass production (crop residue) is managed efficiently as crop residue management practices can significantly improve soil quality, reduce soil erosion and runoff, enhance moisture retention, and lower soil temperature. In addition, incorporation or retention of crop residue on the soil surface increases the soil carbon status and reduces carbon emission into the atmosphere. Zero-tillage practices also improve the nutrient efficiencies of the applied nutrients and prevent the volatilization or denitrification. Therefore, zero tillage, along with residue incorporation and adoption of diversified crop rotation, not only prevents the top soil erosion but also sustains soil fertility and environmental quality (Hobbs et al., 2008; Chauhan et al., 2012). Hence, adoption of conservation agriculture and following its principles may be the best intervention for improving soil health for sustainable production in agriculture.

Balanced and Integrated Nutrient Use

Soil health and OC are greatly affected by balanced application of inorganic fertilizer and organic amendments. Long-term application of balanced fertilizer, either alone or in conjunction with FYM, significantly increase the OC content of the soil in Vertisols. This, in turn, contributes to carbon sequestration within the soil (Hati et al., 2007). The increase in OC content is positively correlated with improved soil aggregate stability, water retention, microporosity, and available water capacity. This correlation underscores the significance and advantages of using balanced fertilizers. Balanced nutrient supply improves soil conditions, leading to enhanced crop production. INM, a customized approach to fertilizer management, optimally utilizes organic and inorganic nutrient sources. INM provides essential macro- and micronutrients, promoting long-term soil health and sustainable, economically viable agriculture (Singh & Wanjari, 2012; Jha et al., 2021; Meena et al., 2021; Dotaniya et al., 2022). Long-term experiments on INM modules have clearly shown that fertilizers' response can be significantly increased with a balanced supply of nutrients along with organic manures (Damodar Reddy et al., 1999; Mandal et al., 2000; Singh and Wanjari 2013; Meena et al., 2019, 2021). Lakaria et al. (2012) reported that application of balanced management and INM considerably improved soil carbon storage, with greater turnover rates of biomass under soybean-wheat cropping system of central India. Similarly, Jha et al. (2014) also reported that the combined use of NPK + FYM was the best option for enhancing the crop yield and SOC sequestration, based on a long-term fertilizer experiment study. Adoption of INM strategy (NPK + FYM) considerably improved SOC compared with only chemical fertilizers in maize-chickpea cropping system in Vertisols (Dotaniya et al., 2022). SOC stock and the carbon sequestration rate significantly improved with adoption of INM modules in Vertisol (Meena et al., 2019). INM strategies in intensive maize-sunflower cropping system enhanced soil carbon sequestration and soil biological properties (Anandakumar et al., 2022).

Efficient Management of Irrigation for Higher Water and Crop Productivity

Scheduling of irrigation based on critical stages of crops or atmospheric demand stimulates optimum plant growth. In Vertisols, when flooding irrigation is applied, a significant amount of irrigation water is lost beyond the root zones due to bypass flow. To reduce this loss, irrigation scheduling based on the irrigation water (IW) and cumulative pan evaporation (CPE)-ratio could be an effective strategy. Deep percolation, which occurs when excess water drains below the crop's root zone, is a common source of water loss in most irrigation systems, especially surface irrigation systems. Additionally, trickle irrigation can help conserve water by reducing evaporative losses. Bandyopadhyay et al. (2009) reported that irrigation scheduling at 0.8 IW/CPE significantly improved the soil water extraction, root length density, and grain yield of wheat over irrigation at

0.6 IW/CPE. In addition, the integrated use of 75% of the recommended doses of NPK to wheat with FYM at 5 Mg ha^{-1}, poultry manure at 1.5 Mg ha^{-1}, or phospho-compost at 5 Mg ha^{-1} to the rainy season crops (e.g., soybean or sorghum) significantly improved the root length density, yield, and water use efficiency of wheat over application of 100% NPK to both crops, leading to saving 25% of fertilizer NPK in both seasons and improving the use efficiency of the fertilizer nutrients. The researchers concluded that to improve the productivity and input use efficiency of wheat in central India, it may be grown under soybean-wheat or soybean + sorghum-wheat system with irrigation at 0.8 IW/CPE along with integrated use of 75% NPK and FYM at 5 Mg ha^{-1}, poultry manure at 1.5 Mg ha^{-1}, or phospho-compost at 5 Mg ha^{-1} to rainy season crops and 75% NPK to wheat.

Organic Farming

In India, agricultural soils have low native fertility and low soil organic matter across the agro-ecological region, except hill and mountain soils under forest cover, with the Indian soils having less than 0.5% OC. Hence, adding organic sources of nutrients, such as FYM, vermicompost, rural and urban compost, crop residues, and green leaf compost, and growing legume-based cropping systems play a crucial role in improving soil health (Meena et al., 2013; Singh et al., 2022).

Organic farming is a major component for improving soil health as it regulates the availability of major and micro nutrients in soil and reflects the overall state of soil health and crop productivity for the long term. Hence, judicious use of organic manures not only improves soil quality parameters but also makes farming sustainable and ecofriendly because manure is the fundamental source of soil organic matter. Soil organic matter guides several processes of the soil environment, namely, nutrient transformation, soil structure formation, carbon storage, water retention, and energy supply to beneficial microorganisms (Lal et al., 2019). The long-term field experiment conducted by Singh et al. (2022) revealed that adoption of organic farming practices significantly improved SOC under organic manures (0.93%) compared with inorganic nutrient-management interventions (0.63%), which was attributed to more carbon being incorporated into organically mediated plots. For that reason, adoption of organic farming is beneficial to sustain soil health and crop productivity by improving soil quality (Peltoniemi et al., 2021; Lori et al., 2022).

Biochar as Soil Amendment

Soil health management intervention with soil amendments are logical solutions for altering soil quality and consequently crop productivity. Biochar can overcome soil problems such as acidity, salinity, drought, and low fertility, among others, and subsequently maintain soil health and agricultural productivity (Laird et al., 2010). However, the primary potential advantages of biochar are carbon sequestration and lower carbon footprint, making it is a very important tool for environmental well-being (Clough et al., 2013).

Water retention and microbial growth within the soil increase considerably with application of biochar to soil as amendment. Moreover, it reduces nutrient leaching and heavy metal contamination of crops (Major et al., 2010), enhances crop growth and yield (Meena et al., 2022), and also reduces anthropogenic greenhouse gas fluxes and increases carbon sequestration (Lal, 2011). Biochar application increases the concentration of macronutrients (phosphorus, potassium, nitrogen, calcium, and magnesium) and micronutrients (copper, zinc, iron, and manganese) in soil, which are required for sustainable agriculture (Major et al., 2010). Biochar addition to hard-setting soil resulted in significant changes in soil properties, including increases in carbon, nitrogen, phosphorus, and pH and reductions in soil strength (Steiner et al., 2008). Crop residue biochar has proven effective in sorbing organic contaminants in soil, and manure-derived biochar is also very effective is sorbing heavy metals and other organic contaminants such as atrazine (Cao et al., 2009). Biochar can reduce nitrogen leaching and retain excess nitrogen, thereby lowering nitrous oxide (N_2O) emissions (Yrjala et al., 2022). Similarly, Mandal et al. (2016) also noted that ammonia (NH_3) volatilization was reduced by 70% with poultry-based biochar application. The decreased NH_3 volatilization with the adoption of biochar is attributed to multiple mechanisms such as ammonia adsorption/immobilization and nitrification. Hence, biochar application to soil has emerged as a potential technology for sequestering carbon, reducing greenhouse gas emissions, and improving soil health.

Conclusions

Vertisols are one of the most important soil orders and cover an area of about 72.9 Mha in central India. They are extensively used for growing of agricultural crops under irrigated and rainfed conditions. Vertisols are low to medium in OC, low in available nitrogen and phosphorus, and rich in exchangeable potassium. They contribute nearly 13% of soil carbon stock of the country and have a good carbon sequestration potential. INM practice (NPK + FYM) was found to be the best strategy for improving physical, chemical, and biological attributes of soil. Adoption of conservation agriculture practices significantly enhanced SOC content and physical properties such as aggregation and saturated hydraulic conductivity of the soil under a no-till system in comparison with a conventional tillage system. Vertisols exhibit wide variation in soil quality due to soil heterogeneity and soil degradation caused by land degradation. Compared with PCA, computation of SQI by combining PCA with expert opinion would be the pragmatic approach for computation of soil quality in Vertisols.

References

Aher, S. B., Lakaria, B. L., Kaleshananda, S., Singh, A. B., Ramana, S., Thakur, J. K., Biswas, A. K., Jha, P., Manna, M. C., & Yashona, D. S. (2018). Soil microbial population and enzyme activities under organic, biodynamic and conventional agriculture in semi-arid tropical conditions of central India. *Journal of Experimental Biology and Agricultural Sciences*, *6*(5), 763–773.

Aher, S. B., Lakaria, B. L., Singh, A. B., & Kaleshananda, S. (2019). Soil aggregation and aggregate associated carbon in a vertisol under conventional, organic and biodynamic agriculture in semi-arid tropics of central India. *Journal Indian Society of Soil Science*, *67*(2), 183–191.

Ahmad, N. (1996). Occurrence and distribution of vertisols. In N. Ahmad & A. Mermut (Eds.), *Vertisols and technologies for their management* (pp. 1–41). Elsevier.

Anandakumar, S., Bakhoum, N., Chinnadurai, C., Malarkodi, M., Arulmozhiselvan, K., Karthikeyan, S., & Balachandar, D. (2022). Impact of long-term nutrient management on sequestration and dynamics of soil organic carbon in a semi-arid tropical Alfisol of India. *Applied Soil Ecology*, *177*, Article 104549

Bairwa, J., Dwivedi, B. S., Rawat, A., Thakur, R. K., & Mahawar, N. (2021). Long-term effect of nutrient management on soil microbial properties and nitrogen fixation in a vertisol under soybean–wheat cropping sequence. *Journal Indian Society of Soil Science*, *69*(2), 171–178.

Bandyopadhyay, K. K., Ghosh, P. K., Hati, K. M., & Misra, A. K. (2009). Efficient utilization of limited available water in wheat through proper irrigation scheduling and integrated nutrient management under different cropping systems in a vertisol. *Journal of the Indian Society of Soil Science*, *57*(2), 121–128.

Behera, U. K. (2009). Organic manuring for soil biological health and productivity of a wheat–soybean cropping system in the vertisols of central India. *Crop and Pasture Science*, *60*(4), 382–393.

Bhat, N. A., Riar, A., Ramesh, A., Iqbal, S., Sharma, M. P., Sharma, S. K., & Bhullar, G. S. (2017). Soil biological activity contributing to phosphorus availability in vertisols under long-term organic and conventional agricultural management. *Frontiers in Plant Science*, *8*, Article 1523

Burford, J. R., & Sahrawat, K. L. (1989). *Nitrogen availability in SAT soils: Environment effects on soil processes. Soil fertility and fertility management in semi-arid tropical India*. In C. B. Christianson (Ed.), *Proc. Colloq. held at ICRISAT* (pp. 53–60). International Fertilizer Development Center.

Cao, X., Ma, L., Gao, B., & Harris, W. (2009). *Dairy-manure derived biochar effectively sorbs lead and atrazine. Environmental Science & Technology, 43*, 3285–3291.

Chaudhary, R. S., Somasundaram, J., Sinha, N. K., Lakaria, B. L., Mohanty, S. R., Singh, A. B., Mohanty, M., Hati, K. M., Singh, R. K., & Patra, A. K. (2022). Participatory soil quality assessment using low-cost tools under contrasting management practices in a vertisol. *Agricultural Research, 11*, 642–651.

Chaudhury, J., Mandal, U. K., Sharma, K. L., Ghosh, H., & Mandal, B. (2005). Assessing soil quality under long-term rice-based cropping system. *Communications in Soil Science and Plant Analysis, 36*(9–10), 1141–1161.

Chauhan, B. S., Singh, R. G., & Mahajan, G. (2012). Ecology and management of weeds under conservation agriculture: A review. *Crop Protection, 38*, 57–65.

Clough, T. J., Condron, L. M., Kammann, C., & Müller, C. (2013). A review of biochar and soil nitrogen dynamics. *Agronomy, 3*, 275–293.

Coumar, M. V., Kundu, S., Rajendiran, S., Saha, J. K., Biswas, A. K., Adhikari, T., & Patra, A. K. (2017). Effect of soil amendments on microbial resilience capacity of acid soil under copper stress. *Bulletin of Environmental Contamination and Toxicology, 99*, 625–632.

Coumar, M. V., Kundu, S., Saha, J. K., Rajendiran, S., Dotaniya, M. L., Karthikeyan, K., & Patra, A. K. (2020). Soil carbon dynamics, carbon budget and its relationship with crop yield under different cropping systems in vertisols of Central India. *Indian Journal of Agricultural Sciences, 90*(1), 123–127.

Dalal, R. C. (1977). Fixed ammonium and carbon–nitrogen ratios of Trinidad soils. *Soil Science, 124*, 323–327.

Damodar Reddy, D., Subba Rao, A., Sammi Reddy, K., & Takkar, P. N. (1999). Yield sustainability and phosphorus utilization in soybean-wheat system on vertisols in response to integrated use of manure and fertilizer phosphorus. *Field Crops Research, 62*, 181–190.

Das, D. K., & Dutta, D. (2000). Role of soil and land quality indicators in sustainable management of land resources. In *Advances in land resources management for 21st century* (pp. 480–486). Soil Conservation Society of India.

Dawe, D., Dobermann, A., Moya, P., Abdulrachman, S., Singh, B., Lal, P., Li, S. Y., Lin, B., Panaullah, G., Sariram, O., Singh, Y., Swarup, A., Tan, P. S., & Zhen, Q. X. (2000). How widespread are yield declines in long-term rice experiments in Asia? *Field Crops Research, 66*, 175–193.

Dhruvanarayana, V. V., & Babu, R. (1983). Estimation of soil erosion in India. *Journal of Irrigation and Drainage Engineering, 109*(4), 419–434.

Dotaniya, C. K., Lakaria, B. L., Sharma, Y., Meena, B. P., Aher, S. B., Shirale, A. O., Pandurang, P. G., Dotaniya, M. L., Biswas, A. K., Patra, A. K., Yadav, S. R., Reager, M. L., Sanwal, R. C., Doutaniya, R. K., & Lata, M. (2022). Performance of chickpea (*Cicer arietinum* L.) in maize-chickpea sequence under various integrated nutrient modules in a vertisol of Central India. *PLoS One*, *17*(2), Article e0262652

Dudal, R. (1965). *Dark clay soils of tropical and subtropical regions* (FAO Agric. Dev. paper 83). FAO.

El-Swaify, S. A., Pathak, P., Rego, T. J., & Singh, S. (1985). Soil management for optimized productivity under rainfed conditions in the semi-arid tropics. In B. A. Stewart (Ed.), *Advances in soil sciences* (Vol. 1, pp. 1–64). Springer.

Fortuna, A., Harwood, R. R., & Paul, E. A. (2003). The effects of compost and crop rotations on carbon turnover and the particulate organic matter fraction. *Soil Science*, *168*, 434–444.

Ghosh, P., Dayal, D., Mandal, K. G., Wanjari, R., & Hati, K. M. (2003). Optimization of fertilizer schedules in fallow and ground-based cropping systems and an assessment of system sustainability. *Field Crops Research*, *80*, 83–98.

Goswami, N. N. (1987). Optimizing soil environment for sustained agricultural production. In T. N. Khoshoo (Ed.), *Perspectives in environmental management* (pp. 95–115). Oxford and IBH Publishing Company Private Limited.

Goswami, N. N. (1995). Soil research in India—A review. *Journal Indian Society of Soil Science*, *43*, 495–503.

Goswami, N. N. (2005). Soil and its quality vis-à-vis sustainability and society: Some random thoughts. In *Indian Society of Soil Science ICSWEQ – Proceedings* (pp. 43–58).

Goswami, N. N., & Rattan, R. K. (1992). Soil health—Key to sustained agricultural productivity. *Fertiliser News*, *37*(12), 53–60.

Hati, K. M., Chaudhary, R. S., Mandal, K. G., Bandyopadhyay, K. K., Singh, R. K., Sinha, N. K., Mohanty, M., Somasundaram, J., & Saha, R. (2015). Effects of tillage, residue and fertilizer nitrogen on crop yields, and soil physical properties under soybean-wheat rotation in vertisols of Central India. *Agricultural Research*, *4*(1), 48–56.

Hati, K. M., Swarup, A., Dwivedi, A. K., Misra, A. K., & Bandyopadhyay, K. K. (2007). Changes in soil physical properties and organic carbon status at the topsoil horizon of a vertisol of central India after 28 years of continuous cropping, fertilization and manuring. *Agriculture, Ecosystems and Environment*, *119*(2), 127–134.

Herrick, J. E., Brown, J. R., Tugel, A. J., Shaver, P. L., & Havstad, K. M. (2002). Application of soil quality to monitoring and management: Paradigms from rangeland ecology. *Agronomy Journal*, *94*, 3–11.

Hobbs, P. R., Sayre, K., & Gupta, R. (2008). The role of conservation agriculture in sustainable agriculture. *Philosophical Transactions of the Royal Society B*, *363*, 543–555.

International Crops Research Institute for Semi-Arid Tropics. (1981). *Farming systems. Annual Report 1979/80* (pp. 182–185). ICRISAT, India.

International Crops Research Institute for Semi-Arid Tropics. (1988). Phosphorus in Indian vertisols (pp. 23–26). Summary Proceedings of a Workshop, August 1988, ICRISAT Center, Patancheru, India.

Jha, P., De, A. , Lakaria, B. L., Biswas, A. K., Singh, M., Reddy, K. S. and Subba Rao, A. (2012) Soil carbon pools, mineralization and fluxes associated with land use change in vertisols of central India. *National Academy Science letters (Springer), 35 (6),* 475–483.

Jha, P., Lakaria, B. L., Biswas, A. K., Saha, R., Mahapatra, P., Agrawal, B. K., Sahi, D. K., Wanjari, R. H., Lal, R., Singh, M., & Rao, A. S. (2014). Effects of carbon input on soil carbon stability and nitrogen dynamics. *Agriculture, Ecosystems and Environment, 189*, 36–42.

Jha, P., Lakaria, B. L., Vishwakarma, A. K., Wanjari, R. H., Mohanty, M., Sinha, N. K., Somasundaram, J., Dheri, G. S., Dwivedi, A. K., Sharma, R. P., Singh, M., Dalal, R. C., Biswas, A. K., Patra, A. K., & Chaudhari, S. K. (2021). Modeling the organic carbon dynamics in long-term fertilizer experiments of India using the Rothamsted carbon model. *Ecological Modelling, 450*, Article 109562

Karlen, D. L., Mausbach, M. J., Doran, J. W., Cline, R. G., & Harris, R. F. (1997). Soil quality: A concept, definition, and framework for evaluation. *Soil Science Society of America Journal, 61*, 4–10.

Katkar, R. N., Kharche, V. K., Sonune, B. A., Wanjari, R. H., & Singh, M. (2012). Long term effect of nutrient management on soil quality and sustainable productivity under sorghum-wheat crop sequence in vertisol of Akola. *Maharashtra. Agropedology, 22*(2), 103–114.

Kumawat, A., Vishwakarma, A. K., Wanjari, R. H., Sharma, N. K., Yadav, D., Kumar, D., & Biswas, A. K. (2022). Impact of levels of residue retention on soil properties under conservation agriculture in vertisols of central India. *Archives of Agronomy and Soil Science, 68*(3), 368–382.

Kundu, S., Coumar, M. V., Saha, J. K., Rajendiran, S., Hati, K. M., Biswas, A. K., Reddy, K. S., & Subba Rao, A. (2012). Assessing soil health of vertisol of AESR 10.1 using selected physical, chemical and biological attributes of soil. *Journal Indian Society of. Soil Science, 60*(4), 281–287.

Kundu, S., Saha, J. K., Adhikari, T., Vassanda Coumar, M., Biswas, A. K., Rajendiran, S., Hati, K. M., Bhattachrya, T., & Subba Rao, A. (2013). Assessment of soil quality and resilience of vertisols of Madhya Pradesh. In S. Kundu, M. C. Manna, A. K. Biswas, R. S. Chaudhary, B. L. Lakaria, & A. Subba Rao (Eds.), *IISS contribution in frontiers area of soil research* (pp. 83–126). Indian Institute of Soil Science.

Kundu, S., Singh, M., Saha, J. K., Biswas, A., Tripathi, A. K., & Acharya, C. L. (2001). Relationship between C addition and storage in a vertisol under soybean-wheat cropping system in sub-tropical central India. *Journal of Plant Nutrition and Soil Science, 164*, 483–486.

Laird, D., Fleming, P., Wang, B., Horton, R., & Karlen, D. (2010). Biochar impact on nutrient leaching from a midwestern agricultural soil. *Geoderma*, *158*, 436–442.

Lakaria, B. L., Adhikari, T., Jha, P., Biswas, A. K., Sammi Reddy, K., Singh, M., Chaudhary, R. S., & Wanjari, R. H. (2017). Characterization of humic and fulvic acid under long-term integrated nutrient management of soybean-wheat cropping system in vertisol. *Journal Indian Society of Soil Science*, *65*(1), 32–41.

Lakaria, B. L., Singh, M., Sammi Reddy, K., Biswas, A. K., Jha, P., Chaudhary, R. S., Singh, A. B., & Subba Rao, A. (2012). Carbon addition and storage under integrated nutrient management in soybean–wheat cropping sequence in a vertisol of central India. *National Academy Science Letters*, *35*(3), 131–137.

Lal, B., Gautam, P., Nayak, A. K., Panda, B. B., Bihari, P., Tripathi, R., Shahid, M., Guru, P. K., Chatterjee, D., Kumar, U., & Meena, B. P. (2019). Energy and carbon budgeting of tillage for environmentally clean and resilient soil health of rice-maize cropping system. *Journal of Cleaner Production*, *226*, 815–830.

Lal, R. (2011). Sequestering carbon in soils of agro-ecosystems. *Food Policy*, *36*, S33–S39.

Lal, R., & Kimble, J. M. (1997). Conservation tillage for carbon sequestration. *Nutrient Cycling in Agroecosystems*, *49*, 243–253.

Lenka, N. K., Biswas, A. K., Chandran, P., Dutta, D., Rajendiran, S., Wanjari, R. H., Singh, A. B., Lakaria, B. L., Sinha, N. K., Vishwakarma, A. K., Meena, B. P., Shirale, A. O., Lenka, S., Patra, A. K., & Chaudhari, S. K. (2022a). *Soil quality mapping and validation of indexing approach for major production region of India* (Technical bulletin 36). ICAR-Indian Institute of Soil Science.

Lenka, N. K., Meena, B. P., Lenka, S., Shirale, A. O., & Wanjari, R. H. (2022b). *Project completion report of the DST-SERB funded project on assessing the impact of imbalanced use of chemical fertilizers on soil health using a soil function based quantitative approach*. ICAR-Indian Institute of Soil Science.

Lingade, S. R., Srivastava, R., Prasad, J., & Saxena, R. K. (2008). Occurrence of sodic vertisols in Nagpur district, Maharashtra. *Journal of the Indian Society of Soil Science*, *56*, 231–232.

Lori, M., Armengot, L., Schneider, M., Schneidewind, U., Bodenhausen, N., Mäder, P., & Krause, H. M. (2022). Organic management enhances soil quality and drives microbial community diversity in cocoa production systems. *Science of the Total Environment*, *834*, Article 155223

Major, J., Rondon, M., Molina, D., Riha, S., & Lehmann, J. (2010). Maize yield and nutrition during 4 years after biochar application to a Colombian Savanna Oxisol. *Plant and Soil*, *333*, 117–128.

Mandal, K. G., Hati, K. M., Misra, A. K., Bandyopadhyay, K. K., & Tripathy, A. K. (2013). Land surface modification and crop diversification for enhancing productivity of a vertisol. *International Journal of Plant Production*, *7*, 455–472.

Mandal, K. G., Misra, A. K., & Hati, K. M. (2000). Effect of combination of NPK and FYM on growth, yield and agronomic efficiency of soybean (*Glycine max*) in vertisol. *Environment and Ecology*, *18*(1), 207–209.

Mandal, K. G., Saha, K. P., Hati, K. M., Singh, V. V., Misra, A. K., Ghosh, P. K., & Bandyopadhyay, K. K. (2005). Cropping systems of central India: An energy and economic analysis. *Journal of Sustainable Agriculture*, *25*(3), 117–140.

Mandal, S., Thangrajan, R., Bolan, N. S., Sarkar, B., Khan, N., Ok, Y. S., & Naidu, R. (2016). Biochar-induced concomitant decrease in ammonia volatilization and increase in nitrogen use efficiency by wheat. *Chemosphere*, *142*, 120–127.

Masto, R. E., Chhonkar, P. K., Singh, D., & Patra, A. K. (2007). Soil quality response to long-term nutrient and crop management on a semi-arid inceptisol. *Agriculture, Ecosystems and Environment*, *118*(1–4), 130–142.

Meena, B. P., Biswas, A. K., Singh, M., Chaudhary, R. S., Singh, A. B., Das, H., & Patra, A. K. (2019). Long-term sustaining crop productivity and soil health in maize–chickpea system through integrated nutrient management practices in vertisols of central India. *Field Crops Research*, *232*, 62–76.

Meena, B. P., Biswas, A. K., Singh, M., Das, H., Chaudhary, R. S., Singh, A. B., Shirale, A. O., & Patra, A. K. (2021). Energy budgeting and carbon footprint in long-term integrated nutrient management modules in a cereal- legume (*Zea mays–Cicer arietinum*) cropping system. *Journal of Cleaner Production*, *314*, Article 127900

Meena, B. P., Jha, P., Ramesh, K., Biswas, A. K., Elanchezhian, R., Das, H., Sathyaseelan, N., Shirale, A. O., & Patra, A. K. (2022). Agronomic management based on multi-split topdressing increases grain yield and nitrogen use efficiency in rainfed maize in Vertisols of India. *Journal of Plant Nutrition*, *45*(6), 828–844.

Meena, B. P., Kumar, A., Meena, S. R., Dhar, S., Rana, D. S., & Rana, K. S. (2013). Effect of sources and levels of nutrients on growth and yield behaviour of popcorn (*Zea mays*) and potato (*Solanum tuberosum*) sequence. *Indian Journal of Agronomy*, *58*(4), 474–479.

Meena, B. P., Shirale, A. O., Dotaniya, M. L., Jha, P., Meena, A. L., Biswas, A. K., & Patra, A. K. (2016). Conservation agriculture: A new paradigm for improving input use efficiency and crop productivity. In J. K. Bisht, V. S. Meena, P. K. Mishra, & A. Pattanayak (Eds.), *Climate change and conservation agriculture in Indian Himalaya*. Springer.

Meshram, N. A., Ismail, S., & Pinjari, S. S. (2018). Long-term effect of FYM and inorganic fertilizers on soil quality and sustainable productivity of soybean (*Glycine max*) and safflower (*Carthamus tinctorius*) in vertisol. *Indian Journal of Agricultural Sciences*, *88*, 845–850.

Mirzavand, J., Asadi-Rahmani, H., & Moradi-Talebbeigi, R. (2020). Biological indicators of soil quality under conventional, reduced, and no-tillage systems. *Archives of Agronomy and Soil Science*, *68*(3), 311–324.

Mohanty, M., Bandyopadhyay, K. K., Painuli, D. K., Ghosh, P. K., Misra, A., & Hati, K. (2007a). Water transmission characteristics of a Vertisol and water use efficiency of rainfed soybean (Glycine max (L.) Merr.) under subsoiling and manuring. *Soil and Tillage Research*, *93*, 420–428.

Mohanty, M., Painuli, D. K., Misra, A. K., & Ghosh, P. K. (2007b). Soil quality effects of tillage and residue under rice-wheat cropping on a vertisol in India. *Soil and Tillage Research*, *92*(1–2), 243–250.

Mohanty, S. R., Parihar, M., Dubey, G., Parmar, R., Jain, D., Singh, M., Patra, A., Chaudhari, S. K., & Kollah, B. (2022). Nitrification under the influence of long-term fertilizer application in a tropical vertisol. *Archives of Agronomy and Soil Science*, *68*(4), 516–530.

Motsara, M. R. (2002). Available NPK status of Indian soils as depicted by soil fertility maps. *Fertiliser News*, *47*, 15–21.

Murthy, A. S. P. (1988). Distribution, properties and management of vertisols of India. *Advances in Soil Science*, *8*, 151–214.

Murthy, I. Y. L. N., Sastry, T. G., Datta, S. C., Narayanasamy, G., & Rattan, R. K. (2002). Phosphate dynamics in Vertisols of different parent material. *Journal of the Indian Society of Soil Science*, *50*, 14–16.

Murthy, R. S., Bhattacharjee, J. C., Landey, R. J., & Pofali, R. M. (1982). *Distribution, characteristics and classification of vertisols*. In *Vertisols and Rice Soils of the Tropics, Symposia Paper II, 12th International Congress of Soil Science, New Delhi* (pp. 3–22). Indian Society of Soil Science.

National Academy of Agricultural Science. (2010). *Degraded and wastelands of India, status and spatial distribution*. Indian Council of Agricultural Research and National Academy of Agricultural Science.

NRSA and Associates. (1996). Mapping of salt affected soils of India, 1:250000 map sheets, legend. NRSA Hyderabad.

Pal, D. K., Dasog, G., & Bhattacharyya, T. (2009). Pedogenetic processes in cracking clay soils (vertisols) in tropical environments of India: A critique. *Journal of the Indian Society of Soil Science*, *57*, 422–432.

Pal, D. K., Lal, S., Bhattacharyya, T., Chandran, P., Ray, S. K., Satyavathi, P. L. A., Raja, P., Maurya, U. K., Durge, S. L., & Kamble, G. K. (2010). *Pedogenic thresholds in benchmark soils under rice-wheat cropping system in a climosequence of the Indo-Gangetic alluvial plains* (Final project report). Division of Soil Resource Studies, NBSS and LUP (ICAR).

Paltineanu, C. (2001). Nutrient leaching in cracked vertisol in Romania. *Agronomie*, *21*, 427–433.

Paroda, R. S. (2018). *Strategy for doubling farmers' income* (Strategy paper). Trust for Advancement of Agricultural Sciences.

Patel, A. M., Patil, R. G., Raman, S., Singh, T. P., & Shukala, P. (2000). *Extent of salt affected and waterlogged soils in Surat Branch of Ukai Kakrapar Command*

(SWM technical bulletin 13). Soil and Water Management Research Unit, Gujarat Agricultural University.

Patil, R. H., & Jawaregowda. (2000). Pre-sowing agronomic practices for disease management. *The Hindu*, (2 Nov 2000).

Patra, A. K., Burford, J. R., & Rego, T. J. (1996). Volatilization losses of surface-applied urea nitrogen from the Vertisols in the Indian semi-arid tropics. *Biology and Fertility of Soils*, *22*, 345–349.

Peltoniemi, K., Velmala, S., Fritze, H., Lemola, R., & Pennanen, T. (2021). Long-term impacts of organic and conventional farming on the soil microbiome in boreal arable soil. *European Journal of Soil Biology*, *104*, Article 103314

Rana, C. H., & Raman, S. (1999). Issues related to water management in Agriculture—An over view. In S. Raman (Ed.), *SWMP pub. 10* (pp. 1–10). Soil and Water Management Research Unit, Gujarat Agricultural University.

Sahrawat, K. L. (1995). Fixed ammonium and carbon-nitrogen ratios of some semi-arid tropical Indian soils. *Geoderma*, *68*, 219–224.

Sharda, V. N., Dogra, P., & Prakash, C. (2010). Assessment of production losses due to water erosion in rainfed areas of India. *Indian Journal of Soil and Water Conservation*, *65*, 79–91.

Sharma, K. L., Chandrika, D. S., Grace, J. K., Maruthi Shankar, G. R., Sharma, S. K., Thakur, H. S., Jain, M. P., Sharma, R. A., Ravindra Chary, G., Srinivas, K., Gajbhiye, P., Venkatravamma, K., Lal, M., Satish Kumar, T., Usha Rani, K., Ramachandran, K., Rao, C. S., Reddy, K. S., & Venkateswarlu, B. (2016). Soil quality assessment under restorative soil management practices in soybean (*Glycine max*) after six years in semi-arid tropical black lands of central India. *Communications in Soil Science and Plant Analysis*, *47*(12), 1465–1475.

Sharma, K. L., Grace, J. K., Mishra, P. K., Venkateswarlu, B., Nagdeve, M. B., Gabhane, V. V., Sankar, G. M., Korwar, G. R., Chary, G. R., Rao, C. S., Gajbhiye, P. N., Madhavi, M., Mandal, U. K., Srinivas, K., & Ramachandran, K. (2011). Effect of soil and nutrient-management treatments on soil quality indices under cotton-based production system in rainfed semi-arid tropical vertisol. *Communications in Soil Science and Plant Analysis*, *42*(11), 1298–1315.

Sharma, K. L., Mandal, U. K., Srinivas, K., Vittal, K. P. R., Mandal, B., Grace, J. K., & Ramesh, V. (2005). Long-term soil management effects on crop yields and soil quality in a dryland Alfisol. *Soil and Tillage Research*, *83*, 246–259.

Singh, A. B., Meena, B. P., Lakaria, B. L., Thakur, J. K., Ramesh, K., Rajput, P. S., & Patra, A. K. (2022). Production potential, soil health and economics of soybean (*Glycine max*)-linseed (*Linum usitatissimum*) cropping system under various nutrient-management protocols. *Indian Journal of Agronomy*, *67*, 102–108.

Singh, M., & Wanjari, R. H. (2012). Yield response to applied nutrients and estimates of N_2 fixation in 33-year-old soybean-wheat experiment on a vertisol. *Experimental Agriculture*, *48*(3), 311–325.

Singh, M., & Wanjari, R. H. (2013). Measures to sustain and restore declined productivity in alfisols under long term fertilizer experiments. *Indian Journal of Fertilizers*, *9*(2), 24–32.

Singh, S. K., Baser, B. L., Shyampura, R. L., & Narain, P. (2003). Phosphorus fractions and their relationship to weathering indices in vertisols. *Journal of the Indian Society of Soil Science*, *51*, 247–251.

Smith, C. K., Coyea, M. R., & Munson, A. D. (2000). Soil carbon, nitrogen and phosphorous stocks and dynamics under disturbed black spurce forests. *Ecological Applications*, *10*, 775–788.

Somasundaram, J., Chaudhary, R. S., Awanish Kumar, D., Biswas, A. K., Sinha, N. K., Mohanty, M., Patra, A. K., & Chaudhari, S. K. (2018). Effect of contrasting tillage and cropping systems on soil aggregation, carbon pools and aggregate-associated carbon in rainfed vertisols. *European Journal of Soil Science*, *69*, 879–891.

Somasundaram, J., Salikram, M., Sinha, N. K., Mohanty, M., Chaudhary, R. S., Dalal, R. C., Mitra, N. G., Blaise, D., Coumar, M. V., Hati, K. M., & Thakur, J. K. (2019). Conservation agriculture effects on soil properties and crop productivity in a semiarid region of India. *Soil Research*, *57*(2), 187–199.

Sonune, B. A., Kharche, V. K., Gabhane, V. V., Jadhao, S. D., Mali, D. V., Katkar, R. N., Kadu, P. R., Konde, N. M., Deshmukh, D. P., & Goramnagar, H. B. (2021). Sustaining soil health and cotton productivity with tillage and integrated nutrient management in vertisols of central India. *Indian Journal of Soil Conservation*, *49*(1), 1–11.

Sparling, G. P., & Schipper, L. A. (2002). Soil quality at a national scale in New Zealand. *Journal of Environmental Quality*, *31*, 1848–1857.

Srinivasarao, C. H., Venkateswarlu, B., Lal, R., Singh, A. K., Vittal, K. P. R., Kundu, S., Singh, S. R., & Singh, S. P. (2012). Long-term effects of soil fertility management on carbon sequestration in a rice-lentil cropping system of the Indo-Gangetic plains. *Soil Science Society of America Journal*, *76*(1), 168–178.

Steiner, C., Glaser, B., Teixeira, W. G., Lehmann, J., Blum, W. E. H., & Zech, W. (2008). Nitrogen retention and plant uptake on a highly weathered central Amazonian Ferralsol amended with compost and charcoal. *Journal of Plant Nutrition and Soil Science*, *171*, 893–899.

Subba Rao, A., Srivastava, S., & Ganeshamurthy, A. N. (2015). Phosphorus supply may dictate food security prospects in India. *Current Sciences*, *108*, 1253–1261.

Tandon, H. L. S., & Kanwar, J. S. (1984). *A review of fertilizer use research on sorghum in India* (Research bulletin no. 8). International Crops Research Institute for the Semi-Arid Tropics.

Thakur, J. K., Mandal, A., Manna, M. C., Jayaraman, S., & Patra, A. K. (2021). Impact of residue burning on soil biological properties. In *Conservation agriculture: A sustainable approach for soil health and food security* (pp. 379–389). Springer.

Vasu, D., Singh, S. K., Ray, S. K., Duraisami, V. P., Tiwary, P., Chandran, P., Nimkar, A. M., & Anantwar, S. G. (2016). Soil quality index (SQI) as a tool to evaluate crop productivity in semi-arid Deccan Plateau, India. *Geoderma*, *282*, 70–79.

Virmani, S. M., Sahrawat, K. L., & Burford, J. R. (1982). Physical and chemical properties of Vertisols and their management. Symposium Papers 1I (pp. 80–91), Transactions of the Twelfth International Congress of Soil Science, New Delhi.

Yrjala, K., Ramakrishnan, M., & Salo, E. (2022). Agricultural waste streams as resource in circular economy for biochar production towards carbon neutrality. *Current Opinion in Environmental Science & Health*, *26*, Article 100339

9

Strategies for Restoring Soil Health in Indian Himalayan Region

Anup Das, Christi B. K. Sangma, Mahasweta Chakraborty, Saurav Saha, Amit Kumar, Bappa Das, Jaynata Layek, Subhash Babu, and Prashant Pandey

Chapter Overview

The fragile ecosystem of the Indian Himalayas is susceptible to catastrophic ecological decline from anthropogenic influences. Poor soil health plays a crucial role in the ecological degradation and disruption in ecosystem functioning, often adversely affecting the livelihood of resource-poor farming communities of the Indian Himalayan region. Changes in land use and associated management practices often exert a strong impact on soil organic carbon dynamics as well as soil health. In this chapter, we evaluate doable strategies and available field-scale technological options for meeting the rising challenges of regional food security and sustaining soil health toward restoration of degraded agro-ecosystem functioning and sustainable crop production in the Indian Himalayan region through recarbonization of depleted soil carbon pools, enhancing soil biodiversity, nutrient recycling, crop rotation, agroforestry systems, small-scale farm mechanization, and crop residue management in combination with minimum soil disturbance. Supportive location-specific future policy planning should aim to facilitate field-scale sustainable technological adaptation for maintaining sustainable agro-ecosystem health in the long run.

Soil Health Series: Volume 4 Soil Health and Sustainability in India, First Edition.
Edited by Ranjan Bhattacharyya.

Introduction

Historically, family-based subsistence farming is practiced in the Himalayan agricultural system. About 50 million people of diverse ethnic backgrounds reside in 10 Indian states/union territories of the Indian Himalayan region (IHR; 533,000 km^2 extending between 21°57′ N to 37°5′ N and 72°40′ E to 97°25′ E). This region accounts for 16.2% of the total geographical area (TGA) of India. The IHR stretches from the western states of Jammu and Kashmir, Himachal Pradesh, and Uttarakhand to the seven hill states of India's northeast and four hill districts (Dima Hasao and Karbi Anglong district of Assam and Darjeeling and Kalimpong district of West Bengal; Figure 9.1). The IHR represents 52.3% of the aerial expansion of the entire Eastern Himalayan region, followed by Myanmar (17.9%), Nepal (16.08%), Bhutan (7.6%), and China (6.26%) (Tse-ring et al., 2010). The IHR also accounts for 19% of the dense forest vegetation of India along with 33% of country's total soil organic carbon (SOC) reserve (Batjes, 1996). In this chapter, we include Nepal Himalaya as a segment of the IHR.

The regional inhabitants of IHR rely heavily on the region's natural resources for their livelihoods and survival (NITI Aayog, 2018). Forest cover accounts >65% of the total IHR geographical area. At present, the future sustainability of human

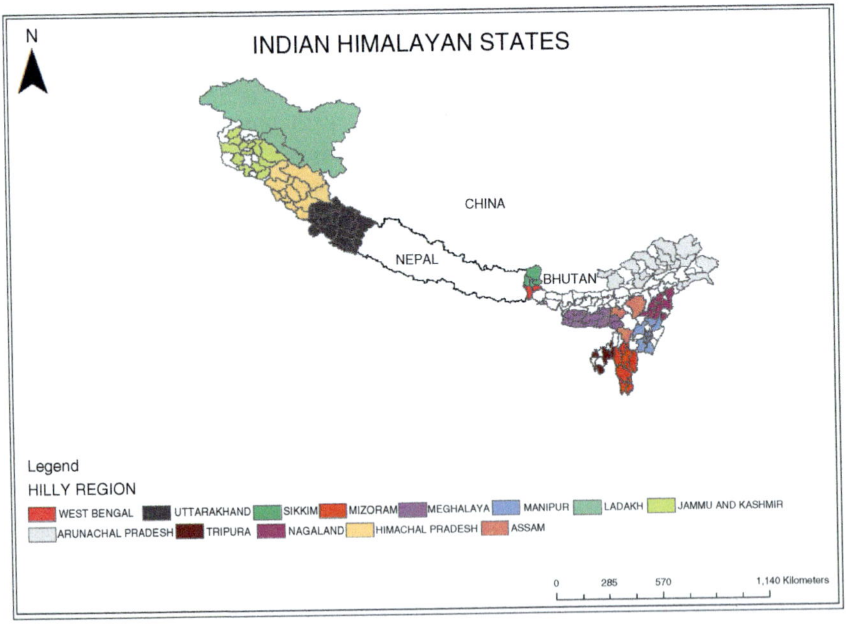

Figure 9.1 Delineated state boundaries of Indian Himalayan region (IHR).

settlements and the progress of civilization are severely affected by an ever-increasing population pressure (2.1% per annum; Tse-ring et al., 2010) and deterioration of soil health from intensified agriculture practices. Despite limited arable land availability, the IHR contributes 15%–18% of the country's total agricultural production with (Nagdeve, 2007). The location-specific soil capacity to sustain agricultural activity and livestock rearing determines the future likelihood of sustainable food production. However, the increased need for improving crop productivity to meet the requirements of a rising population paves the way for overexploitation/improper use of soil resources, followed by the widespread decline in soil quality (i.e., extensive land degradation problems) in the IHR (Das et al., 2011; Lal, 2015).

Moreover, climate change and the poorly planned increase in human activities in the fragile mountains of Indian Himalayas have exacerbated the region's susceptibility to increased ecological disasters, which often arise from natural habitat loss caused by massive deforestation, rapid land use change, deterioration of ecosystem health, and decline in agro-biodiversity across the IHR (ICIMOD, 2010). The projected rise in temperatures across the IHR may be between 1.7 °C and 2.2 °C, with precipitation increasing by 60–206 mm by the 2030s (5%–13% more against 1970 as base year; Anonymous, 2012). In particular, the warming trends across the Kanchenjunga landscape of the IHR vary by 0.01–0.015 °C/yr, which is more prominent during winter months at higher altitudes (>4,000-m elevation; Singh et al., 2011). Several mountain communities in Indian Himalaya are facing the serious problem of more frequent landslide events caused by extreme precipitation driven by climate change. These events adversely affect the local economy, destroy habitats, reduce the scope of agriculture, and narrow other livelihood prospects, particularly for the underprivileged marginal land holders (Dikshit et al., 2020; Saha et al., 2022). Adaptation of ecofriendly and sustainable ecosystem management toward the conservation of natural resources in the IHR has become the foremost policy need for maintaining the social prosperity of ethnic populations in the long run.

One of the most valuable natural resources is soil, which supports life and ecosystem functions. Sustainable soil management is the primary key for maintaining long-term ecosystem productivity and sustainable food grain production through optimized land use (Das et al., 2021). The IHR is highly vulnerable to massive soil losses and plant nutrients as a result of the steeply sloping terrain, variable vegetation dynamics, physiography, and overuse of natural resources such as soils and forests (Sidhu, 2016). In the IHR, widespread land degradation is frequently a key antecedent to soil erosion, a global phenomenon that causes nutrient loss and is a substantial impediment to agricultural development and crop production (Barman et al., 2020; Mishra & Rai, 2013). The majority of the Himalayas in the northern and eastern Indian states are made up of fluvial-deposit-like sandstone, shale, and conglomerates. About 45% of the IHR and Western Ghats area are severely affected by soil erosion (Das et al., 2009; Narayan & Babu, 1983). Water erosion causes the most land degradation (82.57 million ha [Mha]; ~68%), followed by chemical degradation (24.68 Mha), wind erosion (12.4 Mha), and physical degradation (1.070 Mha), which includes waterlogging, mining, and industrial waste disposal (Figure 9.2). Geological erosion, often known as "natural" erosion, is a process whereby rock is gradually eroded away by gravity, water, wind, ice, and other natural forces over time to produce soil and change the ground surface. About

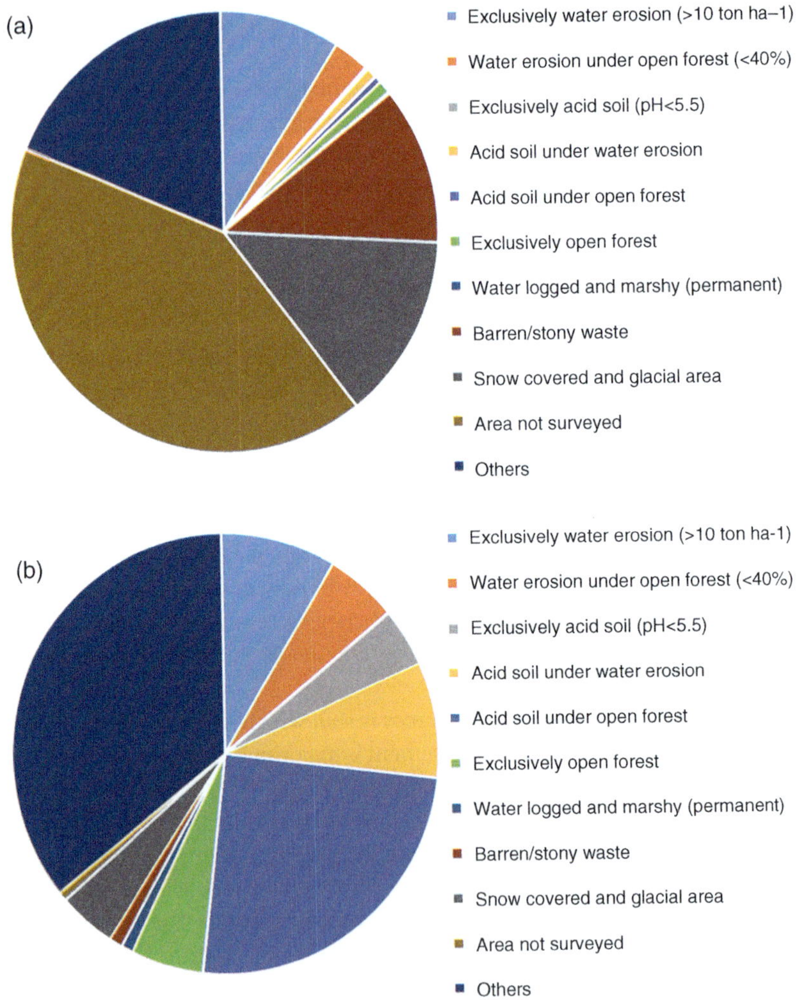

Figure 9.2 Extended of land degradation area (%) in Indian Himalayas (a) Western Himalaya Region - WHR (b) Eastern Himalaya Region – EHR (CSWCR&TI, 2011).

one-third of India's northeastern hill (NEH) region has been degraded as a result of extensive soil erosion by water (Maji et al., 2010; Sehgal & Abrol, 1994).With the rise in anthropogenic disturbances from the expanding human population, the net area under intensified agricultural activities and rapid urbanization has exacerbated the erosion process considerably owing to efforts being made globally to supply the steadily rising need for food, fuel, and wood for daily necessities, including in the IHR (Choudhury et al., 2022; Das et al., 2011).

The net expanse of increased regional food production, fodder availability, and fuel use (timber) may come from the diminishing global forest area, and accelerated land degradation may be adversely affecting ecosystem services across the IHR (Anonymous, 2013; Fróna et al., 2019). In this chapter, we cover the present status of land degradation as a potential threat to existing ecosystem services. We also provide an overview of efficient soil health restoration options under the diverse agro-ecosystems of the IHR.

Land Degradation in the IHR

Worldwide, over 1.9 billion hectares of land is affected by degradation resulting from poor decisions on land use and reduction of soil quality from accelerated soil erosion threatening ecological sustainability (United Nations, 2015). In India, conservative estimates confirmed slightly higher average annual soil erosion rate of ~16.4 t/ha against the global mean varies between 12.0 and 15.0 t/ha (i.e., 0.90–0.95 mm soil; den Biggelaar et al., 2003; Dhruvanarayana & Babu, 1983). Almost 10% of the total eroded soil is deposited in reservoirs, reducing their storage capacity by 1% or 2% annually, while nearly 29% is permanently lost to the sea (Bhattacharyya et al., 2007). The IHR is not a unique case. Extensive soil erosion and heavy sediment load, altered vegetation characters and disappearances of wetlands over past few decades became the effective geo-indicators of anthropogenic land degradation in the IHR (Table 9.1) either through changes in biophysical attributes of earth's surface (land cover) or other human-induced (land use) manipulations (Lal, 2001). Contrary to the ecologically advised guideline of 5–12 ha of forest per hectare of cultivated area, there was only 1.26 ha of forests for every hectare of farming in the IHR with higher annual siltation rates ~2,400 million tonnes (Mt) (Tiwari, 2000). For north East Indian state of Tripura total annual soil loss was estimated to be 15 Mt/yr (Bhattacharyya et al., 2007). Some regions of the IHR are at such high risk of soil erosion from manipulated intensive and inappropriate land use that it became impossible to restore for productive use but left for permanent abandonment, leading to serious land degradation problems (Sharda et al., 2010).

Table 9.1 Brief Status of Reported Soil Loss Across the IHR.

Sl. no.	Location	Annual rainfall (mm)	Methodology and land use	Annual soil loss/ sediment load	References
1	Khasi Hills of Shillong, Meghalaya (1,500 m asl)	1,800–1,900	Erosion plot (shifting cultivation and terrace culture)	49.7 and 56.3 t/ha/yr	Mishra and Ramakrishnan (1983)
2	Northwest part of the central Himalaya (329–2,119 m asl)	—	Erosion plot (forest, *sal*, oak etc.)	0.015–0.057 t/ha/yr	Pathak et al. (1984)
3	Northeastern India	—	Erosion plot (shifting cultivation)	30–170 t/ha/yr	Singh et al. (1992)
4	Nana Kosi Watershed, Central Himalaya (1,200–1,900 m asl)	431.7–1,175	Erosion plot: oak plantation, pine forest, agriculture	0.54–4.77 t/ha/yr	Rawat and Rawat (1994)
5	Umroi watershed, Meghalaya	2,600	Erosion plot (shifting cultivation)	76.6 t/ha/yr	Satapathy (1996)
6	Mamlay watershed, Sikkim (300–2,650 m asl)	1,494–2,100	Discharge study hydro-meteorological stations- large cardamom, fodder fuel, mandarin plantation, forestry, barren, and agriculture	4.18–8.82 t/ha/yr	Rai and Sharma (1998)
7	Pranmati watershed, Chamoli district, Uttarakhand (1,120–4,070 m asl)	1,700	Erosion plot (terrace cultivation)	0.3–64.39 t/ha/yr	Sen et al. (1997)
8	Likhu Khola watershed, middle mountain region of Nepal (780–2,800 m asl)	998–2,839	Universal soil loss equation (USLE)	<1–56 t/ha/yr	Shrestha (1997)
9	Pindar catchment of Nandadevi Biosphere Reserve, Central Himalaya	2,400	Erosion plot (Horse chestnut, Kharsu oak, and Silver fir)	0.009–0.046 t/ha/yr	Negi et al. (1998)

10	Khechepalri Lake, Sikkim (1,700 m asl)	3,837	Erosion plot: forest, cardamom, agroforestry, and barren land	502 $Mg/km^2/yr$	Jain et al. (2000)
11	Balia catchment of Siwalik, Kumaon region (546 and 2,610 m asl)	3,000	Hydro-meteorological stations (grazing, forest, barren, cultivated areas)	22.14–42.30 t/ha/yr	Tiwari (2000)
	Shail Gad watershed in Lesser Himalaya (1,100–1,900 m asl)	1,100–1,900		25.62–47.7 t/ha/yr[a]	
12	Sitlarao subwatershed, Doon valley, Dehra Dun Uttarakhand (440–2,300 m asl)	—	Morgan model and USLE model	20–25 t/ha/yr	Jain et al. (2001)
13	Maw-Ki-Syiem drainage basin of Meghalaya Plateau (1,308–1,390 m asl)	8,000–24,000	$^{210}Pb_{ex}$ and ^{137}Cs activities in soil: gullies, steep grassland, swampy valley, exposed bedrock	0.21 t/ha/yr	Froehlich (2004)
14	Bhetagad sub-watershed (1,090–2,060 m asl)	1,400	Runoff plot: pine forests, tea plantation, rainfed, agriculture, and barren land	0.06–5.47 t/ha/yr	Kothyari et al. (2004)
15	Mamlay watershed, Sikkim Himalaya (300–2,650 m asl)	1,295–1,900	Erosion plot: forests, agroforestry, agriculture, and wastelands	6–7 t/ha/yr	Sharma and Rai (2004)
16	Sitlarao watershed, Doon valley, Uttarakhand (960–1480 m asl)	1,753	USDA Water Erosion Prediction Project (WEPP) model	45 t/ha/yr	Kumar et al. (2005)
17	Dikrong river basin, Arunachal Pradesh (84–1,426 m asl)	1,519–4,169	USLE: forest, shifting agriculture, settlement, fallow, river	19.2–155.5 t/ha/yr	Dabral et al. (2008)
18	Kaneli research watershed, Kumaon Himalaya (1,220–1,540 m asl)	2,840	USDA WEPP model	157.2 t/ha/yr[a]	Ramsankaran et al. (2009)

(Continued)

Table 9.1 (Continued)

Sl. no.	Location	Annual rainfall (mm)	Methodology and land use	Annual soil loss/ sediment load	References
19	Umroi watershed, Meghalaya (882–910 m asl)	2,508–2,843	USDA WEPP watershed model: forest, paddy, barren, and settlements	21.9 t/ha/yr[a]	Singh et al. (2011)
20	Nun Nadi watershed of Siwalik, Uttarakhand (600–2,000 m asl)	2,073.3	Revised universal soil loss equation (RUSLE)	14.19–32.83 t/ha/yr	Naqvi et al. (2012)
21	Meghalaya Plateau (1,820–1,850 m asl), includes Cherrapunji region	2,400–120,00	^{137}Cs technique (upland farming system)	32–79 t/ha/yr	Prokop and Poręba (2012)
			^{137}Cs technique (degraded grassland)	0.5 t/ha/yr	
22	Indian Council of Agricultural Research Complex for NEH Region, Umiam, east Khasi Hills of Meghalaya (980 m asl)	2,439	Runoff plot (agriculture, agri-horti-silvi-pastoral, natural forest, livestock-based land use, natural fallow, and shifting cultivation)	30.2–170.2 t/ha/yr	Saha et al. (2012a, 2012b)
23	Kashmir valley, Jammu and Kashmir (1,500–5,000 m asl; Siwalik hills)	650	ICONA model (forest, agriculture, bare soil, water bodies and wasteland)	3.97–81.71 t/ha/yr	Zaz and Romshoo (2012)
24	Khajuri, sub-catchment of Trijuga river valley, Nepal (165–370 m asl; Siwalik hills)	1,876	Hydro-meteorological stations (forest, agriculture, settlement, floodplain)	0.8–26 t/ha/yr	Ghimire et al. (2013)
25	Dhalai river basin Tripura (16–498 m asl)	2,150	USLE	11–836 t/ha/yr	Ghosh et al. (2013)

26	Pathri Rao sub-watershed in the Siwalik Hills of Haridwar, Uttarakhand (250–900 m asl)	1,044	RUSLE-3D	8.50–134.9 t/ha/yr	Kumar and Kushwaha (2013)
27	Kangra region, Himachal Pradesh (248–5,861 m asl)	—	RUSLE	16.09–60.3 t/ha/yr	Kumar et al. (2014)
28	Likhu Khola Valley, Nepal (530–2,720 m asl)	2,500	Cosmogenic $^{10}Be_{qtz}$ analysis (degraded forests and scrubland)	5–15 t/ha/yr	West et al. (2015)
29	Tripura state	2,000–3,000	USLE: crop cultivation, orchards, shifting cultivation degraded land and fallow.	23.60 t/ha/yr	Tiwary et al. (2015)
30	Teirei watershed, Mizoram (718 m asl)	2,796	USLE: shifting cultivation, forest, etc.	Trace to 40 t/ha/yr	Zonunsanga (2016)
31	Muhuri River Basin, South Tripura	3,352.7	USLE: forest, rubber plantation, agriculture, etc.	0–650 t/ha/yr	Bera (2017)
32	Mid-Himalaya of Tehri Garhwal district, Uttarakhand (248–5,861 m asl)	1,246.3	RUSLE: forest, cropland, scrubland, built-up area, water bodies	15.67–157 t/ha/yr	Gupta and Kumar (2017)
33	Mandakini river basin, Rudraprayag, Uttarakhand	1,000–2,000	USLE and SDSM model: forest, snow cover scrub, barren, and water body	5–20 t/ha/yr	Khare et al. (2017)
34	Papung-Ben Khola watershed, South Sikkim (326–2,600 m asl)	2,600–3,000	Erosion plot: mixed cropping, barren land, terrace cultivation, cardamom and mandarin based agroforestry	0.04–0.21 t/ha/yr	Mishra (2017)
35	Maniyar watershed, Tehri Garhwal district, Uttarakhand (667–2,459 m asl)	1,400	RUSLE: forestry and agriculture	3.24–87.98 Mg/ha/yr	Kalambukattu and Kumar (2017)

(*Continued*)

Table 9.1 (Continued)

Sl. no.	Location	Annual rainfall (mm)	Methodology and land use	Annual soil loss/ sediment load	References
36	Sukhna Lake catchment area of northwestern Siwalik Himalaya	1,121.6	RUSLE: settlement, agriculture, checkdam, etc.	34.27 t/ha/yr	Semwal et al. (2017)
37	Doodhganga watershed of Jammu and Kashmir (1,557–4,700 m asl)	523.13	Morgan Morgan Finney (MMF) and USLE	<1–100 t/ha/yr	Amin and Romshoo (2018)
38	Mago basin, Arunachal Pradesh (2,355–6,436 m asl)	1,139	RUSLE: glacier, barren rocky, grassland, forest	~12 t/ha/yr	Bhadra et al. (2018)
39	Tirap, Arunachal Pradesh (25–1,600 m asl)	3,479	RUSLE: forest shifting cultivation/ settled agriculture, etc.	1.38–59.05 t/ha/yr	Das et al. (2018)
40	Meghalaya plateau	—	RUSLE (shifting cultivation)	36 t/ha/yr	Jena et al. (2018)
41	Nepal Himalaya including, Siwalik and Terai region (60–8,848 m asl)	160–5,500	RUSLE incorporating decadal land use and land cover (LULC) change	6.55–8.76 t/ha/yr (147,181 km^2)	Uddin et al. (2018)
42	Nepal Himalaya including Terai region (60–8,848 m asl)	2,500–3,000	RUSLE	Trace to 273.0 t/ha/yr	Koirala et al. (2019)
43	Catchment of Sitlarao stream, Dehradun, Uttarakhand	1,600–2,000	RUSLE: cropland RUSLE: open scrub	10.61–16.08 t/ha/yr 4.14–26.04 t/ha/yr	Singh et al. (2019a)
44	Tuirial river basin, Mizoram (1,690 m asl)	2,500	RUSLE: shifting cultivation	Soil loss: 115.4 t/ha/yr Sediment load: 6.161 million Mg/yr[a]	Barman et al. (2020)

45	West Kameng watershed, Arunachal Pradesh (115–5,780 m asl)	1,607	RUSLE and analytical hierarchy process approach using LULC map from Sentinel-2 images	124.21 t/ha/yr	Das et al. (2020)
46	Indian Himalaya (Himachal Pradesh, Jammu Kashmir, Sikkim, Arunachal Pradesh and Uttarakhand)	—	RUSLE	Arunachal Pradesh: 150 t/ha/yr Himachal Pradesh: 637 t/ha/yr Jammu and Kashmir: 141 t/ha/yr Sikkim: 659 t/ha/yr Uttarakhand: 632 t/ha/yr	Dawa and Arjune (2021)
47	Sitla Rao watershed, Uttarakhand (920–1,480 m asl)	2,073	APEX model (forest cover, scrub and terrace)	481.8 t/ha/yr	Kumar et al. (2021)
48	Panchnoi River basin, Arunachal Pradesh (55–2,920 m asl)	1,800	RUSLE	12.06–18.30 t/ha/yr	Jaiswal and Amin (2020)
49	Uttarakhand (85–7,817 m asl)	1,500	RUSLE	27.45 t/ha/yr	George et al. (2021)
50	Wokha, Nagaland (1,215 m asl)	1,925–1,996	Erosion plot: shifting cultivation and bench terracing	25.2–70.1 t/ha/yr	Ray et al. (2021)
51	Micro-watersheds of ICAR Research Complex for NEH Region, Umiam, Meghalaya (984–1,020 m asl)	2,394	RUSLE: agri-horti-silvi-pasture, shifting agriculture, forestry, agroforestry, and fodder	11.5–32.7 t/ha/yr	Choudhury et al. (2022)
52	Rani Khola watershed, East Sikkim (311–4,112 m asl)	3,300	Erosion plot: barren land, terrace farming, agroforestry, and mixed farming	3.23–7.73 t/ha/yr	Mishra et al. (2022)

Precipitation characteristics have the primary control in determining erosion rates in the IHR that may be as high as 20–80 t/ha/yr (Garde & Kothyari, 2014; Sharma, 2004). Annual rainfall in the eastern Himalayan region (EHR; 2,000–11,000 mm) is comparatively higher than in the western Himalayan region (WHR; 300–3,500 mm). Thus, the average erosion rate in the EHR is likewise greater (2.9 mm/yr) than in the WHR (2.1 mm/yr; Galy & France-Lanord, 2001). An area is categorized as having moderate to severe degradation severity if the rate of soil erosion fluctuates by more than 5–10 kg/m^2/yr. In contrast to other soil use alternatives, barren regions and alpine pastures are more susceptible to erosion (Koirala et al., 2019; Kumar et al., 2014). Soil exposure from massive deforestation, biomass burning, and dibbling seeds may contribute to a soil loss of ~4.1 t/ha/yr in steep to very steep slopes against the national average of 15.59 t/ha/yr; under shifting cultivation, it may contribute up to 146.6, 170.2, and 30.2 t/ha/yr from first, second, and third year of cultivation, respectively (60%–70% slope; Singh & Singh, 1980; Sharda & Ojasvi, 2016). Soil loss under bench terracing reduced gradually over 4 years, for example, 207.5 t/ha/yr (first year), 136.5 t/ha/yr (second year), 88.9 t/ha/yr (third year), and 78.3 t/ha/yr (fourth year) in Khukhthar and Machi villages, Chandel district, Manipur.

Likewise, *bun* farming practices (i.e., burning of dried vegetation covered with soil) pose a greater environmental threat from higher soil loss than traditional shifting cultivation practiced in the EHR owing to greater soil disturbances (Prokop & Poręba, 2012). The over-estimation of the universal soil loss equation (USLE) annual soil loss output (63.9%–85.7%) is evident, particularly under diverse cropping systems across sub-humid regions of the Indian subcontinent (Ali & Sharda, 2005). The annual erosion rates are mostly higher for Shiwalik region of the WHR (~80 t/ha/yr) and cultivated areas under shifting cultivation across the EHR (~40 t/ha/yr), against the permissible soil loss tolerable value of 11.2 Mg/ha (2.5–12.5 t/ha/yr; Bhattacharyya et al., 2015; Saha et al., 2012a, 2012b; Sharda & Mandal, 2018).

Geologically younger Himalayan regions may tolerate soil loss up to 25 t/ha/yr (Morgan, 1986). About 39% of the area under the EHR and 13.7% of the area under the WHR have erosion rates > 40 t/ha/yr and fall into the very severe category (Mandal & Sharda, 2011; Sharda & Mandal, 2018). Around one-third of the area under the WHR is categorized as having severe or very severe erosion risk (Sharda & Mandal, 2018). In Kashmir Himalaya, 48.27% of the area is under high erosion risk from anthropogenic sources due to reduction in forest cover (~9%) and increased built-up area (Zaz & Romshoo, 2012). Jena et al. (2018) reported that 30% of the area under Jirang block, Ri-Bhoi district of Meghalaya is prone to serious soil erosion problem (>20 t/ha/yr). A total of 41.56% of Uttarakhand is at risk for being in the severe (20–40 t/ha/yr) and very severe (40–80 t/ha/yr) soil loss categories (Mahapatra et al., 2018).

The forest cover of the IHR has been recorded as being the lowest across India (2.1 t/ha/yr; Sharda et al., 2010). The lack of vegetation cover and loose surface soil accelerated soil loss during pre-monsoon months, prominent in Umiam watershed of Meghalaya, North East India (Singh et al., 2011), and the loss of ecosystem services across the IHR (Shrestha et al., 2022). Anthropogenic activities and climate change are expected to accelerate the soil erosion process from the intensified hydrological cycle in the future, with rates predicted to be 80 t/ha/yr by 2100 across the Indian subcontinent, including the IHR (Pal et al., 2021; Tiwari, 2000). Future projections have confirmed that abandoned shifting cultivation lands are more vulnerable to soil erosion processes without any soil water conservation measures being implemented (Choudhury et al., 2022).

In the IHR, substantial crop loss occurs, particularly under *rainfed* hill agriculture, from sheet and rill erosion (Bhattacharyya et al., 2002; Sharda et al., 2010; Shrestha, 1997), along with an estimated SOC loss of between 0.15 and 0.26 t/ha/yr (Sitaula et al., 2004; Tiwary et al., 2015). The subsequent loss of other essential soil nutrients becomes inevitable from the soil loss under the variable slope gradient of the IHR. In the EHR, water erosion has caused 601 Mt of soil loss along with subsequent loss of nitrogen (685.8×10^3 t), phosphorus (99.8×10^3 t), potassium (511.1×10^3 t), manganese (22.6×10^3 t), zinc (14.0×10^3 t), calcium (57.1×10^3 t), and magnesium (43.0×10^3 t) (Central Soil and Water Conservation Research and Training Institute [CSWCRTI], 2011). Under shifting cultivation practice, the annual loss of SOC was estimated to be about 0.7 t/ha, with subsequent loss of 0.00015 t P_2O_5/ha and 0.0075 t K_2O/ha (Hazarika, 2018).

In Sikkim Himalaya (EHR), the maximum loss of nitrogen was reported from barren land and phosphorus loss under mixed cropping with a significant proportion of potassium loss (Figure 9.3; Mishra et al., 2022). Substantial soil nutrient loss has been reported for sugarcane and maize crops (i.e., even more than barren fields), particularly at early crop growth stages (Sharma et al., 2001; Lal & Mishra, 2015). In contrast, alpine forest and tea plantations contribute the maximum losses in the WHR. The magnitude of nutrient loss varied for nitrogen (0.30–21.27 kg/ha/yr), phosphorus (0.14–9.42 kg/ha/yr), and potassium (0.12–11.31 kg/ha/yr) across the WHR (Kothyari et al., 2004).

The ancillary role of soil acidity often facilitates soil degradation across the IHR. In the EHR, the net acreage of soil acidity-induced land degradation varied from one state to another, such as Arunachal Pradesh (21% TGA), Mizoram (55% TGA), Manipur (72% TGA), Nagaland (93% TGA), Meghalaya (46% TGA), Tripura (67% TGA), and Sikkim (9% TGA) (Hazarika, 2018). In particular, widespread aluminum toxicity together with limited availability of essential nutrient elements often limits crop growth under extensive soil acidity coupled with soil erosion across the IHR (Joshi & Negi, 2015; Hazarika, 2018). However, the soil reaction of

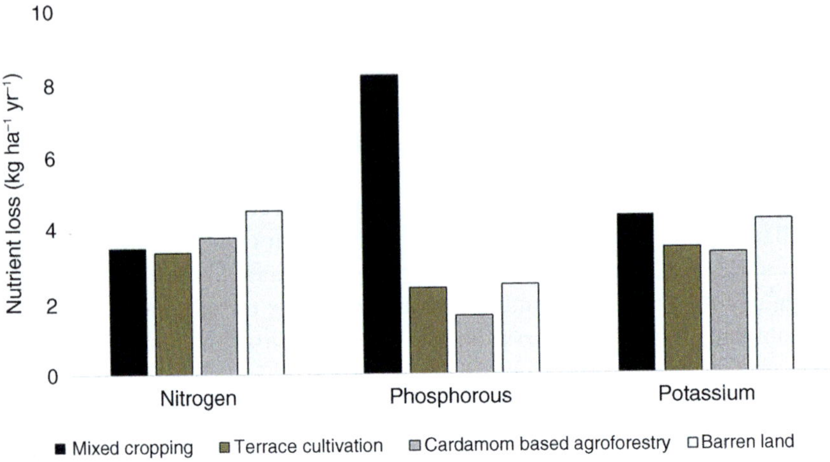

Figure 9.3 Nutrient element lost through soil erosion in Sikkim Himalaya (Mishra et al., 2022).

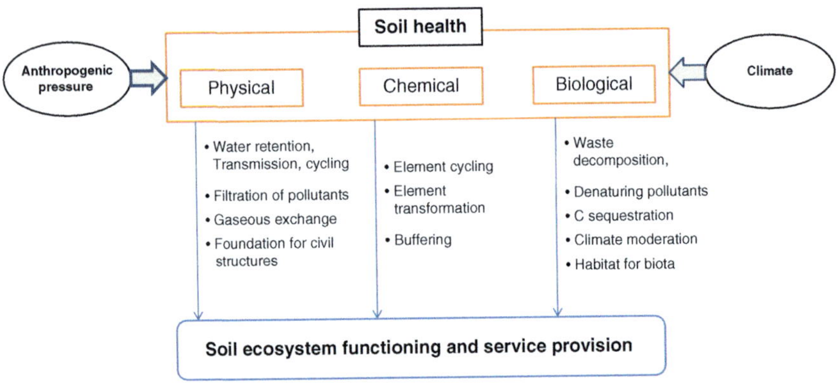

Figure 9.4 Soil health and ecosystem services in IHR.

the Garhwal region and the cold arid regions of the northwestern Himalayas often vary from slightly acidic to slightly alkaline depending on the existing climate regime and parent material, irrespective of their existing land use pattern (Kewlani et al., 2021; Saha et al., 2018).

The regional economy of IHR is severely threatened by periodic crop loss (5%–50% slope) owing to deliberate reduction in soil quality as accelerated by agricultural soil loss-induced land degradation (Sehgal & Abrol, 1994). Prevailing climatic conditions, plant physiological traits, profile nutrient distribution, soil moisture status, rooting depth, and physicochemical properties of subsoil often determine the degree of crop loss under variable soil erosion rates (Lal & Moldenhauer, 1987). For shallow-depth soils of the IHR, yield decline is severe for shallow-rooted cereal crops having high nutrient and water requirements as compared with the deep-rooted legumes. The net impact may be prominent within 2–5 years of intensive cultivation of maize-wheat system under variable management practices in the WHR (Bhattacharyya et al. 2009; Ghosh et al., 2012).

For every 2.8-cm reduction in top soil depth, which formed over 200 to 1,000 years, crop productivity declines by ~7% (Pimentel et al., 1995). In the Shiwalik region, maize yield loss was estimated to be 103 kg/ha for each 1-cm loss of top soils (Sur et al., 1998). In the EHR, the highest crop production loss was recorded in Nagaland (41%), followed by Arunachal Pradesh (35%) across all other states in India (Sharda et al., 2010). Based on soil suitability for crop cultivation, the reduction in crop yield potential may reach 80% in the IHR (Bhattacharyya et al., 2007). For the hilly topography of Sikkim, the estimated crop loss is 25%, followed by Manipur-Mizoram in the Patkai range (~23%) with an average annual monetary loss of >5,000 million INR in the EHR (Sharda et al., 2010). The average estimated annual financial loss in the WHR is almost twice that in the EHR, with agricultural losses of between 10% and 20% in the states of Uttarakhand, Himachal Pradesh, and Jammu and Kashmir. Moreover, the lion's share of total production loss due to water erosion is estimated to be for cereals (358,350 t) followed by pulses (11,093 t) and oilseeds (3,875 t), respectively, with a net worth of Rs. 4,120 million INR across the WHR (Sharda & Mandal, 2018).

Soil Health and Ecosystem Productivity and Functioning in the IHR

The IHR has diverse land uses and vegetative cover with forest, plantation, agro-forestry, and field crops (cereals/vegetables) coexisting and being critical for the livelihood sustenance, especially for inhabitants of the region. From tropical to alpine, the IHR encompasses a succession of vegetation cover, associated with significant differences in climatic variables (mainly temperature, precipitation). These ecosystems have the ability to provide a wide range of ecological services, such as the storage of water and preservation of biodiversity, and have provided food security for millennia.

Kibblewhite et al. (2008) highlighted four ecosystem functions that serve as the foundation for all important soil services. The first is the transformation of carbon by the breakdown of plant residues and other organic matter, as well as the synthetic activities of soil biota, notably soil organic matter (SOM) synthesis. SOM aids in nutrient cycling and soil structure preservation. Soil carbon sequestration also contributes to the regulation of greenhouse gas emissions such as methane and carbon dioxide. The second is the cycling of nutrients such as nitrogen, phosphorous, and sulfur, as well as regulation of nitrous oxide emissions. The third is the maintenance of soil structure via aggregation and particle movement, as well as creation of bio-structures and pore networks at various spatial scales. This function supports the preservation of the soil habitat, the soil-water cycle regulation, and the maintenance of a favorable rooting media for plants. The fourth is the biological regulation of soil populations, which includes organisms known as pests and diseases relevant to plants, animals, and human.

The regional forest cover trends within the Himalayas indicate that this belt is undergoing a severe form of forest degradation. Compared with 2017, the area's forest cover had shrunk by 107 km^2 by 2019 (Forest Survey of India [FSI], 2019). The statistics from the forest matrix suggest that 834 km^2 of very dense forests have grown during this time, but there have been losses of 391 and 432 km^2 of moderately dense and open forests, respectively, compared with 2017. The Garhwal Himalayas experienced annual forest cover loss of 0.22% from 1976 to 1998 and 0.27% from 1998 to 2014 (Batar et al., 2017). In Sikkim Himalaya, the primary forest cover decreased by 16% between 1990 and 2013, and the Western Himalayas lost 1,115 km^2 from 2011 to 2015 (Prasad, 2019). The foothills of the IHR have been subjected to varying anthropogenic pressures, such as shifting cultivation with shorter fallow periods, intensification of agriculture without adequate replenishment of soil reserves, expansion of monoculture plantation, and so forth. In addition to these anthropogenic pressures, extreme climatic variables, often as forms of unprecedented climate change have depleted not only the rich biodiversity and habitat but also the ecosystem services of the IHR, threatening the prosperity of the people.

The soils of the IHR can be largely categorized as acidic soils, with the acidity problem to a great extent being more severe in the NEH region (29% of the TGA) than that in the northwestern hill (NWH) region (0.8% of the TGA). This observed extent of acidic soils in the Himalayan belt may be attributed to the washout of the alkaline/basic ions, especially calcium and sodium, by the surface runoff from the higher altitudes owing to steeper slopes. High rainfall in the NEH region combined with steeply sloped topography results in soils commonly characterized by low pH, often below 5.5 (Ahirwal et al., 2022); whereas, in the northwestern part, soil pH >7.0 has also been observed in some pockets (Bhattacharjya et al., 2017; Saha & Kukal, 2015).

Forests dominate the Himalayan region to a large extent and thus help in sequestering substantial amount of organic carbon in these soils. According to Ahirwal et al. (2021), the top 1 m of soils in the IHR have an average organic carbon stock of 155 Mg C/ha, which is higher than tropical forests in the Andes (Phillips et al., 2019) and lower than successional forests, natural forests, and tree plantations in pantropical regions (Marín-Spiotta & Sharma, 2013). Several regional studies showed wide variation in SOC for the specific regions or particular forest types in the IHR from natural forest to agroforestry to plantation forests (Table 9.2). The variations in carbon stock are mainly due to the different

Table 9.2 Soil Attributes Under Few Prominent Land Uses of the Indian Himalayan Region.

Soil attributes	Land uses				Region/Location	References
	Forest	Agroforestry/ plantation/ orchards	Field crops (Agri/Horti)	Jhum		
pH	4.23–5.87	4.75	—	4.75	Mizoram	Ahirwal et al. (2022)
	4.7–5.8	—	—	4.5–4.9	Wokha, Nagaland	Mishra (2017)
	—	7.19–8.03	6.88–7.7	—	Uttarakhand	Bhattacharjya et al. (2017)
	7.1	—	7.5	—	Lower Himalayan regions of northwest India	Saha and Kukal (2015)
Soil organic carbon (%)	1.41–1.59	—	—	1.19–1.20	Nagaland	Mishra et al. (2019)
	2.02	—	1.53	1.54	Northeastern Himalayan Region	Choudhury et al. (2016)
	2.70	2.36	—	—	—	Saha et al. (2010)
	1.28–1.86	—	0.67–1.09	—	Northwestern hilly region of India	Dadhwal et al. (2011)
	2.3	—	1.0–1.7	—	Shopian, Jammu, and Kashmir	Sofi et al. (2016)

(Continued)

Table 9.2 (Continued)

Soil attributes	Land uses				Region/Location	References
	Forest	Agroforestry/ plantation/ orchards	Field crops (Agri/Horti)	Jhum		
Bulk density (Mg/m^3)	0.95–1.02	—	—	0.92–1.04	Nagaland	Mishra et al. (2019)
	1.05–1.12	1.21	—	1.16	Mizoram	Ahirwal et al. (2022)
	1.09	—	1.24	1.22	Northeastern Himalayan Region	Choudhury et al. (2016)
	0.94	0.98	—	—	—	Saha et al. (2010)
	—	1.28–1.33	1.27–1.45	—	Uttarakhand	Bhattacharjya et al. (2017)
	1.17–1.33	—	1.33–1.52	—	Northwestern hilly region of India	Dadhwal et al. (2011)
Mean weight diameter (mm)	3.13	2.54	—	—	—	Saha et al. (2010)
	0.60	—	0.36	—	Lower Himalayan regions of northwest India	Saha and Kukal (2015)
CEC [cmol (p+)/kg]	2.6–14.1	—	—	5.7–19.2	Wokha, Nagaland	Mishra (2017)
	20.21	—	7.89–17.21	—	Shopian, Jammu, and Kashmir	Sofi et al. (2016)

Available macro-nutrients	N (mg/kg)	184.1	157.3	126.5	127.9	Dimapur, Nagaland	Singh et al. (2014)
	P (mg/kg)	8.5	7.9	7.6	8.2		
	K (mg/kg)	159.9	136.1	107.3	141.2		
	Ex.Ca [cmol (p+)/kg]	3.8	1.4	1.1	1.6		
	Ex. Mg [cmol (p+)/kg]	1.9	0.5	0.6	1.1		
	P (%)	0.007	0.005	0.005	—	Khanikhola watershed, Sikkim	Sharma et al. (2001)
	K (%)	0.51	0.43	0.30	—		
	Ex.Ca [cmol (p+)/kg]	0.26	0.48	—	—		Saha et al. (2010)
	Ex. Mg [cmol (p+)/kg]	0.16	0.60	—	—		
	Ex.Ca [cmol (p+)/kg]	1.0–4.6	—	—	0.5–1.9	Wokha, Nagaland	Mishra (2017)
	Ex. Mg [cmol (p+)/kg]	0.6–1.9	—	—	0.9–4.3		
	P (kg/ha)	2.5–12.10	—	2.7–31.2	—	Northwestern hilly region of India	Dadhwal et al. (2011)
	K (kg/ha)	112–171	—	24.5–175	—		
	N (kg/ha)	380	—	190.0–291.0	—	Shopian, Jammu, and Kashmir	Sofi et al. (2016)
	P (kg/ha)	32.7	—	19.7–29.3	—		
	K (kg/ha)	390.0	—	280.0–330.0	—		

(Continued)

Table 9.2 (Continued)

Soil attributes		Land uses				Region/Location	References
		Forest	Agroforestry/ plantation/ orchards	Field crops (Agri/Horti)	Jhum		
Available micro-nutrients	DTPA-Fe	48.71	—	43.63	40.17	Northeastern Himalayan Region	Choudhury et al. (2021)
	DTPA-Mn	30.46	—	15.12	13.35		
	DTPA-Zn	2.13	—	1.44	1.42		
	DTPA-Cu	1.74	—	0.86	0.81		
	DTPA-Fe	71.2	81.3	88.4	76.9	Dimapur, Nagaland	Singh et al. (2014)
	DTPA-Mn	36.3	34.2	37.7	34.7		
	DTPA-Cu	2.81	2.53	2.53	2.64		
	DTPA-Zn	3.02	2.37	2.43	2.66		
	B	0.40	0.30	0.42	0.29		
	Mo	0.18	0.08	0.18	0.10		
	DTPA-Fe	56.8	—	28.8–36.5	—	Shopian, Jammu and Kashmir	Sofi et al. (2016)
	DTPA-Mn	12.2	—	4.2–7.2	—		
	DTPA-Zn	3.5	—	0.8–1.7	—		
	DTPA-Cu	3.0	—	0.5–1.5	—		

Microbial	MBC (μg/g)	780	527	340	—	—	Sharma et al. (2004)
	MBN (μg/g)	103	55	36	—		
	MBP (μg/g)	35	20	16	—		
	MBC (μg/g)	712.2	252.7	232.5	240.3	Dimapur, Nagaland	Singh et al. (2014)
	MBN (μg/g)	35.2	14.2	18.2	21.1		
	MBP (μg/g)	5.21	1.26	2.11	1.84		
	SR (μg CO_2–C/g/h)	12.2	3.14	3.35	3.29		
	qCO_2 (μg CO_2–C/μg SMBC/h) × 10^{-3}	17.1	12.4	13.8	13.7		
	MBC (μg/g)	—	108.22–236.63	147.61–242.21	—	Uttarakhand	Bhattacharjya et al. (2017)
	SR (μgCO_2–C/g/h)	—	2.41–3.47	2.12–2.55	—		
	qCO_2 [(μg CO_2–C/μg MBC) × 10^{-2}]	—	1.08–4.93	1.12–1.95	—		
	MBC (mg/kg)	1,405.0	—	611.7–861.7	—	Shopian, Jammu, and Kashmir	Sofi et al. (2016)

Abbreviations: CEC, cation exchange capacity; DTPA, diethylenetriaminepentaacetic acid; MBC, microbial biomass carbon; MBN, microbial biomass nitrogen; MBP, microbial biomass phosphorus; qCO_2, metabolic quotient; SR, soil respiration.

composition of tree species in the forests, as well as regional differences including topographical and environmental conditions.

Afforestation increases SOC stock, however, it should be acknowledged that there are several additional factors, such as previous land use (grassland/cropland/degraded land), plantation age, climatic conditions, and soil depth, that play a significant role in influencing the potential contribution of afforestation toward soil carbon stock improvement. Newly planted forests continue to collect soil carbon for 20–50 years or more after establishment in the absence of substantial disturbances, making them a large carbon reservoir (IPCC, 2007). The higher SOC storage under the plantation forests, especially those grown on previously degraded lands, is also attributed to the continuous soil amendments used to revive the health of such soils in order to establish the forests (Brahma et al., 2016).

SOC is the backbone of soil physical, chemical, and biological heath, and thus, the presence of higher organic carbon in soils indicates improvement in its structural stability, microbial activity, and plant nutrients supplying capacity. Organic carbon build-up in forest soils increases the number of micro-and macro-pores, which further helps in lowering the bulk density of soil, increasing the aeration, water holding capacity of soil, and so forth required for optimum plant growth and ecosystem functioning. Plant litter, subsurface decomposition, and microbial biomass all significantly contribute to the active C pools in forest soils.

Among the forests, natural forests have higher values of active C pool, owing mainly to the high biomass and also the quality of litter (litter of natural forests contains more labile carbon than that of regenerated forests; Sahoo et al., 2019). This labile form of carbon acts as a major source of nutrients in topsoil, and the difference in its lability in several forms of forests in the IHR thus leads to wide variation in macro- and micro-nutrient plant-available pools in soils. Available phosphorus and potassium in forest soils (in kg/ha) have been reported to be as low as 2.5 and 112, respectively, to as high as 32.7 and 390.0, respectively, in the NWH region of India (Dadhwal et al., 2011; Sofi et al., 2016). Higher SOM helps in increasing the abundance of chelating agents in forest soils, which further leads to enhanced micronutrient availability. Hence, available macro- and micronutrient pools in soils under forest systems are found to be strongly correlated with organic carbon content in those soils (Choudhury et al., 2021).

Unfortunately, forests of the IHR frequently suffer from land use change that is mostly brought on by industrialization and agricultural growth to feed the burgeoning population. Approximately 86% of the agricultural land in northeast India is under shifting (jhum) farming, the primary source of subsistence for the local population and the second-largest land use in the region after forests (Mishra & Francaviglia, 2021). This practice (i.e., continuous jhumming) and changes in land use negatively affect the SOC stock, thus aggravating the issues related to soil degradation and consequent biodiversity loss. SOC in *jhum* land is substantially lower than the forests because of clearing and burning of vegetation on the hill slopes and continual soil disturbance during cultivation (Choudhury et al., 2022). Although partial burning of biomass, as seen in shifting cultivation, represents a significant supply of organic carbon for soils in the long term, soils are very prone to erosion from the very beginning of the activity because the bare soil surface is exposed to erosive forces. SOC stock in the northeast Indian rainforest has been reduced by 38% (Gogoi et al., 2020), the eastern Himalayas by 63.5% (Gogoi et al., 2022), Mizoram by 56.5% (Sahoo et al., 2019), and Nagaland's SOC stock by 30.8% (Mishra & Francaviglia, 2021). Even in other parts of the world, such as in the Mediterranean hill slopes, surface SOC stock has been reduced by 41.8% with burning of vegetation, which is similar to shifting cultivation practices of the NEH region (Novara et al., 2011).

In addition to changes in SOC stock, conversion of natural forest to *jhum* land also changes the lability of carbon in soil. According to Ahirwal et al. (2022), soils on jhum land have a higher lability index of carbon than those in bamboo forests, indicating that the former are more susceptible to SOC losses. In the north-western part of IHR, soils under various field crops (agricultural/horticultural) are found to have lower carbon stock in them, sometimes even~ 50% lower than that found in forest soils (Dadhwal et al., 2011; Sofi et al., 2016) because of continuous disturbance of soil during cultivation.

The decrease in SOC with deforestation and expansion of arable lands for cultivation comes with a simultaneous deterioration in the physical, chemical, and biological heath of soils of the IHR. Zolfaghari and Hajabbasi (2008) reported that the conversion of forest into cultivated land significantly increased soil bulk density. In the IHR, similar observations were reported by Dadhwal et al. (2011). The mean weight diameter of soil and the water-stable soil macro-aggregates that are prominent soil physical health indicators were higher in agroforestry and grasslands as compared with agricultural lands (Bhowmik et al., 2019). These findings were also highly correlated with the soil microbiota under the various land uses evaluated. Agroforestry land use encourages greater fungal to bacterial ratios and larger atmospheric nitrogen fixing bacterial groups, resulting in more carbon and nitrogen buildup in the soil. Conversion of close canopy forest lands to agricultural lands and wastelands in the EHR (Sharma et al., 2004) was found to dramatically decrease the microbial biomass (with 78%, 73%, and 71% decreases in microbial nitrogen, carbon, and phosphorus, respectively, recorded in wastelands).

These changes in soil microbial biomass are critical components of its turnover, thus contributing to mineralization and release of nutrients in soils. According to a study done in the mountain ecosystem of the Eastern Himalaya, the conversion of forests to upland agriculture significantly reduces the amount of micronutrients that are available in soils (Choudhury et al., 2021). The magnitude of reduction was 17.7%–28.1% for iron, 41.5%–78.8% for manganese, 53.4%–60.9% for copper, and 33%–41% for zinc. In the Lower Shiwalik Foothills of Himalaya, the relative trend of availability of macronutrients (nitrogen, phosphorus, and potassium) and diethylenetriaminepentaacetic acid (DTPA)-extractable micronutrients (iron, manganese, zinc, copper) was found in the order: safeda forest > orange orchard > grassland > wheat-rice > sugarcane land-use system, in both the seasons, which is similar to the trend observed with respect to SOC (Kaur et al., 2021).

Apart from the diverse array of land uses from forests to lowlands, the IHR encompasses significant altitude variations, with the lowest peak at 13 m and the highest at >4,600 m above sea level (m asl; Ahirwal et al., 2021). Altitudinal variation is another major factor that contributes to the wide spatial variation in soil quality attributes in the entire Himalayan belt (including foothills) of the country. Some reports confirm the altitudinal gradient leads variation in climatic variables, vegetation types, and SOC pools as the factors responsible for wide variability in SOC pools in India and also other parts of the world. However, such variation in soil quality attributes with altitudinal gradients in IHR is relatively less explored owing to the obvious complexity of several land uses across the region and their interactions with altitude, climatic variables (rainfall and temperature), and so on (Ahirwal et al., 2021; Choudhury et al., 2016).

SOC concentration at an altitude of 2500–3500 m asl in the north-eastern Himalayan region was found to be almost doubled (91.5% increase) as compared with the SOC concentration at the baseline (Choudhury et al., 2016). Similarly, SOC stock also exhibited an increase with altitudinal gradient (~ 46.3% increase at an elevation of 2,000 m asl, and an additional 33% increase over 2,000 m asl). Decreasing trends in temperature and an increase in total annual rainfall with an increase in altitude have been found to favor accumulation of SOC at higher altitudes. This increase in carbon stock could be attributed to an increase in high-altitude basal area, leaf litter, and canopy cover (Sheikh et al., 2020). Furthermore, steeper terrain at higher elevations may limit access and act as a barrier to human disturbance (forest degradation) and thus result in higher SOC accumulation than at lower altitudes (Spracklen & Righelato, 2014).

In addition to changes in SOC, the available pool of soil micronutrients (iron, manganese, zinc, copper) in the eastern Himalayan region was found to correspond to altitudinal gradients, albeit differently. In general, however, the pool followed a non-linear, sixth-order polynomial trend along the altitudinal gradients from <500 to 4,100 m asl (Choudhury et al., 2021). Such differences in nutrient distributions in soil are mainly associated with differences in SOC content and clay contents with changes in altitude in the entire Himalayan region.

Options for Soil Health Restoration Under Cultivation Systems

Over the years, there has been indiscriminate use of chemicals and use of land beyond its capacity, leading to degradation of soil health and subsequently its quality. Soils of the IHR are usually shallow to medium deep, light to medium textured, and moderate to strongly acidic in hilly regions. Soils in this region are generally rich in SOM but are mostly degraded because of accelerated soil erosion. Therefore, it is essential to preserve and improve soil quality to ensure sustainable agricultural productivity and environmental quality (Das et al., 2012). Although it is impossible to restore soil health to its original form, there are several alternatives for reducing the impact of cultivation practices and anthropogenic activities on soil health.

Re-Carbonization of Depleted Carbon Pools

Soil constitutes the major terrestrial pool of carbon, with an estimated 4,000 petagram (Pg, billion metric tonne) of reserve carbon in a soil depth of up to 3 m, contributing significantly to the global C cycle. There are two types of carbon reserve in soil: (a) SOC in humid, sub-humid, and semi-humid climates that has been derived from biological sources (plants, roots, microorganisms, animals, etc.), and (b) soil inorganic carbon, which comprises lithogenic inorganic carbon (primary carbonates produced from the soil parent material) and pedogenic inorganic carbon (secondary carbonates or leaching of bicarbonates in dry and semi-arid regions and irrigated soils, created by soil processes) (Monger et al., 2015).

Depending on climate, soil, and land use, the rate of SOC and soil inorganic carbon sequestration may range from 0.1 to 1.0 Mg C/ha/yr and from 2 to 5 kg/ha/yr, respectively (Lal, 2018). The organic pool of carbon is the determining factor for the soil quality and governs the ecosystem services. SOC pools in undisturbed ecosystems stay intact, but those under the agro-ecosystems are vulnerable to depletion owing to continuous anthropogenic disturbances (deforestation, cultivation, biomass burning, etc.) and different land use systems. It is estimated that the farmed soils have 25% to 75% less SOC than untouched or natural ecosystems (Lal, 2004). The SOC pools with 1.5% to 2.0% are at a critical level, below which the productivity of soil is low and has low input use efficiency.

Possible options to re-carbonize the depleted carbon pools are based on sequestration with the help of different management techniques. The term "carbonization and recarbonization" refers to the conversion of atmospheric CO_2 into biomass by photosynthesis, as well as the retention of a portion of net primary production (the main carbon input in soil) into the biosphere. The net primary production enters the soil via rhizodeposition and litter decomposition, and the majority of it is transformed back to CO_2 by soil respiration (out of the total gross primary production).

Plants' capacity to photosynthesize atmospheric CO_2 can be used to create a positive carbon budget (C input > C output; carbon input in the form of crop residue or manure and carbon loss in the form of CO_2 through organic matter decomposition). This could enhance ecosystem services as well. Agricultural soils must be managed properly to avoid becoming a major source (10%–12%) of atmospheric CO_2 and N_2O (Smith et al., 2007). At the same time, when management is done properly, it can be used to mitigate and adapt to climate change. Increasing the use efficiency of inputs through practices includes the use of cover crops, agroforestry, controlled grazing, better pastures, integrated nutrient management (INM), biochar, precision farming, and conservation agriculture; these practices can help reduce emissions and capture carbon (Lal, 2018). Recarbonization of the depleted carbon pools requires regular input of biomass carbon and essential elements (i.e., nitrogen, phosphorus, and sulfur) incorporated into the soil. Thus, this restoration can be carried out by minimizing the losses and maximizing the gains in the soil solum.

Management of SOM

SOM is a primary indicator of soil quality, and the amount and quality of SOM influence soil biological activity, soil structure, nutrient dynamics and cycling, and the global carbon cycle. SOM serves as both a source and a sink for major and micronutrients, and it is thus a fundamental component of soil fertility that plays pivotal role in crop production. Crop production itself is one of the primary causes of carbon loss in agricultural soils, depending on the cropping system practice, climatic conditions, and so forth, and other causes of carbon loss include erosion, fire, crop harvest, and leaching, among others (Ciais et al. 2010).

The primary tactic in carbon management is to increase the carbon content and keep it in the soil, preventing its release back into the atmosphere. Retention of the carbon in soil is based on its mean residence time. For stabilization of carbon in soil, stabilization of micro-aggregates (<250 μm) is first and foremost because it prevents the carbon from being easily decomposable by soil microbial processes. The soil flora, roots, microorganisms, ambient factors, and inorganic bonding agents all contribute to the formation of stable aggregates. The binding agents include humified SOM, oxides, and alumino silicates, among others, that can bind primary particles and silt < 20 μm in size. The agents also include fungal hyphae and roots, which are temporary binding agents, and microbial and plant-derived polysaccharides, that bind micro-aggregates of size 20–250 μm and serve as transient binding agents. Plant roots, fungi, and microbial and plant-derived polysaccharides are the main binding agents of macro-aggregates (>250 μm). Another option for protecting carbon from decomposition is to grow crops and plant species with a deep (tap) root system that exudes recalcitrant compounds (Six et al., 2004). Like aggregation, soil pore sizes also play an important role in the mean residence time of carbon, and pores of different sizes are habitats for microorganisms of different sizes. The root-derived particulate organic matter and earthworms' activity and soil biota play important roles in the formation of aggregates and stabilization of C.

Degraded carbon pools or degraded/depleted soils are the largest sinks of carbon. These soils are found in croplands, grazing lands, and eroded lands, and the include soils affected by acidification, salinization, and nutrient depletion. Restoration of these soils is a high priority, and it can be achieved by improving use efficiency of inputs and nutrient management and constantly adding organic matter to the soil through high biomass generation and dead roots improving the carbon status and increasing agronomic productivity.

Strategies to Increase SOM or SOC in Agricultural Soil

Conservation Agriculture

Intensive cropping with conventional tillage practices, removal of crop residues, and imbalanced chemical fertilizer use is the main cause of carbon and nutrient loss. It degrades the soil quality, exacerbates the risk of soil erosion, and ultimately leads to soil infertility. Conventional tillage is also an energy-intensive process, as it involves the consumption of fossil fuel, and a source of carbon emissions owing to the burning of fossil fuel. Conservation agriculture (CA) is an emerging option for carbon sequestration in agriculture. It has become a widely adopted practice in cultivation systems (0.1 billion ha worldwide), with high energy use efficiency and SOC sequestration averaging 0.55×10^{-9} Pg C/ha/yr on a global scale (West & Post, 2002).

CA has four basic principles: (a) retention of crop residue in the field as mulch; (b) adoption of crop sequences and incorporation of a cover crop in the rotational cycle; (c) use of INM involving organic and inorganic fertilizers; and (d) minimal soil disturbance by using minimum/reduced tillage and no-tillage. Collectively these practices induce water-stable aggregates, protect soil from wind and water erosion, enhance carbon gains, and reduce carbon losses by managing the processes that govern the dynamic equilibrium of the SOC pool. Prolonged cultivation of land with tillage and removal of crop residue can accelerate the decomposition of SOM and soil carbon loss, accounting for 20%–67% of the loss

(Yang et al., 2019). Retention of crop residue mulch (10 cm) and incorporation of a cover crop and legumes in the cropping sequence can conserve soil and water and improve carbon and nutrient retention in the surface soils, maintain soil structure, and decrease erosion. Yadav et al. (2021) analyzed the two cropping cycles of maize-maize-field pea with six different tillage combinations and recommended no-tillage raised bed and flatbed methods for improving carbon sequestration (of up to 0.85 Mg/ha/yr cumulative increase) and soil properties in the EHR. Ogle et al. (2012) also reported the increase in carbon by approximately 0.25 t C/ha/yr and 0.29 t C/ha/yr under no-till on sandy and non-sandy soils, respectively.

Nutrient Management and Ameliorations

Regular additions of manures and compost can increase soil carbon content and improve soil physical attributes and nutrient availability, such that plant productivity and residue carbon inputs will increase as well. Manuring can enhance soil aggregation, which will improve soil structure. A decline in soil structure can accentuate the reduction in SOC content and depletion of soil fertility. Unless soil structure is improved, application of fertilizers alone will not be able to restore the soil degradative processes. Ryals et al. (2015) found that after addition of compost for 3 years, carbon storage was increased to 3.3 t C/ha (12.1 CO_2 eq/ha).

Adoption of INM is one of the strategies that represent sustainable management of the SOC pool and its dynamics. It can create a positive soil carbon budget and sequester additional CO_2 from the atmospheric into the SOC pool, and it also decreases soil loss. Das et al. (2016a) observed that the SOC stock of the 0- to 60-cm profile ranged from 67.9 to 83.1 Mg/ha under different nutrient management options, with an unfertilized control having the lowest SOC stock. These findings are consistent with previous research indicating that a rise in SOC stock is directly related to the amount and quality of organic residues, as well as manure application and fertilization (Das et al., 2014; Wang et al., 2012). Himes (1998) and Lal (2005) reported that rapid SOC decline is also due to lack of essential nutrients such as nitrogen, phosphorus, and sulfur.

Liming is the recommended approach for acid soil amelioration and nutrient management. Lime increases soil pH, leading to increased plant nutrient availability and decreased iron and aluminum toxicity. Every year, furrow application of good quality, homogeneous grades/sizes of lime at 250–500 kg/ha can enhance crop yields in acid soils of the NEH region of India. Use of acid-tolerant varieties and application of organic manure also improves productivity of such soils. Low-cost locally available basic slag, a byproduct of the steel industry, was also studied as an ameliorant for acidic red and lateritic soils of West Bengal for mustard-rice cultivation (Bhat et al., 2010).

Biochar

Biochar is prepared from the waste wood of trees, residues, and other plant matter through slow pyrolysis. Biochar is composed of re-arranged macromolecular components with weak colloidal properties having improved resistance to chemical and biological degradation and long residence time (Zimmerman, 2010). Biochar is well known for its acid-neutralizing characteristics in severely degraded acid soils, as well as its potential nutrition supply. The long-term effects of biochar are ensured by its persistency and formation of macro-aggregates. Several researchers have reported that biochar improves the chemical, physical, and biological properties of soil, including bulk density, porosity, penetration resistance, moisture content, hydraulic conductivity, aggregate stability, and erosion resistance (Chan et al., 2007; Jien & Wang, 2013). Biochar incorporation was found to reduce soil loss by 31%, 58%, and 82% at 10, 20, and 30 $t\,ha^{-1}$ application rates, respectively, compared with the control (Adekiya et al., 2020). Biochar was found to be effective in maintaining SOM levels through carbon sequestration (Layek et al. 2022) and increasing fertilizer-use efficiency and crop production, particularly in subtropical and tropical soils (Chan et al., 2007). Hansen et al. (2008) estimated that replacing slash-and-burn agriculture using agricultural and forestry products for biochar production would reduce atmospheric CO_2 by 8 ppm within half a century (equaling approximately 17 Pg C).

Laser-Assisted Precision Land Leveling

Precision land leveling is a technology that employs laser-equipped drag buckets to smooth the ground surface ($\pm$2 cm) from its average height, and it is well recognized for producing better levels of precision in land leveling and has tremendous potential for aggregate-linked SOC. Adoption of laser land leveling technology with organic inputs reduces farm input costs, increases SOC, and improves crop productivity (Naresh et al., 2020). The research suggested that precision land leveling in field crops improved total soil carbon, total inorganic carbon, and total SOC content at surface depth (0–15 cm) by 11.93 and 10.73 g/kg, respectively. However, water-soluble carbon, and microbial biomass carbon were 27.8% and 35.1% in surface soil and 29.2% and 42.9% in sub-surface soil under laser-leveled and unleveled fields, respectively (Cooper et al., 2016; Panettieri et al., 2015).

Soil Biodiversity, Nutrient Cycling, and Resiliency

The concept *soil biodiversity* signifies the number of distinct species (richness) and their proportional abundance (evenness) present in the system. It also includes the functional, structural, or trophic diversity. This unseen diversity in the soil comprises tens of thousands of species of archaea, bacteria, fungi, and protozoa (soil flora including micro, meso, and macro), all of which may occupy different niches in the soil and carry out numerous biogeochemical processes and ecosystem functions. Soil contains by far the greatest diversity of organisms of all the terrestrial ecosystems (Giller, 1996); 4,000 different bacterial genomic units can be present in just 1 g of soil (Torsvik et al., 1990). These microorganisms are highly diverse and constitute approximately 60% of the Earth's biomass. Most of these species are active in the top 2–3 cm of the soil layer, which contains high organic matter, and in the root rhizosphere zone. This soil biodiversity acts as the indicator of soil quality, and a higher species diversity results in greater soil stability. This stability is reflected in the soil's ability to perform key functions such as nutrient cycling (biogeochemical cycles); decomposition; regulation of SOM dynamics, soil temperature, and moisture; maintenance of aggregate stability or modification of soil physical structure; detoxification of xenobiotics and pollutants; and regulation of atmospheric composition. These functions are directly or indirectly responsible for maintaining the ecological balance in the system, making them the "biological engine of the earth" (Rillig & Mummey, 2006). These soil biodiversity functions interconnect with one another, as well as with plants and small animals, creating a network of biological activity that is important for sustainable management of any ecosystem.

Factors Affecting the Soil Biodiversity in Agro-Ecosystems

Previous studies have shown that seasonal changes and management practices influence the size and dynamics of soil biodiversity and the related key soil functions, thereby affecting their ability to supply nutrients. The factors affecting the microbial diversity in soil (Kennedy & Gewin, 1997) are illustrated in Figure 9.5 and described as follows.

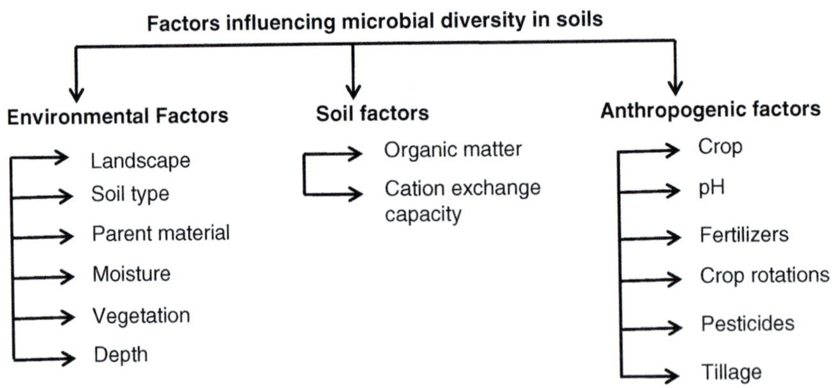

Figure 9.5 Factors affecting soil microbial activities in IHR.

SOM is the primary source of energy, carbon inputs, nutrients, and habitat for the soil biodiversity. Dead organic materials in soil are first incorporated by soil fauna (earthworms, collembolans, mites, nematodes, and protozoa), the first step in decomposition process for any biogeochemical cycle (Paul, 2014). Next, heterotrophic microorganisms (bacteria, yeast etc.) take over and transform the organic material, leading to organic carbon inputs in soil and mineralization of other elemental nutrients in the soil, which enables easy uptake by plants.

Agricultural practices that maintain adequate SOM content encourage the growth of soil biota or soil biodiversity. The moisture content in soil regulates microbial activity and diversity in soil (Griffiths et al., 2003). In addition, systems that boost belowground carbon and nitrogen inputs through the addition of legumes or fibrous-rooted crops in rotations may increase microbial populations and activities, unlike commercial fertilizer application (Lupwayi et al., 1998). A crop-rotation system that includes grass and other compatible plant associations may be able to make the best use of the soil by mobilizing and continually replenishing its biotic potential. Fertilizer application and soil pH both have an impact on the structure of the soil biota. Compared with bacteria, fungi prefer low pH; whereas, high nitrogen concentrations boost bacterial concentrations. Fertilizers interact with soil microbial populations in a variety of ways, either directly by delivering nutrients or indirectly by encouraging plant growth and improving root carbon flow (Buyanovsky & Wagner, 1987). In contrast, fertilizer inputs may cause soil acidity, restricting microbial development and activity (MacRae et al., 1999).

Tillage causes physical disturbance, which is important in determining soil species diversity in the agro-ecosystem. Tillage destroys stratified soil microhabitat, resulting in a decline in the quantity of species that live in such agro-ecosystems. It also aerates the soil, causing fast mineralization of organic materials and, in many cases, significant nutrient loss. Reduced tillage (surface residue placement) produces a more stable environment, encouraging the formation of more diversified decomposer communities and slower nutrient turnover. Evidence shows that in no-till systems, the ratio of fungi to bacteria is greater, but in conventionally tilled areas, bacterial decomposers may predominate (Hendrix et al., 1990). Because plant leftovers decompose on top of the soil, the no-till method favors fungus over bacteria.

Soil resiliency refers to the ability of the soil to recover from a significant disturbance and resume its normal function, and it is increased by soil biodiversity. High elasticity, amplitude, and limited malleability are characteristics of resilient soils. The classes of soil resiliency are listed in Table 9.3. Structure, renewal rate, accretion constant, increase in base saturation, buffering ability, nutrient cycling, biological nitrogen fixation, swell-shrink ability, self-mulching ability, and soil biodiversity are all indicators of soil resiliency.

Table 9.3 Soil Resiliency Classes.

Class	Resilience	Description
0	Highly resilient	Rapid recovery, high buffering
1	Resilient	Recovery with improved management
2	Moderately resilient	Slow recovery with high input
3	Slightly resilient	Slow recovery even with change in land use
4	Non-resilient	No recovery even with change in land use

Site-Specific Nutrient Management (GIS-Based)

Site-specific nutrient management (SSNM) is an approach of supplying organic and inorganic nutrients to plants in the optimum amounts to meet their inherent spatial and temporal needs by using different tools, such as remote sensing, global positioning system (GPS), geographic information system (GIS), variable rate technologies, and yield monitoring. The components of SSNM are described in the following paragraphs.

GPS contains 24 orbiting satellites that emit radio signals that are picked up by GPS receivers to capture a specific location, which enabled having precise information for soil and crop mapping. The precise location is important for data capture and farm operations in precision farming. One example is GPS-based crop dusters (precision GPS helicopters), which can spray an area as small as 4×4 m. Other examples are precise seeding, fertilizer application, and irrigation; insect and disease control; and harvesting and grading of fruits. Remote sensing is a technique of collecting data from distance with data sensors mounted on aircraft or satellites. It is one of the important techniques in precision farming, which helps in different farm operations such as monitoring of soil moisture, crop nutrients, crop pests and diseases, crop growth status, yield estimation of crop, and so forth. GIS is a computer hardware and software system that uses field information to produce various types of maps. The information can be related to yield, soil surveys, and other types of data.

Variable rate technologies are used to adjust the agricultural inputs to meet site-specific requirements in a field. This technique can be manual or automated. Automated ones have a controller mounted on farm machinery, which has a differential GPS receiver to precisely identify the spatial variability in the field and automatically apply the control rate on pre-derived inputs. It is one of the most widely used precision agriculture technologies in recent times. Yield monitoring and mapping are some of the most widely used components in site-specific or precision agriculture.

The objectives of SSNM are (a) balanced fertilization based on crop requirements, (b) plant-based estimation of root nutrient supplies, (c) need-based fertilizer nitrogen management, (c) sustainable phosphorus and potassium management, and (d) increased profitability. The advantages of SSNM include the following: (a) GPS-guided grid sampling delivers more precise soil test results, (b) variable rate fertilizer application can increase effectiveness, (c) starter fertilizer, variety adjustments, and variable rate sowing can adjust for soil characteristics and productivity, (d) crop scouting with new digital technologies enhances field records, and (e) on-the-go yield monitors can instantly measure variability in the field.

With the use of SSNM technology, yield maps, weed distribution maps, soil maps, nutrient status maps, water/moisture availability maps, and pest and disease incidence maps, among other types of maps, can be generated. This technology makes decision-making and planning a lot easier for the cultivator or the farmer. This technique is a new level of management technique in crop production and offers accurate assessment of risk. It will enhance the use efficiency of inputs and cost-effectiveness and enhance the efficiency in crop production as a whole. SSNM will also be profitable in production of high-value crops for which input costs are very high.

Soil Erosion Management in Hill Ecosystems

The topographical features, uncertain rainfall regime, and improper farming practices are the main causal factors for soil erosion, degradation, nutrient loss, and deterioration of soil health in the eastern Himalayan regions, which have undulating topography. Intensive tillage practices in the Himalayan region, which receives high rainfall, can disrupt the macro-aggregates, leading to reductions in the water and nutrient-holding capacity of the soil. On the plot and field scales, erosion can cause yield reduction of 30%–90% depending on the type of crops grown (Lal & Moldenhauer, 1987), and in row crops, erosion can reduce the crop yield by 20%–40% (Fahnestock et al., 1996). As per the estimates of Lal (1995), erosion alone can reduce the total production by 8.2 Mt for cereals, 9.2 Mt for roots and tubers, and 0.6 Mt for pulses. The annual soil fertility depletion rates are 22 kg of nitrogen per hectare, 3 kg of phosphorus per hectare, and 15 kg of potassium per hectare (Stoorvogel et al., 1993). UNEP (1994) estimated that the annual economic losses in South Asia were approximately $600 million due to nutrient loss by erosion and $1,200 million due to soil fertility depletion alone. The common types of erosion and some conservation measures to control such types of soil erosion are presented in Table 9.4.

Table 9.4 Different Types of Erosion and Their Control Measures.

Type of erosion	Depth of erosion	Conservation measures
Sheet, rill, and wind erosion	Surface	Establish ground cover, proper drainage channels, mulching, conservation tillage, rotational cropping, maintain surface soil moisture, etc.
Shallow and deep-seated mass-movement erosion	Shallow and deep landslides, slumps, earthflows (typically of 3–5 m in depth and going up to 10 m)	Spaced planted trees, afforestation, drainage control
Gully erosion	Channeled runoff and large and complex mass-movement–fluvial-erosion features that are typically amphitheater-shaped	Runoff control, afforestation with spaced planted trees, debris or check dams, geotextiles, concrete stacked blocks, etc. Permanent structures: drop spillways (along the gully bed), chute spillways (gully head control) and drop inlet spillways (downstream part for water storage)
Stream bank erosion	Fluvial and mass movement	Plantation of trees at banks, gabion structures, fencing, etc.

Note. Hicks and Anthony (2001).

Table 9.5 Type of Tillage Practices and the Percentage of Runoff Reduction.

Tillage practice	Topography/tillage type	Positive/negative properties	Reference
Conventional tillage	Plowing with moldboard plow with 6–10 tillage operations (primary and secondary tillage is done)	Soil compaction and accelerate erosion	Mehla et al. (2000)
Deep tillage	Plowing with chisel plough	16%–66% runoff reduction with chisel plow	Burcham et al. (2008)
Ridge tillage	Crops are sown on ridges made by ridger		
Reduced tillage or minimum tillage	Less preparatory tillage, practice mostly in flat land	85% reduction in soil loss and 55% runoff reduction	Laufer et al. (2016)

Table 9.5 (Continued)

Tillage practice	Topography/tillage type	Positive/negative properties	Reference
Zero tillage	Only localized soil disturbance at the time of sowing with 30% crop residue	It also increases the SOC content by 12.5%, available P by 14.3%, and K by 29.4% over conventional tillage	
No-tillage	No or very minor soil disturbance, practice both on flat and sloped land	22.5%–64.9% runoff reduced	Ahn and Kim (2016)
Mulch-till	Soil and water preserving by whole surface disturbance and by fair surface cover. Carried out by tine.	50% runoff reduced	Zhang et al. (2016)
Strip-till	Non-inversion tillage techniques that conserves soil moisture and uses crop residue to protect against soil erosion, creates narrow tilled strips which are typically 100–300 mm wide and raised 80–200 mm above the surrounding undisturbed ground. Carried out by specially design tine.	98% soil loss reduction and 92% runoff reduction	Reeder (2000); Laufer et al. (2016)
Down-slope tillage	Practiced on sloped land	10% runoff reduced	Dai et al. (2018)
Cross ridge tillage	Practiced on sloped land	6.11%–64.2% runoff reduced	Guo et al. (2019); Zhang et al. (2016)
Contour ridge tillage	Practiced on sloped land	19%–49% runoff reduced	Dai et al. (2018); Satriawan et al. (2015)
Ripping	Minimum tillage where only parallel furrows are cut using a ripper	8%–40% runoff reduction	Gachene and Kimaru (2003); Chuma et al. (1998)
Tied-ridges	Discontinuous furrows made by cross-ties in the furrow by creating pools to retain water and to promote slow seepage	8%–40% runoff reduction	Gachene and Kimaru (2003); Chuma et al. (1998)

Soil and water conservation measures to control soil erosion are classified into two broad categories: (a) agronomic or biological practices and (b) mechanical conservation. Agronomic or biological practices to combat soil erosion include contour farming or contour hedgerows with vegetative barriers (18%–21% runoff reduction), strip cropping, mulching, and grass waterways, among others. These agronomic practices can reduce runoff from 8% to 40% and soil loss from 6% to 35%. Mechanical conservation practices for soil erosion include bench terracing, contour bunding, contour trenching, and land leveling, among others.

Biological Methods of Erosion Control

Biological methods of erosion control are by far the most widely used method of conservation. It is practiced on 1%–6% slopes.

Contour cultivation, otherwise called *contour farming*, is one of the more common conservation practices used in agriculture. Here, crops are grown across a hillside by adhering to its contour lines of elevation. Contour farming reduces soil erosion rates on agricultural land because the placement of the crops and plants slow the flow of water and thus promote infiltration. It was discovered that contour hedgerow farming in sloped terrain significantly reduced runoff and controlled soil erosion (Fan et al., 2015).

In drylands and hilly areas, respectively, *strip cropping* is a combination of contouring and crop rotation in which consecutive strips of row crops and soil-conserving crops are grown on the same slope, perpendicular to the wind or water flow.

Agrostology is cultivation of suitable grasses to control soil erosion in agricultural lands. Grass hedges such as *Melilotus* hedges and *Pennisetum* hedges in maize fields in sloped land with high-intensity rain can decrease the surface runoff by 27%–72% (Wang et al., 2018). In China, Zhang et al. (2015) studied erosion control with the use of Bahiagrass on bunds/risers with citrus trees on rain-fed terrace beds. The result showed that planting Bahiagrass on the riser and bunds is a useful method for soil erosion conservation on sloping red soil in China. Various researchers (Huang et al., 2014; Pan et al., 2017) used cocksfoot grass as vegetative cover in China's rainfed sloping jujube orchards and in pear-jujube orchards to control soil erosion. In India, Ghosh, Dogra, et al. (2016a) and Ghosh, Meena, et al. (2016b) studied the use of Palmarosa grass and Panicum grass as vegetation strips with minimum tillage for controlling soil erosion and found that they were effective in decreasing runoff and soil erosion.

Mulching, cover crops, intercropping, and grass waterways offer additional biological means of controlling erosion. Organic or inorganic mulch are used in agriculture depending on the land use type and climate of the region. Organic mulch includes living mulch (*Trifolium repens*, *Lolium perenne*, etc.) and crop residues such as paddy straw, sawdust, weeds, bark, wood chips, leaves, grass clippings, plant hulls, and paper. The advantages of organic mulch are numerous. They include improving soil structure, retaining soil moisture levels, releasing nutrients as it slowly decomposes, and keeping the soil cooler. The use of soil amendments and mulch to increase soil cohesiveness from lime or gypsum is known to lessen soil detachment and transport (Won et al., 2016). Thus, it is possible to more effectively minimize soil erosion rates by combining mulching with soil amendments. Inorganic mulches include plastic film (black or white color), stone mulch, and others that do not require regular replacement. The mulching application can reduce runoff by 15.5% and sediment yield by 40.7%, and rice straw mulching on 3%–8% slopes was able to reduce the runoff rate by 9% and sediment yield by 95.9% (Ahn & Kim, 2016; Dai et al., 2018).

Cover crops can also be used as mulch to reduce the impact of rainfall on the soil. Because cover crops boost the soil's ability to hold water and increase infiltration, they can shield the soil from the splashes of rain and allow it to absorb water more effectively (Taguas et al., 2010), which keeps the soil in place. Use of leguminous cover crops has the double benefit of reducing soil erosion and runoff through infiltration and increasing the soil nutrient content, especially the nitrogen content. Between the primary crops, a vegetative filter strip or a row of grass is typically sown as a cover crop. The crops mainly used as cover crops are oat, *Mucuna bracteata,* white clover, and vetch, among others. The deep roots of *M. bracteata* are known to bind soil particles and provide more stability to the slope, reducing the amount of surface water runoff, the peak flow, and the rate of soil erosion (Eshel et al., 2015). Eshel et al. (2015), used different types of cover crops to assess the impact of each in potato cultivation and found that soil erosion and runoff were reduced by 95% and >60%, respectively.

Intercropping is the practice of growing more than one crop in the same field simultaneously. It is usually carried out by planting a combination of nutrient-adding crops (legumes) and nutrient-exhausting crops (cereals, oilseeds, vegetables, etc.). Growing either one row of cowpea or okra in between two rows of maize reported to reduce runoff by about 26% and soil loss by 43% Sharma et al. (2017).

Hedgerow intercropping of nitrogen fixing plant types is another practice in which hedges are grown in bunds to help reduce soil loss and runoff. It also helps in the fixation of atmospheric nitrogen and reduces the leaching losses of mineral nitrogen. The twigs and tender stems of hedge plants can be used as mulch, and owing to their vigorous root systems, the plants can reduce soil loss by 94% and runoff by 78%. Various hedgerow species for plantation suitable for farming in the Eastern Himalayas are *Cajanus cajan*, *Crotalaria tetragona*, *Desmodium rensonii*, *Flemingia macrophylla*, *Indigofera tinctoria*, and *Tephrosia candida*, among others. Grass waterways in agricultural lands are naturally or artificially constructed channels for discharging concentrated runoff from a slope. The grass waterways have been shown to reduce runoff and soil loss by 28%–34% and 34%–60%, respectively (Ghosh et al., 2012). Suitable grass species for this purpose are *Panicum repens*, *Brachiaria mutica*, *Cynodon plectostachyus*, *Cynodon dactylon*, *Paspalum notatum*, *Panicum maximum*, khus (*Vetiveria zizanioides*), bhabhar and munj (*Saccharum spontaneum*), and others.

In the most rapidly eroding locations, *afforestation* is also utilized to control erosion. A broad variety of plant types and species have been employed to control erosion, but the most popular soil conservation plants have been the space-planted poplars and willows because they are suitable for the management of landslides, earthflow, gullies, and stream-bank erosion. Closed-canopy tall woody vegetation often reduces erosion by 90% over a 20-year period and reduces land sliding by 70%–90% during severe storms. In 15 years, space-planted trees can reduce erosion by 70%, and native forest reversion can reduce erosion by 10% (early-stage incomplete canopy closure) to 90% (complete canopy closure) (Dymond et al., 2010; McIvor et al., 2011). Afforestation also has an impact on other ecosystem services by enhancing aquatic habitat and water quality, lowering greenhouse gas emissions by storing carbon and nitrogen, preserving biodiversity, and enhancing soil and nutrient retention.

The cost for *synthetic geo-textiles* is very high compared with *agro-geo-textiles* such as maize straw or stalk mats, *Arundo donax* (giant reed/cane), bamboo mats, borassus mats (*Borassus aethiopum*), buriti mats (*Mauritia flexuosa* L.), and wheat straw mats, as they are found easily and abundantly available. Geo-textiles are used in degraded soils to control soil erosion and improve soil quality (Pilu et al., 2012). Singh et al. (2019a) in India, used agro-geo-textiles (coir) in sloping croplands of the IHR as a soil conservation practice. Their team conducted field experiments on zero-tilled maize and minimum-tilled garden pea and wheat crops planted in rainy, autumn, and winter seasons, with different vegetative filters and agro-geo-textile treatments. Cowpea cover crop and grass weed acted as a vegetative filter between the maize rows. The result showed that conservation tillage together with *A. donax* (giant reed/cane) geo-textile reduced runoff by 24% and soil loss by 8.22 t/ha compared with only conservation tillage practice.

The most popular natural inorganic soil conditioners are *zeolites*, which are considered to improve important soil characteristics such as infiltration rates, water holding capacity, cation exchange capacity, and nutrient retention (Chmielewská, 2014). Zeolites are thought to be an eco-friendly soil conditioner that can enhance the physical and chemical characteristics of agricultural soils. Mirzaei Aminiyan et al. (2015) observed an increase in SOC and water stability of aggregates in soils amended with natural zeolite and plant residues. These results may have been caused by zeolite's propensity to influence nutrient availability by raising cation exchange capacity. Glomalin's ability to maintain its content revealed that the more labile portion of the organic matter, which is primarily composed of refractory aromatic carbon compounds, was influenced by the breakdown of organic carbon. Further studies on zeolite application in combination with a source of organic matter available at the farm, such as winery waste, are being carried out in order to optimize the fertilizer use efficiency and preserve SOM and aggregate stability. Indeed, an increase in dehydrogenase activity and a decrease in SOC were observed in zeolite-treated soils.

Mechanical Methods to Control Soil Erosion

When biological control strategies alone are unable to reduce soil erosion to acceptable levels in agricultural land, mechanical methods are designed to manage soil erosion and runoff (Blanco & Lal, 2010). Development using the mechanical approach will alter the agricultural landscape's features and disturb the soil, but construction using the biological way will not do either.

One of the common mechanical methods of soil erosion control in hilly areas is *terracing*. Terracing (step-like fields) is a common practice on 15%–33% slopes. Terracing divides the long slope into shorter slopes by changing the landforms and reducing the slope gradient. There are three types of terraces: bench terraces (table-top or level-type), outwardly sloping, and inward sloping with mild longitudinal grades for runoff disposal. Terracing reduces the amount of soil erosion and the runoff amount and rate. The average runoff coefficient of the terraced plot (orchard and cropland) was decreased by 47.2% compared with unterraced plot, and runoff and soil loss were reduced by 78.9% and 88%, respectively (Madhu et al., 2018). In comparison with unterraced plots, a terraced plot's average erosion rate is reduced by 83.9%. One of the terracing techniques used to prevent soil erosion on agricultural land is the construction of narrow terraces with a higher slope gradient. On the other side, half moon terraces are used for soil depths greater than 1.0 m. Half-moon terraces are circular, level beds with a 1- to 1.5-m diameter that are shaped like a half-moon and are mostly used for fruit and fodder trees.

Contour bunding is another widely used mechanical measure in soil conservation. It consists of constructing narrow trapezoidal or parabolic-shaped bunds on a gentle slope (0.5%–6.0%) and is generally recommended for low-rainfall areas (<800 mm) with relatively permeable soils. The conservation techniques carried out on 2%–8% slopes can reduced the runoff to 8%–40% and soil loss by 6%–35%. Graded bunds are generally adopted on less permeable soils and recommended on slopes of 6%–10% with annual rainfall >750 mm. Contour trenching is the practice of making trenches or ditches, excavated at the interval of 1.4–3.1 m on slopes >15% to control erosion. Trenches can be divided into four types: continuous contour trench (suitable for areas with annual rainfall of 1,000 mm and <6% slopes), graded contour trench (suitable for areas with annual rainfall of 1,000–1,200 mm, and a slight slope of 0.1%–0.5%), and line trench and staggered contour trench (suitable for high rainfall areas with >1,200 mm annually and steep slopes of 10%–50%). Trenches can reduce runoff loss by 8%–10% and can maintain soil loss below 10 t/ha (Madhu et al., 2018). Other mechanical methods applied in the agricultural area to control soil erosion are basin leaching, sub-soiling (sub-soiling or deep tillage with chisel plow), and ridging, among others.

Tillage operations, or the type of tillage practices, have a great role to play in agriculture field operations through their effect on the soil health and crop yield. Tilling the soil with different types of mechanized plows that have the capability of inverting soil up to 25-cm depth, such as moldboard, rotary, and power tillers, has a short-term benefit of creating favorable soil conditions for crop emergence and nutrient availability. However, in the long run, it can cause soil crusting and compaction, which are the main problems leading to soil erosion and the loss of plant nutrients, organic carbon, and biodiversity from the soil (de Moraes Sá et al., 2015). Thus, adoption of conservation agriculture with 30% residue retention on surface, minimum soil disturbance, and crop rotations is beneficial for improving the soil health.

Zero tillage, minimum tillage, furrow irrigated raised beds, laser leveling, incorporation/retention of crop residues, and so on are major conservation agriculture technologies promoted in India. This has been possible mainly due to the development of the new generation machineries such as zero-till drills, rotary disk drills, happy seeders, laser levelers, and so forth. Modern tillage practices such as reduced/minimum tillage (tillage reduced to 50% with inversion limited to a soil depth of 7–8 cm) and no-tillage have been promoted since the 1990s, and these practices can simultaneously increase carbon sequestration and improve soil health and soil processes. The no-tillage practice is one of the most popular tillage operations for controlling soil erosion and is usually recommended for a high temperature area. The no-tillage method preserves the physical, biological, and chemical characteristics of the soil, which increases infiltration, bulk density, wilting point, and field capacity (Ahn & Kim, 2016). The percentage of the micropores can be increased in no-tillage practices.

Li et al. (2019) experimented with four different tillage operations, including contour drilling (uses a "Lou" plow to create furrows and soil ridges), artificial digging (manual farming using a narrow hoe), manual hoeing (using a broad hoe to cultivate soil), and contour plowing (using a single plow to prepare a fine sowing bed and create no furrows), to assess the surface roughness effect on soil loss. The results showed that the random roughness of surfaces with different tillage practices ranked as contour drilling > artificial digging > manual hoeing > contour plowing > no-tillage control. They also concluded that the artificial digging, artificial hoe, and contour plow were the most suitable as summer tillage practice to control erosion because they delayed the time of runoff, decreased runoff and sediment loss, and increased infiltration. Micro basin tillage, which is a practice of building individual earth blocks along the furrow, is also one of the treatments under a mechanical method to control soil erosion in agricultural land (Sui et al., 2016). With the use of micro basin tillage, the runoff and sediment rate can be reduced by 63% and 96%, respectively.

Restorative Farming and Cropping Systems

Restorative farming, or regenerative farming, is a holistic approach that seeks to rebuild or restore degraded soil by placing an emphasis on soil health. Restorative farming strives to work with the natural ecosystem, not against it. It covers practices such as no-till or minimum tillage, cover crops and mulching, crop rotations and diversification, integrated farming systems (IFSs), and organic farming, among others. According to reports, efficient conservation tillage techniques have a significant impact on soil quality, resource use effectiveness, and productivity of the main agricultural systems in the IHR (Das et al., 2021). Practice of zero-till and minimum tillage, along with recycling of crop residues, has revolutionized agricultural systems by allowing growers to manage greater land areas with reduced energy and machinery inputs; it also helps in forging a better soil environment and thereby stimulates plant growth (Saha et al., 2021). No tillage is also superior to conventional tillage in terms of promoting soil health by preserving organic matter, boosting soil structure, increasing water storage and infiltration, lowering erosion, and lowering greenhouse gas emissions (Thomas et al., 2007). The application of different agro-geo-textiles in a no-tilled rainy season maize crop conducted on two crop rotations at Dehradun on a 4% land slope in the IHR reduced runoff and soil losses better than vegetative filters of cowpea/grass-weed in the same maize crop (Singh et al., 2019a). Adoption of minimum or no tillage along with 50% NPK and weed biomass or green leaf manure in summer rice was found to have multiple benefits not only to the rice but also the succeeding winter pea (Das et al., 2021).

The system of cropping followed in the eastern Himalayan region is mostly mono-cropping where the lands are left fallow after the harvest of the main crop. A change in the cropping system as part of restorative farming could serve as a foundation for improving soil health. Cropping systems used in agriculture, such as crop rotation, intercropping, and other associated agronomic practices, can affect the health and quality of the soil from a variety of spatial and temporal perspectives (Vukicevich et al., 2016). In a cropping system, inclusion of pulses and legumes serves as an effective way of rehabilitating degraded soils and contributes significantly to achieving the twin objectives of increasing productivity and improving the sustainability of crops (Yadav et al., 1998). There are numerous varieties of legumes and pulses cultivated, including adzuki bean (*Vigna angularis*), black gram (*Vigna mungo*), broad bean (*Vicia faba*), cowpea (*Vigna unguiculata*), green gram (*Vigna radiata*), horse gram (*Macrotyloma uniflorum*), lentil (*Lens culinaris*), soybean (*Glycine max*), and rice bean (*Vigna umbellata*). The Northeastern Himalayan region also has a much wider spectrum of crops grown as compared with any other regions, yielding a productivity of about 848 kg/ha, which is higher than that of country overall (743 kg/ha), proving that the region is well suited for production of pulses. Major pulses grown include black gram, broad bean, cowpea, French bean (*Phaseolus vulgaris*), green gram, lablab bean (*Lablab purpureus*), lentil, pea (*Pisum sativum*), pigeon pea (*Cajanus cajan*), rice, bean, and soybean (Das et al., 2018). Other indigenous legumes are moth bean, local khesari, winged bean (*Psophocarpus tetragonolobus*), tree beans (*Parkia roxburghii*), jack bean (*Canavalia ensiformis*), and sword bean (*Canavalia gladiata*). Mucuna/velvet bean (*Mucuna pruriens*) is an important pulse crop of the IHR with potential advantageous medicinal properties (Kala, 2005). It also plays a significant role in soil conservation when incorporated as a cover crop or as green manure (Wabwoba & Mutoro, 2019). In a cereal cropping system, including legumes as a part of a crop rotation is also beneficial. The addition of legumes contributes significantly to restore soil health by minimizing the use of chemicals, increasing the organic matter content, and also acting as cover crops.

Diversification of crops is also designed as an integral part of restorative farming and cropping system to rebuild degraded soil. Crop diversification is often described as the "planned diversity" of cropping systems (Matson et al., 1997). Diversifying crops increases soil health and optimizes crop production by balancing soil biodiversity, enhancing soil nutrient use efficiency, and reducing soil-borne pathogens (Barbieri et al., 2019). With the need to increase production while maintaining the health of soil, agricultural practices are slowly shifting to traditional ways. *Barahnaja* is a type of crop diversification that literally translates to "12 foodgrains" or "cultivating 12 crops in a year." It is usually practiced in the Garhwal Himalayan region of India. The land holding of farmers in the Himalayan regions are generally marginal to small, and traditional mixed cropping system practices such as *Barahnaja* are considered part of the region's heritage and are followed for crop diversification. *Barahnaja* is not confined to the cultivation of cereals, but also includes oilseeds, vegetables, pulses, and other crops. In essence, *Barahnaja* is a mixed farming, companion planting system—a veritable community/society of crops (Gururani et al., 2021). Apart from the obvious advantages that diversified or mixed cropping offers, the *Barahnaja* technique of agriculture also prevents soil loss owing to the diversified root systems from several crops being planted at one time in the same field and the complete coverage of the field soils with the crops. Mandua (finger millet), amaranth, ugal (buckwheat), bhangeera (*Perilla* sp.),

Table 9.6 Present and Possible Future Cropping Systems.

State	Present cropping system	Improved future cropping systems
Arunachal Pradesh	Rice–fallow, maize–fallow	Rice–pea/lentil (lowland), maize + soybean–French bean
Manipur	Rice–fallow, vegetable–fallow, rice–rice	Rice-pea/lentil, rice–broad bean, maize–broad bean
Meghalaya	Rice–fallow, maize–fallow, rice–vegetables, ginger/ turmeric–fallow	Rice–pea/lentil/rapeseed, maize–French bean/black gram/rapeseed, French bean/ carrot/potato–okra/French bean
Mizoram	Rice–fallow, maize–fallow	Rice–pea/lentil (lowlands), maize + soybean–French bean/black gram/green gram
Nagaland	Rice–fallow, maize–fallow	Rice-pea/lentil (lowlands), maize–pulses (French bean/black gram)
Sikkim	Maize–fallow, maize–rice, maize–rapeseed, maize–buckwheat	Maize–pahenlo dal–buckwheat/rapeseed, maize–pea, maize + bean–rapeseed
Tripura	Rice–rice, rice–fallow, vegetable–fallow	Rice–pea/lentil/rapeseed, rice–rice–pea/ lentil, maize–rice–pea/lentil, maize–black gram/green gram

Note. Das et al. (2016a); Pattanayak and Das (2017).

and sesame have roots that hold and conserve soil securely and effectively, preventing the loss of water and nutrients (Gururani et al., 2021). The present and possible cropping systems that can be followed in the NEH region are presented in Table 9.6.

One cannot ignore crop rotation systems when it comes to maintenance and enhancement of soil health; it is an age-old traditional cultivation practice for managing the biodiversity of the agro-ecosystem by enhancing soil health, suppressing pests and disease outbreaks (Barbieri et al., 2019). To fully exploit the efficiency of a crop rotation, several factors are taken into consideration, namely, the type of crops, series of rotation, rotation length, characteristics of soil, and applied frequency of crops (Bainard et al., 2017; Bennett et al., 2011). In the Eastern Himalayan region, crop rotation, conservation-based tillage, and mulching can help address the multifaceted issues of land degradation, deteriorating soil health, and poor crop yield and profitability (Yadav et al., 2021). In two cropping systems, including summer maize-rainy season maize-lentil (SM-RM-L) and summer maize-rainy season maize-mustard (SM-RM-Ma), and with five tillage and mulching practices, including live mulch-based conservation tillage, the overall soil health was improved based on outcomes such as carbon indices and productivity of maize-based cropping systems in the eastern Himalayas (Yadav et al., 2021).

Organic Farming

Organic farming is a type of farming that avoids the use of any types of chemicals and relies heavily upon the use of organic amendments as a source of nutrients and as a tool to combat insect pests and diseases. According to International Federation for Organic Agriculture, the role of organic agriculture, whether in farming, processing, distribution, or consumption, is to sustain and enhance the health of ecosystem and organisms from the smallest in the soil to humans. The concept of organic farming is not new to the farming community of the IHR, particularly the northeastern hill (NEH) region of India. The land is almost virgin in terms of non-organic practices because crops have always been grown organically. Additionally, demand for organic produce and food products is rising globally, and the NEH region has significant potential due to the favorable environmental conditions for growing a variety of crops, the minimal or nonexistent use of synthetic agrochemicals and fertilizers (<12 kg/ha), and the abundance of organic manure, particularly plant biomass (Patel et al., 2014). The region includes a geographical area of 26.2 Mha in the Eastern Himalayas and has a vast opportunity to emerge as a major supplier of organic products (Sanwal et al., 2007).

Organic farming alleviates the negative impact that chemical fertilizers have on the soil and its environment. The use of fertilizers in the NEH region is meager, where nitrogen, phosphorus, and potassium through fertilizer use is only 13.37%, 11.12%, and 11.0% of the crop removal. This necessitates the application of organic sources of nutrition in the domain of soil health management. Organic sources, if pooled together, can supply about 13.07 kg N/ha, 7.18 kg phosphate/ha, and 7.34 kg potash/ha in northeastern India in addition to the adequate micronutrient supply from organic sources. With an adequate supply, biofertilizers can enhance production by 5%–30%. In northeast India, vermicomposting of rural wastes has tremendous promise for reducing soil nutrient deficiency owing to the availability of composting earthworms and need-based training in compost technique. In addition, the fruits of the Green Revolution could be extended to reach and benefit farmers in the region. The system of production in the hills has remained based on low-input, low-risk, low-yield technology, and the average yield of most crops has remained far behind. This provides a huge scope for enhancing productivity and soil health with good organic management. Although the yields of organic farming may not be as high as those supported by chemical fertilizers in the initial years, a bountiful harvest paired with environment safety is a boon in the long run. The highest levels of SOC, accessible phosphorus, and potassium were found in areas where more manure was applied. Over a 150-year period, the manured plots had an increase in SOM and total nitrogen of around 120% (Dhiman et al., 2007).

Agroforestry Options

The IHR, which includes the eight northeastern states and the hills of West Bengal, have a forested area of about 17.50 Mha, which corresponds to about 65.94% of the TGA (FSI, 2013). The need to improve soil health and restore the degraded land makes agroforestry a viable option in sustaining the crop production, while keeping the soil health in good condition. The most serious issue is the region's continuous loss of forest cover from shifting farming and the harvesting of firewood and timber, which leads to unhealthy soil and subsequent environmental degradation (Selvan & Kumar, 2017). In light of such issues, the intervention of agroforestry has huge potential and is proven to have the capacity to reduce soil erosion, conserve SOM, improve soil physical characteristics, enhance nutrient cycling, and sequester carbon (Feliciano et al., 2018; Ollinaho & Kröger, 2021).

In the Northeastern Himalayan region, agri-horticulture, agri-silviculture, agri-horti-silvi-pasture, agri-pisciculuture, and multitier systems are some of the important agroforestry systems (Das et al., 2016b). The traditional agroforestry system in the Northeastern Himalayan region is the popular *jhum* cultivation, which is woven into the regional heritage. Jhumming is, however, not a very sustainable form of agriculture given the short cycles. In the NEH region, tree species such as *Alnus nepalensis*, *Areca catechu*, *Pinus kesiya*, *Prunus domestica*, *Pyrus communis*, *Schima wallichii*, and others are generally integrated with crops such as maize, millet, ginger, pineapple, coffee, and varieties of vegetables.

Tree species can ameliorate the soil through the addition of biomass to the ground. Khasi pine (*P. kesiya*), alder (*A. nepalensis*), tree bean (*P. roxburghii*), Champak (*Magnolia oblonga*), and Gambhar (*Gmelina arborea*) as multipurpose trees lowered the soil bulk density while increasing the organic carbon content as compared to fallow. Soil erodibility decreased up to 23.1%–43.6% with the higher water stability (Saha et al., 2007). An individual tree species alone can also result in better soil structure via various processes such as increasing organic matter and reducing erosion losses (Singh et al., 2004). Similarly, under an alder-based agroforestry system, higher water holding capacity and aggregates were reported as compared to Khasi mandarin (*Citrus reticulata*) and Khasi pine, which in turn facilitated better soil structure and protected the soil from erosion or leaching (Tripathi et al., 2009). The extent of agroforestry benefits to the soil health is not just restricted to their physical and chemical properties, but also to their biological properties such as the microbial population, biomass, and respiration (Panwar et al., 2013; Zornoza et al., 2015). Higher bacterial and fungal counts were observed under traditional agroforestry systems in the NEH region. Consequently, there also existed a strong positive correlation of bacterial population count with the clay and moisture content, while fungal count showed strong positive correlation with total nitrogen in soil (Tangjang & Arunachalam, 2009).

Large cardamom is another successful agroforestry practice that imparts considerable potential in supporting and restoring soil health, particularly in Sikkim Himalaya (Mishra & Rai, 2013). Initially, large cardamom crop was domesticated by the indigenous Lepcha tribe and later on by other communities such as the Bhutias and Nepalis of Sikkim. The nearby Darjeeling district of India, some of Bhutan, and eastern Nepal then caught on with the cultivation technique (Sharma et al., 2000). About 12 indigenous varieties and seven species of wild relatives of the cultivated cardamom have been adapted to the various agroclimatic conditions of the EHR. Some common shade trees for the agroforestry are *A. nepalensis, Engelhardtia acerifolia, Eurya acuminata, Ficus nemoralis, Ficus hookeri, Macaranga pustulata, Maesa chisia, Schima wallichii, Leucosceptrum canum, Litsea polyantha, Nyssa sessiliflora, Osbeckia paniculata, Symplocos theifolia*, and *Viburnum cordifolium* (Sharma et al., 1994). The majority of large cardamom plantations have alder as a shade tree; for example, large cardamom is intercropped as the understory of alder trees (400 trees/ha) since their combination is proven to be sympatric and is both ecologically and economically viable (Das et al., 2012). Soil fertility is also continuously maintained through nutrient cycling from the natural tree cover provided by the Himalayan alder, and owing to its ability to regenerate naturally on sites affected by landslides, the tree provides greater ecological sustainability with cardamom (Sharma et al., 1998).

Being a multipurpose tree, alder is also used as timber, furniture, and fuel, besides improving soil fertility for growing crops. In Nagaland, a lucrative agroforestry method is used to cultivate agricultural crops alongside Himalayan alder trees on around 22,000 ha (Dhyani, 1998). The value of the alder tree was recognized by the tribal farmers long ago, and the evidence of this knowledge lies in trees that aged 200 years or older. Agricultural crops and alder trees combined make a very lucrative agroforestry system.

The tribal farmers of Angami, Chakhesang, Chang, Yimchunger, and Konyak fully utilize the alder tree's capacity to develop and keep soil fertility (Gokhle et al., 1985). In Khonoma village, Nagaland, the alder-based production system is a superb illustration of sustainable land use that has withstood the test of time. In this agroforestry system, alder seedlings are planted on the sloped land intended for cultivation, and the trees grow rapidly for 6–10 years (Dhyani, 1998). At this stage, the trees are pollarded during the months from November until January, the leaves and twigs are burnt, and the resultant ash is mixed into the soil to prepare it for raising crops. Then, the coppice are chopped, with the exception of the 5–6 numbers on top of the main trunk, and a crop schedule is followed, including a fallow period of 2–4 years. This process is repeated once every 4–6 years. Larger branches that have lost their leaves are utilized as firewood, while the tree's roots form nodules in symbiotic association with actinomycetes *Frankia*, a process that fertilizes the soil through nitrogen fixation. Besides fixing atmospheric nitrogen, the litter added to the soil also provides other essential nutrients such as phosphorus, potassium, and calcium (Sharma & Singh, 1994). Jhum fields are typically cultivated for 2 years within a 10-year cycle (1:4 ratio of cropping to fallow), but an alder-based system permits crop harvests in two of every 4 years (1:1 ratio of cropping to fallow). In a normal 4-year cycle, alder-based jhum fields are managed with 2 years of cropping in between the alder trees and 2 years of fallow during which the soil is relaxed and the coppice is allowed to flourish (Das et al., 2012).

In addition to increasing SOC by 96.2%, porosity by 10.9%, aggregate stability by 24.0%, and available soil moisture by 33.2%, multipurpose tree species with greater surface cover, consistent leaf litter fall, and extensive root systems have also been found to decrease bulk density and erosion ratio by 15.9% and 39.5%, respectively (Saha et al., 2007). Decomposition of added leaf litter by six multipurpose tree species, including *A. nepalensis*, *Albizia lebbeck*, *Boehmeria rugulosa*, *Dalbergia sissoo*, *Ficus glomerata*, and *F. roxburghii* in a mixed plantation established on an abandoned agricultural land in Central Himalaya yielded high rates of nitrogen and phosphorus (Semwal et al., 2003). Their addition builds up organic matter and gives better soil aggregation and infiltrability, thus improving soil conservation and health. Alder tree root nodules also fertilize the soil, and the spreading structure of the roots helps to minimize soil erosion on slopes (Das et al., 2012). Agroforestry commonly followed in the NEH region, whether one tiered or multi-tiered, is shown in Table 9.7.

Integrated Farming System Approaches for Small and Marginal Farmers

Farming in the EHR of India suffers a severe setback due to numerous factors, including climatic variation and shifts, undulating topography with accelerated soil erosion in higher hills, floods in the valleys, lack of resources, and limited availability of specialized location-specific improved technologies. Farm holdings in these regions are small (1–2 ha) and marginal (<1 ha). So, the intensive cropping system with selective single crop has a high risk of crop failures due to biotic and abiotic stresses. The IFS and farm diversification through inclusion of different enterprises or components of farming, such as agriculture, horticulture, animal husbandry, aquaculture, and so forth, offer the best options in these regions to fulfill the needs of households, with multiple objectives of food and nutritional security, employment generation, poverty alleviation, and judicious use of natural resources (Kumari et al., 2019).

Table 9.7 Some of the Common Agroforestry Systems in the NEH Region.

States	Agroforestry system/trees/tree species	Associated crops	References
Assam	Arecanut	Coconut, bamboo	Hazarika et al. (2001)
	Aquilaria agallocha (Agar)	*Areca catechu* (Tamul), Piper betel (Pan), *Schima wallichii* (Makarisal), Bamboo species (Banh), and Canes (Bet)	
	Livistona jenkinsiana	*Areca catechu*, Bamboo species, and *Sapindus pinnata* (Ritha)	
	Eri-muga silkworm	*Phoebe goalparensis* (Bonsum), *Ailanthus excelsa*, *Heteropanax fragrans*, Bamboo (Keseru), *Litsea serrata* (Mejankori), *Litsea monopetala* (Sulalo), *Manihot esculenta* (Simolu alu), and *Ricinus communis* (Aragosh)	
	Tea-based agroforestry	*Aquilaria agallocha* (Agar), *Cassia nodosa*, *Cassia siamea*, *Albizia procera*, Piper betle, *Piper longum* (Jaluk), and timbers	
Arunachal Pradesh	Agri-silviculture	*Bambusa tulda*, *Cajanus cajan*, *Gmelina arborea*, *Lagerstroemia* spp., *Mesua ferrea*, *Phaseolus* spp., *Solanum melongena*, and *Zea mays*	Pangging and Singh (2015)
	Agri-horticulture	*Areca catechu*, *Artocarpus heterophyllus*, *Cajanus cajan*, *Citrus* spp., *Solanum melongena*, *Mangifera indica*, *Psidium guajava*, *Musa* spp., and *Zea mays*	
	Agri-horti-silviculture	*Areca catechu*, *Artocarpus heterophyllus*, *Bambusa* spp., *Cajanus cajan*, *Mangifera indica*, *Mesua ferrea*, and *Psidium guajava*	
	Aqua-forestry	*Labeo rohita*, *Catla catla*, *Ctenopharyngodon idellus* Silvicultural crops: *Bambusa* spp., *Melia azedarach*	
	Bamboo	Pine	Tangjang and Nair (2016)
Manipur	Pond-based agroforestry	Banana, arecanut, and vegetable garden	Deb et al. (2009)

Meghalaya	Khasi mandarin	Maize, turmeric, cauliflower, cabbage, ginger, groundnut, and soybean	Bhatt et al. (2005)
	Pyrus communis	Maize, cauliflower, cabbage	
	Musa paradisiaca	Pineapple, broom grass	
	Areca catechu	Pineapple, betel vine	
Mizoram	Teak	Rice	Lalramnghinglova and Jha (1996)
Nagaland	Alder (*Alnus nepalensis*)	Jhum rice, tapioca, colocasia, large cardamom and turmeric	Kehie (2017)
	Oak (*Quercus serrate*)	Jhum rice	Singh and Teron (2019)
Sikkim	*Schima wallichii*, *Engelhardtia acerifolia*, *Eurya acuminata*, *Leucosceptrum canum*, *Maesa chisia*, *Symplocos theifolia*, *Ficus nemoralis*, *F. hookeri*, *Nyssa sessiliflora*, *Osbeckia paniculata*, *Viburnum cordifolium*, *Litsea polyantha*, *Macaranga pustulata*, and *Alnus nepalensis*	Large cardamom	Sharma et al. (1994)
	Mandarin (*Citrus reticulata*)	Ginger, spices, cereals and vegetables	Singh et al. (1991)
Tripura	Pond-based agroforestry	*Leucaena leucocephala* and *Moringa oleifera*	Deb et al. (2009)
	Arecanut	Pineapple, banana, mango, jackfruit and coconut	Mathur and Bhattacharya (2022)

In an IFS, the farmer and the farming family at the center are the decision makers, and the decisions are influenced by the family's priorities and household requirements. A minimum of 1 ha of land is required for the farm to support the family (Figure 9.6), and the successful execution of the IFS is dependent on the micro-climatic condition of the area, accurate planning, economic and socio-cultural aspects, and other variables. Before adoption, it is essential to know the present cropping system practice and the food demand and options in the region, with pre-planning of proper marketing strategies. After the adoption, the household and its resources and the resource flow should be planned such that each enterprise incorporated in the system functions interdependently as one unit. In this way, this method of farming reduces farmers' risks, increases productivity and profit, and maximizes the use of organic manure and crop wastes locally (Kumar et al., 2018). IFS also helps in utilizing or recycling the by-products of different components of the system and helps in decreasing air pollution, improving water quality, enhancing soil fertility, and finally increasing the total output in a sustainable manner. In short, IFS is one of the cutting-edge farming methods that offers resistance to climatic irregularities and is regarded as one of the appropriate solutions for climate-resilient agriculture (Das et al., 2015). There are several approaches and types of IFS, and some of the productive and widely adopted IFS models are as follows.

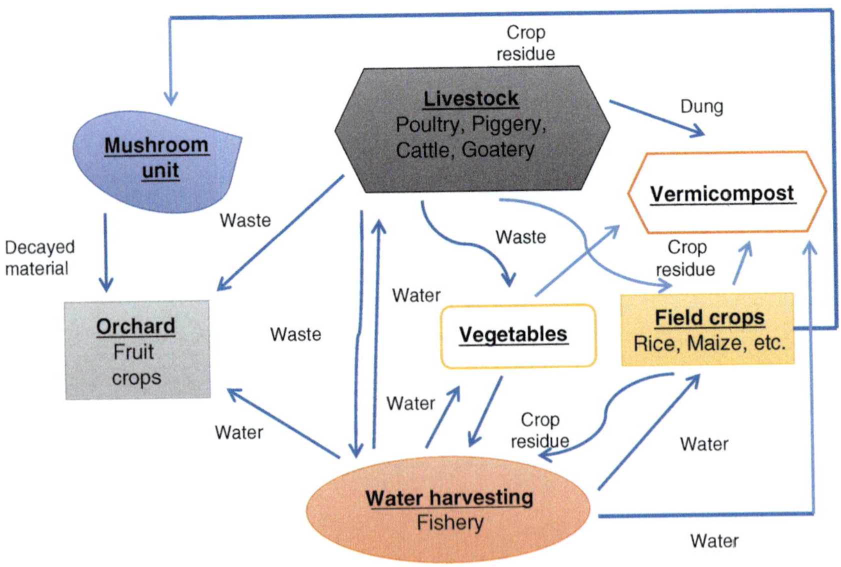

Figure 9.6 Flow diagram of Integrated Farming System (IFS).

Pond-Based IFS

In a pond-based IFS, the fishery unit or the fish is the major commodity of the farm as a whole. This type of IFS aims to integrate fish farming in dugout ponds either in a corner or center of the farm, as per land topography, along with livestock, crops, vegetables, fruits, and so forth, with provisions for on-farm resource recycling. This is practiced mostly in regions with ample rainfall during the monsoon, with 1,000–2,000 mm precipitation, and dry spells from October to April. The dugout ponds are used for rainwater harvesting for fish culture, and they can also be used for lifesaving irrigation during dry spells. The IFS model types under this system can be of different types based on the integration of different enterprises:

- Pond (fishery) + Rice cultivation + Vegetables cultivation + Piggery unit
- Pond (fishery) + Vegetable cultivation + Fruit crops + Piggery unit
- Pond (fishery) + Vegetable cultivation + Poultry unit + Fruit crops
- Pond (fishery) + Rice cultivation + Vegetable cultivation + Duckery unit
- Pond (fishery) + Vegetable cultivation+ Goatery unit integration, etc.
- Pond (fishery) + Vegetable cultivation + Mushroom unit + Agroforestry system + Fruit crops
- Pond (fishery) + Cereals+ Goatery unit + Vermicompost unit + Fruit crops + Spice production + Agroforestry system

This pond-based IFS has the ability to significantly raise the farm family's resource income and offer year-round food, nutrition, employment opportunities, and revenue. When farming components are chosen based on regional demand, resource availability, and agro-climatic conditions, this sort of IFS is also an economically sensible, environmentally viable, and socially acceptable system (Babu et al., 2019; Biswas et al., 2013).

Crop-Based IFS

In a crop-based IFS, either agricultural or horticultural crops are the major commodity in the farm. This IFS model combines crops, animals, multi-storied horticultural crops, seasonal vegetables, vermicomposting unit, a water pond for irrigation, and a boundary plantation with fruit crops or an agroforestry unit. This IFS model can be cereal based with subsistence livestock, cereal and fodder crops with livestock, and so on.

Balanced and Efficient Use of Agrochemicals

Plants extract required nutrients from the soil and use them for different metabolic processes, and the nutrients are later returned to the soil after the plant decays or through litter fall. In the context of the Himalayan agro-ecosystems, this nutrient cycling is an important process that enhances the organic matter deposition in soil, which results in the sequestration of carbon stocks (Bargali et al., 2015). While the majority of inputs used in modern agriculture include pesticides, chemical fertilizers, reliable irrigation, better seeds, and herbicides, it is a well-known fact that excessive usage causes soil acidification, eutrophication, and disruption of microbial activities. These factors can degrade environmental quality and soil productivity, which is a matter of concern (Dar & Bhat, 2020). The unbalanced use of inputs is also a major cause of yield gaps and food insecurity, particularly in regions with smallholder farms (Dutta et al., 2020; Giller et al., 2021). The adverse consequences are also observed in the Himalayan agro-ecosystems and attributed to the problems of increased erosion, decreased soil fertility, loss of biodiversity, water pollution, and alteration of atmospheric and climate processes (Shahi, 2020). The growth of pesticide-resistant weed and insect strains, ecological instability, soil and water pollution, toxicity to people and other organisms, and the application of pesticides and fertilizers on a regular basis all pose significant dangers. The management of soil health is necessary to promote sustainable agricultural output and to maintain biodiversity despite the negative impacts of chemical fertilizers. Because soils are the primary source of nutrients and microbial activity, managing them carefully is essential to meeting the world's food demand. Farming in the NEH region is usually done following organic farming systems and thereby create very little opportunity for the use of chemical fertilizers and pesticides (as presented in Table 9.8), which helps to maintain the soil health.

Table 9.8 Consumption of Fertilizers and Pesticides in the NEH Region.

	Fertilizer, 2020–2021	**Pesticides, 2017**
States	**kg/ha**	**tonnes**
Arunachal Pradesh	NIl	17
Assam	56.81	183
Manipur	40.39	30
Meghalaya	0.0	9
Mizoram	8.78	4
Nagaland	NA	16
Sikkim	0	3
Tripura	30.5	266

Source: North Eastern Council Secretariat (2024), Devi et al., (2017).

Crop Residue Management

The regular supply of organic amendments to soil increases the SOM content and also contributes to soil. Crop residues are the parts of plants that are left in the field after harvest. This residue derived from crops is considered "the greatest source of soil organic matter" (Tisdale et al., 1985) for agricultural soils. Crop residues are not necessarily "waste"; they are instead one of the several ways to restore soil health when managed in the right way. The method of retaining crop residues on the soil surface helps in maintaining the physical, chemical, and biological properties of soils (Wilhelm et al., 2007). Therefore, farmers are encouraged to return the residues back into the soil rather than burn them. Among the cereal crops, maize (*Zea mays*), wheat (*Triticum aestivum*), rice (*Oryza sativa*), and sorghum (*Sorghum bicolor*) produce the largest amount of crop residues (Blanco-Canqui & Lal, 2009). Crop residues are not added as is to the soil if one desires to reap maximum benefits. The types of residues, return ratios, return techniques, tillage intensity, fertilization rate, climate conditions, and soil features all have an impact on how healthy the soil is (Su et al., 2020). Therefore, to enhance crop yield and soil health, crop leftovers must be introduced along with the proper management techniques. Crop residues affect soil water movement, runoff, and infiltration because they add a significant amount of nutrients to the soil for crop production.

Crop residue management usually refers to protecting the soil surface from runoff or erosion and losses of nutrients. The practice not only protects soil from erosion, but also conserves soil moisture and improves infiltration and aeration. When crop residue is properly managed, it also aida in building up SOM and provides food for soil microbes (Shan & Yan, 2013). Additionally, crop residue returned to the soil increases the OC content and the available nitrogen, phosphorus, and potassium and prevents the loss of nutrients (Rasool et al., 2020; Zhang et al., 2021). Addition of crop residues can influence the availability of nitrogen to the crop depending on the type of crops. For instance, adding residue from legumes with a low C/N ratio can lead to nitrogen mineralization, whereas adding residue from cereals with a high C/N ratio can temporarily immobilize nitrogen during the decomposition process (Govaerts et al., 2007). The losses of mineral nitrogen fertilizers through denitrification is also greater when residues are left on the surface due to higher soil moisture content and if fertilizers are not properly incorporated. Similarly, addition of crop residues also increases the concentration of phosphorus in the top soil (Zibilske et al., 2002).

Using Weeds as an Alternate Source for Nutrient Management

Depending on the crop variety, weeds can significantly reduce agricultural yield by 22%–80%. The losses caused by weed infestation are also far more than the combined losses from insects, pests, and diseases (Sinha et al. 2018; Zannat et al. 2014). Weeds compete with crops for nutrients, and the existing weed control measures are no longer feasible, owing to labor shortages. In addition, chemical control measures based on the use of herbicides raise questions about the sustainability of the agro-ecosystems and concerns about the health hazards. Furthermore, excessive use of herbicides for weed control causes weeds to gain resistance to the chemicals. An emerging technique in weed management in recent years is the use of remote sensing, modeling, and robotics as a primary component in precision weed management. Another technique for weed management under conservation agriculture systems is the use of thermal weed management techniques and the implementation of modified tillage. Weed mat as the mulch is also a possible alternative in weed management. It is more cost efficient for mulching and has the benefit of moisture conservation and soil temperature management (Strik et al., 2020).

Another possible alternative in weed management is the use of weeds as the source of nutrients in crop production systems. Composting is an age-old technique that can also be used for weeds. It enables removal of the weeds' allelopathic effects and turn them into soil conditioner and organic fertilizer. One such technique is the composting of *Parthenium hysterophorus*, which has global significance since it causes various problems for biodiversity and for human and animal health, including eczema, asthma, and bronchitis. In addition, crop yield reduction caused by *P. hysterophorus* can be up to 40%. Composting of *Parthenium* can convert this weed to a useful material for use as a soil conditioner. The usual method of composting of *Parthenium* takes a long time and is less eco-friendly. As a result, a different method has been developed for preparing compost of *Parthenium* that takes comparatively less time (60 days only). This method has been named *Vishishta* composting (meaning something special). The quantity of matured compost obtained by this method is greater than other methods, and it is also suitable for destroying the weed seeds during the composting process. Application of composted material as soil conditioners improves the soil physical, chemical, and biological properties of soil and provides a source of nutrients and organic matter to the soil. The composting process can also be used to transform various weeds, such as *Chenopodium album*, *Chenopodium murale*, *Rumex dentatus*, *Asphodelus tenuifolius*, *Melilotus indica*, *Amaranthus albus*, *Avena fatua*, and others, into nutrient sources.

Conclusion

Intensive farming for higher food production in the IHR is often linked to a number of negative effects, including an increase in soil erosion, loss in SOM, decline in soil biodiversity, an overall deterioration of the soil health, and eventually a decline in factor productivity. Anthropogenic pressures on land, combined with unprecedented climate change, exacerbates these problems, threatening ecosystem services of the IHR and prosperity of the inhabitants. All these problems signal an urgent need to act more quickly to stop degradation and protect the natural wealth of the vulnerable region through adoption of different management and reclamation procedures. Although in the present century, there is immense need to improve economic growth and livelihood security of the peoples of the Himalayan region, but it cannot come at the cost of its rich biodiversity and environment. Thus, integrated approaches that are conservation effective, farmer oriented, and environmentally friendly should be taken into consideration. The Himalayan region is made up of plains, valleys, and steep slopes, and the issues are therefore also varied and demand site-specific management strategies to tackle them. To promote integrated natural resource management, including IFS, INM, and agroforestry-based systems, as well as the adoption of conservation agriculture and rainwater management principles for sustainable food production in the ecologically vulnerable IHR, developing strong policies and legislation is imperative.

References

Adekiya, A. O., Agbede, T. M., Olayanju, A., Ejue, W. S., Adekanye, T. A., Adenusi, T. T., & Ayeni, J. F. (2020). Effect of biochar on soil properties, soil loss, and cocoyam yield on a tropical Sandy loam Alfisol. *The Scientific World Journal, 2020*, Article 391630. https://doi.org/10.1155/2020/9391630

Ahirwal, J., Gogoi, A., & Sahoo, U. K. (2022). Stability of soil organic carbon pools affected by land use and land cover changes in forests of eastern himalayan region India. *Catena, 215*, Article 106308

Ahirwal, J., Nath, A., Brahma, B., Deb, S., Sahoo, U. K., & Nath, A. J. (2021). Patterns and driving factors of biomass carbon and soil organic carbon stock in the Indian Himalayan region. *Science of the Total Environment, 770*, Article 145292

Ahn, S.-R., & Kim, S.-J. (2016). The effect of rice straw mulching and no-tillage practice in upland crop areas on nonpoint-source pollution loads based on HSPF. *Water, 8*(3), Article 106. https://doi.org/10.3390/w8030106

Ali, S., & Sharda, V. N. (2005). Evaluation of the universal soil loss equation (USLE) in semi-arid and sub-humid climates of India. *Applied Engineering in Agriculture, 21*(2), 217–225. https://doi.org/10.13031/2013.18156

Amin, M., & Romshoo, S. A. (2018). Comparative assessment of soil erosion modelling approaches in a Himalayan watershed. *Modeling Earth Systems and Environment, 5*(1), 175–192. https://doi.org/10.1007/s40808-018-0526-x

Anonymous. (2012). *Uttarakhand state action plan for climate change, transforming crisis into opportunity* (pp. 1–226) [Review of *Uttarakhand State Action Plan for Climate Change, Transforming Crisis into Opportunity*]. Government of Uttarakhand.

Anonymous. (2013). Press note on poverty estimates 2011–12. Planning Commission, Government of India.

Babu, S., Das, A., Mohapatra, K. P., Yadav, G. S., Singh, R., Tahashildar, M., Devi, M. T., Das, S., Panwar, A. S., & Prakash, N. (2019). Pond dyke utilization: An innovative means for enhancing productivity and income under Integrated Farming System in North East Hill Region of India. *The Indian Journal of Agricultural Sciences*, *89*(1), 117–122. https://doi.org/10.56093/ijas.v89i1.86190

Bainard, L. D., Navarro-Borrell, A., Hamel, C., Braun, K., Hanson, K., & Gan, Y. (2017). Increasing the frequency of pulses in crop rotations reduces soil fungal diversity and increases the proportion of fungal pathotrophs in a semiarid agroecosystem. *Agriculture, Ecosystems & Environment*, *240*, 206–214. https://doi.org/10.1016/j.agee.2017.02.020

Barbieri, P., Pellerin, S., Seufert, V., & Nesme, T. (2019). Changes in crop rotations would impact food production in an organically farmed world. *Nature Sustainability*, *2*(5), 378–385. https://doi.org/10.1038/s41893-019-0259-5

Bargali, S. S., Shukla, K., Singh, L., Ghosh, L., & Lakhera, M. L. (2015). Leaf litter decomposition and nutrient dynamics in four tree species of dry deciduous forest. *Tropical Ecology*, *56*(2), 191–200.

Barman, B. K., Srinivasa Rao, K., Sonowal, K., Zohmingliani, Prasad, N. S. R., & Kumar Sahoo, U. (2020). Soil erosion assessment using revised universal soil loss equation model and geo-spatial technology: A case study of upper Tuirial river basin. *Mizoram India, AIMS Geosciences*, *6*(4), 525–544. https://doi.org/10.3934/geosci.2020030

Batar, A. K., Watanabe, T., & Kumar, A. (2017). Assessment of land-use/land-cover change and forest fragmentation in the Garhwal Himalayan region of India. *Environments*, *4*(2), Article 34

Batjes, N. H. (1996). Total carbon and nitrogen in the soils of the world. *European Journal of Soil Science*, *47*(2), 151–163. https://doi.org/10.1111/j.1365-2389.1996.tb01386.x

Bennett, A. J., Bending, G. D., Chandler, D., Hilton, S., & Mills, P. (2011). Meeting the demand for crop production: the challenge of yield decline in crops grown in short rotations. *Biological Reviews*, *87*(1), 52–71. https://doi.org/10.1111/j.1469-185x.2011.00184.x

Bera, A. (2017). Estimation of soil loss by USLE model using GIS and remote sensing techniques: A case study of Muhuri River basin, Tripura India. *Eurasian Journal of Soil Science*, *6*(3), 206–215. https://doi.org/10.18393/ejss.288350

Bhadra, A., Lalramnghaki, H., Kiba, L. G., & Bandyopadhyay, A. (2018). *Temporal variation in water induced soil erosion by RUSLE model using remote sensing and GIS*. https://doi.org/10.29007/271c

Bhat, J. A., Kundu, M. C., Hazra, G. C., Santra, G. H., & Mandal, B. (2010). Rehabilitating acid soils for increasing crop productivity through low-cost liming material. *Science of the Total Environment, 408*(20), 4346–4353. https://doi.org/10.1016/j.scitotenv.2010.07.011

Bhatt, B. P., Sachan, M. S., & Singh, K. (2005). Production potential of traditional agroforestry systems of Meghalaya: a case study [review of *Production potential of traditional agroforestry systems of Meghalaya: a case study*]. In B. P. Bhatt & K. M. Bujarbaruah (Eds.), *Agroforestry in north east India: Opportunities and challenges*.

Bhattacharjya, S., Bhaduri, D., Chauhan, S., Chandra, R., Raverkar, K. P., & Pareek, N. (2017). Comparative evaluation of three contrasting land use systems for soil carbon, microbial and biochemical indicators in North-Western Himalaya. *Ecological Engineering, 103*, 21–30.

Bhattacharyya, R., Ghosh, B., Mishra, P., Mandal, B., Rao, C., Sarkar, D., Das, K., Anil, K., Lalitha, M., Hati, K., & Franzluebbers, A. (2015). Soil degradation in India: Challenges and potential solutions. *Sustainability, 7*(4), 3528–3570. https://doi.org/10.3390/su7043528

Bhattacharyya, R., Prakash, V., Kundu, S., Srivastva, A. K., & Gupta, H. S. (2009). Soil aggregation and organic matter in a sandy clay loam soil of the Indian Himalayas under different tillage and crop regimes. *Agriculture, Ecosystems & Environment, 132*(1–2), 126–134. https://doi.org/10.1016/j.agee.2009.03.007

Bhattacharyya, T., Babu, R., Sarkar, D., Mandal, C., Dhyani, B. L., & Nagar, A. P. (2007). Soil loss and crop productivity model in humid subtropical India. *Current Science, 93*(10), 1397–1403. https://www.currentscience.ac.in/Volumes/93/10/1397.pdf

Bhattacharyya, T., Babu, R., Sarkar, D., Mandal, C., & Nagar, A. P. (2002). *Soil erosion of Tripura, a model for soil conservation and crop performance* (NBSS publication no. 97). NBSS & LUP.

Bhowmik, A., Kukal, S. S., Saha, D., Sharma, H., Kalia, A., & Sharma, S. (2019). Potential indicators of soil health degradation in different land use-based ecosystems in the Shiwaliks of northwestern India. *Sustainability, 11*(14), Article 3908

Biswas, S., Goswami, B., Sahu, N. C., & Dinajpur, D. (2013). Fish-duck and dyke vegetable cultivation practices in rural integrated farming system. *Indian Research Journal of Extension Education, 13*, 72–76.

Blanco, H., & Lal, R. (2010). *Principles of soil conservation and management*. Springer.

Blanco-Canqui, H., & Lal, R. (2009). Crop residue removal impacts on soil productivity and environmental quality. *Critical Reviews in Plant Sciences, 28*(3), 139–163. https://doi.org/10.1080/07352680902776507

Brahma, B., Nath, A. J., & Das, A. K. (2016). Managing rubber plantations for advancing climate change mitigation strategy. *Current Science, 110*(10), 2015–2019.

Burcham, M., R. O. Maguire, M. Alley, W. Thomason, & B. Jones. (2008). Injecting dairy manure to decrease nitrogen and phosphorous in runoff. SERA17 Annual Conference, Grasonville, MD.

Buyanovsky, G. A., & Wagner, G. H. (1987). Carbon transfer in a winter wheat (*Triticum aestivum*) ecosystem. *Biology and Fertility of Soils*, *5*, 76–82.

Central Soil and Water Conservation Research and Training Institute. (2011). *CSWCRTI vision 2030*. http://www.cswcrtiweb.org/Training/CSWCRTI%20Vision%202030.pdf

Chan, K. Y., Van Zwieten, L., Meszaros, I., Downie, A., & Joseph, S. (2007). Agronomic values of greenwaste biochar as a soil amendment. *Australian Journal of Soil Research*, *45*(8), 629–634. https://doi.org/10.1071/sr07109

Chmielewská, E. (2014). Zeolitic adsorption in course of pollutants mitigation and environmental control. *Journal of Radioanalytical and Nuclear Chemistry*, *299*, 255–260.

Choudhury, B. U., Ansari, M. A., Chakraborty, M., & Meetei, T. T. (2021). Effect of land-use change along altitudinal gradients on soil micronutrients in the mountain ecosystem of Indian (eastern) Himalaya. *Scientific Reports*, *11*(1), Article 14279

Choudhury, B. U., Fiyaz, A. R., Mohapatra, K. P., & Ngachan, S. (2016). Impact of land uses, agrophysical variables and altitudinal gradient on soil organic carbon concentration of north-eastern Himalayan region of India. *Land Degradation & Development*, *27*(4), 1163–1174.

Choudhury, B. U., Nengzouzam, G., & Islam, A. (2022). Evaluation of climate change impact on soil erosion in the integrated farming system based hilly micro-watersheds using revised universal soil loss equation. *Catena*, *214*, Article 106306. https://doi.org/10.1016/j.catena.2022.106306

Chuma, E., Benites, J., Fowler, R., Kienzle, J., Molapong, K., Nyagumbo, I., . . . & Manu, J. (1998, June 22–27). Proceedings.-Pt. 1: Workshop report.-Pt. 2: Annexes. International Workshop on Conservation Tillage for Sustainable Agriculture, Harare, Zimbabwe. GTZ.

Ciais, P., Wattenbach, M., Vuichard, N., Smith, P., Piao, S. L., Don, A., Luyssaert, S., Janssens, I. A., Bondeau, A., Dechow, R., Leip, A., Smith, P., Beer, C., van der Werf, G. R., Gervois, S., Van Oost, K., Tomelleri, E., Freibauer, A., & Schulze, E. D. (2010). The European carbon balance. Part 2: Croplands. *Global Change Biology*, *16*(5), 1409–1428. https://doi.org/10.1111/j.1365-2486.2009.02055.x

Cooper, J., Baranski, M., Stewart, G., Nobel-de Lange, M., Bàrberi, P., Fließbach, A., Peigné, J., Berner, A., Brock, C., Casagrande, M., Crowley, O., David, C., De Vliegher, A., Döring, T. F., Dupont, A., Entz, M., Grosse, M., Haase, T., Halde, C., & Hammerl, V. (2016). Shallow non-inversion tillage in organic farming maintains crop yields and increases soil C stocks: A meta-analysis. *Agronomy for Sustainable Development*, *36*(1), Article 22. https://doi.org/10.1007/s13593-016-0354-1

CSWCRTI. (2011). CSWCRTI Vision 2030 Dr. P. Dogra, (Ed.). http://www.cswcrtiweb.org/Training/CSWCRTI%20Vision%202030.pdf

Dabral, P. P., Baithuri, N., & Pandey, A. (2008). Soil erosion assessment in a hilly catchment of north eastern India using USLE, GIS and remote sensing. *Water Resources Management*, *22*(12), 1783–1798. https://doi.org/10.1007/s11269-008-9253-9

Dadhwal, K. S., Mandal, D., & Shrimali, S. S. (2011). Soil quality index for different land use systems in northwestern hilly region of India. *Journal of the Indian Society of Soil Science*, *59*(2), 169–176.

Dai, C., Liu, Y., Wang, T., Li, Z., & Zhou, Y. (2018). Exploring optimal measures to reduce soil erosion and nutrient losses in southern China. *Agricultural Water Management*, *210*, 41–48. https://doi.org/10.1016/j.agwat.2018.07.032

Dar, S. A., & Bhat, R. A. (2020). Aquatic pollution stress and role of biofilms as environment cleanup technology. In H. Qadri, R. A. Bhat, M. A. Mehmood, & G. H. Dar (Eds.), *Fresh Water Pollution Dynamics and Remediation* (pp. 293–318). Springer. https://doi.org/10.1007/978-981-13-8277-2_16

Das, A., Babu, S., Yadav, G. S., & Ansari, M. A. (2016a). Status and strategies for pulses production for food and nutritional security in north eastern region of India. *Indian Journal of Agronomy*, *61*, 43–57.

Das, A., Ghosh, P. K., Choudhury, B. U., Patel, D. P., Munda, G. C., Ngachan, S. V., & Chowdhury P. (2009) *Climate change in north East India: Recent facts and events—Worry for agricultural management* (pp. 32-37), Proceedings of the Workshop on Impact of Climate Change on Agriculture.

Das, A., Jayanta, L., Yadav, G. S., Lal, R., Saha, S., Singh, R. P., Ansari, M. A., Babu, S. S., Mohapatra, K. P., Devi, S., Kandpal, B. K., & Ghosh, P. (2021). Managing soil organic carbon in croplands of the eastern Himalayas, India. In R. Lal (Ed.), *Soil organic matter and feeding the future* (pp. 279–304). CRC Press. https://doi.org/10.1201/9781003102762-12

Das, A., Munda, G. C., Ghosh, P. K., Ramkrushna, G. I., Patel, D. P., Choudhury, B. U., Mandal, S., Ngachan, S. V., & Jayanta, L. (2011). Managing degraded land and soil organic carbon for climate resilient agriculture in north East India. In C. Srinivasa Rao, B. Venkateswarlu, K. Srinivas, S. Kundu, & A. K. Singh (Eds.), *Soil carbon sequestration for climate change mitigation and food security* (pp. 241–265). Central Research Institute for Dryland Agriculture.

Das, A., Ramkrushna, G. I., Choudhury, B. S., Munda, G. C., Patel, D., Ngachan, S. V., Ghosh, P., Tripathi, A. K., Das, S., & Sharma, M. (2012). Natural resource conservation through indigenous farming systems: Wisdom alive in North East India. *Indian Journal of Traditional Knowledge*, *11*(3), 505–513.

Das, A., Ramkrushna, G. I., Makdoh, B., Sarkar, D., Layek, J., Mandal, S., & Lal, R. (2016b). Lower Himalayas: Soil management. In *Encyclopedia of Soil Science* (3rd ed). CRC Press.

Das, A., Ramkrushna, G. I., Yadav, G., Layek, J., Debnath, C., Choudhury, B. S., Mohaptara, K. P., Ngachan, S. V., & Das, S. (2015). Capturing traditional practices of rice based farming systems and identifying interventions for resource conservation and food security in Tripura, India. *Applied Ecology and Environmental Sciences*, *3*(4), 100–107. https://doi.org/10.12691/aees-3-4-2

Das, B., Bordoloi, R., Thungon, L. T., Paul, A., Pandey, P. K., Mishra, M., & Tripathi, O. P. (2020). An integrated approach of GIS, RUSLE and AHP to model soil erosion in West Kameng watershed, Arunachal Pradesh. *Journal of Earth System Science*, *129*(1), Article 94. https://doi.org/10.1007/s12040-020-1356-6

Das, B., Chakraborty, D., Singh, V. K., Aggarwal, P., Singh, R., Dwivedi, B. S., & Mishra, R. P. (2014). Effect of integrated nutrient management practice on soil aggregate properties, its stability and aggregate-associated carbon content in an intensive rice–wheat system. *Soil and Tillage Research*, *136*, 9–18. https://doi.org/10.1016/j.still.2013.09.009

Das, B., Paul, A., Bordoloi, R., Tripathi, O. P., & Pandey, P. K. (2018). Soil erosion risk assessment of hilly terrain through integrated approach of RUSLE and geospatial technology: A case study of Tirap District, Arunachal Pradesh. *Modeling Earth Systems and Environment*, *4*(1), 373–381. https://doi.org/10.1007/s40808-018-0435-z

Dawa, D., & Arjune, V. (2021). Identifying potential erosion-prone areas in the Indian Himalayan region using the revised universal soil loss equation (RUSLE). *Asian Journal of Water, Environment and Pollution*, *18*(1), 15–23. https://doi.org/10.3233/ajw210003

de Moraes Sá, J. C., Séguy, L., Tivet, F., Lal, R., Bouzinac, S., Borszowskei, P. R., Briedis, C., dos Santos, J. B., da Cruz Hartman, D., Bertoloni, C. G., Rosa, J., & Friedrich, T. (2015). Carbon depletion by plowing and its restoration by no-till cropping systems in Oxisols of subtropical and tropical agro-ecoregions in Brazil. *Land Degradation & Development*, *26*(6), 531–543. https://doi.org/10.1002/ldr.2218

Deb, S., Arunachalam, A., & Das, A. (2009). Indigenous knowledge of Nyishi tribes on traditional agroforestry systems. *Indian Journal of Traditional Knowledge*, *8*(1), 41–46.

den Biggelaar, C., Lal, R., Eswaran, H., Breneman, V. E., & Reich, P. F. (2003). Crop yield losses to soil erosion at regional and global scales: Evidence from plot-level and GIS data. https://econpapers.repec.org/RePEc:elg:eechap:2478_12

Devi, P. I., Thomas, J., & Raju, R. K. (2017). Pesticide consumption in India: A spatiotemporal analysis. *Agricultural Economics Research Review*, *30*(1), 163. https://doi.org/10.5958/0974-0279.2017.00015.5

Dhiman, S. D., Hari, O., & Singh, D. (2007). Effect of long term nutrient management on soil properties and productivity in rice (*Oryza sativa*)-wheat (*Triticum aestivum*) sequence. *Indian Journal of Agronomy*, *32*, 112–120.

Dhruvanarayana, V. V., & Babu, R. (1983). Estimation of soil erosion in India. *Journal of Irrigation and Drainage Engineering*, *109*(4), 419–434.

Dhyani, S. K. (1998). Tribal alders: An agroforestry system in the north eastern hills of India. *Agroforestry Today*, *10*(4), 14–15.

Dikshit, A., Sarkar, R., Pradhan, B., Segoni, S., & Alamri, A. M. (2020). Rainfall induced landslide studies in Indian Himalayan region: A critical review. *Applied Sciences*, *10*(7), Article 2466. https://doi.org/10.3390/app10072466

Dutta, S., Chakraborty, S., Goswami, R., Banerjee, H., Majumdar, K., Li, B., & Jat, M. L. (2020). Maize yield in smallholder agriculture system—An approach integrating socio-economic and crop management factors. *PLoS One*, *15*(2), Article e0229100. https://doi.org/10.1371/journal.pone.0229100

Dymond, J. R., Betts, H. D., & Schierlitz, C. S. (2010). An erosion model for evaluating regional land-use scenarios. *Environmental Modelling & Software*, *25*(3), 289–298. https://doi.org/10.1016/j.envsoft.2009.09.011

Eshel, G., Egozi, R., Goldwasser, Y., Kashti, Y., Fine, P., Hayut, E., Kazukro, H., Rubin, B., Dar, Z., Keisar, O., & DiSegni, D. M. (2015). Benefits of growing potatoes under cover crops in a Mediterranean climate. *Agriculture, Ecosystems & Environment*, *211*, 1–9. https://doi.org/10.1016/j.agee.2015.05.002

Fahnestock, P., Lal, R., & Hall, G. F. (1996). Land use and erosional effects on two Ohio Alfisols: II. crop yields. *Journal of Sustainable Agriculture*, *7*(2–3), 85–100. https://doi.org/10.1300/j064v07n02_09

Fan, F., Xie, D., Wei, C., Ni, J., Yang, J. J., Tang, Z., & Zhou, C. (2015). Reducing soil erosion and nutrient loss on sloping land under crop-mulberry management system. *Environmental Science and Pollution Research*, *22*(18), 14067–14077. https://doi.org/10.1007/s11356-015-4608-2

Feliciano, D., Ledo, A., Hillier, J., & Nayak, D. R. (2018). Which agroforestry options give the greatest soil and above ground carbon benefits in different world regions? *Agriculture, Ecosystems & Environment*, *254*, 117–129.

Forest Survey of India. (2013). *India State of Forest Reports (2013)*. Ministry of Environment, Forest and Climate Change, Government of India.

Forest Survey of India. (2019). *India State of Forest Report (2019)*. Ministry of Environment, Forest and Climate Change, Government of India.

Froehlich, W. (2004). *Soil erosion, suspended sediment sources and deposition in the Maw-Ki-Syiem drainage basin, Cherrapunji, northeastern India* (IAHS publ. 288). Sediment Transfer through the Fluvial System [Proceedings of a symposium held in Moscow, August 2004]. https://iahs.info/uploads/dms/12877.138-146-288-Froehlich.pdf

Fróna, D., Szenderák, J., & Harangi-Rákos, M. (2019). The challenge of feeding the world. *Sustainability*, *11*(20), Article 5816. https://doi.org/10.3390/su11205816

Gachene, C. K., & Kimaru, K. G. (2003). Soil fertility and Land Productivity Technical Handbook No. 30.

Galy, A., & France-Lanord, C. (2001). Higher erosion rates in the Himalaya: Geochemical constraints on riverine fluxes. *Geology*, *29*(1), 23–26.

Garde, R. J., & Kothyari, U. C. (2014). Sediment yield estimation. *Water and Energy International*, *44*(3), 97–123.

George, K. J., Kumar, S., & Hole, R. M. (2021). Geospatial modelling of soil erosion and risk assessment in Indian Himalayan region—A study of Uttarakhand state. *Environmental. Advances*, *4*, Article 100039. https://doi.org/10.1016/j.envadv.2021.100039

Ghimire, S., Higaki, D., & Bhattarai, T. (2013). Estimation of soil erosion rates and eroded sediment in a degraded catchment of the Siwalik Hills. *Nepal. Land*, *2*(3), 370–391. https://doi.org/10.3390/land2030370

Ghosh, B. K., Dogra, P., Sharma, N., Alam, N. M., Singh, R. P., & Mishra, P. C. (2016a). Effects of resource conservation practices on productivity, profitability and energy budgeting in maize–wheat cropping system of Indian sub-Himalayas. *Proceedings of the National Academy of Sciences, India Section B: Biological Sciences*, *86*(3), 595–605. https://doi.org/10.1007/s40011-015-0492-2

Ghosh, B. N., Dogra, P., Sharma, N. K., & Dadhwal, K. S. (2012). Soil erosion-productivity relationship assessment in sloping lands of north-West Himalayas. *Indian Journal of Agricultural Sciences*, *82*(12), 1068–1071.

Ghosh, B. N., Meena, V. S., Alam, N. M., Dogra, P., Bhattacharyya, R., Sharma, N. K., & Mishra, P. K. (2016b). Impact of conservation practices on soil aggregation and the carbon management index after seven years of maize–wheat cropping system in the Indian Himalayas. *Agriculture, Ecosystems & Environment*, *216*, 247–257. https://doi.org/10.1016/j.agee.2015.09.038

Ghosh, K., Kumar De, S., Bandyopadhyay, S., & Saha, S. (2013). Assessment of soil loss of the Dhalai River basin, Tripura, India using USLE. *International Journal of Geosciences*, *4*(1), 11–23. https://doi.org/10.4236/ijg.2013.41002

Giller, K. E., Delaune, T., Silva, J. V., van Wijk, M., Hammond, J., Descheemaeker, K., van de Ven, G., Schut, A. G. T., Taulya, G., Chikowo, R., & Andersson, J. A. (2021). Small farms and development in sub-Saharan Africa: Farming for food, for income or for lack of better options? *Food Security*, *13*, 1431–1454. https://doi.org/10.1007/s12571-021-01209-0

Giller, P. S. (1996). The diversity of soil communities, the 'poor man's tropical rainforest'. *Biodiversity & Conservation*, *5*, 135–168.

Gogoi, A., Ahirwal, J., & Sahoo, U. K. (2022). Evaluation of ecosystem carbon storage in major forest types of eastern Himalaya: Implications for carbon sink management. *Journal of Environmental Management*, *302*, Article 113972

Gogoi, A., Sahoo, U. K., & Saikia, H. (2020). Vegetation and ecosystem carbon recovery following shifting cultivation in Mizoram-Manipur-Kachin rainforest eco-region, Southern Asia. *Ecological processes*, *9*, 1–13.

Gokhle, A. M., Zeliang, D. K., Kevichusa, R., Angami, T., & Bendangnugsang, S. (1985). *The use of alder tree*. State Council of Education Research and Training, Government of Nagaland.

Govaerts, B., Sayre, K. D., Lichter, K., Dendooven, L., & Deckers, J. (2007). Influence of permanent raised bed planting and residue management on physical and chemical soil quality in rain fed maize/wheat systems. *Plant and Soil, 291*(1–2), 39–54. https://doi.org/10.1007/s11104-006-9172-6

Griffiths, R. I., Whiteley, A. S., O'Donnell, A. G., & Bailey, M. J. (2003). Physiological and community responses of established grassland bacterial populations to water stress. *Applied and Environmental Microbiology, 69*(12), 6961–6968. https://doi.org/10.1128/aem.69.12.6961-6968.2003

Guo, S., Zhai, L., Liu, J., Liu, H., Chen, A., Wang, H., Wu, S., & Lei, Q. (2019). Cross-ridge tillage decreases nitrogen and phosphorus losses from sloping farmlands in southern hilly regions of China. *Soil and Tillage Research, 191*, 48–56. https://doi.org/10.1016/j.still.2019.03.015

Gupta, S., & Kumar, S. (2017). Simulating climate change impact on soil erosion using RUSLE model—A case study in a watershed of mid-Himalayan landscape. *Journal of Earth System Science, 126*(3), Article 43. https://doi.org/10.1007/s12040-017-0823-1

Gururani, K., Sood, S., Kumar, A., Joshi, D. C., Pandey, D., & Sharma, A. R. (2021). Mainstreaming Barahnaja cultivation for food and nutritional security in the Himalayan region. *Biodiversity and Conservation, 30*(3), 551–574. https://doi.org/10.1007/s10531-021-02123-9

Hansen, J., Sato, M., Kharecha, P., Beerling, D., Berner, R., Delmotte, V. M., Pagani, M., Raymo, M., Royer, D. L., & Zachos, J. C. (2008). Target atmospheric CO_2: where should humanity aim? [review of *Target atmospheric CO2: Where should humanity aim?*]. *The Open Atmospheric Science Journal, 2*, 217–231.

Hazarika, P., Kakati, N., & Kalita, R. K. (2001). *Indigenous knowledge in relation to conservation and management of forest biodiversity of Assam*. In B. P. Bhatt, K. M. Bujarbaruah, Y. P. Sharma, & Patiram (Eds.), *Approaches for increasing agricultural productivity in hill and mountain ecosystem* (pp. 149–152). ICAR Research Complex for NEH Region.

Hazarika, S. (2018). Approaches for managing degraded lands with special reference to north East India. In A. Das, K. P. Mohapatra, S. V. Ngachan, A. S. Panwar, D. J. Rajkhowa, G. I. Ramkrushna, & J. Layek (Eds.), *Conservation agriculture for advancing food security in changing climate* (Vol. 2, pp. 649–665). Today & Tomorrow's Printers and Publishers.

Hendrix, P. F., Crossley, D. A. Jr., Blair, J. M., Coleman, D. C., 1990. Soil biota as components of sustainable agroecosystems. In: Edwards, C.A., Lal, Rattan, Madden, Patrick, Miller, Robert H., House, Gar (Eds.), Sustainable Agricultural Systems. *Soil and Water Conservation Society*, IA, pp. 637–654

Hicks, D. L., & Anthony, T. (2001). *Soil conservation technical handbook*. New Zealand Resource Management Association, Ministry for the Environment.

Himes, F. L. (1998). Nitrogen, sulfur and phosphorus and the sequestering of carbon. In R. Lal, J. M. Kimble, R. Follett, & B. A. Stewart (Eds.), *Soil processes and the carbon cycle*. CRC Press.

Huang, J., Wang, J., Zhao, X., Wu, P., Qi, Z., & Li, H. (2014). Effects of permanent ground cover on soil moisture in jujube orchards under sloping ground: A simulation study. *Agricultural Water Management*, *138*, 68–77. https://doi.org/10.1016/j.agwat.2014.03.002

ICIMOD. (2010). *Biodiversity in the Eastern Himalayas: Status, trends and vulnerability to climate change climate change impact and vulnerability in the Eastern Himalayas* (Technical report 2).

Jain, A., Rai, S. C., & Sharma, E. (2000). Hydro-ecological analysis of a sacred lake watershed system in relation to land-use/cover change from Sikkim Himalaya. *Catena*, *40*(3), 263–278. https://doi.org/10.1016/s0341-8162(00)00086-2

Jain, S. K., Kumar, S., & Varghese, J. (2001). Estimation of soil erosion for a Himalayan watershed using GIS technique. *Water Resources Management*, *15*(1), 41–54. https://doi.org/10.1023/a:1012246029263

Jaiswal, M. K., & Amin, N. (2020). The impact of land use dynamics on the soil erosion in the Panchnoi River basin, Northeast India. *Journal of the Geographical Institute Jovan Cvijic SASA*, *70*(1), 1–14. https://doi.org/10.2298/ijgi2001001j

Jena, R. K., Padua, S., Ray, P., Ramachandran, S., Bandyopadhyay, S., Deb Roy, P., Obi Reddy, G. P., Singh, S. K., & Ray, S. K. (2018). Assessment of soil erosion in sub tropical ecosystem of Meghalaya, India using remote sensing, GIS and RUSLE. *Indian Journal of Soil Conservation*, *46*(3), 273–282.

Jien, S.-H., & Wang, C.-S. (2013). Effects of biochar on soil properties and erosion potential in a highly weathered soil. *Catena*, *110*, 225–233. https://doi.org/10.1016/j.catena.2013.06.021

Joshi, G., & Negi, G. C. S. (2015). Physico-chemical properties along soil profiles of two dominant forest types in Western Himalaya. *Current Science*, *109*(4), 798–803. http://www.jstor.org/stable/24905743

Kala, C. P. (2005). Ethnomedicinal botany of the Apatani in the eastern Himalayan region of India. *Journal of Ethnobiology and Ethnomedicine*, *1*(1), Article 11. https://doi.org/10.1186/1746-4269-1-11

Kalambukattu, J., & Kumar, S. (2017). Modelling soil erosion risk in a mountainous watershed of mid-Himalaya by integrating RUSLE model with GIS. *Eurasian Journal of Soil Science*, *6*(2), 92–105. https://doi.org/10.18393/ejss.286442

Kaur, T., Sehgal, S. K., Singh, S., Sharma, S., Dhaliwal, S. S., & Sharma, V. (2021). Assessment of seasonal variability in soil nutrients and its impact on soil quality under different land use systems of lower Shiwalik foothills of Himalaya, India. *Sustainability*, *13*(3), 1398.

Kehie, M. (2017). Indigenous alder based farming practices in Nagaland, India: A sustainable agricultural model. *Journal of Traditional and Folk Practices*, *5*(2). https://doi.org/10.25173/jtfp.2017.5.2.75

Kennedy, A. C., & Gewin, V. L. (1997). Soil microbial diversity: Present and future considerations. *Soil Science, 162*(9), 607–617. https://doi.org/10.1097/00010694-199709000-00002

Kewlani, P., Negi, V. S., Bhatt, I. D., Rawal, R. S., & Nandi, S. K. (2021). Soil nutrients concentration along altitudinal gradients in Indian Western Himalaya. *Scandinavian Journal of Forest Research, 36*(2–3), 98–104. https://doi.org/10.1080/02827581.2020.1871065

Khare, D., Mondal, A., Kundu, S., & Mishra, P. K. (2017). Climate change impact on soil erosion in the Mandakini River Basin, North India. *Applied Water Science, 7*, 2373–2383.

Kibblewhite, M. G., Ritz, K., & Swift, M. J. (2008). Soil health in agricultural systems. Philosophical transactions of the Royal Society of London. *Series B, Biological sciences, 363*(1492), 685–701. https://doi.org/10.1098/rstb.2007.2178

Koirala, P., Thakuri, S., Joshi, S., & Chauhan, R. (2019). Estimation of soil erosion in Nepal using a RUSLE modeling and geospatial tool. *Geosciences, 9*, 4, Article 147. https://doi.org/10.3390/geosciences9040147

Kothyari, B. P., Verma, P. K., Joshi, B. K., & Kothyari, U. C. (2004). Rainfall–runoff-soil and nutrient loss relationships for plot size areas of bhetagad watershed in Central Himalaya, India. *Journal of Hydrology, 293*(1–4), 137–150. https://doi.org/10.1016/j.jhydrol.2004.01.011

Kumar, A., Devi, M., & Deshmukh, B. (2014). Integrated remote sensing and geographic information system based RUSLE modelling for estimation of soil loss in Western Himalaya, India. *Water Resources Management, 28*(10), 3307–3317. https://doi.org/10.1007/s11269-014-0680-5

Kumar, S., & Kushwaha, S. P. S. (2013). Modelling soil erosion risk based on RUSLE-3D using GIS in a Shivalik sub-watershed. *Journal of Earth System Science, 122*(2), 389–398. https://doi.org/10.1007/s12040-013-0276-0

Kumar, S., Murthy, D. K., Gumma, M. K., Khan, P. E., Khatri-Chhetri, A., Aggarwal, P. K., Murthy, C. S. R., Chakravarthy, K., & Whitbread, A. M. (2018). *Towards climate-smart agricultural policies and investments in Telangana* (CCAFS info note). CGIAR Research Program on Climate Change, Agriculture and Food Security. https://hdl.handle.net/10568/90627

Kumar, S., Singh, R. P., & Kalambukattu, J. G. (2021). Modeling daily surface runoff, sediment and nutrient loss at watershed scale employing arc-APEX model interfaced with GIS: A case study in lesser Himalayan landscape. *Environmental Earth Sciences, 80*, Article 498. https://doi.org/10.1007/s12665-021-09791-4

Kumar, S., Sterk, G., & Dadhwal, V. K. (2005). *Process based modeling for simulating surface runoff and soil erosion at watershed basis*. Commission VI, WG VI/4. Indian Institute of Remote Sensing.

Kumari, C. P., Goverdhan, M., Sridevi, S., Reddy, G. K., Pasha, M. L., Ramana, M. V., & Rani, B. (2019). Profitability of cropping systems module for different farming systems in southern Telangana zone of Telangana state. *Journal of Pharmacognosy and Phytochemistry*, *8*(3), 153–157.

Lal, M., & Mishra, S. (2015). Characterization of surface runoff, soil erosion, nutrient loss and their relationship for agricultural plots in India. *Current World Environment*, *10*(2), 593–601. https://doi.org/10.12944/cwe.10.2.24

Lal, R. (1995). Erosion-crop productivity relationships for soils of Africa. *Soil Science Society of American Journal*, *59*, 661–667. https://doi.org/10.2136/sssaj1995.03615995005900030004x

Lal, R. (2001). Soil degradation by erosion. *Land Degradation & Development*, *12*(6), 519–539. https://doi.org/10.1002/ldr.472

Lal, R. (2004). Soil carbon sequestration impacts on global climate change and food security. *Science*, *304*(5677), 1623–1627. https://doi.org/10.1126/science.1097396

Lal, R. (2005). Forest soils and carbon sequestration. *Forest Ecology and Management*, *220*(1–3), 242–258.

Lal, R. (2015, December 5–7). International Year of Soil and the Sustainable Development Goals of the United Nation. National Seminar on Sustaining Hill Agriculture in Changing Climate, Agartala, Tripura.

Lal, R. (2018). Digging deeper: A holistic perspective of factors affecting soil organic carbon sequestration in agroecosystems. *Global Change Biology*, *24*(8), 3285–3301.

Lal, R., & Moldenhauer, W. C. (1987). Effects of soil erosion on crop productivity. *Critical Reviews in Plant Sciences*, *5*(4), 303–367. https://doi.org/10.1080/07352688709382244

Lalramnghinglova, J. H., & Jha, L. K. (1996). Prominent agroforestry systems and important multipurpose trees in farming system of Mizoram. *Indian Forester*, *122*, 604–609. https://doi.org/10.36808/if/1996/v122i7/6395

Laufer, D., Loibl, B., Märländer, B., & Koch, H.-J. (2016). Soil erosion and surface runoff under strip tillage for sugar beet (*Beta vulgaris* L.) in Central Europe. *Soil and Tillage Research*, *162*, 1–7. https://doi.org/10.1016/j.still.2016.04.007

Layek, J., Das, A., Ghosh, P. K., Rangappa, K., Lal, R., Idapuganti, R. G., Nath, C. P., & Dey, U. (2022). Double no-till and rice straw retention in terraced sloping lands improves water content, soil health and productivity of lentil in Himalayan foothills. *Soil and Tillage Research*, *221*, Article 105381

Li, T., Zhao, L., Duan, H., Yang, Y., Wang, Y., & Wu, F. (2019). Exploring the interaction of surface roughness and slope gradient in controlling rates of soil loss from sloping farmland on the loess plateau of China. *Hydrological Processes*, *34*(2), 339–354. https://doi.org/10.1002/hyp.13588

Lupwayi, N. Z., Rice, W. A., & Clayton, G. W. (1998). Soil microbial diversity and community structure under wheat as influenced by tillage and crop rotation.

Soil Biology and Biochemistry, *30*(13), 1733–1741. https://doi.org/10.1016/s0038-0717(98)00025-x

Macrae, C. N., Bodenhausen, G. V., Schloerscheidt, A. M., Milne A. B. Tales of the unexpected: executive function and person perception. *J Pers Soc Psychol.* 1999 Feb;76(2):200–13. doi: 10.1037//0022-3514.76.2.200. PMID: 10074705.

Madhu, U., Begum, M., Salam, A., & Sarkar, S. K. (2018). Influence of sowing date on the growth and yield performance of wheat (Triticum aestivum L.) varieties. *Archives of Agriculture and Environmental Science*, *3*(1), 89–94. https://doi.org/10.26832/24566632.2018.0301014

Mahapatra, S. K., Obi Reddy, G. P., Nagdev, R., Yadav, R. P., Singh, S. K., & Sharda, V. N. (2018). Assessment of soil erosion in the fragile Himalayan ecosystem of Uttarakhand, India using USLE and GIS for sustainable productivity. *Current Science*, *115*(1), 108–121. https://doi.org/10.18520/cs/v115/i1/108-121

Maji, A. K., Reddy, G. O., & Sarkar, D. (2010). *Degraded and wastelands of India: Status and spatial distribution*. Indian Council of Agricultural Research and National Academy of Agricultural Science.

Mandal, D., & Sharda, V. N. (2011). Appraisal of soil erosion risk in the eastern Himalayan region of India for soil conservation planning. *Land Degradation & Development*, *24*(5), 430–437. https://doi.org/10.1002/ldr.1139

Marín-Spiotta, E., & Sharma, S. (2013). Carbon storage in successional and plantation forest soils: a tropical analysis. *Global Ecology and Biogeography*, *22*(1), 105–117.

Mathur, I., & Bhattacharya, P. (2022). Transition from shifting cultivation to agroforestry: A case study of regrouped villages in Tripura, India. *Environmental. Challenges*, *7*, Article 100471. https://doi.org/10.1016/j.envc.2022.100471

Matson, P. A., Parton, W. J., Power, A. G., & Swift, M. J. (1997). Agricultural intensification and ecosystem properties. *Science*, *277*(5325), 504–509.

McIvor, I., Douglas, G., Dymond, J., Eyles, G., & Marden, M. (2011). Pastoral hill slope erosion in New Zealand and the role of poplar and willow trees in its reduction. In D. Godone & S. Stanchi (Eds.), *Soil erosion issues in agriculture* (pp. 257–278).

Mehla, R. S., Verma, J. K., Gupta, R. K., & Hobbs, P. R. (2000). *Stagnation in the productivity of wheat in the Indo-Gangetic Plains: Zero-till-seed-cum-fertilizer drill as an integrated solution*. Rice-Wheat Consortium.

Mirzaei Aminiyan, M., Safari Sinegani, A. A., & Sheklabadi, M. (2015). Aggregation stability and organic carbon fraction in a soil amended with some plant residues, nanozeolite, and natural zeolite. *International Journal of Recycling of Organic Waste in Agriculture*, *4*(1), 11–22. https://doi.org/10.1007/s40093-014-0080-0

Mishra, B. K., & Ramakrishnan, P. S. (1983). Slash and burn agriculture at higher elevations in north-eastern India. I. Sediment, water and nutrient losses. *Agriculture, Ecosystems & Environment*, *9*(1), 69–82. https://doi.org/10.1016/0167-8809(83)90007-5

Mishra, G., Jangir, A., & Francaviglia, R. (2019). Modeling soil organic carbon dynamics under shifting cultivation and forests using Rothc model. *Ecological Modelling, 396*, 33–41.

Mishra, P. K. (2017). Assessment of run-off and soil loss under different land-use practices in a Himalayan watershed, India. *American Research Journal of Humanities and Social Sciences, 3*, 1–10.

Mishra, P. K., Rai, A., Abdelrahman, K., Rai, S. C., & Tiwari, A. (2021). Analysing challenges and strategies in land productivity in Sikkim Himalaya, India. *Sustainability, 13*(19), Article 11112. https://doi.org/10.3390/su131911112

Mishra, P. K., Rai, A., Abdelrahman, K., Rai, S. C., & Tiwari, A. (2022). Land degradation, overland flow, soil erosion, and nutrient loss in the eastern Himalayas, India. *Land, 11*(2), 179. https://doi.org/10.3390/land11020179

Mishra, P. K., & Rai, S. C. (2013). Use of indigenous soil and water conservation practices among farmers in Sikkim Himalaya. *Indian Journal of Traditional Knowledge, 12*(3), 454–464. http://nopr.niscpr.res.in/handle/123456789/19434

Monger, H. C., Kraimer, R. A., Khresat Sa'eb, Cole, D. R., Wang, X. and Wang J. (2015). Sequestration of inorganic carbon in soil and groundwater. *Geology 2015, 43*(5): 375–378. https://doi.org/10.1130/G36449.1

Morgan, R. P. C. (1986). *Soil erosion and conservation*. Longman Group.

Nagdeve, D. A. (2007). Population growth and environmental degradation in India. *Asian Pacific Journal on Environment and Development, 14*, 41–63.

Naqvi, H. R., Mallick, J., Devi, L. M., & Siddiqui, M. A. (2012). Multi-temporal annual soil loss risk mapping employing revised universal soil loss equation (RUSLE) model in Nun Nadi watershed, Uttrakhand (India). *Arabian Journal of Geosciences, 6*(10), 4045–4056. https://doi.org/10.1007/s12517-012-0661-z

Narayana, D. V., & Babu, R. (1983). Estimation of soil erosion in India. *Journal of Irrigation and Drainage Engineering, 109*, 419–434. https://doi.org/10.1061/(ASCE)0733-9437(1983)109:4(419)

Naresh, R. K., Kumar, Y., Tomar, S. S., Kumar, M., Chandra, M. S., Rajput, P., Alam, M. S., & Singh, P. K. (2020). Measuring and monitoring the impact of precision land levelling and arable cropping systems on aggregate associated carbon fractions and soil carbon stock in sub-tropical ecosystems. *International Research Journal of Pure and Applied Chemistry, 21*(24), 197–220.

Negi, S., Rikhari, H. C., & Garkoti, S. C. (1998). The hydrology of three high-altitude forests in central Himalaya, India: a reconnaissance study. *Hydrological Processes, 12*(2), 343–350. https://doi.org/10.1002/(sici)1099-1085(199802)12:2%3C343::aid-hyp578%3E3.0.co;2-l

NITI Aayog. (2018). *A summary report—Contributing to sustainable development in the Indian Himalayan Region: Key messages from five thematic reports and way forward*. NITI Aayog.

North Eastern Council Secretariat. (2000). The rural sector: Basic crops, cash crops, rural development, non-farm rural employment, irrigation and flood control.

In *Northeastern region vision 2020* (pp. 47–98). Ministry of Development of North Eastern Region.

North East Organization. (2024). NEC Yearbook 2024. North Eastern Council. https://necouncil.gov.in/sites/default/files/NEC%20year%20book%202024_0.pdf

Novara, A., Gristina, L., Saladino, S. S., Santoro, A., & Cerdà, A. (2011). Soil erosion assessment on tillage and alternative soil managements in a Sicilian vineyard. *Soil and Tillage Research*, *117*, 140–147.

Ogle, S. M., Swan, A., & Paustian, K. (2012). No-till management impacts on crop productivity, carbon input and soil carbon sequestration. *Agriculture, Ecosystems & Environment*, 149, 37–49.

Ollinaho, O. I., & Kröger, M. (2021). Agroforestry transitions: The good, the bad and the ugly. *Journal of Rural Studies*, *82*, 210–221. https://doi.org/10.1016/j.jrurstud.2021.01.016

Pal, S. C., Chakrabortty, R., Roy, P., Chowdhuri, I., Das, B., Saha, A., & Shit, M. (2021). Changing climate and land use of 21st century influences soil erosion in India. *Gondwana Research*, *94*, 164–185. https://doi.org/10.1016/j.gr.2021.02.021

Pan, D., Song, Y., Dyck, M., Gao, X., Wu, P., & Zhao, X. (2017). Effect of plant cover type on soil water budget and tree photosynthesis in jujube orchards. *Agricultural Water Management*, *184*, 135–144. https://doi.org/10.1016/j.agwat.2017.01.009

Panettieri, M., Berns, A. E., Knicker, H., Murillo, J. M., & Madejón, E. (2015). Evaluation of seasonal variability of soil biogeochemical properties in aggregate-size fractioned soil under different tillages. *Soil and Tillage Research*, *151*, 39–49.

Pangging, G., & Singh, S. L. (2015). Traditional agroforestry system in eastern Himalaya: A case study of fringe villages of Doimukh forest range, Arunachal Pradesh, India. *International Journal of Current Research in Biosciences and Plant Biology*, *2*(10), 23–28.

Panwar, P., Pal, S., Chakravarty, S., & Alam, M. (2013). Soil quality and production of low land paddy under agrisilviculture systems in acid soil of West Bengal, India. *Range Management and Agroforestry*, *34*(1), 51–57.

Patel, D. P., Das, A., Kumar, M., Munda, G. C., Ngachan, S. V., Ramkrushna, G. I., Layek, J., Pongla, N., Buragohain, J., & Somireddy, U. (2014). Continuous application of organic amendments enhances soil health, produce quality and system productivity of vegetable-based cropping systems in subtropical eastern Himalayas. *Experimental Agriculture*, *51*(1), 85–106. https://doi.org/10.1017/s0014479714000167

Pathak, P. C., Pandey, A. N., & Singh, J. S. (1984). Overland flow, sediment output and nutrient loss from certain forested sites in the central Himalaya. *India. Journal of Hydrology*, *71*(3–4), 239–251. https://doi.org/10.1016/0022-1694(84)90099-4

Pattanayak, A., & Das, A. (2017, February). *Present and future cropping system of north East India in changing climatic conditions.* In SOUVENIR, conference on "farmers first for conserving soil and water resources in north eastern region (FFCSWR 2017) (pp. 44–48), Guwahati, Assam, India.

Paul, E. (2014). *Soil microbiology, ecology and biochemistry*. Academic Press.

Phillips, J., Ramirez, S., Wayson, C., & Duque, A. (2019). Differences in carbon stocks along an elevational gradient in tropical mountain forests of Colombia. *Biotropica, 51*(4), 490–499.

Pilu, R., Bucci, A., Badone, F. C., & Landoni, M. (2012). Giant reed (*Arundo donax* L.): A weed plant or a promising energy crop? *African Journal of Biotechnology, 11*(38), 9163–9174. https://doi.org/10.5897/ajb11.4182

Pimentel, D., Harvey, C., Resosudarmo, P., Sinclair, K., Kurz, D., McNair, M., Crist, S., Shpritz, L., Fitton, L., Saffouri, R., & Blair, R. (1995). Environmental and economic costs of soil erosion and conservation benefits. *Science, 267*(5201), 1117–1123. https://doi.org/10.1126/science.267.5201.1117

Prasad, S. (2019). Diminishing forest area in the Western Himalaya. *Current Science, 116*(2), 170–171.

Prokop, P., & Poręba, G. J. (2012). Soil erosion associated with an upland farming system under population pressure in Northeast India. *Land Degradation & Development, 23*(4), 310–321. https://doi.org/10.1002/ldr.2147

Rai, S. C., & Sharma, E. (1998). Comparative assessment of runoff characteristics under different land use patterns within a Himalayan watershed. *Hydrological Processes, 12*(13–14), 2235–2248. https://doi.org/10.1002/(sici)1099-1085(19981030)12:13/14%3C2235::aid-hyp732%3E3.0.co;2-5

Ramsankaran, R., Kothyari, U. C., & Rawat, J. S. (2009). Simulation of surface runoff and sediment yield using the water erosion prediction project (WEPP) model: A study in Kaneli watershed, Himalaya, India/Simulation de ruissellement de surface et d'érosion à l'aide du modèle WEPP: cas du bassin versant de Kaneli, Himalaya. *Inde. Hydrological Sciences Journal, 54*(3), 513–525. https://doi.org/10.1623/hysj.54.3.513

Rasool, G., Guo, X., Wang, Z., Ali, M. U., Chen, S., Zhang, S., Wu, Q., & Ullah, M. S. (2020). Coupling fertigation and buried straw layer improves fertilizer use efficiency, fruit yield, and quality of greenhouse tomato. *Agricultural Water Management, 239*, Article 106239. https://doi.org/10.1016/j.agwat.2020.106239

Rawat, J. S., & Rawat, M. S. (1994). Accelerated erosion and denudation in the Nana Kosi watershed, central Himalaya, India. Part I: Sediment load. *Mountain Research and Development, 14*(1), 25–38. https://doi.org/10.2307/3673736

Ray, S. K., Chatterjee, D., Chowdhury, P., Deka, B. C., Bihari, P., & Saha, S. (2021). Comparison of indigenous and mechanical conservation technologies for shifting cultivation agro-ecology of north-eastern Himalaya. *Journal of the Indian Society of Soil Science, 69*(1), 60–71. https://doi.org/10.5958/0974-0228.2021.00021.9

Reeder, R. (2000). *Conservation tillage systems and management*. Iowa State University.

Rillig, M. C., & Mummey, D. L. (2006). Mycorrhizas and soil structure. *New Phytologist, 171*(1), 41–53. https://doi.org/10.1111/j.1469-8137.2006.01750.x

Ryals, R., Hartman, M. D., Parton, W. J., DeLonge, M. S., & Silver, W. L. (2015). Long-term climate change mitigation potential with organic matter management on grasslands. *Ecological Applications*, *25*(2), 531–545. https://doi.org/10.1890/13-2126.1

Saha, D., & Kukal, S. S. (2015). Soil structural stability and water retention characteristics under different land uses of degraded lower Himalayas of north-West India. *Land Degradation & Development*, *26*(3), 263–271.

Saha, R., Chaudhary, R. S., & Somasundaram, J. (2012a). Soil health management under hill agroecosystem of north east India. *Applied and Environmental Soil Science*, *2012*, Article 696174. https://doi.org/10.1155/2012/696174

Saha, R., Ghosh, P. K., Mishra, V. K., Majumdar, B., & Tomar, J. M. S. (2010). Can agroforestry be a resource conservation tool to maintain soil health in the fragile ecosystem of north-east India? *Outlook on Agriculture*, *39*(3), 191–196.

Saha, R., Mishra, V. K., & Khan, S. K. (2012b). Soil erodibility characteristics under modified land-use systems as against shifting cultivation in hilly ecosystems of Meghalaya, India. *Journal of Sustainable Forestry*, *30*(4), 301–312. https://doi.org/10.1080/10549811.2011.531992

Saha, R., Tomar, J. M. S., & Ghosh, P. K. (2007). Evaluation and selection of multipurpose tree for improving soil hydro-physical behaviour under hilly eco-system of north East India. *Agroforestry Systems*, *69*(3), 239–247.

Saha, S., Chakraborty, D., Hazarika, S., Shakuntala, I., Das, B., Chhabra, A., Sadhu, S., Chakraborty, D., Mukherjee, J., Singson, L., & Mishra, V. K. (2022). Spatiotemporal variability of weather extremes over eastern India: evidences of ascertained long-term trend persistence and effective global climate controls. *Theoretical and Applied Climatology*, *148*(1–2), 643–659. https://doi.org/10.1007/s00704-022-03949-1

Saha, S., Rajwar, G. S., & Kumar, M. (2018). Soil properties along altitudinal gradient in Himalayan temperate forest of Garhwal region. *Acta Ecologica Sinica*, *38*(1), 1–8. https://doi.org/10.1016/j.chnaes.2017.02.003

Saha, S., Yadav, D., Layak, J., Babu, S., Ansari, M. A., Das, B., Choudhary, S., Singh, R., Rajkhowa, D. J., Ghosh, P. K., & Das, A. (2021). Potential of conservation agriculture in hill ecosystems: Impact on resource sustainability. *Indian Journal of Agronomy*, 66 (5th 515 IAC Special issue), 516, 128–141.

Sahoo, U. K., Singh, S. L., Gogoi, A., Kenye, A., & Sahoo, S. S. (2019). Active and passive soil organic carbon pools as affected by different land use types in Mizoram, Northeast India. *PLoS One*, *14*(7), Article e0219969

Sanwal, S. K., Yadav, R. K., & Singh, P. K. (2007). Effect of organic manure on growth, yield and quality parameters of ginger (*Zingiber officinale*). *Indian Journal of Agriculture Sciences*, *77*(2), 67–72.

Satapathy, K. K. (1996). Shifting cultivation in north eastern region: An overview. *Indian Journal Soil Water Conservation*, *34*, 171–179.

Satriawan, H., Harahap, E., Rahmawaty, & Karim, A. (2015). Effectiveness of soil conservation to erosion control on several land use types. *Agriculture (Pol'nohospodárstvo), 61*(2), 61–68. https://doi.org/10.1515/agri-2015-0011

Sehgal, J. L., & Abrol, I. P. (1994). *Soil degradation in India: Status and impact.* Oxford & IBH Publishing Co.

Selvan, T., & Kumar, S. (2017). Agroforestry in the north-eastern Himalayas. In *Agroforestry for increased production and livelihood security* (pp. 23–40).

Semwal, P., Khobragade, S. D., & Nainwal, H. C. (2017). Modelling of recent erosion rates in a lake catchment in the north-western Siwalik Himalayas. *Environmental Processes, 4*(2), 355–374. https://doi.org/10.1007/s40710-017-0234-y

Semwal, R. L., Maikhuri, R. K., Rao, K. S., Sen, K. K., & Saxena, K. G. (2003). Leaf litter decomposition and nutrient release patterns of six multipurpose tree species of central Himalaya. *India. Biomass and Bioenergy, 24*(1), 3–11.

Sen, K. K., Rao, K. S., & Saxena, K. G. (1997). Soil erosion due to settled upland farming in the Himalaya: A case study in Pranmati watershed. *International Journal of Sustainable Development & World Ecology, 4*(1), 65–74. https://doi.org/10.1080/13504509709469942

Shahi, C. (2020). *Agroecosystems of Kumaun Himalaya: Ecology, productivity, nutrient dynamics and energy budget*. Kumaun University.

Shan, J., & Yan, X. (2013). Effects of crop residue returning on nitrous oxide emissions in agricultural soils. *Atmospheric Environment*, 71, 170–175.

Sharda, V. N., Dogra, P., & Prakash, C. (2010). Assessment of production losses due to water erosion in rainfed areas of India. *Journal of Soil and Water Conservation, 65*(2), 79–91. https://doi.org/10.2489/jswc.65.2.79

Sharda, V. N., & Mandal, D. (2018). Prioritization and field validation of erosion risk areas for combating land degradation in North Western Himalayas. *Catena, 164*, 71–78. https://doi.org/10.1016/j.catena.2017.12.037

Sharda, V. N., & Ojasvi, P. R. (2016). A revised soil erosion budget for India: Role of reservoir sedimentation and land-use protection measures. *Earth Surface Processes and Landforms, 41*(14), 2007–2023. https://doi.org/10.1002/esp.3965

Sharma, E., R. Sharma, K.K. Singh & G. Sharma. (2000). A boon to mountain populations. Large cardamomfarming in the Sikkim Himalaya. *MountainResearch and Development* 20: 108–111.

Sharma, E., Rai, S. C., & Sharma, R. (2001). Soil, water and nutrient conservation in mountain farming systems: case-study from the Sikkim Himalaya. *Journal of Environmental Management, 61*(2), 123–135. https://doi.org/10.1006/jema.2000.0386

Sharma, E., Sharma, R., & Pradhan, M. (1998). Ecology of Himalayan alder (*Alnus nepalensis* D. Don). *Proceedings Indian National Science Academy Part B, 64*, 59–78.

Sharma, N. K., Singh, R. J., Mandal, D., Kumar, A., Alam, N. M., & Keesstra, S. (2017). Increasing farmer's income and reducing soil erosion using intercropping in rainfed maize-wheat rotation of Himalaya, India. *Agriculture, Ecosystems & Environment, 247*, 43–53.

Sharma, P., & Rai, S. (2004). Streamflow, sediment and carbon transport from a Himalayan watershed. *Journal of Hydrology, 289*(1–4), 190–203. https://doi.org/10.1016/j.jhydrol.2003.11.023

Sharma, P., Rai, S. C., Sharma, R., & Sharma, E. (2004). Effects of land-use change on soil microbial C, N and P in a Himalayan watershed. *Pedobiologia, 48*(1), 83–92.

Sharma, P. D. (2004). Managing natural resources in Indian Himalayas. *Journal of the Indian Society of Soil Science, 52*, 314–331.

Sharma, R., Sharma, E., & Purohit, A. N. (1994). Dry matter production and nutrient cycling in agroforestry systems of cardamom grown under *Alnus* and natural forest. *Agroforestry Systems, 27*(3), 293–306. https://doi.org/10.1007/bf00705063

Sharma, U. C., & Singh, K. A. (1994, November). *Agriculture with Alder—A potential indigenous farming system of Nagaland*. Proceedings of the National Seminar on Developments in Soil Science (pp. 570–571).

Sheikh, M. A., Kumar, M., Todaria, N. P., & Pandey, R. (2020). Biomass and soil carbon along altitudinal gradients in temperate *Cedrus deodara* forests in central Himalaya, India: Implications for climate change mitigation. *Ecological Indicators, 111*, Article 106025

Shrestha, B., Zhang, L., Sharma, S., Shrestha, S., & Khadka, N. (2022). Effects on ecosystem services value due to land use and land cover change (1990–2020) in the transboundary Karnali River basin, Central Himalayas. *SN. Applied Sciences, 4*(5), Article 137. https://doi.org/10.1007/s42452-022-05022-y

Shrestha, D. P. (1997). Assessment of soil erosion in the Nepalese Himalaya: A case study in Likhu Khola Valley, Middle Mountain Region. *Land Husbandry, 2*(1), 59–80.

Sidhu, G. S. (2016). Soil conservation of northwestern Himalayas (NWH): Their constraints and potentials for sustainable hill agriculture. In J. K. Bisht, V. Meena, P. Mishra, & A. Pattanayak (Eds.), *Conservation agriculture* (pp. 315–338). Springer. https://doi.org/10.1007/978-981-10-2558-7_12

Singh, A., & Singh, M. D. (1980). Effect of various stages of shifting cultivation on soil erosion from steep hill slopes. *Indian Forester, 106*(2), 116–121.

Singh, A. K., Bordoloi, L. J., Kumar, M., Hazarika, S., & Parmar, B. (2014). Land use impact on soil quality in eastern Himalayan region of India. *Environmental Monitoring and Assessment, 186*, 2013–2024.

Singh, A. K., Kumar, S., Naithani, S., & George, J. K. (2019a). Soil erosion and sediment yield estimation using remote sensing data and GIS in a Sitlarao watershed of north-western Himalayan region. *Indian Journal of Soil Conservation, 47*(3), 244–254.

Singh, A. N., Raghubanshi, A. S., & Singh, J. S. (2004). Comparative performance and restoration potential of two Albizia species planted on mine spoil in a dry tropical region. *India. Ecological Engineering*, *22*(2), 123–140. https://doi.org/10.1016/j.ecoleng.2004.04.001

Singh, G., Babu, R., Narain, P., Bhushan, L. S., & Abrol, I. P. (1992). Soil erosion rates in India. *Journal of Soil and Water Conservation*, *47*(1), 97–99.

Singh, K. A., Rai, R. N., & Pradham, I. P. (1991). Agroforestry systems in Sikkim hills. *Indian Farming*, *41*(3), 7–10.

Singh, R. K., Panda, R. K., Satapathy, K. K., & Ngachan, S. V. (2011). Simulation of runoff and sediment yield from a hilly watershed in the eastern Himalaya, India using the WEPP model. *Journal of Hydrology*, *405*(3–4), 261–276. https://doi.org/10.1016/j.jhydrol.2011.05.022

Singh, A. B., & Teron, R. (2019). Traditional oak agroforestry practices in Mima village of Khonoma district, Nagaland state, Northeast India. *Indian Journal of Hill Farming*, *32*(2), 203–215.

Sinha, T., Paul, S. K., & Sarkar, A. R. (2018). Effect of age of seedlings at staggered transplanting and weed management on the growth and yield of aromatic Boro rice (cv. BRRI dhan50). *Journal of the Bangladesh Agricultural University*, *16*(1), 5–11.

Sitaula, B. K., Bajracharya, R. M., Singh, B. R., & Solberg, B. (2004). Factors affecting organic carbon dynamics in soils of Nepal/Himalayan region—A review and analysis. *Nutrient Cycling in Agroecosystems*, *70*(2), 215–229. https://doi.org/10.1023/b:fres.0000048474.85331.7d

Six, J., Bossuyt, H., Degryze, S., & Denef, K. (2004). A history of research on the link between (micro) aggregates, soil biota, and soil organic matter dynamics. *Soil and Tillage Research*, *79*(1), 7–31. https://doi.org/10.1016/j.still.2004.03.008

Smith, P., Martino, D., Cai, Z., Gwary, D., Janzen, H., Kumar, P., McCarl, B., Ogle, S., O'Mara, F., & Rice, C. (2007). Policy and technological constraints to implementation of greenhouse gas mitigation options in agriculture. *Agriculture, Ecosystems & Environment*, *118*(1–4), 6–28. https://doi.org/10.1016/j.agee.2006.06.006

Sofi, J. A., Bhat, A. G., Kirmai, N. A., Wani, J. A., Lone, A. H., Ganie, M. A., & Dar, G. I. H. (2016). Soil quality index as affected by different cropping systems in northwestern Himalayas. *Environmental Monitoring and Assessment*, *188*, 1–13.

Solomon, S., Qin, D., Manning, M., Chen, Z., Marquis, M., Averyt, K., ... & Miller, H. (2007). IPCC fourth assessment report (AR4). *Climate change*, 374.

Spracklen, D. V., & Righelato, R. (2014). Tropical montane forests are a larger than expected global carbon store. *Biogeosciences*, *11*(10), 2741–2754. https://doi.org/10.5194/bg-11-2741-2014

Stoorvogel, J. J., Smaling, E. M. A., & Janssen, B. H. (1993). Calculating soil nutrient balances in Africa at different scales. *Fertilizer Research*, *35*(3), 227–235. https://doi.org/10.1007/bf00750641

Strik, B. C., Davis, A. J., Bryla, D. R., & Orr, S. T. (2020). Individual and combined use of sawdust and weed mat mulch in a new planting of northern highbush blueberry I. Impacts on plant growth and soil and canopy temperature. *HortScience*, *55*(8), 1280–1287. https://doi.org/10.21273/hortsci15122-20

Su, Y., Yu, M., Xi, H., Lv, J., Ma, Z., Kou, C., & Shen, A. (2020). Soil microbial community shifts with long-term of different straw return in wheat-corn rotation system. *Scientific Reports, 10*(1), Article 6360. https://doi.org/10.1038/s41598-020-63409-6

Sui, Y., Ou, Y., Yan, B., Xu, X., Rousseau, A. N., & Zhang, Y. (2016). Assessment of micro-basin tillage as a soil and water conservation practice in the black soil region of northeast China. *PLoS One, 11*(3), e0152313. https://doi.org/10.1371/journal.pone.0152313

Sur, H. S., Kukal, S. S., & Maskina, M. S. (1998). *Indigenous technical knowledge for soil and water conservation and crop management in Kandi area of NW India [Research bulletin]*. Zonal Research Station for Kandi Area, Punjab Agricultural University.

Taguas, E. V., Peña, A., Ayuso, J. L., Pérez, R., Yuan, Y., & Giráldez, J. V. (2010). Rainfall variability and hydrological and erosive response of an olive tree microcatchment under no-tillage with a spontaneous grass cover in Spain earth surface processes and landforms. *The Journal of the British Geomorphological Research Group, 35*(7), 750–760.

Tangjang, S., & Arunachalam, K. (2009). Microbial population dynamics of soil under traditional agroforestry systems in Northeast India. *Research Journal of Soil Biology, 1*(1), 1–7. https://doi.org/10.3923/rjsb.2009.1.7

Tangjang, S., & Nair, R. (2016). Integrated bamboo + pine homegardens: A unique agroforestry system in Ziro valley of Arunachal Pradesh, India. *International Journal of Environmental & Agriculture Research, 2*(2), 25–34.

Thomas, G. A., Titmarsh, G. W., Freebairn, D. M., & Radford, B. J. (2007). No tillage and conservation farming practices in grain growing areas of Queenslanda a review of 40 years of development. *Australian Journal of Experimental Agricultural, 47*, 887–898.

Tisdale, S. L., Nelson, W. L., Beaton, J. D., & Havlin, J. L. (1985). *Soil fertility and fertilizers*. Macmillan Publishing Company.

Tiwari, P. C. (2000). Land-use changes in Himalaya and their impact on the plains ecosystem: Need for sustainable land use. *Land Use Policy, 17*(2), 101–111. https://doi.org/10.1016/s0264-8377(00)00002-8

Tiwary, P., Bhattacharyya, T., Mandal, C., Dasgupta, D., & Telpande, B. (2015). Pedometric mapping of soil organic carbon loss using soil erosion maps of Tripura. *Current Science, 108*(7), 1326–1339. http://www.jstor.org/stable/24905496

Torsvik, V., Goksøyr, J., & Daae, F. L. (1990). High diversity in DNA of soil bacteria. *Applied and Environmental Microbiology, 56*(3), 782–787. https://doi.org/10.1128/aem.56.3.782-787.1990

Tripathi, O. P., Pandey, H. N., & Tripathi, R. S. (2009). Litter production, decomposition and physicochemical properties of soil in 3 developed agroforestry systems of Meghalaya, Northeast India. *African Journal of Plant Science, 3*(8), 160–167.

Tse-ring, K., Sharma, E., Chettri, N., & Shrestha, A. (2010). *Climate change vulnerability of mountain ecosystems in the Eastern Himalayas; Climate change impact an vulnerability in the Eastern Himalayas–Synthesis report*. ICIMOD.

Uddin, K., Abdul Matin, M., & Maharjan, S. (2018). Assessment of land cover change and its impact on changes in soil erosion risk in Nepal. *Sustainability, 10*(12), Article 4715

UNEP (1994) Land Degradation in South Asia: Its Severity, Causes and Effects upon the People. USDA-ARS.

United Nations. (2015). *Transforming our world: The 2030 agenda for sustainable development* (Report no. A/RES/70/1). United Nations General Assembly.

Vukicevich, E., Lowery, T., Bowen, P., Urbez-Torres, J. R., & Hart, M. (2016). Cover crops to increase soil microbial diversity and mitigate decline in perennial agriculture. A review. *Agronomy for Sustainable Development, 36*, Article 48

Wabwoba, M. S., & Mutoro, K. (2019). Promoting Mucuna beans production for soil rehabilitation, incomes, food and nutrition security in Kenya. *Global Journal of Nutrition & Food Science, 2*(4). http://dx.doi.org/10.33552/GJNFS.2019.02.000543

Wang, K., Zhang, C., & Li, W. (2012). Comparison of geographically weighted regression and regression kriging for estimating the spatial distribution of soil organic matter. *GIScience & Remote Sensing, 49*(6), 915–932.

Wang, S., Lai, Z. Q., Hai, J. S., Xin Xin, J., & Wang, Q. B. (2018). Spatial variations of soil organic carbon stocks in a coastal hilly area of China. *Geoderma, 314*, 8–19.

West, A. J., Arnold, M., Auma Ître, G., Bourlès, D. L., Keddadouche, K., Bickle, M., & Ojha, T. (2015). High natural erosion rates are the backdrop for present-day soil erosion in the agricultural Middle Hills of Nepal. *Earth Surface Dynamics, 3*(3), 363–387. https://doi.org/10.5194/esurf-3-363-2015

West, T. O., & Post, W. M. (2002). Soil organic carbon sequestration by tillage and crop rotations. *Soil Science Society of America Journal, 66*, 1930–1946.

Wilhelm, W. W., Johnson, J. M. F., Karlen, D. L., & Lightle, D. T. (2007). Corn stover to sustain soil organic carbon further constrains biomass supply. *Agronomy Journal, 99*, 1665–1667.

Won, C., Shin, M., Lee, S., Park, Y., Lee, Y., Shin, Y., & Choi, J. (2016). NPS pollution reduction from alpine fields using surface cover material and soil amendments. *Irrigation and Drainage, 65*, 193–199.

Yadav, G. S., Das, A., Babu, S., Mohapatra, K. P., Lal, R., & Rajkhowa, D. (2021). Potential of conservation tillage and altered land configuration to improve soil properties, carbon sequestration and productivity of maize based cropping system in eastern Himalayas, India. *International Soil and Water Conservation Research, 9*(2), 279–290.

Yadav, R. L., Dwivedi, B. S., Gangwar, K. S., & Prasad, K. (1998). Overview and prospects for enhancing residual benefits of legumes in rice and wheat cropping systems in India. In J. V. D. K. Kumar Rao & C. Johansen (Eds.), *Residual effects of legumes in rice and wheat cropping systems of the Indo-Gangetic Plain* (pp. 207–226). ICRISAT, Oxford & IBH Publishing Co. Pvt. Ltd.

Yang, Y., Tilman, D., Furey, G., & Lehman, C. (2019). Soil carbon sequestration accelerated by restoration of grassland biodiversity. *Nature. Communications, 10*, Article 718

Zannat, S. T., Paul, S. K., & Salam, M. A. (2014). Effect of weeding regime and nitrogen management on the performance of trans-plant aromatic Aman rice (cv. Binadhan-9). *Bangladesh Journal of. Seed Science & Technology, 18*(1&2), 29–34.

Zaz, S. N., & Romshoo, S. A. (2012). Assessing the geoindicators of land degradation in the Kashmir Himalayan region. *India. Natural Hazards, 64*(2), 1219–1245. https://doi.org/10.1007/s11069-012-0293-3

Zhang, G. H., Xie, C. B., Pi, X. Y., & Zuo, C. Q. (2015). Effectiveness of soil conservation methods in preventing red soil erosion in southern China. *Journal of Mountain Science, 12*, 1281–1291.

Zhang, Q., Liu, D., Cheng, S., & Huang, X. (2016). Combined effects of runoff and soil erodibility on available nitrogen losses from sloping farmland affected by agricultural practices. *Agricultural Water Management, 176*, 1–8.

Zhang, S. J., Zhang, G., Wang, D. J., & Liu, Q. (2021). Abiotic and biotic effects of long-term straw retention on reactive nitrogen runoff losses in a rice–wheat cropping system in the Yangtze Delta region. *Agriculture, Ecosystems & Environment, 305*, Article 107162

Zibilske, L. M., Bradford, J. M., & Smart, J. R. (2002). Conservation tillage induced changes in organic carbon, total nitrogen and available phosphorus in a semi-arid alkaline subtropical soil. *Soil and Tillage Research, 66*, 153–163.

Zimmerman, A. R. (2010). Abiotic and microbial oxidation of laboratory-produced black carbon (biochar). *Environmental Science & Technology, 44*(4), 1295–1301. https://doi.org/10.1021/es903140c

Zolfaghari, A., & Hajabbasi, M. (2008). Effect of Different Land Use Treatments on Soil Structural Quality and Relations with Fractal Dimensions. *International Journal of Soil Science, 3*(2), 101–108. https://doi.org/10.3923/ijss.2008.101.108

Zonunsanga, R. (2016). Estimation of soil loss in Teirei watershed of Mizoram by using Universal Soil Loss Equation model. *Journal of Science and Technology, 4*(1), 43–47.

Zornoza, R., Acosta, J. A., Bastida, F., Dominguez, S. G., Toledo, D. M., & Faz, A. (2015). Identification of sensitive indicators to assess the interrelationship between soil quality, management practices and human health. *Soil, 1*, 173–185.

10

Strategies for Restoring and Enhancing Soil Health in India: Management Practices to Restore Soil Health in Arid, Semi-Arid, and Plateau Regions

Avijit Ghosh, Pramod Jha, Sukanya Misra, and Ranjan Bhattacharyya

Chapter Overview

Restoring and enhancing soil health in India's arid and semi-arid regions is vital for sustainable agriculture and food security. These regions face significant challenges such as low organic matter, high salinity, soil erosion, and water scarcity, which undermine agricultural productivity. Effective management practices must address these issues holistically. Improving organic matter is a critical step, achieved by incorporating organic inputs like compost, biochar, and farmyard manure to enhance soil structure, microbial activity, and nutrient retention. Green manuring and crop residue management also contribute to organic carbon replenishment, promoting a resilient soil ecosystem. Conservation agriculture practices, such as minimum tillage, crop rotation, and agroforestry, further enhance soil health by reducing erosion, improving water retention, and increasing biodiversity. Water management is equally important in these regions, where rainfall is erratic, and water resources are limited. Rainwater harvesting structures, including check dams and farm ponds, help capture and store water, while techniques like mulching and drip irrigation optimize moisture conservation and usage. Addressing salinity and alkalinity is crucial in arid soils, and the application of gypsum and the cultivation of salt-tolerant crops like barley and millets can reclaim degraded land. Vegetative barriers, such as vetiver grass, reduce soil erosion, while contour plowing minimizes water runoff and promotes in-situ moisture conservation. Biofertilizers and microbial inoculants, including Rhizobium and mycorrhizal fungi, offer sustainable solutions for enhancing soil fertility and nutrient cycling. Policymakers and agricultural institutions must

Soil Health Series: Volume 4 Soil Health and Sustainability in India, First Edition.
Edited by Ranjan Bhattacharyya.

support these practices through awareness campaigns, farmer training, and the expansion of soil health card programs. Subsidies for soil testing and nutrient management tools can empower farmers to adopt location-specific interventions effectively. The adoption of these strategies yields numerous benefits, including enhanced soil fertility, improved water efficiency, and greater resilience to climate change. By integrating traditional knowledge with scientific innovations, these approaches can transform the arid and semi-arid regions of India into productive landscapes. They not only improve agricultural productivity but also ensure long-term environmental sustainability and better livelihoods for farmers. Such a comprehensive strategy is essential to secure India's agricultural future in the face of growing challenges.

Introduction

The arid zone of India covers about 15.8% of the country's geographical area, including 31,900 million km^2 of hot desert located in parts of Rajasthan (61%), Gujarat (20%), Punjab and Haryana (9%), and Andhra Pradesh and Karnataka (10%). The Great Indian Desert, or Thar Desert, lies in western Rajasthan and covers an area of 196,150 km^2. In addition, an area of 78,300 km^2 of cold desert is located in the states of Jammu and Kashmir. The climate in the arid regions has erratic rainfall, frequent droughts, high evaporation, intense heat, and high-velocity winds. The human population in the Indian arid zone is high and includes nomadic people who maintain many livestock. The population density of both humans and livestock is moderately higher than the national average, which places severe stress on the scarce natural resources of this zone. Owing to the growing season being very short in the arid zone, millets and short-duration pulses dominate the cropping systems. Livestock farming forms an integral part of this ecosystem. There are six National Parks and Wildlife Sanctuaries in the deserts covering16,076 km^2 or 7.45% of the bio-geographical area of the country.

The semi-arid zone is a transition zone between desert and dense forests of Western Ghats. The semi-arid zone comprises 970,530 km^2, which is 37% of the total geographical area of the country. The semi-arid zone is distributed in the states of Punjab, Haryana, Rajasthan, Uttar Pradesh, Madhya Pradesh, Gujarat, Maharashtra, Karnataka, Andhra Pradesh, and Tamil Nadu. In the semi-arid zones, the crops and cropping systems can be quite diverse depending upon the soil types and the length of growing season. Sorghum, soybean, groundnut, and pulses are the major crops grown in this zone. The dominant grass and palatable shrub layer in this zone support a high wildlife biomass. The semi-arid zone covers 15,302 km^2 or 2.8% of the bio-geographical area of the country.

Characteristics of an Arid Ecosystem and Its Soil Constraints

High wind speeds, frequent droughts, persistent water shortages, and faster evaporation rates than precipitation are all characteristics of arid areas. A loss of biodiversity can happen when desertification occurs, and formerly non-arid areas became arid. Aridity is defined as the difference between the quantity of water that is accessible and the total amount that is consumed. Due to the stability of the atmosphere, dry regions frequently experience cold winters and precipitation that is less than evapo-transpiration. Aridity is frequently caused by dry, stable air masses that resist convective currents. Aridity can also result from the absence of storm systems, which bring about convergence, create unstable conditions, and provide the upward air movement needed for precipitation. Strong convectional showers, which seldom cover huge areas, make up most of the precipitation in hot deserts, making widespread rain practically unknown there. Low-latitude deserts typically have clear skies and a lot of sunshine. The yearly temperature range of a low-latitude desert is higher than that of any other tropical environment. Because the soils in dry areas cannot hold all the rain that falls during intense storms, water is frequently lost as runoff (Brooks et al., 2013). In other instances, a large portion of the precipitation that falls over arid territory evaporates. In dry environments, up to 90% of rainfall evaporates back into the sky, leaving only 10% for useful transpiration (Molden & Oweis, 2007). Low precipitation and high temperatures result in high evaporation in arid regions, which also leads to aridic and xeric soil moisture regimes.

Scarcity of Water

In dry regions, a lack of water is one of the main factors restricting agricultural productivity. In dry regions, there is a shortage of high-quality water due to low precipitation and excessive evapotranspiration. Global climate change, which includes catastrophic weather events and extended dry seasons, exacerbates this issue. Improving soil, water, and product quality requires agricultural practices that are optimized for productivity and water management (Morison et al., 2008). Water shortage causes the soils in dry areas to have low soil fertility and poor natural primary output. The accumulation of potassium, salt, calcium, and other minerals in arid soils results in an alkaline pH and can be harmful to plant development. Additionally, certain dry soils generate gypsic crusts that might harbor specialized plant species. Therefore, in dry soils, poor soil fertility adversely affects plant development and biodiversity (Dlamini et al., 2014). Husein et al. (2021) concluded that conventional systems have a low potential for adoption in arid areas owing to the existence of environmental conditions that were not considered in soil fertility assessment systems developed by researchers and scientists for temperate and humid areas. Soil fertility evaluation techniques made specifically for dry regions have to be created to ensure their broad use.

Soil Organic and Inorganic Carbon

The fundamental factor affecting both the composition and structure of soil is soil organic carbon (SOC). Soil organic matter (SOM) also helps dry areas be more resilient to drought and increases crop production. Climate restrictions result in soils in dry areas having a naturally limited supply of organic carbon. They also contain a significant amount of inorganic carbon, primarily in the form of soil carbonates (Naorem et al., 2022). Lal (2004) found that the amount of soil inorganic carbon sequestered by dryland ecosystems was typically between 0.1 and 0.2 Mg/ha/yr. Inorganic carbon in soil, in addition to serving as a sink for atmospheric CO_2, may also benefit soil aggregation through the interaction of carbonates with SOM. Additionally, SOM regulates the beneficial impact of carbonates on soil structure (Bronick & Lal, 2005).

In dry regions, it has long been understood that vegetal covers are crucial for limiting soil erosion and conserving SOC. Conventional management techniques, such as extensive tillage, the demand for animal feed, and excessive grazing, make it difficult to provide enough protection for the soil surface (Álvaro-Fuentes et al., 2008; Hoffmann et al., 2008; López et al., 2003). The extractive nature of employing animal dung as a cooking fuel and crop leftovers as cow feed in developing nations in Asia and Africa threatens soil quality and long-term food security (Lal, 2006). The generation of more agricultural leftovers is necessary to stop the loss of SOC in these countries. However, because of their highly weathered character, certain developing locations, such as West Africa, require fertilization to promote significant biomass production.

Salinity

Soil salinization has become a significant environmental problem in dry areas, endangering food security, agricultural sustainability, and soil productivity (Cuevas et al., 2019). The soil turns salty when it retains salts that are soluble in water. It can happen because of human activities, most frequently those connected to agricultural operations, or it may happen spontaneously. Low precipitation and dry conditions prevent extra salts from being rinsed out of the soil, leading to soil salinization. Salts can accumulate on the ground surface where there are high evaporation rates. In addition, poor drainage or waterlogging can lead to insufficient water conveyance, which means salts cannot be cleared. The soil salt concentration also rises when the soil is irrigated with water that has a high salt content. The water table is raised when deep-rooted vegetation is removed. Furthermore, saltwater submergence followed by water evaporation, saltwater intrusion into groundwater, coastal winds bringing salty air masses inland, seawater submergence, and incorrect fertilizer application are all factors in soil salinization. Currently, saline soils cover a total area of 932.2 million hectares (Mha) worldwide and are present in at least 100 different nations (Cuevas et al., 2019). This encompasses several countries in Central and Western Asia, as well as along the Mediterranean coast, and notably sizable portions in Pakistan, China, the United States, India, Argentina, and the Sudan (Cuevas et al., 2019). According to Wichelns and Qadir (2015), salinization of agricultural soils causes the agriculture sector to lose $27.3 million annually. According to Bridges and Oldeman (1999), between 10 and 20 Mha of irrigated land becomes unproductive each year because of secondary salinization. This equates to 3 ha of arable land becoming unproductive globally per minute. Soil salinization will continue to be a significant obstacle to food production and meeting food needs as the world's population rises to 9.8 billion by 2050 (Shabala & Munns, 2017).

Nutrient Imbalances

Even though desert soils already have a restricted supply of carbon, nitrogen, and phosphorus, Jiao et al. (2016) noted that aridity and drought can further diminish the nutrients' availability. The decrease in soil water availability and vegetation cover brought on by aridity, both of which have direct or indirect effects on processes involving carbon and nitrogen, particularly biological processes such as photosynthesis, atmospheric nitrogen fixation, and the activity of microorganisms and soil enzymes. This leads to soil carbon and nitrogen depletion. In addition to speeding up soil drying, the combination of greater temperatures and reduced plant cover can cause erosion, which can wash away tiny, nutrient-rich particles such as clay (Reynolds et al., 2007). Aridity-related soil nitrogen loss can still occur even in locations with low annual precipitation and high evaporation rates that might otherwise prevent available nutrients from being washed out of the soil profile. Since nitrogen availability and photosynthesis are highly correlated, if nitrogen availability decreases owing to drought, plants will not be able to produce as much food or reduce atmospheric CO_2 (Field & Mooney, 1986).

Because a rise in the absorption of chlorine corresponds to a decrease in the quantity of NO_3 in the shoots of many crops, salinity can restrict nitrogen uptake in agricultural plants. Salinity also decreases rhizobia density and root hair colonization, which can obstruct symbiotic N_2-fixation. Symbiosis, on the other hand, increases the quantity of photosynthate delivered to the nodule, which declines because of the host plant's restricted growth (Fageria et al., 2011).

Phosphorus is abundant in soil but has a low bioavailability, like nitrogen. Only 0.1% of the typical soil phosphorus content (0.05% w/w) is thought to be exploitable by plants. Because of this, phosphorus-based fertilizers are frequently used; nevertheless, shortly after application, a quantity equal to 75%–90% of the given phosphorus precipitates with metal–cation complexes, quickly becoming fixed in soil (Feng et al., 2004). Rhizobacteria that promote plant development can increase the availability of vital minerals such as nitrogen, phosphors, and iron, which can help plants grow faster. Here, phosphate-solubilizing bacteria may serve as a long-term and environmentally safe way to boost the amount of phosphorus that is accessible to plants, particularly in phosphorus-poor soils. The solubilization of inorganic phosphorus is facilitated by the synthesis of low molecular weight organic acids such as gluconic and citric acid, the liberation of protons, and the creation of chelating molecules (Figure 10.1). In addition, the enzymes phosphomonoesterase, phosphodiesterase, and phosphotriesterase, which are produced by microscopic organisms during the mineralization of organic phosphorus, catalyze the hydrolysis of phosphoric esters (Alori et al., 2017). In general, microorganisms isolated from severe soil conditions (such as saline/alkaline soils and soil with nutrient deficit) are more effective in solubilizing phosphate than those originating from soils with moderate conditions (Zhu et al., 2011).

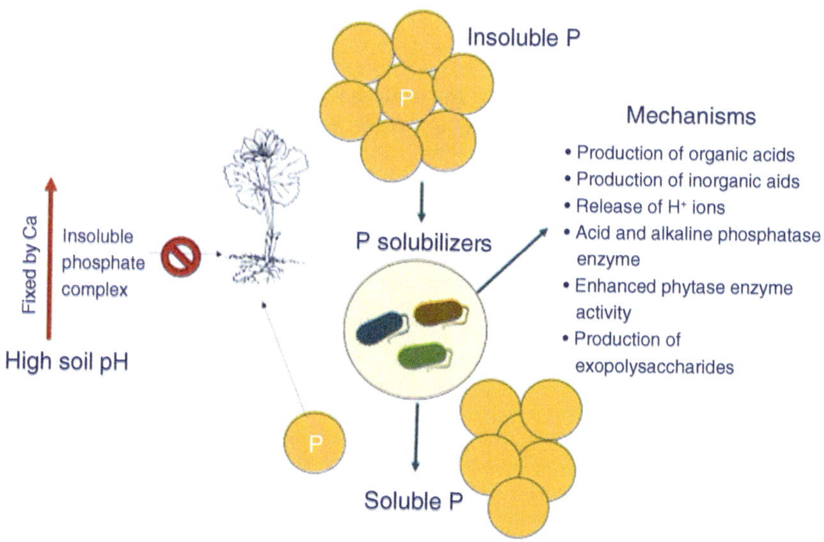

Figure 10.1 Results of phosphorus-solubilizing microbes for improving phosphorus use efficiency in arid soils.

Sandy Soils

Silt and clay size fractions can prevent SOC from breaking down, which has a significant influence on carbon storage owing to soil texture and aggregation (Hassink, 1997). When organic matter attaches to inorganic substances such as silt and clay during decomposition, aggregates are created (Churchman, 2018). In the soil that Hassink (1997) investigated, there was no association between the total carbon and clay + silt concentration, but there was a correlation between the clay + silt content and the quantity of carbon held in the 20-μm size fraction (Hassink, 1997). Low amounts of soil carbon were found in the sand size fraction, in contrast to the high levels of soil carbon found in the silt and clay size fractions, as reported by Gabarrón-Galeote et al. (2015) and Tiessen and Stewart (1983).

Soil Loss Through Erosion

Large expanses of deteriorated soils or soil loss due to erosion are additional characteristics of the dry environment (Table 10.1). The likelihood of increasing irrigation in most dry agro-ecosystems is quite low; as a result, new strategies for optimizing land use must be developed (Hall & Richards, 2013). Arid locations are particularly susceptible to soil erosion owing to a lack of significant soil cover, low soil structural stability with a limited quantity of SOC, and high human pressure. In dry regions, it is well recognized that vegetative cover significantly contributes to avoiding soil erosion and long-term SOM preservation. Intensive tillage, excessive grazing, a high demand for feedstock for bioenergy, and the need for feed for livestock production all work against the availability of a high SOC content in arid soils, even though adequate soil surface protection is required in these areas (Miner et al., 2013).

Table 10.1 Extent of Soil Degradation in Arid Soils of Different Continents.[a]

Sl no.	Continent	Total area of degraded soil (Mha)
1	Asia	150.7
2	Africa	172.5
3	Australia	48.9
4	Europe	4.8
5	South America	7.5
6	North America	7.9

[a] *Source:* Sharma (2002).

To reduce soil erosion, engineering, chemical, and biological methods are presently applied. Engineering methods, includingusing checkerboards constructed of organic materials such as straw or tree branches, and chemical methods, includingusing the usage of organic or inorganic substances, are effective in reducing soil erosion (Chi et al., 2020). Additionally gaining popularity as a possible biotechnology tactic is the use of inoculation-based tactics for controlling the soil microbial community (Zhao & Wang, 2019). Inoculating soil with cyanobacteria has been shown to improve soil quality in several studies, including those investigating ways to stabilize and fertilize Mediterranean desert soils, fix moving sand in Chinese arid lands, stabilize dried-up sand lakebeds, manage runoff on damaged soils in northwest Iran, and restore post-mining soils in South Australia (Kheirfam & Asadzadeh, 2020; Zhao & Wang, 2019). By using conservation tillage (no-tillage) techniques and leaving crop leftovers on the field, soil erosion can be significantly decreased (Delgado et al., 2013). For instance, the government is promoting the adoption of conservation tillage techniques in the vast desert regions of Northern China owing to their ability to slow down soil degradation (Wang et al., 2007).

In arid locations, no-till farming can benefit from new technology. For instance, the use of tool attachments called stripper-headers has great potential for reducing the risks of soil erosion brought on by no-tillage in the cultivation of field crops. This technological breakthrough helps farmers save money while simultaneously lessening their impact on the environment by lowering CO_2 emissions during harvest (Spokas & Steponavičius, 2011). Stripper-headers leave much of a crop's stalks standing, which lessens the impact of wind on the soil. Thus, soil carbon is less likely to be carried away by the wind and less water will evaporate. It has long been known that utilizing conservation tillage techniques followed by leaving crop residues on the soil's surface greatly reduces soil erosion (Delgado et al., 2013). Unfortunately, no-till farming and its benefits are not thoroughly established in all of the world's dry agricultural regions (Plaza-Bonilla et al., 2015). Due to the intense competition from other residue uses, farmers in dry locations are less inclined to embrace conservation farming practices (Lahmar, 2010).

Hardsetting and Hardpan

Hardsetting soils are those that dehydrate into a solid, unstructured mass that is challenging or impossible to deal with until the profile is rehydrated. In agriculture, hardsetting soils can lead to a number of issues, such as a shorter tilling window and greater physical obstacles to strong root development. Hardsetting is frequently associated with soil degradation processes such erosion, compaction, and crusting (Mullins, 1997). Farmers frequently express unhappiness with these soils because of the low yield and high cost of farming. Hardsetting soil is unstable because it shrinks more when it dries and collapses structurally when it gets wet. As a result, there are extremely few or no fractures and very little pore space in the soil layer, which causes it to become "massive" (Lima et al., 2006). Typically, this "massive" design has low infiltration, high soil strength, and a little water-holding capacity. High exchangeable sodium percentage soil aggregates tend to spread, which may cause hardsetting. When traffic or soil genetic characteristics produce hard layers, horizons with a high density or cemented soil particles are generated (Hamza & Anderson, 2005). These horizons have high penetration resistance that obstructs root growth and lowers water and air circulation. The ability of plants to absorb water and nutrients is hampered by inadequate root growth. Rain and irrigation cannot seep into the soil and be used later if water flow is impeded. Lack of airflow inhibits microorganisms and plants from obtaining the oxygen they require to survive. These restrictions cause crop yields to decrease.

Gypic Soils

Gypsic soil predominates in sparsely inhabited, dry or semi-arid regions and makes up a tiny part of the world's total soil area (Casby-Horton et al., 2015). In locations with more comprehensive management, the presence of gypsum in agricultural soils raises potential issues with pipework dissolving, particularly in reaction to irrigation. Additionally, poor soil frequently loses nutrients and water. Farmers will have to begin farming more marginal soil landscapes, such as those found in gypsic soils, to fulfill the food demands of a growing population. For gypsic soils now managed for agriculture or rangeland, enhancing productivity requires an understanding of the possible land use and function of this special substrate.

According to calculations made by Eswaran and Zi-Tong (2015), around 60,000 km^2 of soil has petrogypsic layers and 2.01 million km^2 has gypsic horizons. Gypsum has been shown to be present in soils in dry and semi-arid areas at higher proportions than any other sulfate mineral. Micro-crystalline gypsum, a weathering by-product of rock gypsum, may be transported under aridic soil moisture regimes because of mass movement and runoff, followed by dissolution and re-precipitation or leaching. Gypsic soil development in areas with little precipitation, little leaching, and shallow wetting depths is mostly controlled by the parent materials. Gypsum-rich parent materials, such as rock gypsum, alluvial sediments weathered from rock gypsum and transported in runoff or mass movement, gypsum-rich lacustrine deposits, eolian deposits derived from weathered gypsum crystals, and gypsum-rich coastal precipitates, are the direct sources of pedogenic gypsum in gypsic soils. In aridic, xeric, and ustic soil moisture regimes, gypsiferous and gypsic soils are produced in addition to the parent materials. The climate has a considerable impact on this process. Gypsum is a semi-soluble mineral that is quickly leached under moist soil conditions; hence it is unusual to find gypsum-rich soils in humid regions. In aridic or xeric soils, the balance between leaching and evaporation supports the accumulation and stability of gypsum.

Calcareous Soils

Calcareous, calcisols, and calcids are terms used to describe soils that are rich in calcium carbonates or calcium and magnesium carbonates. Calcareous soils are naturally found in dry and semi-arid regions owing to low levels of leaching (Brady & Weil, 2007). In humid and semi-arid areas, parent material can also be discovered if it is fresh and has experienced little weathering (e.g., limestone, shells, or calcareous glacial tills; Monier et al., 2019). Calcareous soils develop in regions with little rainfall and must be irrigated for crops to be grown (Pearce et al., 1999). As a result, one of the main obstacles to production in dry locations is the availability of water for irrigation. To guarantee long-term viability while farming on calcareous soils, it is crucial that the irrigation water is of a high caliber.

Inorganic salts are dissolved in almost all irrigation fluids. These salts can build up in the soil profile, changing the soil's structure, reducing water permeability, and seriously impairing plant development. Another important problem is the surface crusting. These particular soils have an organic matter level that is less than 0.4% and frequently much lower. Prior to the initial irrigation, these soils seem to be in good physical condition, but once water is added, major chemical changes occur. The conversion of carbonates to bicarbonates and subsequent precipitation of the latter upon drying promote crust growth, which is regulated by elements such as texture and salt dominance. Crusting of the surface may impede infiltration and soil aeration, which may obstruct the germination of seeds.

The underlying layers' cementing conditions may hinder root development and water flow characteristics. Available nitrogen is often insufficient in soils that are mostly rich in calcium carbonate. Owing to the formation of apatite and other insoluble calcium phosphates in high-pH conditions (lime-induced chlorosis), micronutrients such as zinc and iron may be less readily available. Magnesium and potassium deficits can also result from calcium insufficiency. Many macro- and micro-elements' chemistry and availability are changed by $CaCO_3$. To lessen the likelihood of ammonium-N volatilization, nitrogen fertilizers should be added to calcareous soils (Brady & Weil, 2007).

Soil Biodiversity

Actinobacteria are more prevalent in dry soils than Proteobacteria, Cyanobacteria, or Planctomycetes, demonstrating that microbial structure appears to be strongly influenced by mean annual precipitation and mean annual temperature and not by pH in desert soils. A meta-analysis by Ren et al. (2018) revealed that lower rainfall significantly lowered the overall soil microbial biomass. Rather than reductions in rainfall duration, percentage reductions (size) in rainfall were primarily important for explaining the directions and magnitudes. According to several methods for reducing rainfall, it has been suggested that substrate and microbial collisions are controlled by the availability of water (Beier et al., 2012; Manzoni et al., 2014).

Additionally, large decreases in rainfall decreases relative to small reductions may have an impact on microbes' access to substrates, which severely limits the proliferation of soil microorganisms. In their meta-analysis, Schimel et al. (2007) emphasized the necessity of taking into consideration the extent of rainfall decreases and aridity levels to correctly assess and predict soil carbon dynamics.

Biological soil crusts, also known as bio-crusts, are complex communities of bacteria found on soil surfaces. They are often found in dry and semi-arid regions, where plant cover is frequently low (Rodriguez-Caballero et al., 2018). Owing to their ecological and microbiological characteristics and the widespread presence of bio-crusts in dry environments, bio-crusts have become a significant and expanding topic of study (Bowker et al., 2006). Numerous scientific topics are addressed by these bio-crusts, including stress survival, ecological assembly in difficult conditions, and biogeochemical effect on the soils they are in. Each form of bio-crust has its own unique bacterial and fungal communities that are different from the soil surrounding it. Abiotic stressors such as UV and moisture may be less of a concern for below-crust soils, but they are critical for bio-crusts (Mueller et al., 2015). Reduced microbial diversity beneath the crusts of arid soils may be caused by strong selection of species that can endure extremely dry settings with high UV indices.

Potential Management Strategies for Restoration

To illustrate the mechanisms underpinning restoration's impact on soil fertility, we describe three potential management systems that differ in intensity, impact, and time to full restoration. In doing so, we broadly illustrate three overarching uses: climate change mitigation (bioenergy cropping systems), animal protein production (intensive rotational grazing), and biodiversity restoration (conservation plantings). All have great potential to restore soil fertility on degraded lands.

Bioenergy Cropping Systems

Whether cellulosic bioenergy is used to produce liquid fuel, electricity, or both, it is essential to all Intergovernmental Panel on Climate Change (IPCC) mitigation strategies that the increase in global temperature by the end of the century remains below 1.5 °C (IPCC, 2018). A significant portion of cellulosic feedstocks must originate from specially produced biomass crops, namely perennial grasses and short-rotation trees (Robertson et al., 2017). Some cellulose feedstocks will come from agricultural and industrial by-products.

Thus, there are two benefits to producing cellulosic bioenergy on damaged areas across the world. First, it prevents indirect land use changes (ILUC) impacts so that cellulosic bioenergy's full climate advantages may be used for climate mitigation without benefits needing to be reduced to offset the extra greenhouse gases created when natural lands elsewhere are converted to food production. Targets from Sustainable Development Goal 13 pertaining to climate action are also addressed here. Second, it offers a method for restoring degraded lands to a more fertile state, enabling a greater productive capacity to gradually decrease the amount of land required for bioenergy production, which in turn will allow these lands to eventually return to a more robust natural state to support global biodiversity goals (Brondizio et al., 2019) or returned to food crop production to achieve Sustainable Development Goal 2 related to hunger. Once global temperatures have stabilized sometime after 2100, a rising fraction of bioenergy lands with restored soil fertility would be available for food cultivation.

Most of the the ecological nutrient management methods, such as, fertility restoration—soil carbon accretion, nitrogen conservation, and more effective phosphorus cycling—are used to restore soil fertility under perennial bioenergy crops. Because soil carbon accumulation is crucial to the climatic benefits of bioenergy crops, we have the greatest knowledge about it. Nitrogen is significant for both its beneficial effects on biomass production and its potential drawbacks, which include the ability to increase the reactive nitrogen load in the biosphere and to negate the beneficial effects of bioenergy production on the climate through the creation of fertilizer and nitrous oxide. Additionally, phosphorus may restrict the growth of biomass, particularly in tropical soils with severe weathering. As mentioned earlier, soil carbon gains under perennial crops come from root biomass and stabilized soil C, whether they are grasses such as switchgrass (*Panicum virgatum*) or short-rotation trees such as hybrid poplar (*Populus* spp.), or even complex semi-natural communities such as restored prairie (Gelfand et al., 2020; Tilman et al., 2006). More specifically, these soil carbon gains come from root exudates and decomposition by-products in soil environments that support the persistence of stable forms of soil carbon, particularly carbon associated with aggregates and mineral surfaces (Garten & Wullschleger, 2000; Tiemann & Grandy, 2015). However, the climatic benefit would probably be less than if the biomass carbon were entirely converted to energy production to counterbalance the use of fossil fuels (Paustian et al., 2016). Some researchers have also advised adding biochar to bioenergy crops.

The fact that soil carbon may build up underneath bioenergy crops even after all the aboveground biomass has been harvested highlights the significance of roots as sources of stabilized soil carbon. Perennial legumes used as fodder crops are long renowned for their ability to store soil carbon. There are some good examples in the United States that perennial fodder crops have great potential for SOC sequestration and soil health improvement. Such measures may be taken up in India. A continuous lucerne stand in southwest Michigan, United States, gained 1.0 Mg C/ha/yr in the Ap horizon during its first 12 years, as others have shown (Kumar et al., 2018), and at nearly three times this rate (2.9 Mg C/ha/yr when considering the entire profile to 1 m), according to Syswerda et al. (2011). In a 5-year study of harvested switchgrass fields on farms in Nebraska, United States, Schmer et al. (2011) observed rates of SOC acquisition between 1.4 and 3.3 Mg C/ha/yr to 1.2 m depth. Soil carbon also accumulates under grasses, although frequently more slowly. Not all, but some researchers (Chimento et al., 2016; Frank et al., 2004), have discovered increases of a comparable size immediately after switchgrass introduction. Under short-rotation tree crops, soil carbon also builds up (Chimento et al., 2016), although post-harvest SOC losses can significantly offset soil carbon gains when the soil is exposed to erosive conditions and moisture and temperature conditions that speed up decomposition when the soil is exposed before the next crop's canopy closes.

Apart from using current farmland currently utilized for food production, the two biggest dangers of large-scale bioenergy production are additional biodiversity loss owing to unsuitable land cover alteration and increased nitrogen loading of the biosphere. Both dangers are reduced when appropriate crops are grown on damaged ground. To start, using the 864–951 Mha of uncultivated land that has been abandoned worldwide prevents the destruction of wetlands, old-growth forests, and other natural areas crucial to the preservation of biodiversity (see Introduction; Campbell et al., 2008). Avoiding areas with considerable forest cover is crucial to preventing long-term carbon debt that undermines climate benefits (Robertson et al., 2017); areas that are now undergoing regeneration are already helping to reduce global warming (Griscom et al., 2017). The conservation value of most degraded lands, which are often dominated by non-native invasive species with lower biodiversity value, will be improved by planting native grasses and short-rotation trees that are native to the area. Mixed-species native plantings ought to be even more advantageous (Werling et al., 2014).

Additionally, the release of reactive nitrogen into the environment can be prevented by avoiding crops with high nitrogen needs or poor nitrogen usage efficiency. Perennial grasses and short-rotation trees are examples of non-leguminous biomass crops with low nitrogen needs and high nitrogen usage efficiency. If gathered after senescence, these crops remove very little nitrogen during harvest. Mechanisms in particular persistent roots that take up a significant portion of the available soil volume, pre-harvest nitrogen transfer from aboveground biomass to roots, and relatively little nitrogen allocated to reproductive biomass such as seeds, resulting in high C:N ratios at harvest, create a relatively closed ecosystem nitrogen cycle. When planting species that can obtain the majority or all of their own nitrogen through biological N fixation (BNF), this is much truer. The potential for associative nitrogen fixation (Roley et al., 2019, 2021) and empirical evidence for minimal nitrogen losses from perennial biomass cropping systems are both growing (Hussain et al., 2019; Ruan et al., 2016). Additionally, perennial biomass crops could prevent a significant amount of current nitrogen loss from cropland (Basso et al., 2019) planted in low-performing, unprofitable subfield areas used for biodiversity conservation, bioenergy, or both (Brandes et al., 2018; Schulte et al., 2017).

Intensive Rotational Grazing

Another important management strategy for rehabilitating degraded lands is the management of perennial forage as grazing pastures. Here we refer to grazing management, which is intended to maintain perennial forage supply in a sustainable manner by rotating animals over the landscape, in contrast to present grazing practices that frequently result in overgrazing that can preserve or worsen degraded soils. Various configurations of rotational grazing are possible. With 2–14 pastures per herd, simple rotational grazing rotates animals at modest intensities (Briske et al., 2011; Roche et al., 2015). Adaptive multi-paddock grazing (AMP) frequently comprises 40+ pastures per herd, whereas management intense grazing moves animals at greater intensities with 16+ pastures per herd (Barton et al., 2020; Mosier et al., 2021; Teague et al., 2015). While more intense systems are more flexible and tend to move the animals based on forage growth, low-intensity systems often have more set, scheduled animal movements (Undersander et al., 2002). The soil and perennial plants are given grazing-free intervals by rotating the animals, which helps to preserve and enhance soil fertility and perennial productivity (Kemp et al., 2000).

In comparison with grazing systems that heavily rely on annually planted crops for year-round forage, pastures with diverse, perennial vegetation offer forage that can frequently be more productive (Minns et al., 2001; Moore et al., 2004; Skinner et al., 2004) and available for a greater portion of the growing season (Ferchaud et al., 2016). The greater drought and stress tolerance of many perennial grasses may also contribute to the higher production of grazed perennial fodder (Skinner et al., 2006; Tilman & Downing, 1994). The absence of yearly tillage and dramatic harvesting episodes in permanent pastures also means that greater above- and belowground biomass survives grazing activities, particularly in AMP grazing systems, which seek to leave 50% of fodder uneaten (Teague et al., 2013). These areas can host more animals with greater stocking rates without suffering the detrimental impacts of overgrazing thanks to the higher productivity of perennial fodder (Jakoby et al., 2015; Teague et al., 2015). The vegetation that is accessible for grazing is further diversified by perennialization, which frequently results in more nutritious feed and enhances the well-being of cattle (Teague et al., 2016). In India, rotational grazing at a lower stocking rate was found to sustain the SOC and availability of other nutrients (Table 10.2; Ghosh et al., 2022).

There are several instances of perennial pastures that are grazed in rotation improving soil fertility. Because manure produced on-site offers organic carbon, nitrogen, and phosphorus inputs that can increase nutrient retention and availability, animals maintain nutrient cycles in a more closed, conservative, and cost-effective way when grazing perennial pastures (Elser & Bennett, 2011; Mosier et al., 2021). The usage of fertilizers and other inputs is reduced or even avoided in AMP grazing systems; for instance, Mosier et al. (2021) discovered that unfertilized AMP pastures had more soil nitrogen than non-AMP pastures that were fertilized yearly. Physical, chemical, and biological indices of soil health have all been demonstrated to be enhanced by rotationally grazing perennial systems (Teague et al., 2011; Byrnes et al., 2018), which also leads to increased water retention and infiltration (McCallum et al., 2004; Teague et al., 2011). Humid environments, such as the southern United States (Machmuller et al., 2015; Mosier et al., 2021), and semi-arid rangelands in Africa and the western United States (Badini et al., 2007; McDonald et al., 2019; Teague et al., 2015) have both demonstrated the ability of perennial pastures to restore degraded land.

Rotational grazing in perennial pastures can lessen certain greenhouse gas emissions in addition to its benefits for soil fertility. Through AMP grazing, additional reductions in greenhouse gas emissions, particularly methane, can be made. Shrestha et al. (2020) discovered that AMP grazing systems boosted methane absorption 1.5 times more than other grazing systems. Through improved carbon absorption and decreased external feed requirements, rotationally grazed perennial systems can also lower the overall quantity of CO_2 emissions associated with production when compared with other traditionally grazed systems (Bosch et al., 2008). Although these emissions were mitigated by greater levels of soil carbon accretion, other studies have reported that N_2O emissions were enhanced in these rotating systems because to higher stocking rates (Bosch et al., 2008; Rowntree et al., 2020). The Sustainable Development Goals related to climate action will also be addressed via reductions in net greenhouse gas emissions.

As shown for regions in the U.S. Midwest (Riley et al., 1997), the costs involved in converting degraded croplands into perennial forage systems that are rotationally grazed, such as the need for more fencing, are relatively low and easily offset by increased productivity and economic returns. A further cost-saving benefit is

Table 10.2 Total Nitrogen (N), Available Phosphorus (P), Available Potassium (K), and Available Sulfur (S) in Surface (0–15 cm) and Subsurface (15–30 cm) Soil Layers as Influenced by Different Grazing Intensity in a Semi-Arid Tropical Inceptisol.[a]

	Stocking rate			
Attributes	**Low grazing**	**Medium grazing**	**Heavy grazing**	**Control**
Total N (g/kg)				
0–15 cm	0.63 ± 0.04Ca	0.55 ± 0.03Da	0.70 ± 0.06Ba	0.98 ± 0.07Aa
15–30 cm	0.42 ± 0.06Cc	0.41 ± 0.04Cb	0.53 ± 0.06Bb	0.60 ± 0.05Ab
30–60 cm	0.52 ± 0.05Ab	0.41 ± 0.04Bb	0.39 ± 0.4Bc	0.51 ± 0.04Ac
Available P (kg/ha)				
0–15 cm	9.76 ± 0.08Aa	9.26 ± 0.09Ba	10.10 ± 0.09Aa	8.82 ± 0.06Cb
15–30 cm	9.21 ± 0.09Bb	8.45 ± 0.08Cb	9.68 ± 0.10ABb	10.45 ± 0.08Aa
30–60 cm	9.80 ± 0.08Ba	9.24 ± 0.08Ba	9.51 ± 0.08Bb	10.52 ± 0.09Aa
Available K (kg/ha)				
0–15 cm	278.45 ± 12.6Ca	256.19 ± 20.1 Da	299.52 ± 22.4Ba	370.88 ± 24.2Aa
15–30 cm	228.91 ± 14.1Bb	222.56 ± 17.3Bb	296.85 ± 20.1Ab	289.04 ± 21.2Ab
30–60 cm	191.25 ± 13.6Ac	205.17 ± 14.5Bb	192.48 ± 16.4Bc	260.80 ± 20.6Ac
Available S (mg/kg)				
0–15 cm	26.38 ± 1.64Ba	27.55 ± 2.19Ba	27.35 ± 2.27Ba	30.07 ± 2.64Aa
15–30 cm	24.82 ± 1.37Ab	26.38 ± 2.13Aa	27.55 ± 2.82Aa	24.82 ± 2.66Ab
30–60 cm	23.85 ± 2.20Cb	26.18 ± 2.06Ba	24.05 ± 1.97BCb	29.49 ± 2.54Aa

[a] *Source:* Ghosh et al. (2022). Means with similar uppercase letters within a row are not significantly different at $P < 0.05$ according to Tukey's HSD test. Means with similar lower-case letters within a column are not significantly different at $P < 0.05$ according to Tukey's HSD test.

that animals in these settings become less dependent on outside supplies of feed owing to better perennial production and longer growing seasons (Wang et al., 2018). The conversion of underperforming croplands to perennial pastures for grazing is appealing, especially in arid areas, especially when ranchers consider predicted increases in drought intensity (Wang et al., 2021).

Compared with degraded grazing lands and underperforming cropland, perennial pasture systems that employ intensive rotational grazing have the potential to provide an improved suite of ecosystem services while also restoring soil fertility to provide an increasing base of productive land to meet future food needs. The urge to convert untouched natural areas to pastureland will be proportionately reduced by increased production, which will have a large indirect impact for biodiversity.

Conservation Plantings

The practice of turning damaged croplands into conservation plantings in order to restore soil fertility is at the low end of the range of perennial management intensity. As mentioned previously in the Introduction, the simplest form of it is the fallow phase of bush-fallow or shifting cultivation agriculture, which entails only letting the land undergo biological succession with the restoration of SOM and nutrient availability. However, in many situations today, soil degradation has progressed past the threshold of quick recovery, and because invasive species are so common, the vegetation that does rebound may not resemble the original native community or provide the same ecological advantages. Frequently, more direct conservation management is necessary, with certain species being chosen specifically to promote ecosystem health or biodiversity.

For conservation plants, similar patterns in soil carbon recovery have been recorded for Europe and Asia. In Germany, Breuer et al. (2006) found that areas that were converted to perennial grassland species had higher soil carbon and nitrogen reserves than locations that had continuous agriculture. In croplands that have been replanted with natural grass flora in Russia and China, multiple authors have demonstrated a high potential for soil carbon stock recovery. Regenerated grasslands in Russia had soil carbon accretion rates that were over 50% greater than those of the surrounding woods (Shvidenko & Nilsson, 2003; Vuichard et al., 2008). In a similar vein, abandoned croplands that were turned into grasslands in the Chinese Loess Plateau absorbed more soil carbon than did restored forested areas (Jin et al., 2014; Wei et al., 2012;). In addition, increased soil nitrogen, phosphorus, and total soil fertility were measured by Zhang and Shao (2018) after maize and wheat croplands were converted to perennial grassland.

Many species, including those that make up soil microbial communities, benefit from the conversion of degraded land to conservation plantings. For instance,

Matamala et al. (2008) demonstrated the recovery of microbial communities in a restored grassland that had been converted from farmland, albeit at a significantly slower rate than soil carbon and nitrogen stocks (Matamala et al., 2008). Similar results were found by Baer et al. (2000) on Conservation Reserve Program fields planted to native perennial grasses, which demonstrated a moderate but steady rise in microbial biomass. Zhang et al. (2012) found that once farmland was converted to perennial grassland in China's Loess Plateau, microbial species richness and biomass increased, along with microbial activity. The majority of publications link recovery to perennial rooting systems and corresponding rises in carbon inputs through root turnover and exudation.

The ecological services provided by conservation plants are substantial, with the exclusion of direct financial gain or food production (Figure 10.2). Many of the environmental costs of agriculture are instantly eliminated when land and its related inputs are taken out of production. This also starts the process of restoring supporting services such as biodiversity and soil fertility as well as regulatory functions such as flood management. Most provisioning services, or the direct economic benefits associated with the production of food, fiber, or fuel, are

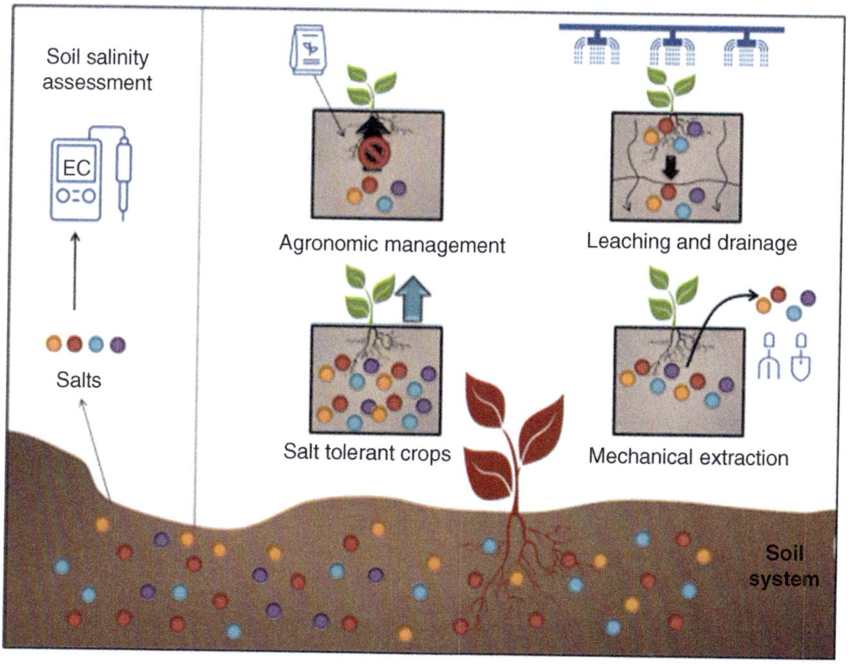

Figure 10.2 The main approaches for ameliorating saline soils in arid regions. Different colors depict the presence of different forms of salt ions in the soil system.

nonetheless eliminated. However, with the possibility of ecosystem service payments (Swinton et al., 2007) that may cover compensation for biodiversity benefits such as pollination and naturally occurring pest suppression (Asbjornsen et al., 2014; Landis & Gage, 2015), direct financial gains may become a new ecosystem service offered by conservation plantings (Figure 10.2). In the end, returning these areas to food production would increase future food security since degraded lands may once again support food crops.

Return to Food Production After Soil Fertility Restoration With Perennials

Any effort to restore soil must include recovery as a major component to stop the cycle of deterioration. It does not make much sense to devote decades to management that increases a system's capacity for production—whether through bioenergy production, intensive rotational grazing, conservation plantings, or any other number of techniques—only to watch the system deteriorate once it is put back into use for food production. Therefore, the manufacturing system put in place after recovery needs to be sustainable. This problem is embodied in Sustainable Development Goal 2, which focuses on ensuring sustainable food production systems and putting in place resilient agricultural practices that boost productivity and production; support ecosystem preservation; strengthen capacity for adaptation to climate change, extreme weather, drought, flooding, and other disasters; and gradually improve land and soil quality.

The ecological services provided by conservation plants are substantial, with the exclusion of direct financial gain or food production (Figure 10.2). Many of the environmental costs of agriculture are instantly eliminated when land and its related inputs are taken out of production. This also starts the process of restoring supporting services such as biodiversity and soil fertility as well as regulatory functions such as flood management. Most provisioning services, or the direct economic benefits associated with the production of food, fiber, or fuel, are nonetheless eliminated. Agricultural systems can never have closed cycles since they export carbon, nitrogen, phosphorus, and other elements with harvest. However, constant green cover and crops that fix nitrogen and mobilize plant-available phosphorus can result in high system-wide nutrient usage efficiency. By intercropping as well as by employing mixed-species cover crops and forages, within-crop variety can also aid in generating and maintaining bigger stabilized carbon reserves and enhance the efficiency of nutrient utilization.

Other important practices to adopt include no-till to improve soil health; variable rate nutrient management and subfield conservation plantings to prevent over-fertilizing low-yielding areas of a given field; minimal, precisely targeted pesticide use to maintain intact and fully functional soil food webs; and animal integration through periodic grazing or manure return. Perennial crops might be included on cereal croplands if they prove to be economically viable in the coming decades. Perennials are therefore an essential component of sustainable, regenerative soil management and should be included in post-recovery rotations.

Barriers to Adoption

Farmers and landowners have a wide range of options for incorporating perennials into their production and land management strategies, from the establishment of perennial cropping systems like those for cellulosic bioenergy, rotational grazing, and conservation plantings noted above, to more complex crop rotations that include perennial crops such as forage grasses and legumes (King & Blesh, 2018). What prevents these methods from being used more frequently? Most farmers have the information and expertise necessary to successfully build and manage restorative perennial cropping systems, hence barriers are typically unrelated to gaps in our understanding. Instead, socioeconomic factors such as international commerce and national policies that support the status quo present hurdles. In particular, three obstacles are particularly prominent.

The first is the ongoing strain from the world's food demand. The demand for food is predicted to increase by 30%–50% by the middle of the century because of population and income development (Food Security Information Network, 2017). This will put constant pressure on farmers to expand and improve food production. There will be pressure to use degraded lands for food production with their inherent productivity and nutrient conservation restrictions to the degree that increased output cannot be satisfied through intensification.

Policies that encourage and reward yearly cultivation on damaged farms come in second. Crop insurance in the United States encourages cultivating on even unproductive cropland since farmers get rewarded for low yearly yields, which happen more frequently on such areas and lead to a downward spiral of crop yields. Crop insurance in the United States and other countries' subsidies, which are only offered for a small number of commodities and do not encourage ecological management (Archer et al., 2003), directly discourage crop diversification and reward low-diversity, high-input production systems. There is no worldwide market for soil fertility or farmland diversification, and because existing regulations erect financial barriers that deter regenerative practices, land abandonment is all too frequently the end outcome, as stated by Pascual and Perrings (2007).

The absence of markets for some of the most promising restorative perennial agricultural systems is a last significant hurdle. In the examples, neither the conservation plantings nor the cellulosic bioenergy crops have current markets, nor are there sufficient regional markets for a variety of grain crops. Even in the American Midwest, crops such as canola (*Brassica rapa*), which is widely grown, cannot be grown due to the lack of nearby processing facilities. So, not only are there frequently disincentives for leaving damaged farms, but there are also no immediate incentives—indeed, there are sometimes disincentives—for implementing regenerative practices. Removing perverse incentives that encourage land degradation and compensating farmers for the provision of ecosystem services are two viable alternatives, particularly in more affluent nations (Robertson et al., 2014).

Producers might be rewarded for the management of crop diversification for ecological nutrient management and the related soil restoration by extending agricultural subsidy payments, whether direct or indirect, to include new annual and perennial crops. A more stable food supply, tolerance to adverse weather, and avoiding yield penalties associated with continuous rotations are also co-benefits of diversity (Bowles et al., 2020; Renard & Tilman, 2019; Seifert et al. 2017).

Second, payments for ecosystem services offer a way for society to directly reward landowners and operators for agricultural practices they would not otherwise embrace (Swinton et al., 2007). Markets still in their infancy are compensating farmers for soil carbon sequestration (Climate Action Reserve, 2020) and more accurate nitrogen management (Millar et al., 2012), but rewards are still insufficient to encourage widespread adoption. However, current proposals in the United States for compensation for climate mitigation practices might provide incentives for soil carbon sequestration through more varied agricultural practices, such as perennial crops.

References

Alori, E. T., Glick, B. R., & Babalola, O. O. (2017). Microbial phosphorus solubilization and its potential for use in sustainable agriculture. *Frontiers in Microbiology*, *8*. https://doi.org/10.3389/fmicb.2017.00971

Álvaro-Fuentes, J., Arrúe, J. L., Gracia, R., & López, M. V. (2008). Tillage and cropping intensification effects on soil aggregation: Temporal dynamics and controlling factors under semiarid conditions. *Geoderma*, *145*(3–4), 390–396. https://doi.org/10.1016/j.geoderma.2008.04.005

Archer, D. W., Pikul, J. L. J., & Riedell, W. E. (2003). Analyzing risk and risk management in cropping systems. In J. D. Hanson & J. M. Krupinsky (Eds.), *Proceedings of the Dynamic Cropping Systems: Principles, Processes and Challenges* (pp. 155–164).

Asbjornsen, H., Hernandez-Santana, V., Liebman, M., Bayala, J., Chen, J., Helmers, M., . . . Schulte, L. A. (2014). Targeting perennial vegetation in agricultural landscapes for enhancing ecosystem services. *Renewable Agriculture and Food Systems*, *29*(2), 101–125. https://doi.org/10.1017/S1742170512000385

Badini, O., Stöckle, C. O., Jones, J. W., Nelson, R., Kodio, A., & Keita, M. (2007). A simulation-based analysis of productivity and soil carbon in response to time-controlled rotational grazing in the west African Sahel region. *Agricultural Systems, 94*(1), 87–96. https://doi.org/10.1016/j.agsy.2005.09.010

Baer, S. G., Rice, C. W., & Blair, J. M. (2000). Assessment of soil quality in fields with short-, and long-term enrollment in the CRP. *Journal of Soil and Water Conservation, 55*, 142–146.

Barton, E., Bennett, D. E., & Burnidge, W. (2020). Holistic perspectives—Understanding rancher experiences with holistic resource management to bridge the gap between rancher and researcher perspectives. *Rangelands, 42*(5), 143–150. https://doi.org/10.1016/j.rala.2020.05.003

Basso, B., Shuai, G., Zhang, J., & Robertson, G. P. (2019). Yield stability analysis reveals sources of large-scale nitrogen loss from the US Midwest. *Scientific Reports, 9*(1), Article 5774. https://doi.org/10.1038/s41598-019-42271-1

Beier, C., Beierkuhnlein, C., Wohlgemuth, T., Penuelas, J., Emmett, B., Körner, C., . . . Hansen, K. (2012). Precipitation manipulation experiments—Challenges and recommendations for the future. *Ecology Letters, 15*(8), 899–911. https://doi.org/10.1111/j.1461-0248.2012.01793.x

Bosch, D. J., Stephenson, K., Groover, G., & Hutchins, B. (2008). Farm returns to carbon credit creation with intensive rotational grazing. *Journal of Soil and Water Conservation, 63*(2), 91–98. https://doi.org/10.2489/jswc.63.2.91

Bowker, M. A., Belnap, J., Davidson, D. W., & Goldstein, H. (2006). Correlates of biological soil crust abundance across a continuum of spatial scales: Support for a hierarchical conceptual model. *Journal of Applied Ecology, 43*(1), 152–163. https://doi.org/10.1111/j.1365-2664.2006.01122.x

Bowles, T. M., Mooshammer, M., Socolar, Y., Calderón, F., Cavigelli, M. A., Culman, S., Deen, W., Drury, C. F., Garcia, A. G., Gaudin, A., Harkcom, W., Lehman, R., Osborne, S., Robertson, G., Salerno, J., Schmer, M., Strock, J., & Grandy, A. (2020). Long-term evidence shows that crop-rotation diversification increases agricultural resilience to adverse growing conditions in North America. *One Earth, 2*(3), 284–293. https://doi.org/10.1016/j.oneear.2020.02.007

Brady, N. C., & Weil, R. R. (2007). *The nature and properties of soils* (14th ed.). Pearson.

Brandes, E., McNunn, G. S., Schulte, L. A., Muth, D. J., Vanloocke, A., & Heaton, E. A. (2018). Targeted subfield switchgrass integration could improve the farm economy, water quality, and bioenergy feedstock production. *GCB Bioenergy, 10*(3), 199–212. https://doi.org/10.1111/gcbb.12481

Breuer, L., Huisman, J. A., Keller, T., & Frede, H. (2006). Impact of a conversion from cropland to grassland on C and N storage and related soil properties: Analysis of a 60-year chronosequence. *Geoderma, 133*(1–2), 6–18. https://doi.org/10.1016/j.geoderma.2006.03.033

Bridges, E. M., & Oldeman, L. R. (1999). Global assessment of human-induced soil degradation. *Arid Soil Research and Rehabilitation*, *13*(4), 319–325. https://doi.org/10.1080/089030699263212

Briske, D. D., Sayre, N. F., Huntsinger, L., Fernandez-Giminez, M., Budd, B., & Derner, J. D. (2011). Origin, persistence, and resolution of the rotational grazing debate: Integrating human dimensions into rangeland research. *Rangeland Ecology & Management*, *64*(4), 325–334. https://doi.org/10.2111/REM-D-10-00084.1

Brondizio, E. S. (2019). In J. Settele, S. Díaz, & H. T. Ngo (Eds.), *Global assessment report on biodiversity and ecosystem services*. IPBES. https://ipbes.net/global-assessment

Bronick, C. J., & Lal, R. (2005). Soil structure and management: A review. *Geoderma*, *124*(1–2), 3–22. https://doi.org/10.1016/j.geoderma.2004.03.005

Brooks, K. N., Ffolliott, P. F., & Magner, J. A. (2013). *Hydrology and the management of watersheds: Brooks/hydrology and the management of watersheds* (4th ed.). Wiley-Blackwell.

Byrnes, R. C., Eastburn, D. J., Tate, K. W., & Roche, L. M. (2018). A global meta-analysis of grazing impacts on soil health indicators. *Journal of Environmental Quality*, *47*(4), 758–765. https://doi.org/10.2134/jeq2017.08.0313

Campbell, J. E., Lobell, D. B., Genova, R. C., & Field, C. B. (2008). The global potential of bioenergy on abandoned agriculture lands. *Environmental Science & Technology*, *42*(15), 5791–5794. https://doi.org/10.1021/es800052w

Casby-Horton, S., Herrero, J., & Rolong, N. A. (2015). Gypsum soils—Their morphology, classification, function, and landscapes. In *Advances in agronomy* (pp. 231–290). Elsevier.

Chi, Y., Li, Z., Zhang, G., Zhao, L., Gao, Y. H., Wang, D., Liu, L., Cai, D., & Wu, Z. (2020). Inhibiting desertification using aquatic cyanobacteria assisted by a nanocomposite. *ACS Sustainable Chemistry & Engineering*, *8*(8), 3477–3486. https://doi.org/10.1021/acssuschemeng.0c00233

Chimento, C., Almagro, M., & Amaducci, S. (2016). Carbon sequestration potential in perennial bioenergy crops: The importance of organic matter inputs and its physical protection. *GCB Bioenergy*, *8*(1), 111–121. https://doi.org/10.1111/gcbb.12232

Churchman, G. J. (2018). Game changer in soil science. Functional role of clay minerals in soil. *Journal of Plant Nutrition and Soil Science*, *181*(1), 99–103. https://doi.org/10.1002/jpln.201700605

Climate Action Reserve. (2020). *Soil enrichment protocol*. Version 1.0. https://www.climateactionreserve.org/%0Ahow/protocols/soil-enrichment

Cuevas, J., Daliakopoulos, I. N., del Moral, F., Hueso, J. J., & Tsanis, I. K. (2019). A review of soil-improving cropping systems for soil salinization. *Agronomy*, *9*(6), Article 295. https://doi.org/10.3390/agronomy9060295

Delgado, J. A., Nearing, M. A., & Rice, C. W. (2013). Conservation practices for climate change adaptation. In *Advances in agronomy* (pp. 47–115). Elsevier.

Dlamini, P., Chivenge, P., Manson, A., & Chaplot, V. (2014). Land degradation impact on soil organic carbon and nitrogen stocks of sub-tropical humid grasslands in South Africa. *Geoderma*, *235–236*, 372–381. https://doi.org/10.1016/j.geoderma.2014.07.016

Elser, J., & Bennett, E. (2011). A broken biogeochemical cycle. *Nature*, *478*(7367), 29–31. https://doi.org/10.1038/478029a

Eswaran, H., & Zi-Tong, G. (2015). Properties, genesis, classification, and distribution of soils with gypsum. In W. D. Nettleton (Ed.), *Occurrence, characteristics, and genesis of carbonate, gypsum, and silica accumulations in soils* (pp. 89–119). Soil Science Society of America.

Fageria, N. K., Gheyi, H. R., & Moreira, A. (2011). Nutrient bioavailability in salt affected soils. *Journal of Plant Nutrition*, *34*(7), 945–962. https://doi.org/10.1080/01904167.2011.555578

Feng, K., Nadeem, M., Ashiq, W., Zaeem, M., Gilani, S. S. M., Rajabi-Khamseh, S., Pham, T. H., Kavanagh, V., Thomas, R., & Cheema, M. (2004). Effect of organic ligands on biological availability of inorganic phosphorus in soils. *Pedosphere*, *14*, 85–92.

Ferchaud, F., Vitte, G., & Mary, B. (2016). Changes in soil carbon stocks under perennial and annual bioenergy crops. *GCB Bioenergy*, *8*(2), 290–306. https://doi.org/10.1111/gcbb.12249

Field, C. B., & Mooney, H. A. (1986). The photosynthesis-nitrogen relationship in wild plants. In T. J. Givnish (Ed.), *On the economy of plant form and function* (pp. 25–55). Cambridge University Press.

Food Security Information Network. (2017). *Global report on food crises 2017*. https://www.wfp.org/publications/global-reportfood- crisis-2017

Frank, A. B., Berdahl, J. D., Hanson, J. D., Liebig, M. A., & Johnson, H. A. (2004). Biomass and carbon partitioning in Switchgrass. *Crop Science*, *44*(4), 1391–1396. https://doi.org/10.2135/cropsci2004.1391

Gabarrón-Galeote, M. A., Trigalet, S., & van Wesemael, B. (2015). Effect of land abandonment on soil organic carbon fractions along a Mediterranean precipitation gradient. *Geoderma*, *249–250*, 69–78. https://doi.org/10.1016/j.geoderma.2015.03.007

Garten, C. T., & Wullschleger, S. D. (2000). Soil carbon dynamics beneath switchgrass as indicated by stable isotope analysis. *Journal of Environmental Quality*, *29*(2), 645–653. https://doi.org/10.2134/jeq2000.00472425002900020036x

Gelfand, I., Hamilton, S. K., Kravchenko, A. N., Jackson, R. D., Thelen, K. D., & Robertson, G. P. (2020). Empirical evidence for the potential climate benefits of decarbonizing light vehicle transport in the U. S. with bioenergy from purpose-grown biomass with and without BECCS. *Environmental Science & Technology*, *54*(5), 2961–2974. https://doi.org/10.1021/acs.est.9b07019

Ghosh, A., Mahanta, S. K., Manna, M. C., Singh, S., Bhattacharyya, R., Tyagi, V. C., Singh, J. B., Ram, S. N., Srinivasan, R., Singh, A. K., & Gupta, A. (2022). Long-term grazing mediates soil organic carbon dynamics by reorienting enzyme activities and elemental stoichiometry in semi-arid tropical inceptisol. *Journal of Soil Science and Plant Nutrition, 22*(2), 1422–1433.

Griscom, B. W., Adams, J., Ellis, P. W., Houghton, R. A., Lomax, G., Miteva, D. A., . . . Fargione, J. (2017). Natural climate solutions. *Proceedings of the National Academy of Sciences, 114*(44), 11645–11650. https://doi.org/10.1073/pnas.1710465114

Hall, A. J., & Richards, R. A. (2013). Prognosis for genetic improvement of yield potential and water-limited yield of major grain crops. *Field Crops Research, 143*, 18–33. https://doi.org/10.1016/j.fcr.2012.05.014

Hamza, M. A., & Anderson, W. K. (2005). Soil compaction in cropping systems. *Soil and Tillage Research, 82*(2), 121–145. https://doi.org/10.1016/j.still.2004.08.009

Hassink, J. (1997). The capacity of soils to preserve organic C and N by their association with clay and silt particles. *Plant and Soil, 191*, 77–87. https://doi.org/10.1023/A:1004213929699

Hoffmann, C., Funk, R., Li, Y., & Sommer, M. (2008). Effect of grazing on wind driven carbon and nitrogen ratios in the grasslands of Inner Mongolia. *Catena, 75*(2), 182–190. https://doi.org/10.1016/j.catena.2008.06.003

Husein, H. H., Lucke, B., Bäumler, R., & Sahwan, W. (2021). A contribution to soil fertility assessment for arid and semi-arid lands. *Soil Systems, 5*(3), Article 42. https://doi.org/10.3390/soilsystems5030042

Hussain, M. Z., Bhardwaj, A. K., Basso, B., Robertson, G. P., & Hamilton, S. K. (2019). Nitrate leaching from continuous corn, perennial grasses, and poplar in the US Midwest. *Journal of Environmental Quality, 48*(6), 1849–1855. https://doi.org/10.2134/jeq2019.04.0156

Intergovernmental Panel on Climate Change. (2018). Global warming of 1.5°C. In V. Masson-Delmotte, et al. (Eds.), *An IPCC special report on the impacts of global warming of 1.5°C above pre-industrial levels and related global greenhouse gas emission pathways, in the context of strengthening the global response to the threat of climate change, sustainable development*. https://www.ipcc.ch/sr15

Jakoby, O., Quaas, M. F., Baumgärtner, S., & Frank, K. (2015). Adapting livestock management to spatio-temporal heterogeneity in semi-arid rangelands. *Journal of Environmental Management, 162*, 179–189. https://doi.org/10.1016/j.jenvman.2015.07.047

Jiao, F., Shi, X. R., Han, F. P., & Yuan, Z. Y. (2016). Increasing aridity, temperature and soil pH induce soil C-N-P imbalance in grasslands. *Scientific Reports, 6*(1), Article 19601. https://doi.org/10.1038/srep19601

Jin, Z., Dong, Y., Wang, Y., Wei, X., Wang, Y., Cui, B., & Zhou, W. (2014). Natural vegetation restoration is more beneficial to soil surface organic and inorganic carbon sequestration than tree plantation on the loess plateau of China. *Science of the Total Environment, 485–486*, 615–623. https://doi.org/10.1016/j.scitotenv.2014.03.105

Kemp, D. R., Michalk, D. L., & Virgona, J. M. (2000). Towards more sustainable pastures: Lessons learnt. *Australian Journal of Experimental Agriculture*, *40*(2), Article 343. https://doi.org/10.1071/EA99001

Kheirfam, H., & Asadzadeh, F. (2020). Stabilizing sand from dried-up lakebeds against wind erosion by accelerating biological soil crust development. *European Journal of Soil Biology*, *98*, Article 103189. https://doi.org/10.1016/j.ejsobi.2020.103189

King, A. E., & Blesh, J. (2018). Crop rotations for increased soil carbon: Perenniality as a guiding principle. *Ecological Applications*, *28*(1), 249–261. https://doi.org/10.1002/eap.1648

Kumar, S., et al. (2018). Role of legumes in soil carbon sequestration. In *Legumes for soil health and sustainable management* (pp. 109–138). Springer. https://doi.org/10.1007/978-981-13-0253-4_4

Lahmar, R. (2010). Adoption of conservation agriculture in Europe. *Land Use Policy*, *27*(1), 4–10. https://doi.org/10.1016/j.landusepol.2008.02.001

Lal, R. (2004). Carbon sequestration in dryland ecosystems. *Environmental Management*, *33*(4). https://doi.org/10.1007/s00267-003-9110-9

Lal, R. (2006). Enhancing crop yields in the developing countries through restoration of the soil organic carbon pool in agricultural lands. *Land Degradation & Development*, *17*(2), 197–209. https://doi.org/10.1002/ldr.696

Landis, D. A., & Gage, S. H. (2015). Arthropod diversity and pest suppression in agricultural landscapes. In S. K. Hamilton, J. E. Doll, & G. P. Robertson (Eds.), *The ecology of agricultural landscapes: Long-term research on the path to sustainability.* Oxford University Press.

Lima, H. V., Silva, A. P., Santos, M. C., Cooper, M., & Romero, R. E. (2006). Micromorphology and image analysis of a hardsetting Ultisol (Argissolo) in the state of Ceará (Brazil). *Geoderma*, *132*(3–4), 416–426. https://doi.org/10.1016/j.geoderma.2005.06.006

López, M. V., Moret, D., Gracia, R., & Arrue, J. L. (2003). Tillage effects on barley residue cover during fallow in semiarid Aragon. *Soil and Tillage Research*, *72*(1), 53–64. https://doi.org/10.1016/S0167-1987(03)00047-3

Machmuller, M. B., et al. (2015). Emerging land use practices rapidly increase soil organic matter. *Nature Communications*, *6*(1), Article 6995. https://doi.org/10.1038/ncomms7995

Manzoni, S., Schaeffer, S. M., Katul, G., Porporato, A., & Schimel, J. P. (2014). A theoretical analysis of microbial eco-physiological and diffusion limitations to carbon cycling in drying soils. *Soil Biology and Biochemistry*, *73*, 69–83. https://doi.org/10.1016/j.soilbio.2014.02.008

Matamala, R., Jastrow, J. D., Miller, R. M., & Garten, C. T. (2008). Temporal changes in C and N stocks of restored prairie: Implications for C sequestration strategies. *Ecological Applications*, *18*(6), 1470–1488. https://doi.org/10.1890/07-1609.1

McCallum, M. H., Kirkegaard, J. A., Green, T. W., Cresswell, H. P., Davies, S. L., Angus, J. F., & Peoples, M. B. (2004). Improved subsoil macroporosity following perennial pastures. *Australian Journal of Experimental Agriculture*, *44*(3). Article 299. https://doi.org/10.1071/EA03076

McDonald, S. E., Reid, N., Smith, R., Waters, C., Rader, R., & Hunter, J. (2019). Rotational grazing management achieves similar plant diversity outcomes to areas managed for conservation in a semi-arid rangeland. *The Rangeland Journal*, *41*(2), Article 135. https://doi.org/10.1071/RJ18090

Millar, N., Robertson, G. P., Diamant, A., Gehl, R. J., Grace, P. R., & Hoben, J. P. (2012). *Methodology for quantifying nitrous oxide (N2O) emissions reductions by reducing nitrogen fertilizer use on agricultural crops*. American Carbon Registry, Winrock International.

Miner, G. L., Hansen, N. C., Inman, D., Sherrod, L. A., & Peterson, G. A. (2013). Constraints of no-till dryland agroecosystems as bioenergy production systems. *Agronomy Journal*, *105*(2), 364–376. https://doi.org/10.2134/agronj2012.0243

Minns, A., Finn, J., Hector, A., Caldeira, M., Joshi, J., Palmborg, C., Schmid, B., Scherer-Lorenzen, M., & Spehn, E. (2001). The functioning of European grassland ecosystems: Potential benefits of biodiversity to agriculture. *Outlook on Agriculture*, *30*(3), 179–185. https://doi.org/10.5367/000000001101293634

Molden, D., & Oweis, T. Y. O. (2007). Pathways for increasing agricultural water productivity. In D. Molden (Ed.), *Water for food, water for life*. Earthscan International Water Management Institute.

Monier, M. W., Fawkia, L., & Alaa, Z. (2019). Management of calcareous soils in arid region. *International Journal of Environmental Pollution and Environmental Modelling*, *2*, 248–258.

Moore, K. J., White, T., Hintz, R., Patrick, P. K., & Brummer, E. C. (2004). Sequential grazing of cool- and warm-season pastures. *Agronomy Journal*, *96*(4), 1103–1111. https://doi.org/10.2134/agronj2004.1103

Morison, J. I., Baker, N. R., Mullineaux, P. M., & Davies, W. J. (2008). Improving water use in crop production. *Philosophical Transactions of the Royal Society, B: Biological Sciences*, *363*(1491), 639–658. https://doi.org/10.1098/rstb.2007.2175

Mosier, S., Apfelbaum, S., Byck, P., Calderon, F., Teague, R., Thompson, R., & Cotrufo, M. F. (2021). Adaptive multi-paddock grazing enhances soil carbon and nitrogen stocks and stabilization through mineral association in southeastern U. S. grazing lands. *Journal of Environmental Management*, *288*, Article 112409. https://doi.org/10.1016/j.jenvman.2021.112409

Mueller, R. C., Belnap, J., & Kuske, C. R. (2015). Soil bacterial and fungal community responses to nitrogen addition across soil depth and microhabitat in an arid shrubland. *Frontiers in Microbiology*, *6*. https://doi.org/10.3389/fmicb.2015.00891

Mullins, C. E. (1997). Hardsetting. In R. Lal, et al. (Eds.), *Methods for assessment of soil degradation: Vol. 19. Advances in soil science* (pp. 109–128). CRC Press.

Naorem, A., Jayaraman, S., Dalal, R. C., Patra, A., Rao, C. S., & Lal, R. (2022). Soil inorganic carbon as a potential sink in carbon storage in dryland soils—A review. *Agriculture*, *12*(8), Article 1256. https://doi.org/10.3390/agriculture12081256

Pascual, U., & Perrings, C. (2007). Developing incentives and economic mechanisms for in situ biodiversity conservation in agricultural landscapes. *Agriculture, Ecosystems & Environment*, *121*(3), 256–268. https://doi.org/10.1016/j.agee.2006.12.025

Paustian, K., Lehmann, J., Ogle, S., Reay, D., Robertson, G. P., & Smith, P. (2016). Climate-smart soils. *Nature*, *532*(7597), 49–57. https://doi.org/10.1038/nature17174

Pearce, R. C., Li, Y., & Bush, L. P. (1999). Calcium and bicarbonate effects on the growth and nutrient uptake of burley tobacco seedlings: Float system 1. *Journal of Plant Nutrition*, *22*(7), 1079–1090. https://doi.org/10.1080/01904169909365697

Plaza-Bonilla, D., Arrúe, J. L., Cantero-Martínez, C., Fanlo, R., Iglesias, A., & Álvaro-Fuentes, J. (2015). Carbon management in dryland agricultural systems. *A review. Agronomy for Sustainable Development*, *35*(4), 1319–1334. https://doi.org/10.1007/s13593-015-0326-x

Ren, C., Chen, J., Lu, X., Doughty, R., Zhao, F., Zhong, Z., . . . Ren, G. (2018). Responses of soil total microbial biomass and community compositions to rainfall reductions. *Soil Biology and Biochemistry*, *116*, 4–10. https://doi.org/10.1016/j.soilbio.2017.09.028

Renard, D., & Tilman, D. (2019). National food production stabilized by crop diversity. *Nature*, *571*(7764), 257–260. https://doi.org/10.1038/s41586-019-1316-y

Reynolds, J. F., Smith, D. M., Lambin, E. F., Turner, B. L., 2nd, Mortimore, M., Batterbury, S. P., Downing, T. E., Dowlatabadi, H., Fernández, R. J., Herrick, J. E., Huber-Sannwald, E., Jiang, H., Leemans, R., Lynam, T., Maestre, F. T., Ayarza, M., & Walker, B. (2007). Global desertification: Building a science for dryland development. *Science*, *316*(5826), 847–851. https://doi.org/10.1126/science.1131634

Riley, B., Nelson, C., Strohbehn, D., BreDahl, R., Peterson, B., & Sprague, R. (1997). *An economic comparison of rotational grazing systems to eight crop alternatives and the CRP option for highly erodible land in Southwest Iowa*. Iowa State University Animal Industry Report 1(1).

Robertson, G. P., Gross, K. L., Hamilton, S. K., Landis, D. A., Schmidt, T. M., Snapp, S. S., & Swinton, S. M. (2014). Farming for ecosystem services: An ecological approach to production agriculture. *Bioscience*, *64*(5), 404–415. https://doi.org/10.1093/biosci/biu037

Robertson, G. P., Hamilton, S. K., Barham, B. L., Dale, B. E., Izaurralde, R. C., Jackson, R. D., . . . Tiedje, J. M. (2017). Cellulosic biofuel contributions to a sustainable energy future: Choices and outcomes. *Science*, *356*(6345), Article eaal2324. https://doi.org/10.1126/science.aal2324

Roche, L. M., Cutts, B. B., Derner, J. D., Lubell, M. N., & Tate, K. W. (2015). On-ranch grazing strategies: Context for the rotational grazing dilemma. *Rangeland Ecology & Management*, *68*(3), 248–256. https://doi.org/10.1016/j.rama.2015.03.011

Rodriguez-Caballero, E., Belnap, J., Büdel, B., Crutzen, P. J., Andreae, M. O., Pöschl, U., & Weber, B. (2018). Dryland photoautotrophic soil surface communities endangered by global change. *Nature Geoscience, 11*(3), 185–189. https://doi.org/10.1038/s41561-018-0072-1

Roley, S. S., Ulbrich, T. C., & Robertson, G. P. (2021). Nitrogen fixation and resorption efficiency differences among twelve upland and lowland Switchgrass cultivars. *Phytobiomes Journal, 5*(1), 97–107. https://doi.org/10.1094/PBIOMES-11-19-0064-FI

Roley, S. S., Xue, C., Hamilton, S. K., Tiedje, J. M., & Robertson, G. P. (2019). Isotopic evidence for episodic nitrogen fixation in switchgrass (*Panicum virgatum* L.). *Soil Biology and Biochemistry, 129*, 90–98. https://doi.org/10.1016/j.soilbio.2018.11.006

Rowntree, J. E., Rowntree, J., Windisch, W., Waters, S. M., Shalloo, L., & Manzano, P. (2020). Ecosystem impacts and productive capacity of a multi-species pastured livestock system. *Frontiers in Sustainable Food Systems, 4*. https://doi.org/10.3389/fsufs.2020.544984

Ruan, L., Bhardwaj, A. K., Hamilton, S. K., & Robertson, G. P. (2016). Nitrogen fertilization challenges the climate benefit of cellulosic biofuels. *Environmental Research Letters, 11*(6), Article 064007. https://doi.org/10.1088/1748-9326/11/6/064007

Schimel, J., Balser, T. C., & Wallenstein, M. (2007). Microbial stress-response physiology and its implications for ecosystem function. *Ecology, 88*(6), 1386–1394. https://doi.org/10.1890/06-0219

Schmer, M. R., Liebig, M. A., Vogel, K. P., & Mitchell, R. B. (2011). Field-scale soil property changes under switchgrass managed for bioenergy. *GCB Bioenergy, 3*(6), 439–448. https://doi.org/10.1111/j.1757-1707.2011.01099.x

Schulte, L. A., Niemi, J., Helmers, M. J., Liebman, M., Arbuckle, J. G., James, D. E., Kolka, R. K., O'Neal, M. E., Tomer, M. D., Tyndall, J. C., Asbjornsen, H., Drobney, P., Neal, J., Van Ryswyk, G., & Witte, C. (2017). Prairie strips improve biodiversity and the delivery of multiple ecosystem services from corn–soybean croplands. *Proceedings of the National Academy of Sciences, 114*(42), 11247–11252. https://doi.org/10.1073/pnas.1620229114

Seifert, C. A., Roberts, M. J., & Lobell, D. B. (2017). Continuous corn and soybean yield penalties across hundreds of thousands of fields. *Agronomy Journal, 109*(2), 541–548. https://doi.org/10.2134/agronj2016.03.0134

Shabala, S., & Munns, R. (2017). Salinity stress: Physiological constraints and adaptive mechanisms. In *Plant stress physiology* (pp. 24–63). CABI.

Shrestha, B. M., Bork, E. W., Chang, S. X., Carlyle, C. N., Ma, Z., Döbert, T. F., . . . Boyce, M. S. (2020). Adaptive multi-paddock grazing lowers soil greenhouse gas emission potential by altering extracellular enzyme activity. *Agronomy, 10*(11), Article 1781. https://doi.org/10.3390/agronomy10111781

Shvidenko, A., & Nilsson, S. (2003). A synthesis of the impact of Russian forests on the global carbon budget for 1961–1998. *Tellus B: Chemical and Physical Meteorology*, *55*(2), 391–415. https://doi.org/10.1034/j.1600-0889.2003.00046.x

Skinner, R. H., Gustine, D. L., & Sanderson, M. A. (2004). Growth, water relations, and nutritive value of pasture species mixtures under moisture stress. *Crop Science*, *44*(4), 1361–1369. https://doi.org/10.2135/cropsci2004.1361

Skinner, R. H., Sanderson, M. A., Tracy, B. F., & Dell, C. J. (2006). Above- and belowground productivity and soil carbon dynamics of pasture mixtures. *Agronomy Journal*, *98*(2), 320–326. https://doi.org/10.2134/agronj2005.0180a

Spokas, L., & Steponavičius, D. (2011). Fuel consumption during cereal and rape harvesting and methods of its reduction. *Journal of Food, Agriculture and Environment*, *9*, 257–263.

Swinton, S. M., Lupi, F., Robertson, G. P., & Hamilton, S. K. (2007). Ecosystem services and agriculture: Cultivating agricultural ecosystems for diverse benefits. *Ecological Economics*, *64*(2), 245–252. https://doi.org/10.1016/j.ecolecon.2007.09.020

Syswerda, S. P., Corbin, A. T., Mokma, D. L., Kravchenko, A. N., & Robertson, G. P. (2011). Agricultural management and soil carbon storage in surface vs. deep layers. *Soil Science Society of America Journal*, *75*(1), 92–101. https://doi.org/10.2136/sssaj2009.0414

Teague, W. R., Dowhower, S., Baker, S. A., Haile, N., DeLaune, P., & Conover, D. (2011). Grazing management impacts on vegetation, soil biota and soil chemical, physical and hydrological properties in tall grass prairie. *Agriculture, Ecosystems and Environment*, *141*(3–4), 310–322. https://doi.org/10.1016/j.agee.2011.03.009

Teague, R., Grant, B., & Wang, H.-H. (2015). Assessing optimal configurations of multi-paddock grazing strategies in tallgrass prairie using a simulation model. *Journal of Environmental Management*, *150*, 262–273. https://doi.org/10.1016/j.jenvman.2014.09.027

Teague, R., Provenza, F., Kreuter, U., Steffens, T., & Barnes, M. (2013). Multi-paddock grazing on rangelands: Why the perceptual dichotomy between research results and rancher experience? *Journal of Environmental Management*, *128*, 699–717. https://doi.org/10.1016/j.jenvman.2013.05.064

Teague, W. R., Teague, W. R., Apfelbaum, S. I., Lal, R., Kreuter, U. P., Rowntree, J., Davies, C. A., Conser, R. E., Rasmussen, M. A., Hatfeld, J., Wang, T., Wang, F., & Byck, P. (2016). The role of ruminants in reducing agriculture's carbon footprint in North America. *Journal of Soil and Water Conservation*, *71*(2), 156–164. https://doi.org/10.2489/jswc.71.2.156

Tiemann, L. K., & Grandy, A. S. (2015). Mechanisms of soil carbon accrual and storage in bioenergy cropping systems. *GCB Bioenergy*, *7*(2), 161–174. https://doi.org/10.1111/gcbb.12126

Tiessen, H., & Stewart, J. W. B. (1983). Particle-size fractions and their use in studies of soil organic matter: II. Cultivation effects on organic matter composition in size fractions. *Soil Science Society of America Journal, 47*(3), 509–514. https://doi.org/10.2136/sssaj1983.03615995004700030023x

Tilman, D., Hill, J., & Lehman, C. (2006). Carbon-negative biofuels from low-input high-diversity grassland biomass. *Science, 314*(5805), 1598–1600. https://doi.org/10.1126/science.1133306

Tilman, D., & Downing, J. A. (1994). Biodiversity and stability in grasslands. *Nature, 367*(6461), 363–365. https://doi.org/10.1038/367363a0

Undersander, D., Albert, B., Cosgrove, D., Johnson, D., & Peterson, P. (2002). *Pastures for profit: A guide to rotational grazing* (No. A3529). University of Wisconsin Extension.

Vuichard, N., Ciais, P., Belelli, L., Smith, P., & Valentini, R. (2008). Carbon sequestration due to the abandonment of agriculture in the former USSR since 1990. *Global Biogeochemical Cycles, 22*(4). https://doi.org/10.1029/2008GB003212

Wang, T., Jin, H., Kreuter, U., & Teague, R. (2021). Expanding grass-based agriculture on marginal land in the U. S. Great Plains: The role of management intensive grazing. *Land Use Policy, 104*, Article 105155. https://doi.org/10.1016/j.landusepol.2020.105155

Wang, T., Teague, W. R., & Seong, P. (2018). Evaluating long-term economic and ecological consequences of continuous and multi-paddock grazing—A modeling approach. *Agricultural Systems, 165*, 197–207. https://doi.org/10.1016/j.agsy.2018.06.012

Wang, X. B., Cai, D. X., Hoogmoed, W. B., Oenema, O., & Perdok, U. D. (2007). Developments in conservation tillage in rainfed regions of North China. *Soil and Tillage Research, 93*(2), 239–250. https://doi.org/10.1016/j.still.2006.05.005

Wei, J., Cheng, J., Li, W., & Liu, W. (2012). Comparing the effect of naturally restored Forest and grassland on carbon sequestration and its vertical distribution in the Chinese loess plateau. *PLoS One, 7*(7), Article e40123. https://doi.org/10.1371/journal.pone.0040123

Werling, B. P., Dickson, T. L., Isaacs, R., Gaines, H., Gratton, C., Gross, K. L., Liere, H., Malmstrom, C. M., Meehan, T. D., Ruan, L., Robertson, B. A., Robertson, G. P., Schmidt, T. M., Schrotenboer, A. C., Teal, T. K., Wilson, J. K., & Landis, D. A. (2014). Perennial grasslands enhance biodiversity and multiple ecosystem services in bioenergy landscapes. *Proceedings of the National Academy of Sciences, 111*(4), 1652–1657. https://doi.org/10.1073/pnas.1309492111

Wichelns, D., & Qadir, M. (2015). Achieving sustainable irrigation requires effective management of salts, soil salinity, and shallow groundwater. *Agricultural Water Management, 157*, 31–38. https://doi.org/10.1016/j.agwat.2014.08.016

Zhang, C., Guobin, L., Sha, X., & Changsheng, Z. (2012). Rhizosphere soil microbial properties on abandoned croplands in the loess plateau China during vegetation succession. *European Journal of Soil Biology, 50*, 127–136. https://doi.org/10.1016/j.ejsobi.2012.01.002

Zhang, Q., & Shao, M. (2018). Soil fertility increases rapidly during the 6–10 yr following conversion of cropland to grassland in China's loess plateau region.

Canadian Journal of Soil Science, *98*(3), 531–541. https://doi.org/10.1139/cjss-2017-0107

Zhao, Y., & Wang, J. (2019). Mechanical sand fixing is more beneficial than chemical sand fixing for artificial cyanobacteria crust colonization and development in a sand desert. *Applied Soil Ecology*, *140*, 115–120. https://doi.org/10.1016/j.apsoil.2019.04.008

Zhu, F., Qu, L., Hong, X., & Sun, X. (2011). Isolation and characterization of a phosphate-solubilizing halophilic bacterium *Kushneria* sp. YCWA18 from Daqiao Saltern on the coast of Yellow Sea of China. *Evidence-Based Complementary and Alternative Medicine*, *2011*, Article 615032. https://doi.org/10.1155/2011/615032

11

Enhancing Soil Health by Mitigating Salinity in India

Ashim Datta, Nirmalendu Basak, Bholanath Saha, Piu Basak, Arvind Kumar Rai, and Rajender Kumar Yadav

Chapter Overview

Salt-affected soils are constraints for agricultural crop production, particularly in arid and semi-arid regions. To feed the growing population of India, these marginal lands should be reclaimed and brought back under productive cultivation with adoption of innovative approaches. This chapter discusses the extent of salt–affected soils and their physical, chemical, and biological constraints, and synthesizes existing management technologies and researchable issues for reclaiming saline and sodic soils to enhance soil health in India.

Climate change, increasing temperature, and shifting rainfall or greater evapotranspiration demands for crops increase soil salinization and aggravate the sodicity. Therefore, monitoring soil and water bodies at frequent intervals is imperative. The primary indicators of soil health assessment (physical, chemical, and microbiological properties) represent the soil processes and can be early indicators of salinity. These tools need to be tested to benchmark salt–affected soils in India. Application of different organic and inorganic amendments, imposing innovative agronomic managements, and alternate land use systems for reclamation of sodic and saline soils helps in soil quality improvement with organic carbon enrichment in soils, thereby serving many ecosystem functions in the present context of climate change. Finally, salt–affected lands should not be considered as wastelands. With the adoption of innovative reclamation management practices, these soils can support biodiversity, secure the livelihoods of resource-poor farmers, and generate revenue for the nation.

Soil Health Series: Volume 4 Soil Health and Sustainability in India, First Edition.
Edited by Ranjan Bhattacharyya.

Introduction

Globally about 1 billion hectares of land is estimated to be affected by salinity (Shaygan & Baumgartl, 2022) owing to natural and anthropogenic activities such as mining, improper agricultural activities, and inappropriate irrigation practices. Climate, soil type, irrigation water quality and/or management, groundwater level and salinity, and soil management practices influence salinity development. Arid and semi-arid environments, where the amount of rainfall is not sufficient to leach salts from surface soil, are conducive for salinity development.

Climate change and rapidly expanding populations in areas with salt-affected soils are major threats to sustainable agricultural production and ecosystem sustainability owing to the effect of salinity on crop growth. In India, about 6.73 million hectares (Mha) of land is salt affected, posing a serious concern for food production. Annual losses of 230 billion Indian Rupees (INR) or US$2,780,010,000 are expected as salt-affected areas may increase to 16.2 Mha by 2050 in India (Central Soil Salinity Research Institute [CSSRI], 2015). In India, the arable area is constant at 144 Mha, but to feed a steadily increasing population, many salt-affected areas need to be reclaimed. Salinity and sodicity both hamper successful crop production because of their negative effects on soil physical, chemical, and biological properties and processes (i.e., soil quality). Soil organic matter (SOM) turnover is also affected by both salinity and sodicity, thereby contributing significantly to greenhouse gas emissions.

Salt-affected soils have been shown to be able to act as a sink for soil organic carbon (SOC) sequestration (Datta et al., 2019b). Therefore, management practices that enhance SOC also improve the soil quality of salt-affected soils. Understanding SOC dynamics is essential for coping with the global warming effects (Setia et al., 2013). In this chapter, the area and the extent of salt-affected soils in India, different soil quality constraints, and various innovative management practices for enhancing SOC and soil quality of these soils are discussed.

Extent of Salt-Affected Soils in India

Worldwide, salt-affected soils occupy about 952 Mha and exhibit various degrees of deterioration (Table 11.1). About 23% of the world's cultivated lands (1.5×10^7 km^2) are reported to be saline, while saline and sodic soils cover about 10% of the total potentially arable land (Szabolcs, 1989). India is no different from others, having 6.73 Mha of salt-affected lands (Mandal et al., 2009; Table 11.2).

Table 11.1 Worldwide Salt-Affected Soils[a].

	Salinit	Alkalinity	Total
Continent		Mha	
North America	6.2	9.6	15.8
Mexico and Central America	1.9	–	1.9
South America	69.4	59.6	129.0
Europe	0	50.8	50.8
Africa	53.5	26.9	80.4
South Asia	83.3	1.8	85.1
North and Central Asia	91.6	120.1	211.7
Southeast Asia	19.9	—	19.9
Australia	17.4	340	357.4
Total	343.2	608.8	952.0

[a] *Source:* FAO/UNESCO (1974).

Table 11.2 Occurrence of Salt-Affected Soils in India by State[a].

	Saline	Sodic/alkali	Total
State		ha	
Andhra Pradesh	77,598	196,609	274,207
Andaman & Nicobar	77,000	0	77,000
Bihar	47,301	105,852	153,153
Gujarat	1,680,570	541,430	2,222,000
Haryana	49,157	183,399	232,556
Karnataka	1,893	148,136	150,029
Kerala	20,000	—	20,000
Madhya Pradesh	—	139,720	139,720
Maharashtra	184,089	422,670	606,759
Odisha	147,138	—	147,138
Punjab	—	151,717	151,717
Rajasthan	195,571	179,371	374,942
Tamil Nadu	13,231	354,784	368,015
Uttar Pradesh	21,989	1,346,971	1,368,960
West Bengal	441,272	—	441,272
Total	2,956,809	3,770,659	6,727,468

[a] *Source:* Mandal et al. (2009).

Types of Salt-Affected Soils

Salt–affected soils, based on their origin, nature, characteristics and plant growth relationships, are classified into three groups: saline soils, sodic/alkali soils, and saline–alkali soils.

Saline Soil

Soil with sufficient soluble salt to impair its productivity is known as a saline soil or *solonchak*. The salts are mainly chlorides and sulfates of sodium, calcium, and magnesium. In arid climatic regions, saline soils may also have excessive amounts of boron, fluoride, and/or nitrates. Sometimes, appreciable amounts of sparingly soluble gypsum are also observed in patches within saline areas.

The U.S. Salinity Laboratory Staff (1954) identified the distinguishing features of saline soils from an agricultural standpoint as those that contain sufficient neutral soluble salts to adversely affect the growth of most crop plants. They stated saline soils must have a saturated soil extract electrical conductivity (EC_e) exceeding 4.0 dS/m at 25 °C, pH_s of less than 8.5, and an exchangeable sodium percentage (ESP) of less than 15.0. However, Gupta and Abrol (1990) defined saline soils as those having pH_s values of less than 8.2, EC_e values exceeding 4.0 dS/m, and a preponderance of chlorides and sulfates of sodium, calcium, and magnesium. Soil salinity classes, based primarily on their saturated extract conductivity and effects on crop plants, are given in Table 11.3 (Jehangir et al., 2013).

Saline soils form through salinization, which is a natural process occurring through weathering of minerals or deposits of fossil salts. Low rainfall in arid and semi–arid climates provides insufficient water to leach the soluble weathered products, causing them to deposit as salts on the soil surface and thus cause.

Table 11.3 Soil Salinity Classes Based on Electrical Conductivity of the Saturated Extract (EC_e) and Their Effects on Crop Plants.[a]

Soil salinity class	EC_e (dS/m)	Effect on crop plants
Non saline	0–2	Salinity effects negligible
Slightly saline	2–4	Yield of sensitive crops may be restricted
Moderately saline	4–8	Yield of many crops are restricted
Strongly saline	8–16	Only tolerant crops yield satisfactorily
Very strongly saline	>16	Only a few very tolerant crops yield satisfactorily

[a] *Source:* Jehangir et al. (2013).

salinity hazards. Some human activities that accelerate the salinization process are excessive use of basic fertilizers, such as sodium nitrate or basic slag; removal of native vegetation; and growing of shallow–rooted annual crops. Excessive irrigation can also increase leaching of salts into the groundwater system. Use of saline groundwater for irrigation and a rise of the water table brings salt into the root zone and to the soil surface, where it accelerates both salt mobilization and accumulation.

In summary, owing to the many different mechanisms through which salinization can occur, salinity is often classified in varying ways based on the broad cause. These classifications include dryland salinity, irrigation–induced salinity, urban salinity, and industrial salinity.

Sodic Soils

Non–saline soil contains sufficient exchangeable sodium to adversely affect crop production and soil structure. Carbonates and bicarbonates of sodium are the dominant salts, and the concentration of neutral salts is very low. Soluble gypsum is nearly absent or is present in only small amounts.

The pH_s of sodic soils usually varies between 8.5 and 10.0, EC_e is less than 4.0 dS/m, and ESP and the sodium adsorption ratio (SAR) are more than 15 and 13, respectively. In India, alkali soils are distinguished as those having pH_s less than 8.2 and a ratio of $[Na^+]/([Cl^-]+[SO_4^{2-}])$ in soil solution of more than 1.0 (Bajwa & Swarup, 2002).

Sodic soils are generally found in low -lying areas with annual rainfall of 550–900 mm. This amount of precipitation is insufficient to leach the weathering products, and therefore, sodium becomes the dominant cation and increases SAR (>13). The sodium ions replace calcium and magnesium ions on clay surfaces, hence, the solubility of calcium carbonate, calcium sulfate, and magnesium carbonates limits and precipitates. As a result, ESR increases (>15).

$$Ca[Clay]+2Na^+ \leftrightharpoons [Clay]^{Na}_{Na}+Ca^{2+}$$

An increase in the pH is associated with hydrolysis of Na_2CO_3, resulting in high alkalinity of soils.

$$Na_2CO_3+2HOH \rightarrow 2Na^+ +2OH^- +H_2CO_3$$

The release of OH^- ions results in the maintenance of high soil pH_s (i.e., >8.5).

Historically, sodic soils were popularly known as black alkali soils because with high pH values, organic matter gets dissolved and deposited on the soil surface forming a thin film over the soil particles. This happens when water is transported to the soil surface through capillary action and then evaporates. These soils are mainly found in the arid and semi-arid regions of Punjab, Haryana, Uttar Pradesh, Bihar, and Rajasthan.

Saline–Sodic Soil

Saline–sodic soil is defined as a soil having a saturated extract conductivity greater than 4.0 dS/m and an ESP greater than 15. The pH_s is variable but usually above 8.5 depending on the relative amounts of exchangeable sodium and soluble salts. These soils form because of the combined processes of salinization and alkalization. If the excess soluble salts within these soils are leached downward, the soil properties may change markedly, causing the soil to become similar to a sodic soil.

Soil Quality Constraints of Salt-Affected Soils

Saline Soils

Increasing greenhouse gas emissions raise the earth's projected surface temperature, warm the seawater temperatures, and shift rainfall patterns. Pedogenic weathering, increased soluble electrolyte concentrations, eolian process, and intrusion of and/or inundation with seawater paired with its evaporation can increase salt concentrations in soils (Gupta & Mathur, 2011). Shortage of rainfall, increased evaporation demand, and higher summer temperature cause the inflow of secondary salt that accumulates in root zone. Excessive and poor irrigation practices can increase underground water intrusion and expedite salinity development (Cai et al., 2010). The Indo–Gangetic Plain in India and Pakistan (Basak et al., 2022; Fishman, 2018; Hayat et al., 2020), Australia (Ranatunga et al., 2010), China (Xu et al., 2013), Iran (Maleki et al., 2022), Egypt (Abdel–Fattah et al., 2020), Argentina (Sione et al., 2017), and the United States (Hansen et al., 2018) exhibit this condition because they have been supporting an extensive irrigation network to increase settlement, food productivity, and the livelihood of people living there.

The increased abundance of Na^+, Ca^{2+}, and Mg^{2+} salts available in irrigation water when applied in root zone causes soil salinization. For coastal ecosystems, the increasing volume of sea–water associated with temperature rise, glacial melting, and ingression of surface and subsurface brackish or brine water are the primary threats for increased salinization (Dasgupta et al., 2015; Mandal et al., 2022). Wider cation ratios (i.e., Na^+: Ca^{2+}, Mg^{2+}:Ca^{2+}, or Cl^-:HCO_3^-) due to inundated sea–water ingressing into tube–wells can degrade the quality of soils in coastal and adjacent areas. The influx of saline seawater shows the risk of groundwater contamination by saline water (Sappa et al., 2019). Reclamation and rehabilitation of contaminated aquifers serving as the source of groundwater is a recurring problem in coastal ecosystems. Inundation and submergence of low-lying coastal lines with brackish water are often accelerated by intensive storms and/or surges. Areas with repeated seawater intrusion have gradually transitioned to shellfish or shrimp farming instead of paddy rice agriculture (Blankespoor et al., 2017).

Salinity leads to the presence of undesirable electrolytes in soil water solution during the plant growing season. The resulting deficiency or toxicity of essential elements creates an unfavorable environment for the survival of microorganisms because of the desiccating action of salts. The result is a shift in microbial community leading to reduced substrate utilization efficiency (Zhang et al., 2023).

SOC accumulation and storage in salt-affected soil varies from low to medium in range, while inorganic C effects are even more variable (Datta et al., 2019b). Salt-affected soils have low vegetation growth, which minimizes organic input (Bhardwaj et al., 2019). The dynamic nature of salinity stress also affects SOM decomposition. As salinity increases, moisture stress also increases as rising temperatures cause dissolved organic carbon to decline (Liu et al., 2017). Since each increment of SOC in salt-affected soils prefer to bond with Ca^{2+} rather than Na^+, saline soil carries a disproportionately larger quantity of Cl^- and SO_4^{2-} than CO_3^{2-} and HCO_3^-. The physical health of saline soil is generally not impeded, but chemical equilibria are largely affected (Rai et al., 2021). In waterlogged soils, saline conditions reduce nitrification, causing more NH_4^+ to accumulate and consequently be lost through volatilization (Bandyopadhyay & Sen, 1992). Soil phosphorus dynamics are affected in salt-affected soil due to changes of phosphorus sorbent sites owing to higher amounts of Na^+, Ca^{2+}, Mg^{2+}, and SO_4^{2-} in saline soil (Meena et al., 2016).

Addition of organic mulch can reduce the fixation of water-soluble phosphorus owing to presence of ligand substances and facilitated mineralization of organic phosphorus. Therefore, adoption of management practices that provide easily decomposable organic substrate (e.g., applying mature compost or implementing reduced and/or zero tillage) favors nitrogen mineralization rates owing to the intensification of soil microbial activity as well as the growth of aboveground and belowground biomass (Mitran et al., 2021; Soni et al., 2021).

Sodic Soils

Constraints of Soil Physical Properties

Soil physical attributes play a major role in supporting optimum root growth with both positive and negative changes requiring time to be detected (Andrews et al., 2004). Soil bulk density has often been identified as an indicator of soil quality (Imaz et al., 2010) because it influences most soil state and rate variables, and its measurement is both easy and reproducible. Soil bulk density has specific importance in intensively cultivated agro–ecosystems. Soil structural stability fails upon sodification because net negative charges on clay particle exchange sites repulse aggregation. This causes aggregates to slake and disperse with subsequent migration of clay minerals and clogging of water–conducting soil pores. This phenomenon reduces pore size and ultimately causes poor water storage in sodic soils (Ahmad et al., 2012; Chaganti & Crohn, 2015; Levy & Shainberg, 2005).

The poor infiltration capacity of sodic soil makes them prone to waterlogging, impedes air and water entry, results in anoxia, and deflocculates soil particles resulting in a structural degradation. The "hard-settled" or dispersed soils impede tillage operations and often have hard crusts on the surface layer that hinder seedling emergence and result in poor plant establishment (Dagar et al., 2022; Mandal et al., 2008). The unfavorable soil conditions for crop establishment limit the economic return to producers. The failure in hydraulic conductivity as sodicity increases results in swelling that leads to restricted pores in Vertisols (Ali et al., 2022; Bhardwaj et al., 2007). Clay dispersion and migration can also cause pore blocking in coarser soils (Shainberg & Letey, 1984), causing crops to suffer most from waterlogging problems. Surface crusting after rain or the application of irrigation water can cause this situation to worsen because of the beating action of raindrops (Bennett et al., 2019). Soil dispersion and aggregate failure caused by irrigation with sodic water reduces infiltration, decreases post–infiltration soil profile water storage, and results in the need for longer times to redistribute profile water because of poor drainage (Quirk et al., 1986). Clay mineralogy and soil texture control inherent infiltration capacities and modify factors including the sodicity level of the soil, its electrolytes release patterns, and the quantity and quality of SOM (Minhas, 2010; Suarez, 2012). Decreased aggregate stability in sodic soils reduces protection to the SOM and can result in further soil degradation. Intensive rice–wheat cropping systems on heavy–textured soils are more prone to sodification and structural failure when irrigated with sodic waters (Minhas et al., 2019). The hardness of water, alkalinity, SAR (>10), appearance of residual CO_3^{2-} and HCO_3^-, and the presence of high Mg^{2+} to Ca^{2+} ratios, also degrade soil physical health (Balpande et al., 1996; Qadir et al., 2018; Vaidya & Pal, 2002).

Constraints of Soil Chemical Properties and Plant Nutrients

Numerous multi–phase equilibria concurrently persist in salt–affected soil because of soluble electrolytes and exchangeable bases. Plant roots in sodic soil encounter dilute to concentrated electrolyte concentrations that may possess quasi–stable or more complex salts. This includes the toxic appearance of CO_3^{2-}, HCO_3^-, and Na^+; unfavorable moisture regimes; decreased ability to infiltrate precipitation; increased runoff events; and temperature changes.

Cations, such as Na^+, Ca^{2+}, Mg^{2+}, and K^+; anions, such as Cl^-, SO_4^{2-}, CO_3^{2-}, HCO_3^- and SiO_x^{-n}; and minor amounts of other electrolytes [e.g., $(H)_xPO_4^{-n}$ and NO_3^-] are generally present in sodic soils. Several chemical reactions, such as ion exchange and organo–macromolecules, are facilitated by the presence of electrolytes (Rai et al., 2020). Other auxiliary reactions such as complexation, precipitation–dissolution, occlusion by soil particles, formation of ion pairs, speciation(s), and solubility in soil water solution ultimately distress nutrient availability, uptake, and root development (Rengasamy, 2016) and are critical issues. These reactions are regulated by the cations and anions present in soils, as well as the supply of sodic/high SAR irrigation water if any, and the application of sodicity–neutralizing chemical amendments for rehabilitation (Sheoran et al., 2021; Sundha et al., 2020). The salt–affected soils show higher values of Gapon's selectivity coefficient (K_G) with greater Na^+ content and poor organic carbon storage (Basak et al., 2015, 2021; Bhardwaj et al., 2019). However, the external supply of Ca^{2+} or organic amendments mobilizes the native $CaCO_3$ dissolution and causes declines in Gapon's selectivity coefficient (K_G) (Sundha et al., 2020).

Ionic imbalance and poor nutrient availability hinder plant growth, slow microbial activity, and impede soil structure in salt–affected soils (Bhardwaj et al., 2019; Datta et al., 2019b). Decomposition of simple organic materials is strongly affected by low microbiological activity in high alkaline soil (Dendooven et al., 2010). The solubility of SOM at alkaline pH may lead to greater N mineralization. Nitrification rather than ammonification is affected by high pH. Mobility and erosion of SOM, wasteful transformation of supplied nitrogen, and low symbiotic nitrogen fixation all contribute to the deficiency of nitrogen in sodic soils (Sundha et al., 2017). Sodic soil also remains dispersed upon wetting and vulnerable for decomposition of soil microbial biomass. A wasteful and slow transformation of N occurs within sodic soil, and volatilization losses of NH_4^+ are significant (Bharadwaj & Abrol, 1978). Symbiotic nitrogen fixation is also slow in alkaline soil. Phosphate remains soluble because of alkaline hydrolysis of $Na_2CO_3/NaHCO_3$ or Na–clay. The native insoluble calcium phosphate is converted to soluble sodium phosphate with excess appearance of HCO_3^{2-} and CO_3^{2-} (Sundha et al., 2022, 2023a, 2023b). The occurrence of micaceous minerals in sodic soils of alluvial plain fulfill the K demand of crop even though the high ESP reduces the K uptake and imbalances Na^+/K^+ ratio in plant cell (Chhabra, 1985). In sodic soil, soluble calcium precipitates as insoluble $CaCO_3$ (varying from trace to 40%).

Soils with high exchangeable magnesium are found scattered throughout several areas with salt–affected soils and groundwater-irrigated basins owing to the high presence of hydrous silicates of magnesium and very high magnesium to calcium ratios, serpentine clay minerals, and magnesium-rich water (Qadir et al., 2018). The presence of soluble cations (Na^+, Ca^{2+}, Mg^{2+}) and SO_4^{2-} in a sodic environment targets phosphorus adsorption sites and affects soil phosphorus dynamics

(Meena et al., 2018; Sundha et al., 2022). The total iron concentration is not limited in sodic soils, but plant available iron concentrations are limited due to high pH, sorption onto $CaCO_3$, soil oxidation status, and low SOM. Zinc availability in alkali soils is largely affected by the presence of free HCO_3^-, CO_3^{2-}, $CaCO_3$, high soluble phosphorus, and low organic matter. The solubility of zinc is drastically reduced by immediate formation of less soluble $Zn(OH)_2$, $ZnCO_3$, and the retention of zinc on surfaces of freshly precipitated $CaCO_3$ (Swarup, 1985). The presence of soluble Na^+ and formation of highly soluble $NaBO_2$ and Na_2MoO_4 at alkaline pH (Gupta & Chandra, 1972) increases availability of boron and molybdenum (Pasricha & Randhawa, 1971). Boron availability sometime reaches a toxic level owing to formation of highly soluble sodium metaborate as pH_s and ESP increase. In saline–sodic soil, both the salt and alkalinity increase H_3BO_3 dissociation. Like boron, the solubility of molybdenum increases incrementally as soil pH increases. Neutralization of alkalinity, increased biomass production, and its return to the soil are favorable strategies for increasing SOC and reclaiming these soils.

Constraints of Soil Biological Properties

Increasing sodicity causes Na^+ salts with HCO_3^-, CO_3^{2-}, Cl^-, or SO_4^{2-} to increase, affecting crop, grass, and tree growth and development (Wong et al., 2008, 2010) as well as mineralization of SOM (Rengasamy, 2006, 2008; Wong et al., 2008). Under high sodic conditions, microbes require greater metabolic energy with the presence of concentrated electrolyte concentrations. At extremely alkaline pH, nutrient assimilation by microbes is hindered, and consequently, microbial cells lyse and soil microbial activities slow. The microbial population remains stressed under a sodic environment, resulting in a considerable loss of SOC through respiration rather than being incorporated into cells ultimately developing humus.

Sodicity is detrimental to microorganisms and shows a low microbial quotient. It is uncertain whether SOM mineralization is hampered in the presence of electrolytes and high pH. The structural instability in a sodic environment makes the organic matter more accessible for microbial action, resulting in greater carbon mineralization (CO_2 emission), and also increases microbial decomposition owing to the lack of soil moisture and presence of electrolytes (Mitran et al., 2021). However, the instability may also decrease mineralization owing to anaerobic conditions. Both the ESP and the SAR_e affect the physical and chemical environment of soil because of modified anionic movement. This is reflected in the significant negative relationships between microbial biomass carbon and the soil microbial community (Rietz & Haynes, 2003; Singh et al., 2012).

Activities of extracellular soil enzymes are regulated by soil pH (Kaur et al., 2008). In soil with a high ESP, clays disperse, and the enzymes become vulnerable and more prone to denaturation by proteolysis (Pathak & Rao, 1998). An increase in soil sodicity linearly inhibits β–glucosidase, alkaline phosphatase, and arylsulfatase activities (Rietz & Haynes, 2003). Therefore, assay values of soil enzymes have been postulated to be potential indicators of effective land use management and severity of land degradation (Singh et al., 2012; Soni et al., 2021).

The soil microbial biomass and population generally decrease with increasing soil salinity and sodicity, thus limiting efficient nitrogen cycling. In general, soil carbon, nitrogen, available phosphorus, salinity, and soil reactions play critical roles in shaping the soil microbial community structure. Increased soil salinization, or soil salt content, decreases soil fertility, thus reducing soil bacteria and fungi abundance, richness, and diversity (Zong et al., 2022).

Soil Organic Carbon and Soil Quality in Salt-Affected Soils

SOC performs a central role in soil functioning and is a principal indicator of it, and there is a need for a better understanding of the factors influencing the behavior and persistence of carbon in soils (Kopittke et al., 2022). It is even more important for the salt-affected soils, which are often very low in SOC owing to lower carbon input from plants (Basak et al., 2023). Low SOC in salt-affected soils results in poor soil quality in terms of physical, chemical, and biological soil health. Therefore, any management practices that enhance SOC can improve soil quality of these marginal lands. SOC sequestration is a key sustainable development strategy with a likely strong potential to increase the resistance and the resilience of societies toward future crises, in addition to being beneficial for climate change mitigation and adaptation and food security (Rumpel et al., 2022). Supplying soluble Ca^{2+} through gypsum and mobilization of native $CaCO_3$ checked soil sodification when alkali water was used for irrigation (5–7 me/L) and improved SOC storage compared with the unamended control (Figure 11.1) (Basak et al., 2021).

Improving Soil Quality of Salt-Affected Soils

Improving Soil Quality of Sodic Soils: Impact of Inorganic and Organic Reclamation Technology

Inorganic Amendments

Calcium-based amendments (gypsum, phosphogypsum) and calcium mobilizers (elemental sulfur, sulfuric acids, pyrites, fly ash, aluminum chloride, agricultural industry by-product, press mud and acid-formers, and bio-augmented materials) are recommended for soil sodicity reclamation, a corrective measure based on the application of specific amounts of reclaiming agents, provided that irrigation water quality is safe for crop production (CSSRI, 2006; Rai et al., 2021; Sundha et al., 2023a, 2023b; Rai et al., 2024). Rice-based cropping systems usually require

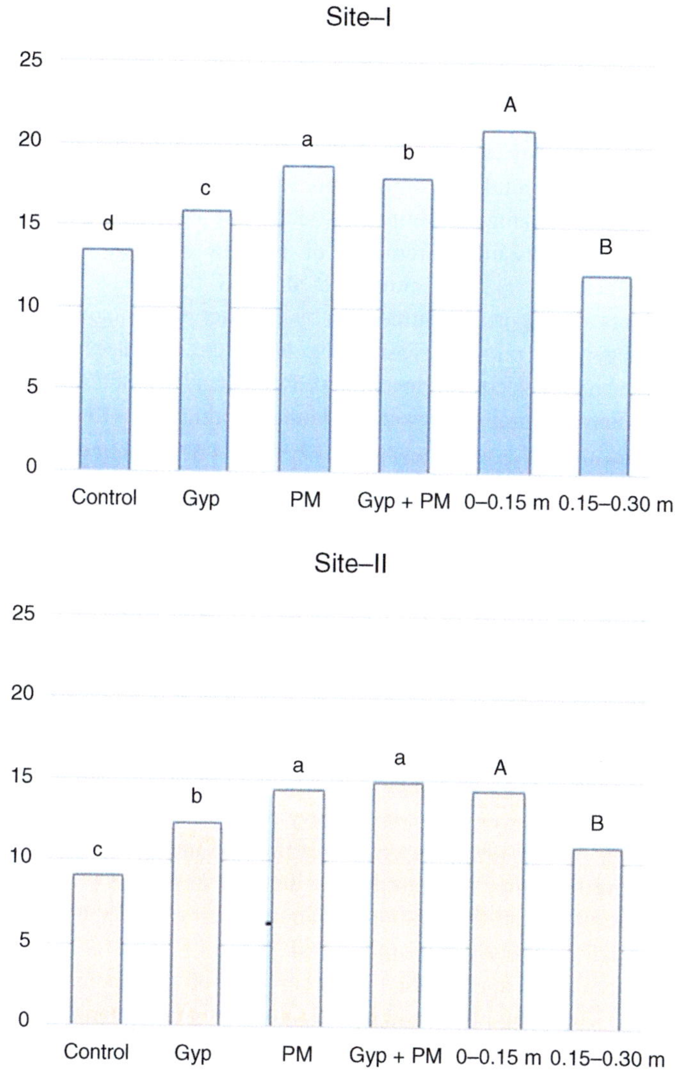

Figure 11.1 The improvemet in soil organic carbon storage on adoption of chemical and intrgratedamendment strategies along depth (values followed by different uppercase letters in different treatment soil depths differ significantly by DMRT ($P \leqslant 0.05$); thseseinteraction between different treatments and soil depth significance at $P \leqslant 0.01$, respectively (adopted from Basak et al., 2021).

3–4 years after the adoption of reclamation technology to reach their full productivity potential because carbonates and bicarbonates must be reclaimed and large amounts of Na^+ must be removed from sodic soils. During reclamation, an exchange reaction occurs [Na^+–Soil–$Na^+ + CaSO_4 \rightarrow$ Soil–$Ca^{2+} + Na_2SO_4$]

(Abrol & Bhumbla, 1979; Zhang et al., 2021). When incipient sodicity is a consequence of water-borne alkalinity, then recurring applications of amendments are recommended to overcome water-borne sodicity (residual alkalinity), maintain soil quality, and sustain crop yield (Murtaza et al., 2017; Sheoran et al., 2021). Organic amendments, such as city waste compost, sewage sludge, calcium/sulfur-enriched compost, and press mud in conjunction with gypsum application, improve grain yield by neutralizing alkalinity and decreasing ESP, which slows precipitation of Ca^{2+} and accelerates removal of Na^{+} in drainage waters (Choudhary et al., 2011; Grigg et al., 2006; Sundha et al., 2018, 2020).

Phosphoric acid plants are important sources of a by-product phosphogypsum. Compared with mined gypsum, phosphogypsum dissolves faster, produces acidic reactions and generates higher electrolyte leaching solutions, and promotes particle aggregation and therefore improves soil hydraulic conductivity (Keren & Shainberg, 1981). Application of phosphogypsum at 10 Mg/ha resulted in greater yields of both rice and wheat than other treatments (Nayak et al., 2013). Application of phosphogypsum is also desirable to reduce the deleterious effect of magnesium-rich water (brackish water) and was found to be effective in additional reduction of the exchangeable magnesium per-cent in a cotton-based cropping system in the semi-arid Kazakhstan (Karimov et al., 2009). However, the potential presence of heavy metals in phosphogypsum may reduce its applicability for sodicity reclamation (Mishra et al., 2019). Combined use of fly ash and organic amendments, such as farmyard manure, press mud, paper factory sludge, sewage sludge, crop residues, and organic compost, can also improve degraded/marginal soil (Lokeshappa & Dikshit, 2011). The beneficial effects of combined use of fly ash and organic manures can reduce heavy metal availability and kill pathogens in the sludge (Wong, 1995); improve soils through increased nutrient concentrations, improved texture, decreased bulk density, improved porosity, increased moisture retention, and fine-grained minerals (Shen et al., 2008); and enhance biological activity in the soil (Kumpiene et al., 2007).

Elemental sulfur can also be applied for the reclamation of calcareous sodic soils (Ganjegunte et al., 2018). In-house and participatory research programs led by CSSRI and other national laboratories have claimed improvements in soil quality through reclamation (Sharma et al., 2019). The sodic land of the western Indo-Gangetic Plain maintained an optimum grain yield for rice-wheat cropping system by applying high doses (up to 150%) of chemical fertilizer and organic amendments. There was an appreciable reduction of $pH_{1:2}$ and $EC_{1:2}$ and improvements in SOC, Olsen's P, NH_4OAc-extractable K, and DTPA-extractable micronutrients by adopting the aforementioned practice (Swarup & Yaduvanshi, 2000). Decreases in pH, EC, ESP, and SAR_e, as well as improvements in SOC storage, porosity, void ratio, and saturated hydraulic conductivity, were documented in sodic soil at 10 locations in the central Indo-Gangetic Plain upon adoption of green manuring in participatory rice–wheat cropping system experiments (Singh et al., 2014). A balanced amount of flue-gas desulfurization gypsum and leaching water also improved pore structure and increased hydraulic conductivity and infiltration, while neutralizing soil sodicity (Yu et al., 2014; Zhao et al., 2021; Sundha et al., 2023b).

Organic Amendments

Press mud is a solid waste by-product of sugar mills, accounting for about 3% of the total quantity of cane crushed. Press mud is a good source of organic carbon, nitrogen, phosphorus, potassium, and micronutrients. Sulfidation of press mud successfully decreased EC and SAR in calcareous saline–sodic soils (Yaduvanshi & Yadav, 1990). Combined use of gypsum (5 Mg/ha) and a mixture of farmyard manure (FYM) and press mud (at 10 Mg/ha) was effective for sustaining productivity in areas having sodic groundwater for irrigation (Yaduvanshi & Swarup, 2005).

A recent large-scale field demonstration concluded that applying press mud as a soil sodicity ameliorant is a safer disposal vis-à-vis a potential alternative to gypsum in the long run. These approaches are also applicable to larger areas undergoing soil sodification induced by high alkali water, a process further mitigating soil and environment degradation in salt-affected soil of India and similar agro-ecosystems (Sheoran et al., 2021, 2022).

The application of farmyard manure and *Sesbania* green manure at 20 Mg/ha plus wheat straw at 6 Mg/ha solubilized calcium from inherent and precipitated $CaCO_3$ in calcareous soils and reduced the dependency on gypsum in sodic water irrigation for achieving sustainable yields (Choudhary et al., 2011). Bio-methanated spentwash is another potential amendment that contains a considerable amount of organic carbon as well as calcium and potassium, which may prove to be beneficial for reclamation and for increasing the fertility of sodic soils. Several laboratory and field trials are being carried out to examine the effect of phosphogypsum and distillery spent wash on salt accumulation and mobility in sodic soil reclamation (Khandkar et al., 2017). Application of organic amendments, such as crushed cotton gintrash (60 Mg/ha), cattle manure (60 Mg/ha), and composted chicken manure (18 Mg/ha) in sodic soils, significantly decreased ESP (by about 50%), compared with the unamended soil (Ghosh et al., 2010; Tejada et al., 2006). Biochar prepared from poultry manure can also manage sodicity by reducing soil pH and EC and increasing soil Ca^{2+}, Mg^{2+}, and available phosphorus (Sappor et al., 2017).

The concept of environmental management is based on the recycling, reuse, and composting of waste organic sources for reclamation and rehabilitation of degraded soils. Gypsum and municipal city waste compost-amended saline–sodic soil, when leached with different water quality, revealed that reclamation efficiency increased with application of 20 Mg/ha of city waste amendments in conjunction with 25% of the gypsum requirement (GR25). Likewise, the percentage of cation mass for calcium was greater in GR25 + city waste compost than other inorganic–organic combinations, when low SAR water was used for leaching. Similarly, higher doses of organic amendments also showed significantly greater leaching of calcium when high SAR water was used for leaching (Sundha et al., 2018). Acidified rice straw and dicer wood chip biochars can be used to provide larger quantities of Ca^{2+} and Mg^{2+} to facilitate exchange reactions and promote leaching of Na^{+} from the surface of the soil colloids and deeper with the soil profile (Sadegh-Zadeh et al., 2018).

Improving Quality of Saline Soil

Minimum tillage, deficit– or need–based irrigation, surface mulching, cultivation of crops with low water demand, and adoption of salt–tolerant varieties for water conservation and reduced salt load in applied irrigation measures are approaches for productive use of saline soils (Rai et al., 2022; Singh, 2009). Frequent drought and water scarcity can be overcome with the application of deficit saline irrigation (Jiang et al., 2012; Nagaz et al., 2012). Soil moisture evaporation loss was reduced, crop productivity was increased, and soil quality was restored by applying paddy straw mulch in a saline soil (Purakayastha et al., 2019; Soni et al., 2021). Those practices permit the water transmission and leaching of salt (Grigg et al., 2006; Yuan et al., 2019; Zhang et al., 2021) and thus provide a nature–based solution in salt–affected areas (Soni et al., 2021; Visser et al., 2019). Surface soil salinization may be checked to some extent by mulching because the evaporation rate and capillary rise of underground saline water are reduced.

Conservation agriculture with no till or minimum till is another potential choice particularly for acid sulfate coastal soils, because those practices prevent exposure of potential acid sulfate soil to air. Soil fertility improvement, increasing SOC, and restoring soil structure are the benefits of practicing zero tillage (Mandal et al., 2022; Wang et al., 2014).

Soil microbial biomass carbon, nitrogen, and basal soil respiration are adversely affected by salinity, while metabolic quotient is positively affected by salinity (Egamberdieva et al., 2011; Iwai et al., 2012; Mahajan et al., 2016; Tripathi et al., 2006). Saline irrigation has been shown to increase secondary soil salinity, reduce microbial biomass carbon (Meena et al., 2016), and alter microbial community structure (Tripathi et al., 2006). Organic matter transformation is then impeded by salinity because of low microbial biomass and slow microbial metabolism (Iwai et al., 2012; Mamilov et al., 2004). The toxic concentration of Na^{+} slows the growth of soil microorganisms and reduces microbial biomass carbon content rather than the carbon input (Egamberdieva et al., 2011; Mavi & Marschner, 2013; Rietz & Haynes, 2003).

In general, greater soil enzyme activity was noted with low electrolyte concentration and *vice versa* (Mitran et al., 2021). Leaching with seasonal rain, river/canal water, or best available underground water is the option to leach excess electrolytes, which is accomplished by ponding water in a properly leveled field. The amount of salt leaching depends on the quantity and quality of irrigation water and soil texture. Surface flushing with irrigation is suggested to leach surface-deposited salts (Nayak et al., 2008). This practice is recommended for soil with slow permeability that is prone to crust formation.

Scraping the soil surface can also be used to reduce the salt load, but scraping is a recurring practice for salt removal and maintaining crop production in salinity-affected landscapes. For fruitful utilization of saline landscapes and saline water, the saline soil needs to be leveled, and good agronomic practices, such as minimum tillage, crop residue mulching, need-based application of saline irrigation in a mixing/cycling mode, irrigation with best available water at crop sensitive stages (germination and seedling arising stages), pre-sowing irrigation for *kharif* (rainy season) crop, and adaptation of pressurize drip/sprinkler irrigation that facilitate the washing of root zone salinity, are recommended to sustain crop production.

The application of poultry or city waste compost to a saline soil can significantly improve the soil biological and chemical environment by increasing the cation exchange capacity and soluble and exchangeable potassium, limiting the entry of Na^+ into the exchange complex. These approaches reduce the harmful effects of soluble salts and soluble sodium (SAR; Meena et al., 2016). Plant transpiration is also a useful technology for managing excess soil water and shallow water tables. Slow drainage problems can be managed by implementing bio-drainage (Dagar et al., 2016).

Tillage

Soil physical and hydraulic properties, such as bulk density, porosity, and water holding capacity, are influenced by tillage (Strudley et al., 2008), which also facilitates leaching of salts in sodic soils prone to surface crusting and clogging of pores (Sadiq et al., 2007). Improvement in soil structure and fertility of sodic soils with tillage were also reported by Qadir et al. (2001) and Wong et al. (2010). Tillage favors the mineralization of SOC in sodic soils, thereby modifying distribution of organic carbon in the soil profile (Wong et al., 2010). Medium- to long-term tillage (5 and 9 years of tillage) enhanced SOC by modulating crop residues distribution (Singh et al., 2016) in sodic soils. Dalal (1989) also reported that SOM and microbial activity at surface layers and movement of salt to lower soil depths were influenced by tillage and crop residue management in a Vertisol. Singh et al. (2016) showed that SOC was increased by 35% and 39% at surface and subsurface soil depth, respectively, following 9 years of continuous tillage compared with untilled sodic lands. Fodder sorghum yield was maintained with reduced tillage when the wheat crop was taken with mulching and irrigation with saline water (EC_{iw} 8.0 dS/m) at a deficit of 60% of the crop water

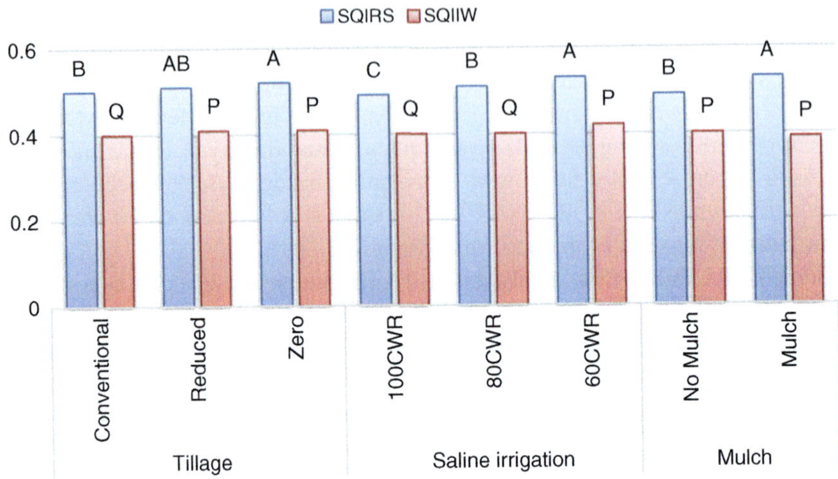

Figure 11.2 Improvement in soil quality index (SQI; SQIRS and SQIIW: after sorghum and wheat) on adoption oftillage, mulch and deficit saline irrigation (different uppercase letters (A to C; and P and Q) denote significant differences ($P < 0.05$, Tukey's HSD test); CWR: wheat water requirement.

requirement. This integrated practice is also suitable for improving Olsen's P, NH_4OAc–K, and seasonal soil quality indices, reducing the salinity stress over conventional tillage under both crops (Figure 11.2; Rai et al., 2022; Soni et al., 2021).

Irrigation

The amelioration strategies for saline and sodic soils can have significant impacts on SOC concentrations. For example, Chandel et al. (2021b) reported that organic carbon and total soil carbon were 15.2% and 22.4% higher with alternate application of saline and freshwater irrigation compared with continuous application of saline or fresh water. Less labile and non-labile carbon pools were increased under four seed spice crops. Chandel et al. (2021a) identified EC_e, SAR, potassium, and pH_s as the important indicators upon alternate application of fresh and saline water irrigation and better explained soil salinity under seed spices. They recommended cyclic use of saline and fresh water in growing seed spice crops to prevent soil degradation.

Alternate application of saline and fresh water also increased both the active and passive pools of carbon. SOC dynamics in salt-affected soils treated with wastewater irrigation were studied by Ganjegunte et al. (2018). They found SOC concentrations at 0- to 60-cm soil depth were increased by on an average of 3–5 Mg C/ha for different treatments after 6 years. Setia et al. (2014) showed that SOC decomposition in saline soil was enhanced more by Na^+ than K^+ ions. Divalent cations, such as Ca^{2+}, play an important role in binding organic matter as well as C stabilization (Rowley et al., 2018). Partial neutralization of alkali water (residual sodium carbonate from ~10 to ~5 me/L) with gypsum or sulfurous amendments was not sufficient for sustaining long-term soil quality (Singh et al., 2022). In Rann of Kutch, Gujarat, India, higher SOC in water was also reported by Datta et al. (2020).

Organic Amendments

Application of compost decreased the proportion of Na^+ in the exchange complex by its Ca^{2+} ions and facilitated the leaching of exchangeable Na^+, which led to flocculation of clay minerals, improved soil structure with the formation of a large number of soil aggregates, and thereby controlled erosion in saline soils (Oo et al., 2013; Qadir & Oster, 2004). Combined application of different organic amendments decreased EC by 87% and ESP by 71% and increased SOC by 96% in a coastal saline soil (Wang et al., 2014). However, types of organic matter applied also influenced carbon and nitrogen mineralization patterns in saline soil (Walpola & Arunakumara, 2010). Municipal solid waste can be used as an organic amendment for restoration of fertility of saline soils. Sole or combined application of municipal solid waste compost, gypsum-enriched compost, and rice straw compost with mineral fertilizers as reclamation agents and their effect on chemical and biological properties of saline soil have been studied in a mustard–pearl millet cropping system (Meena et al., 2016). Significantly higher SOC was observed under integrated use of 25% recommended dose of fertilizer along with rice straw compost (at 3.5 Mg/ha) + gypsum enrich compost (at 3.5 Mg/ha) + municipal city waste compost (at 4 Mg/ha) over a control plot after harvesting of mustard and pearl millet (Meena et al., 2016). Rice–sunflower cropping performed better than rice–chili/tomato and integrated use of mineral fertilizer with organic amendments ([green leaf and farmyard] manure + vermicompost) improved soil quality in coastal saline soils (Mitran et al., 2021). Similar management practice (NPK and farmyard manure/vermicompost/*Gliricidia* and *Eupatorium*/water hyacinth/*Sesbania*) also maintained sustainable yield index in lowland rice in west coast of India (Mahajan et al., 2021). Therefore, organic material application in soil can be an effective low-input alternative to offset the detrimental effects of salinization and improving carbon storage and soil quality of saline soils (Mandal et al., 2022).

Alternate Land Use Systems in Salt-Affected Soils

Carbon sequestration in soil is significantly influenced by land use change. Cultivation of appropriate crops, shrubs, and trees has potential to increase soil carbon through biomass production in sodic and saline-sodic soils (Garg, 1998; Kaur et al., 2002). Higher SOC was observed in soils under shisham

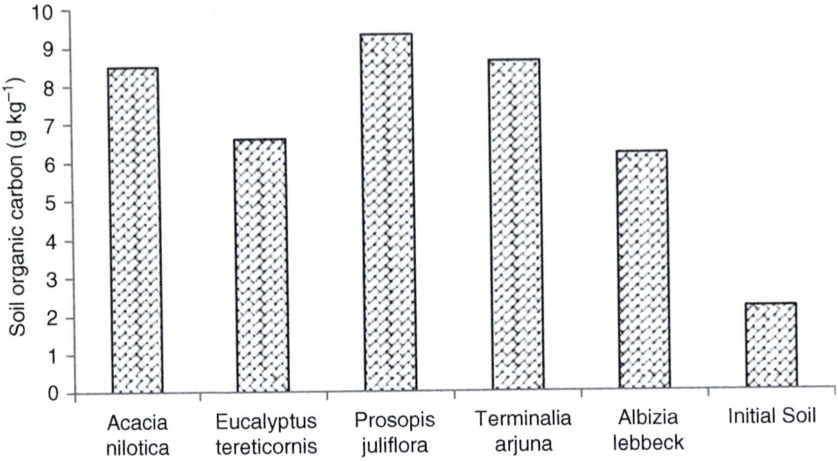

Figure 11.3 Ameliorative effect of 20-year-old tree plantations on soil organic carbon enrichment at 0–15 cm depth of an alkali soil (Singh and Gill, 1990).

(*Dalbergia sissoo* Roxb. ex DC.) and mesquite [*Prosopis juliflora* (Sw.) DC.] compared with others owing to increased activity of microorganisms in the upper soil depth facilitating decomposition of leaf litter and root decay, leading to accumulation of humus (Garg, 1998). Singh and Gill (1990) found maximum carbon sequestration in soil with *P. juliflora* (9.3 g/kg) after 20 years of planting in an alkali soil, and the carbon sequestration rate was 0.2–0.8 Mg C/ha/yr (Figure 11.3).

Datta et al. (2015) observed maximum SOC storage (133 Mg C/ha) including passive pool carbon (76 Mg C/ha) in the soils of guava orchard compared to mango, jamun, litchi, eucalyptus, and prosopis. Datta et al. (2019a) showed that precise measurements of salinity and sodicity parameters are essential for formulating management practices under different land uses. Adoption of these management practices not only rehabilitate the salt-affected soils, but also sequester atmospheric CO_2.

Conclusions and the Way Forward

Spatio-temporal variability of salt-affected soil is the principal constraint for formulating proper management practices. Therefore, frequent and detailed or narrow-scale and in-depth sampling is necessary for the correct diagnosis of salinity, sodicity, or both. Salinity, sodicity, or both may be present in the surface or subsurface soil or in soils at a lower depth. Sometimes, the problem may arise because of long-time continuous irrigation with saline or sodic water. The dynamics of electrolytes in the root zone varies with seasonal temperature, precipitation, evapotranspiration, vegetation cover, and supply of external irrigation and regional drainage. The greater concentration of electrolytes hampers biogeochemical cycling of essential nutrients and promotes the occurrence of an oligohaline zone. Changes in climate, rising atmospheric temperature, and shifting rainfall events or increments in evapotranspiration demands for crops increase the chances of salinization and aggravate the sodicity problem. Therefore, frequent monitoring of soil and water bodies is imperative for real-time assessment of problems. In such cases, analytical values of principal indicators may be representative of a large amount of soil or water data. The principal indicators must reflect soil processes, be cross–functioning, provide early warning of salinity, and be stakeholder friendly. These attempts need to address for different benchmark salt–affected soil of the country.

Because of the multi–dimensional problems in salt–affected soils, crops growing in these soils experience physiological stress(s). Therefore, plan–led indicators may be complementary to certain salt–affected soil for capturing the soil/water–borne constraints. The problem of salinity can be tackled by adopting good agronomic practices, using efficient/judicious irrigation, and cultivating salt–tolerant cultivars. The problem of sodicity may be reclaimed on application of the calcium–based amendments or mobilizing native $CaCO_3$ to neutralize alkalinity by applying a mineral acid or acid–forming agent or through afforestation of sodic land by sodicity–tolerant trees. The limited availability, slow rate of reclamation, and associated environmental hazards limit the application of certain conventional amendments (phosphogypsum, fly ash, pyrites, etc.).

Recent advances in formulation science have led to the development of materials such as polymers, nanomaterials, acidified biochars, elemental sulfur, sulfuric acid generator, and flue gas desulfurization gypsum having the potential to replace gypsum as an amendment. Future application of these alternatives for sustainable and profitable use in sodic soils will not be practicable unless efficacy and efficiency in sodic soil reclamation is established. Reclamation of salt–affected soils make them suitable for plant growth, leading to higher productivity. This causes not only higher aboveground biomass but also a considerable amount of belowground biomass. As a result, SOC content is enhanced, which subsequently improves other soil properties and thus leads to overall soil quality improvement in these marginal lands.

We suggest that the following issues need further assessment:

- Assessment of soil quality indices for different types of salt–affected soil and their relationships with different management practices.
- The assessment of how rehabilitation processes affect carbon cycling and carbon stocks, and how to maximize the accumulation of carbon stocks in areas where SOC stocks are very small.

- The consideration of salinity and sodicity drivers when assessing soil carbon turnover.
- Estimation of carbon stocks in salt-affected areas to ascertain the amount of both SOC and soil inorganic carbon that has been lost or gained as these areas become degraded.

References

Abdel-Fattah, M. K., Abd-Elmabod, S. K., Aldosari, A. A., Elrys, A. S., & Mohamed, E. S. (2020). Multivariate analysis for assessing irrigation water quality: A case study of the Bahr Mouise canal, eastern Nile delta. *Water, 12*(9), Article 2537. https://doi.org/10.3390/w12092537

Abrol, I. P., & Bhumbla, D. R. (1979). Crop responses to differential gypsum applications in a highly sodic soil and the tolerance of several crops to exchangeable sodium under field conditions. *Soil Science, 127*, 79–86.

Ahmad, S., Chaudhari, S. K., Dagar, J. C., & Basak, N. (2012). Soil aggregates as indicator of soil health in waterlogged sodic soil. *Journal of Soil Salinity and Water Quality, 4*(2), 92–96.

Ali, A., Bennett, J. M., Biggs, A. J. W., & Marchuk, A. (2022). Incorporating solution alkalinity into a hydraulic reduction model to account for disaggregation and dispersion. *Geoderma, 413*, Article 115742. https://doi.org/10.1016/j.geoderma.2022.115742

Andrews, S. S., Karlen, D. L., & Cambardella, C. A. (2004). The soil management assessment framework. *Soil Science Society of America Journal, 68*(6), 1945–1962. https://doi.org/10.2136/sssaj2004.1945

Bajwa, M. S., & Swarup, A. (2002). Soil salinity and alkalinity. In N. N. Goswami, R. K. Rattan, G. Dev, G. Narayanasamy, D. K. Das, S. K. Sanyal, D. K. Pal, & D. L. N. Rao (Eds.), *Fundamentals of soil science* (pp. 329–347). Indian Society of Soil Science.

Balpande, S. S., Deshpande, S. B., & Pal, D. K. (1996). Factors and processes of soil degradation in vertisols of the Purna Valley. *Maharashira India, Land Degradation and Development*, 7(4), 313–324. https://doi.org/10.1002/(SICI)1099-145X(199612)7:4<313::AID-LDR236>3.0.CO;2-#

Bandyopadhyay, B. K., & Sen, H. S. (1992). Effect of excess soil water conditions for a short period on growth and nutrition of crops on coastal saline soil. *Journal of the Indian Society of Soil Science, 40*, 823–827.

Basak, N., Chaudhari, S. K., & Sharma, D. K. (2015). Influence of water quality on exchange phase-solution phase behaviour of texturally different salt-affected soils. *Journal of the Indian Society of Soil Science, 63*(4), 365–372.

Basak, N., Rai, A. K., Krishnamurty, S. L., Singh, R. K., Sundha, P., Patel, S., Bedwal, S., & Sharm, P. C. (2023). Nutrients availability in salt–affected soils varies with the ionic composition of solid and solution phases of inland and coastal agroecosystems of India. *Environmental Earth Sciences*, *82*, 489. https://doi.org/10.1007/s12665-023-11167-9

Basak, N., Rai, A. K., Sundha, P., Meena, R. L., Bedwal, S., Yadav, R. K., & Sharma, P. C. (2022). Assessing soil quality for rehabilitation of salt-affected agroecosystem: A comprehensive review. *Frontiers in Environmental. Science*, *10*, Article 935785. https://doi.org/10.3389/fenvs.2022.93578

Basak, N., Sheoran, P., Sharma, R., Yadav, R. K., & Singh, R. K. (2021). Gypsum and pressmud amelioration improve soil organic carbon storage and stability in sodic agroecosystems. *Land Degradation and Development*, *32*(15), 4430–4444. https://doi.org/10.1002/ldr.4047

Bennett, J. M., Marchuk, A., Marchuk, S., & Raine, S. R. (2019). Towards predicting the soil-specific threshold electrolyte concentration of soil as a reduction in saturated hydraulic conductivity: The role of clay net negative charge. *Geoderma*, *337*, 122–131. https://doi.org/10.1016/j.geoderma.2018.08.030

Bharadwaj, K. K. R., & Abrol, I. P. (1978). Nitrogen management in alkali soils. In S. P. Sen, et al. (Eds.), *Nitrogen assimilation and crop productivity* (pp. 83–86). Associated Publishing Company.

Bhardwaj, A. K., Mishra, V. K., Singh, A. K., Arora, S., & Srivastava, S. (2019). Soil salinity and land use-land cover interactions with soil carbon in a salt-affected irrigation canal command of Indo-Gangetic plain. *Catena*, *180*, 392–400. https://doi.org/10.1016/j.catena.2019.05.015

Bhardwaj, A. K., Shainberg, I., Goldstein, D., Warrington, D. N., & Levy, G. J. (2007). Water retention and hydraulic conductivity of cross-linked polyacrylamides in sandy soils. *Soil Science Society of America Journal*, *71*(2), 406–412. https://doi.org/10.2136/sssaj2006.0138

Blankespoor, B., Dasgupta, S., & Lange, G. M. (2017). Mangroves as a protection from storm surges in a changing climate. *Ambio*, *46*(4), 478–491. https://doi.org/10.1007/s13280-016-0838-x

Cai, X., Sharma, B. R., Matin, M. A., Sharma, D., & Gunasinghe, S. (2010). *An assessment of crop water productivity in the Indus and Ganges River Basins: Current status and scope for improvement* (Research report 140). International Water Management Institute.

Central Soil Salinity Research Institute. (2006). *CSSRI: A journey to excellence (1969–2006)*. Central Soil Salinity Research Institute.

Central Soil Salinity Research Institute. (2015). *Vision 2050*. Central Soil Salinity Research Institute.

Chaganti, V. N., & Crohn, D. M. (2015). Evaluating the relative contribution of physiochemical and biological factors in ameliorating a saline–sodic soil amended with composts and biochar and leached with reclaimed water. *Geoderma*, *259–260*, 45–55. https://doi.org/10.1016/j.geoderma.2015.05.005

Chandel, S., Datta, A., & Yadav, R. K. (2021a). Soil salinity indicators and salinity build up on saline water irrigation in seed spices. *Crop and Pasture Science*, *73*(6), 663–678. https://doi.org/10.1071/CP21585

Chandel, S., Datta, A., Yadav, R. K., & Dheri, G. S. (2021b). Does saline water irrigation influence soil carbon pools and nutrient distribution in soil under seed spices? *Journal of Soil Science and Plant Nutrition*, *21*, 949–966. https://doi.org/10.1007/s42729-021-00413-3

Chhabra, R. (1985). Crop responses to phosphorus and potassium fertilization of a sodic soil. *Agronomy Journal*, *77*, 699–702.

Choudhary, O. P., Ghuman, B. S., Bijay-Singh, Thuy, N., & Buresh, R. J. (2011). Effects of long-term use of sodic water irrigation, amendments and crop residues on soil properties and crop yields in rice–wheat cropping system in a calcareous soil. *Field Crops Research*, *121*(3), 363–372. https://doi.org/10.1016/j.fcr.2011.01.004

Dagar, J. C., Lal, K., Ram, J., Kumar, M., & Chaudhari, S. K. (2016). Eucalyptus geometry in agroforestry on waterlogged saline soils influences plant and soil traits in north-west India. *Agriculture, Ecosystem and Environment*, *233*, 33–42. https://doi.org/10.1016/j.agee.2016.08.025

Dagar, J. C., Rai, A. K., Basak, N., & Yadav, R. K. (2022). Soil alkalinity/sodicity: Degradation processes, constraints for crop production, and their management. In Y. Dang, N. Menzies, & R. Dalal (Eds.), *Soil constraints on crop production* (pp. 116–138). Cambridge Scholars Publishing.

Dalal, R. C. (1989). Long-term effects of no-tillage, crop residue, and nitrogen application on properties of a vertisol. *Soil Science Society of America Journal*, *53*(5), 1511–1515. https://doi.org/10.2136/sssaj1989.03615995005300050035x

Dasgupta, S., Hossain, M. M., Huq, M., & Wheeler, D. (2015). Climate change and soil salinity: The case of coastal Bangladesh. *Ambio*, *44*(8), 815–826. https://doi.org/10.1007/s13280-015-0681-5

Datta, A., Basak, N., Chaudhari, S. K., & Sharma, D. K. (2015). Soil properties and organic carbon distribution under different land uses in reclaimed sodic soils of north-west India. *Geoderma Regional*, *4*, 134–146. https://doi.org/10.1016/j.geodrs.2015.01.006

Datta, A., Mahato, A. K. R., Choudhary, M., Nisha, P., Jat, H. S., & Sharma, P. C. (2020). Soil organic carbon pools and microbial population in extremely saline soils: A case study in Salt Desert of Rann of Kachchh, India. *European Journal of Environment and Earth Sciences*, *1*(3). https://doi.org/10.24018/ejgeo.2020.1.3.24

Datta, A., Mandal, A. K., & Yadav, R. K. (2019a). Proper measurement of electrical conductivity and other parameters influence profile salinity and sodicity under different land uses. *Ecological Indicators*, *101*, 1004–1006.

Datta, A., Setia, R., Barman, A., Guo, Y., & Basak, N. (2019b). Carbon dynamics in salt-affected soils. In J. C. Dagar, R. K. Yadav, & P. C. Sharma (Eds.), *Research developments in saline agriculture* (pp. 369–389). Springer Singapore.

Dendooven, L., Alcántara-Hernández, R. J., Valenzuela-Encinas, C., Luna-Guido, M., & Perez-Guevara, F. (2010). Dynamics of carbon and nitrogen in an extreme alkaline saline soil: A review. *Soil Biology and Biochemistry, 42*(6), 865–877. https://doi.org/10.1016/j.soilbio.2010.02.014

Egamberdieva, D., Kucharova, Z., Davranov, K., Berg, G., & Makarova, N. (2011). Bacteria able to control foot and root rot and to promote growth of cucumber in salinated soils. *Biology and Fertility of Soils, 47*(2), 197–205. https://doi.org/10.1007/s00374-010-0523-3

FAO - UNESCO. (1974). *Soil map of the world 1: 5,000,000. Volume I Legend.* UNESCO Paris.

Fishman, R. (2018). Groundwater depletion limits the scope for adaptation to increased rainfall variability in India. *Climatic Change, 147*(1), 195–209. https://doi.org/10.1007/s10584-018-2146-x

Ganjegunte, G. K., Clark, J. A., Parajulee, M. N., Enciso, J., & Kumar, S. (2018). Salinity management in pima cotton fields using sulfur burner. *Agrosystems, Geosciences and Environment, 1*(1), Article 180006. https://doi.org/10.2134/age2018.04.0006

Garg, V. K. (1998). Interaction of tree crops with a sodic soil environment: Potential for rehabilitation of degraded environments. *Land Degradation and Development, 9*, 81–93.

Ghosh, S., Lockwood, P., Hulugalle, N., Daniel, H., & Kristiansen, P. (2010). Changes in properties of sodic Australian vertisols with application of organic waste products. *Soil Science Sociaty of America Journal, 74*(1), 153–160. https://doi.org/10.2136/sssaj2008.0282

Grigg, A. H., Sheridan, G. J., Pearce, A. B., & Mulligan, D. R. (2006). The effect of organic mulch amendments on the physical and chemical properties and revegetation success of a saline-sodic minespoil from Central Queensland Australia. *Soil Research, 44*(2), 97–105.

Gupta, I. C., & Chandra, H. (1972). Effect of gypsum in reducing boron hazards of saline waters and irrigated soil. *Annals of Arid Zone, 11*, 228–230.

Gupta, R. K., & Abrol, I. P. (1990). Salt-affected soils: Their reclamation and management for crop production. *Advances in Soil Science, 11*, 223–228.

Gupta, R. N., & Mathur, A. (2011). Origin, distribution and classification of the salt-affeted soils of India. In R. K. Rattan & A. K. Singh (Eds.), *Bulletin of the Indian Society of Soil Science* (pp. 1–28). Indian Society of Soil Science.

Hansen, J. A., Jurgens, B. C., & Fram, M. S. (2018). Quantifying anthropogenic contributions to century-scale groundwater salinity changes, San Joaquin Valley, California, USA. *Science of the Total Environment, 642*, 125–136. https://doi.org/10.1016/j.scitotenv.2018.05.333

Hayat, K., Bundschuh, J., Jan, F., Menhas, S., & Hayat, S. (2020). Combating soil salinity with combining saline agriculture and phytomanagement with salt-accumulating plants. *Critical Reviews in Environmental Science and Technology, 50*(11), 1085–1115. https://doi.org/10.1080/10643389.2019.1646087

Imaz, M. J., Virto, I., Bescansa, P., Enrique, A., Fernandez-Ugalde, O., & Karlen, D. L. (2010). Soil quality indicator response to tillage and residue management on semi-arid Mediterranean cropland. *Soil and Tillage Research, 107*(1), 17–25. https://doi.org/10.1016/j.still.2010.02.003

Iwai, C. B., Oo, A. N., & Topark-ngarm, B. (2012). Soil property and microbial activity in natural salt affected soils in an alternating wet–dry tropical climate. *Geoderma, 189–190*, 144–152. https://doi.org/10.1016/j.geoderma.2012.05.001

Jehangir, I. A., Khan, M. H., & Bhat, Z. A. (2013). Strategies of increasing crop production and productivity in problem soils. *International Journal of Forest, Soil and Erosion, 3*(2), 73–78.

Jiang, J., Huo, Z., Feng, S., & Zhang, C. (2012). Effect of irrigation amount and water salinity on water consumption and water productivity of spring wheat in Northwest China. *Field Crops Research, 137*, 78–88. https://doi.org/10.1016/j.fcr.2012.08.019

Karimov, A., Qadir, M., Noble, A., Vyshpolsky, F., & Anzelm, K. (2009). Development of magnesium-dominant soils under irrigated agriculture in southern Kazakhstan. *Pedosphere, 19*(3), 331–343. https://doi.org/10.1016/S1002-0160(09)60124-7

Kaur, B., Gupta, S. R., & Singh, G. (2002). Bioamelioration of a sodic soil by silvopastoral systems in northwestern India. *Agroforestry Systems, 54*, 13–20.

Kaur, J., Choudhary, O. P., & Bijay-Singh. (2008). Microbial biomass carbon and some soil properties as influenced by long-term sodic-water irrigation, gypsum, and organic amendments. *Soil Research, 46*(2), 141–151. https://doi.org/10.1071/SR07108

Keren, R., & Shainberg, I. (1981). Effect of dissolution rate on the efficiency of industrial and mined gypsum in improving infiltration of a sodic soil. *Soil Science Society of America Journal, 45*(1), 103–107. https://doi.org/10.2136/sssaj1981.03615995004500010022x

Khandkar, U., Tiwari, S., Sharma, R., Verma, S., & Meena, R. (2017). Relative efficiency of amendments for reclamation of sodic Vertisols and their effects on crop production. *Journal of Soil Salinity and Water Quality, 9*(2), 187–193.

Kopittke, P. M., Berhe, A. A., Carrillo, Y., Cavagnaro, T. R., Chen, D., Chen, Q. L., Dobarco, M. R., Dijkstra, F. A., Field, D. J., Grundy, M. J., He, J. Z., Hoyle, F. C., Kögel-Knabner, I., Lam, S. K., Marschner, P., Martinez, C., McBratney, A. B., McDonald-Madden, E., Menzies, N. W., . . . Minasny, B. (2022). Ensuring planetary survival: The centrality of organic carbon in balancing the multifunctional nature of soils. *Critical Reviews in Environmental Science and Technology, 52*(23), 4308–4324. https://doi.org/10.1080/10643389.2021.2024484

Kumpiene, J., Lagerkvist, A., & Maurice, C. (2007). Stabilization of Pb- and Cu-contaminated soil using coal fly ash and peat. *Environmental Pollution, 145*(1), 365–373. https://doi.org/10.1016/j.envpol.2006.01.037

Levy, G. J., & Shainberg, I. (2005). Sodic soils. In D. Hillel (Ed.), *Encyclopedia of Soils in the Environment* (pp. 504–513). Elsevier. https://doi.org/10.1016/B0-12-348530-4/00218-6

Liu, X., Ren, Y., Gao, C., Yan, Z., & Li, Q. (2017). Compensation effect of winter wheat grain yield reduction under straw mulching in wide-precision planting in the North China plain. *Scientific Reports, 7*(1), Article 213. https://doi.org/10.1038/s41598-017-00391-6

Lokeshappa, B., & Dikshit, A. K. (2011). *Disposal and management of flyash.* In *International Conference on Life Science and Technology IPCBEE* (pp. 11–14). IACSIT Press.

Mahajan, G. R., Manjunath, B. L., Latare, A. M., D'Souza, R., & Vishwakarma, S. (2016). Microbial and enzyme activities and carbon stock in unique coastal acid saline soils of Goa. *Proceedings of the National Academy of Science-India Section B Biological Sciences, 86*(4), 961–971. https://doi.org/10.1007/s40011-015-0552-7

Mahajan, G. R., Manjunath, B. L., Morajkar, S., Desai, A., Das, B., & Paramesh, V. (2021). Long-term effect of various organic and inorganic nutrient sources on rice yield and soil quality in west coast India using suitable indexing techniques. *Communications in Soil Science and Plant Analysis, 52*(15), 1819–1833. https://doi.org/10.1080/00103624.2021.1900221

Maleki, S., Fathizad, H., Karimi, A., Taghizadeh-Mehrjardi, R., & Pourghasemi, H. R. (2022). Monitoring of spatiotemporal changes of soil salinity and alkalinity in eastern and central parts of Iran. In H. R. Pourghasemi (Ed.), *Computers in earth and environmental sciences* (pp. 547–561). Elsevier.

Mamilov, A., Dilly, O. M., Mamilov, S., & Inubushi, K. (2004). Microbial eco-physiology of degrading Aral Sea wetlands: Consequences for C-cycling. *Soil Science and Plant Nutrition, 50*(6), 839–842. https://doi.org/10.1080/00380768.2004.10408544

Mandal, A. K., Sharma, R. C., & Singh, G. (2019). Assessment of salt affected soils in India using GIS. *Geocarto International, 24*(6), 437–456. https://doi.org/10.1080/10106040902781002

Mandal, B., Dash, B., Jaison, M., Das, A., & Ghosh, S. (2022). Soil quality assessment for coastal agroecosystem—Problems and perspectives. In T. D. Lama, D. Burman, U. K. Mandal, S. K. Sarangi, & H. S. Sen (Eds.), *Transforming coastal zone for sustainable food and income security* (pp. 607–619). Springer International Publishing.

Mandal, U. K., Bhardwaj, A. K., Warrington, D. N., Goldstein, D., & Bar Tal, A. (2008). Changes in soil hydraulic conductivity, runoff, and soil loss due to irrigation with different types of saline–sodic water. *Geoderma*, *144*(3), 509–516. https://doi.org/10.1016/j.geoderma.2008.01.005

Mavi, M. S., & Marschner, P. (2013). Salinity affects the response of soil microbial activity and biomass to addition of carbon and nitrogen. *Soil Research*, *51*(1), 68–75. https://doi.org/10.1071/SR12191

Meena, M. D., Joshi, P. K., Narjary, B., Sheoran, P., & Jat, H. S. (2016). Effects of municipal solid waste compost, rice-straw compost and mineral fertilisers on biological and chemical properties of a saline soil and yields in a mustard–pearl millet cropping system. *Soil Research*, *54*(8), 958–969. https://doi.org/10.1071/SR15342

Meena, M. D., Narjary, B., Sheoran, P., Jat, H. S., & Joshi, P. K. (2018). Changes of phosphorus fractions in saline soil amended with municipal solid waste compost and mineral fertilizers in a mustard-pearl millet cropping system. *Catena*, *160*, 32–40. https://doi.org/10.1016/j.catena.2017.09.002

Minhas, P. S. (2010). A re-look at the diagnostic criteria for salt-affected soils in India. *Journal of the Indian Scoiety of Soil Science*, *58*(1), 12–24.

Minhas, P. S., Qadir, M., & Yadav, R. K. (2019). Groundwater irrigation induced soil sodification and response options. *Agricultural Water Management*, *215*, 74–85. http://doi.org/10.1016/j.agwat.2018.12.030

Mishra, V. K., Jha, S. K., Damodaran, T., Singh, Y. P., & Srivastava, S. (2019). Feasibility of coal combustion fly ash alone and in combination with gypsum and green manure for reclamation of degraded sodic soils of the indo-Gangetic Plains: A mechanism evaluation. *Land Degradation and Development*, *30*(11), 1300–1312. https://doi.org/10.1002/ldr.3308

Mitran, T., Basak, N., Mani, P. K., Tamang, A., & Singh, D. K. (2021). Improving crop productivity and soil quality through soil management practices in coastal saline agro-ecosystem. *Journal of Soil Science and Plant Nutrition*, *21*, 3514–3529. https://doi.org/10.1007/s42729-021-00624-8

Murtaza, G., Murtaza, B., Kahlon, U. Z., & Yaseen, M. (2017). A comparative study of different amendments on amelioration of saline-sodic soils irrigated with water having different EC: SAR ratios. *Communication in Soil Science and Plant Analysis*, *48*(22), 2630–2641. https://doi.org/10.1080/00103624.2017.1416130

Nagaz, K., Masmoudi, M. M., & Ben Mechlia, N. (2012). Impacts of irrigation regimes with saline water on carrot productivity and soil salinity. *Journal of the Saudi Society of Agricultural Sciences*, *11*(1), 19–27. https://doi.org/10.1016/j.jssas.2011.06.001

Nayak, A. K., Mishra, V. K., Sharma, D. K., Jha, S. K., & Singh, C. S. (2013). Efficiency of phosphogypsum and mined gypsum in reclamation and productivity of rice–wheat cropping system in sodic soil. *Communication in Soil Science and Plant Analysis*, *44*(5), 909–921. https://doi.org/10.1080/00103624.2012.747601

Nayak, A. K., Sharma, D. K., Mishra, V. K., Minhas, P. S., & Verma, C. L. (2008). Reclamation of saline-sodic soil under a rice–wheat system by horizontal surface flushing. *Soil Use and Management*, *24*(4), 337–343. https://doi.org/10.1111/j.1475-2743.2008.00162.x

Oo, A. N., Iwai, C. B., & Saenjan, P. (2013). Soil properties and maize growth in saline and nonsaline soils using cassava-industrial waste compost and vermicompost with or without earthworms. *Land Degradation and Development*, *26*, 300–310.

Pasricha, N. S., & Randhawa, N. S. (1971). *Available molybdenum status of some recently reclaimed saline-sodic soils and its effect on concentration of molybdenum, copper, sulphur, and nitrogen in barseem (Trifolium alexandrinum) grown on these soils*. Proceedings of the International Symposium on Soil Fertility Evaluation (pp. 1017–1026), Hisar, India.

Pathak, H., & Rao, D. L. N. (1998). Carbon and nitrogen mineralization from added organic matter in saline and alkali soils. *Soil Biology and Biochemistry*, *30*(6), 695–702. https://doi.org/10.1016/S0038-0717(97)00208-3

Purakayastha, T. J., Pathak, H., Kumari, S., Biswas, S., & Chakrabarty, B. (2019). Soil health card development for efficient soil management in Haryana, India. *Soil and Tillage Research*, *191*, 294–305. https://doi.org/10.1016/j.still.2018.12.024

Qadir, M., & Oster, J. D. (2004). Crop and irrigation management strategies for saline-sodic soils and waters aimed at environmentally sustainable agriculture. *Science of the Total Environment*, *323*, 1–19.

Qadir, M., Schubert, S., Ghafoor, A., & Murtaza, G. (2001). Amelioration strategies for sodic soils: A review. *Land Degradation and Development*, *12*, 357–386.

Qadir, M., Schubert, S., Oster, J. D., Sposito, G., & Minhas, P. S. (2018). High-magnesium waters and soils: Emerging environmental and food security constraints. *Science of the Total Environment*, *642*, 1108–1117. https://doi.org/10.1016/j.scitotenv.2018.06.090

Quirk, J., Pereira, C., & Tanton, T. W. (1986). Soil permeability in relation to sodicity and salinity. *Philosophical Transactions of the Royal Society of London, series A*, *316*(1537), 297–317.

Rai, A. K., Basak, N., Chandra, P., Sundha, P., Kaur, H., Bedwal, S., Patel, S., Kumar, S., Choubey, K., Meena, R. L., Sharma, P. C., & Yadav, R. K. (2024). Occurrence of Sphingomonas olei with elemental S oxidation capability in sodic soil: Potential role in sodicity reclamation and plant growth promotion. *Ecotoxicology and Environmental Safety*, *278*, 116396. https://doi.org/10.1016/j.ecoenv.2024.116396

Rai, A. K., Basak, N., & Sundha, P. (2020). Chemistry of salt-affected soils. In P. S. Minhas, R. K. Yadav, & P. C. Sharma (Eds.), *Salinity management in India*. Indian Council of Agricultural Research (ICAR).

Rai, A. K., Basak, N., & Sundha, P. (2021). Saline and sodic ecosystems in the changing world. In A. Rakshit, S. K. Singh, P. C. Abhilash, & A. Biswas (Eds.), *Soil science: Fundamentals to recent advances* (pp. 175–190). Springer Singapore. https://doi.org/10.1007/978-981-16-0917-6_9

Rai, A. K., Basak, P., Soni, P. G., Kumar, S., & Sundha, P. (2022). Bioenergy sorghum as balancing feedback loop for intensification of cropping system in salt-affected soils of the semi–arid region: Energetics, biomass quality and soil properties. *European Journal of Agronomy*, *134*, Article 126452. https://doi.org/10.1016/j.eja.2021.126452

Ranatunga, K., Nation, E. R., & Barodien, G. (2010). Potential use of saline groundwater for irrigation in the Murray hydrogeological basin of Australia. *Environmental Modelling Software*, *25*(10), 1188–1196. https://doi.org/10.1016/j.envsoft.2010.03.028

Rengasamy, P. (2006). World salinization with emphasis on Australia. *Journal of Experimental Botany*, *57*(5), 1017–1023. https://doi.org/10.1093/jxb/erj108

Rengasamy, P. (2008). Salinity in the landscape: A growing problem in Australia. *Geotimes*, *53*, 34–39.

Rengasamy, P. (2016). Soil chemistry factors confounding crop salinity tolerance—A review. *Agronomy*, *6*(4), Article 53. https://doi.org/10.3390/agronomy6040053

Rietz, D. N., & Haynes, R. J. (2003). Effects of irrigation-induced salinity and sodicity on soil microbial activity. *Soil Biology and Biochemistry*, *35*(6), 845–854. https://doi.org/10.1016/S0038-0717(03)00125-1

Rowley, M. C., Grand, S., & Verrecchia, E. P. (2018). Calcium-mediated stabilisation of soil organic carbon. *Biogeochemistry*, *137*, 27–49. https://doi.org/10.1007/s10533-017-0410-1

Rumpel, C., Amiraslani, F., Bossio, D., Chenu, C., Henry, B., Espinoza, A. F., Koutika, L. S., Ladha, J., Madari, B., Minasny, B., Olaleye, A. O., Shirato, Y., Sall, S. N., Soussana, J. F., & Varela-Ortega, C. (2022). The role of soil carbon sequestration in enhancing human resilience in tackling global crises including pandemics. *Soil. Security*, *8*, Article 100069. https://doi.org/10.1016/j.soisec.2022.100069

Sadegh-Zadeh, F., Parichehreh, M., Jalili, B., & Bahmanyar, M. A. (2018). Rehabilitation of calcareous saline-sodic soil by means of biochars and acidified biochars. *Land Degradation and Development*, *29*(10), 3262–3271. https://doi.org/10.1002/ldr.3079

Sadiq, M., Hassan, G., Mehdi, S. M., Hussain, N., & Jamil, M. (2007). Amelioration of saline-sodic soils with tillage implements and sulfuric acid application. *Pedosphere*, *17*, 182–190.

Sappa, G., Iacurto, S., Ferranti, F., & De Filippi, F. M. (2019). Groundwater quality assessment in a karst coastal region of the west Aurunci Mountains (Central Italy). *Geofluids*, *2019*, Article 3261713. https://doi.org/10.1155/2019/3261713

Sappor, D. K., Osei, B. A., & Ahmed, M. R. (2017). Reclaiming sodium affected soil: The potential of organic amendments. *International Journal of Plant and Soil Science*, *16*(2), 1–11. http://doi.org/10.9734/IJPSS/2017/33410

Setia, R., Gottschalk, P., Smith, P., Marschner, P., Baldock, J., Setia, D., & Smith, J. (2013). Soil salinity decreased global soil organic carbon stocks. *Science of the Total Environment*, *465*, 267–272.

Setia, R., Rengasamy, P., & Marschner, P. (2014). Effect of mono- and divalent cations on sorption of water-extractable organic carbon and microbial activity. *Biology and Fertility Soils*, *50*, 727–734.

Shainberg, I., & Letey, J. (1984). *Response of soils to sodic and saline conditions. Hilgardia.*

Sharma, P. C., Singh, A., & Choudhary, M. (2019). *Annual report*. ICAR-Central Soil Salinity Research Institute.

Shaygan, M., & Baumgartl, T. (2022). Reclamation of salt-affected land: A review. *Soil Systems*, *6*, Article 61. https://doi.org/10.3390/soilsystems6030061

Shen, J., Zhou, X., Sun, D., Fang, J., & Liu, Z. (2008). Soil improvement with coal ash and sewage sludge: A field experiment. *Environment and Geology*, *53*(8), 1777–1785. https://doi.org/10.1007/s00254-007-0783-x

Sheoran, P., Basak, N., Kumar, A., Yadav, R. K., & Singh, R. (2021). Ameliorants and salt tolerant varieties improve rice-wheat production in soils undergoing sodification with alkali water irrigation in Indo–Gangetic plains of India. *Agricultural Water Management*, *243*, Article 106492. https://doi.org/10.1016/j.agwat.2020.106492

Sheoran, P., Sharma, R., Kumar, A., Singh, R. K., & Barman, A. (2022). Climate resilient integrated soil–crop management (CRISCM) for salt affected wheat Agri–food production systems. *Science of the Total Environment*, *837*, Article 155843. https://doi.org/10.1016/j.scitotenv.2022.155843

Singh, A., Kumar, A., Yadav, R. K., Minhas, P. S., & Saini, U. (2022). Long-term effect of alkali and partially neutralized irrigation water on soil quality. *Journal of Soil Science and Plant Nutrition*, *22*(2), 1252–1266. https://doi.org/10.1007/s42729-021-00728-1

Singh, G. (2009). Salinity-related desertification and management strategies: Indian experience. *Land Degradation and Development*, *20*(4), 367–385. https://doi.org/10.1002/ldr.933

Singh, G. B., & Gill, H. S. (1990). Raising trees in alkali soils. *Wasteland News*, *6*, 15–18.

Singh, K., Mishra, A. K., Singh, B., Singh, R. P., & Patra, D. D. (2016). Tillage effects on crop yield and physicochemical properties of sodic soils. *Land Degradation and Development*, *27*, 223–230.

Singh, K., Singh, B., & Singh, R. R. (2012). Changes in physico-chemical, microbial and enzymatic activities during restoration of degraded sodic land: Ecological suitability of mixed forest over monoculture plantation. *Catena*, *96*, 57–67. https://doi.org/10.1016/j.catena.2012.04.007

Singh, R., Yaduvanshi, N. P. S., Singh, K. N., Kumar, S., & Mishra, V. K. (2014). Bio-chemical amelioration effects on physico-chemical dynamics of sodic soils under rice (*Oryza sativa*)–wheat (*Triticum aestivum*) cropping system. *Indian Journal of Agricultural Sciences*, *84*(3), 349–355.

Sione, S. M. J., Wilson, M. G., Lado, M., & González, A. P. (2017). Evaluation of soil degradation produced by rice crop systems in a vertisol, using a soil quality index. *Catena*, *150*, 79–86. https://doi.org/10.1016/j.catena.2016.11.011

Soni, P. G., Basak, N., Rai, A. K., Sundha, P., & Narjary, B. (2021). Deficit saline water irrigation under reduced tillage and residue mulch improves soil health in sorghum-wheat cropping system in semi-arid region. *Scientific Reports*, *11*(1), Article 1880. https://doi.org/10.1038/s41598-020-80364-4

Strudley, M. W., Green, T. R., & Ascough, I. I. J. C. (2008). Tillage effects on soil hydraulic properties in space and time: State of the science. *Soil and Tillage Research*, *99*, 4–48.

Suarez, D. L. (2012). Irrigation water quality assessments. In W. W. Wallender & K. K. Tanji (Eds.), *ASCE manual and reports on engineering practice no. 71 agricultural salinity assessment and management* (2nd ed., pp. 343–370). ASCE.

Sundha, P., Basak, N., Rai, A. K., Yadav, R. K., & Sharma, D. K. (2017). N and P release pattern in saline-sodic soil amended with gypsum and municipal solid waste compost. *Journal of Soil Salinity and Water Quality*, *9*, 145–155.

Sundha, P., Basak, N., Rai, A. K., Yadav, R. K., & Sharma, D. K. (2018). Utilization of municipal solid waste compost in reclamation of saline-sodic soil irrigated with poor quality water. *Journal of the Indian Society of Soil Science*, *66*(1), 28–39.

Sundha, P., Basak, N., Rai, A. K., Yadav, R. K., & Sharma, P. C. (2023a). Irrigation water quality, gypsum, and city waste compost addition affect P dynamics in saline-sodic soils. *Environmental Research*, *216*, Article 114559. https://doi.org/10.1016/j.envres.2022.114559

Sundha, P., Basak, N., Rai, A. K., Yadav, R. K., Sharma, P. C., & Sharma, D. K. (2020). Can conjunctive use of gypsum, city waste composts and marginal quality water rehabilitate saline-sodic soils? *Soil and Tillage Research*, *200*, Article 104608. https://doi.org/10.1016/j.still.2020.104608

Sundha, P., Mukhopadhyay, R., Basak, N., Rai, A. K., Bedwal, S., Patel, S., Kumar, S., Kaur, H., Chandra, P., Sharma, P. C., Saxena, S. K., Parihar, S. S., & Yadav, R. K. (2023b). Characterization of flue gas desulphurized (FGD) gypsum of a coal-fired plant and its relevant risk of associated potential toxic elements in sodic soil reclamation. *Scientific Reports*, *13*, 19787. https://doi.org/10.1038/s41598-023-45706-y

Sundha, P., Rai, A. K., Basak, N., Yadav, R. K., & Sharma, P. C. (2022). P solubility and release kinetics in the leachate of saline–sodic soil: Effect of reclamation strategies and water quality. *Soil and Tillage Research*, *222*, Article 105440. https://doi.org/10.1016/j.still.2022.105440

Swarup, A. (1985). Effect of exchangeable sodium percentage and presubmergence on yield and nutrition of rice under field conditions. *Plant and Soil*, *85*, 279–288.

Swarup, A., & Yaduvanshi, N. P. S. (2000). Effects of integrated nutrient management on soil properties and yield of rice in alkali soils. *Journal of the Indian Society of Soil Science, 48*(2), 279–282.

Szabolcs, I. (1989). *Salt-affected soils*. CRC Press.

Tejada, M., Garcia, C., Gonzalez, J. L., & Hernandez, M. T. (2006). Use of organic amendment as a strategy for saline soil remediation: Influence on the physical, chemical and biological properties of soil. *Soil Biology and Biochemistry, 38*(6), 1413–1421. https://doi.org/10.1016/j.soilbio.2005.10.017

Tripathi, S., Kumari, S., Chakraborty, A., Gupta, A., & Chakrabarti, K. (2006). Microbial biomass and its activities in salt-affected coastal soils. *Biology and Fertility of Soils, 42*(3), 273–277. https://doi.org/10.1007/s00374-005-0037-6

U.S. Salinity Laboratory Staff. (1954). *Agriculture handbook* (Vol. No. 60). U.S. Department of Agriculture.

Vaidya, P. H., & Pal, D. K. (2002). Microtopography as a factor in the degradation of Vertisols in Central India. *Land Degradation and Development, 13*(5), 429–445. https://doi.org/10.1002/ldr.525

Visser, S., Keesstra, S., Maas, G., de Cleen, M., & Molenaar, C. (2019). Soil as a basis to create enabling conditions for transitions towards sustainable land management as a key to achieve the SDGs by 2030. *Sustainability, 11*(23). https://doi.org/10.3390/su11236792

Walpola, B. C., & Arunakumara, K. K. I. U. (2010). Effect of salt stress on decomposition of organic matter and nitrogen mineralization in animal manure amended soils. *Journal of Agricultural Sciences, 5*, 9–18.

Wang, Q., Lu, C., Li, H., He, J., & Sarker, K. K. (2014). The effects of no-tillage with subsoiling on soil properties and maize yield: 12-year experiment on alkaline soils of northeast China. *Soil and Tillage Research, 137*, 43–49. https://doi.org/10.1016/j.still.2013.11.006

Wong, J. W. C. (1995). The production of artificial soil mix from coal fly ash and sewage sludge. *Environmental Technology, 16*(8), 741–751. https://doi.org/10.1080/09593331608616313

Wong, V. N. L., Dalal, R. C., & Greene, R. S. B. (2008). Salinity and sodicity effects on respiration and microbial biomass of soil. *Biology and Fertility of Soils, 44*, 943–953. https://doi.org/10.1007/s00374-008-0279-1

Wong, V. N. L., Greene, R. S. B., Dalal, R. C., & Murphy, B. W. (2010). Soil carbon dynamics in saline and sodic soils: A review. *Soil Use and Management, 26*(1), 2–11. https://doi.org/10.1111/j.1475-2743.2009.00251.x

Xu, X., Huang, G., Sun, C., Pereira, L. S., & Ramos, T. B. (2013). Assessing the effects of water table depth on water use, soil salinity and wheat yield: Searching for a target depth for irrigated areas in the upper Yellow River basin. *Agricultural Water Management, 125*, 46–60. https://doi.org/10.1016/j.agwat.2013.04.004

Yaduvanshi, N. P. S., & Swarup, A. (2005). Effect of continuous use of sodic irrigation water with and without gypsum, farmyard manure, pressmud and fertilizer on soil properties and yields of rice and wheat in a long term experiment. *Nutrient Cycling in Agroecosystems*, *73*(2), 111–118. https://doi.org/10.1007/s10705-005-3361-1

Yaduvanshi, N. P. S., & Yadav, D. V. (1990). Effects of sulphitation press mud and nitrogen fertilizer on biomass, nitrogen economy and plant composition in sugarcane and on soil chemical properties. *Journal of Agricultural Science*, *114*, 259–263.

Yu, H., Yang, P., Lin, H., Ren, S., & He, X. (2014). Effects of sodic soil reclamation using flue gas desulphurization gypsum on soil pore characteristics, bulk density, and saturated hydraulic conductivity. *Soil Science Society of America Journal*, *78*(4), 1201–1213. https://doi.org/10.2136/sssaj2013.08.0352

Yuan, C., Feng, S., Huo, Z., & Ji, Q. (2019). Effects of deficit irrigation with saline water on soil water-salt distribution and water use efficiency of maize for seed production in arid Northwest China. *Agricultural Water Management*, *212*, 424–432. https://doi.org/10.1016/j.agwat.2018.09.019

Zhang, L., Su, X., Meng, H., Men, Y., & Liu, C. (2023). Cotton stubble return and subsoiling alter soil microbial community, carbon and nitrogen in coastal saline cotton fields. *Soil and Tillage Research*, *226*, Article 105585. https://doi.org/10.1016/j.still.2022.105585

Zhang, W., Zhao, Y., Wang, S., Li, Y., & Zhuo, Y. (2021). Soil salinity and sodicity changes after a one time application of flue gas desulphurization gypsum to paddy fields. *Land Degradation and Development*, *32*(15), 4193–4202. https://doi.org/10.1002/ldr.4025

Zhao, Y. G., Wang, S. J., Liu, J., Zhuo, Y. Q., & Li, Y. (2021). Fertility and biochemical activity in sodic soils 17 years after reclamation with flue gas desulfurization gypsum. *Journal of Integrated Agriculture*, *20*(12), 3312–3322. https://doi.org/10.1016/S2095-3119(20)63446-3

Zong, R., Tan, M., Han, Y., Zou, R., & Wang, Z. (2022). Changes in soil microbial community following the conversion of wasteland to cotton land in saline-sodic region of northwest China. *Applied Soil Ecology*, *174*, Article 104424. https://doi.org/10.1016/j.apsoil.2022.104424

12

Mitigating Soil Acidity and Waterlogging Impacts on Soil Health in India

Ruma Das, Samrat Ghosh, Khushboo Rani, Amit Phonglosa, and B. N. Ghosh

Chapter Overview

Soil acidity and waterlogging are two major constrains of sustainable crop production in India. Nearly 30 million hectares land of India falls under strongly to moderately acidic soils having pH < 5.5. Soil acidity accelerates land degradation and deterioration of soil physical, chemical and biological health and reduces crop productivity by 10%–80%. Application of conventional as well as alternative cheap source liming materials, organic ammendments and carbon-rich biochar are important to mitigate soil acidity. Proper management of applied nutrient fertilizers, inoculation of soil and seed with different acid-tolerant soil microbes, planting of acid tolerant crops, grasses and trees and different engineering soil conservation measures are some approaches to restore soil acidity. Development of acid tolerant crop varieties by genetic engineering is the new breakthrough for ensuring sustainable production in acid soil. Although acid soils are considered as degraded soils, they provide various ecosystem services for overall development of mankind. Waterlogging of acid soil leads to the formation of acid sulphate soil which adversely affects crops and ecosystems. Besides, waterlogging affects the chemical, geochemical, and biological reactions in soil and accelerates land degradation and makes the soil unproductive. Thus, soil acidity and waterlogged conditions adversely affect the soil health and hinder the goal of sustainable crop production. Therefore, this chapter emphasizes the different aspects of acid soil,

Soil Health Series: Volume 4 Soil Health and Sustainability in India, First Edition.
Edited by Ranjan Bhattacharyya.

acid sulfate soil, waterlogging and their impacts on soil health and nutrient dynamics. In addition, this chapter describes the different mitigation approaches, important observations and areas for future research to manage soil acidity and waterlogged conditions for better soil health and crop production.

Introduction

Soil acidification causes soil pH to decrease over time due to high acid strength and increased base-neutralizing capacity. Formation of acid soils is a natural pedogenic process accelerated by poor soil management and other anthropogenic decisions. In India, nearly 93 million hectares (Mha) of land, covering 28% of the total geographical area and 48 agro-ecological subregions (AESRs), displays various degrees of soils acidity. Acidic soils have been classified based on the strength of their acidity, measured by their pH (Figure 12.1). Strongly acidic soils, characterized by a pH of <4.5, covers 6.0 Mha; moderately acidic soils, with pH 4.5–5.5, cover 24 Mha; and slightly acidic soils, with pH 5.5–6.5, cover 63 Mha (Maji et al., 2012; Mandal et al., 2019). Any soils with pH > 6.5 are considered nonacidic, and development of acidity in soil is a sign of chemical degradation. In India, acid soils mainly occur in the hilly areas of the northeast and northwest Himalayan Region, Eastern Plateau, and the Western Coastal Plains, where the mean annual rainfall exceeds 1,200 mm (Figure 12.2). In northeast India, more than 80% of the land is under hilly terrain with a soil pH of <5.5 and >50% of the total area is

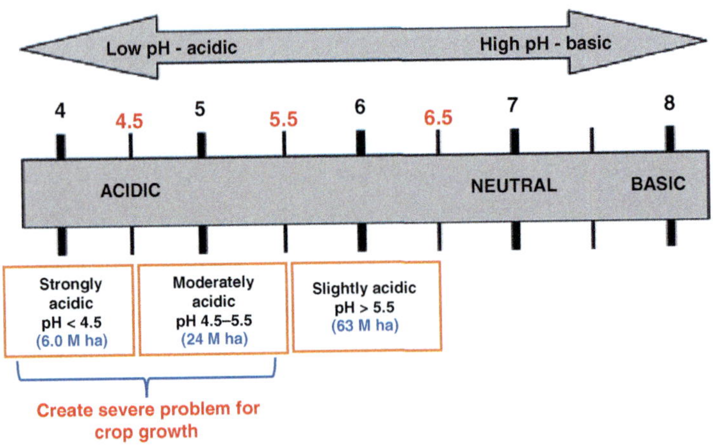

Figure 12.1 Soil acidity based on pH range.

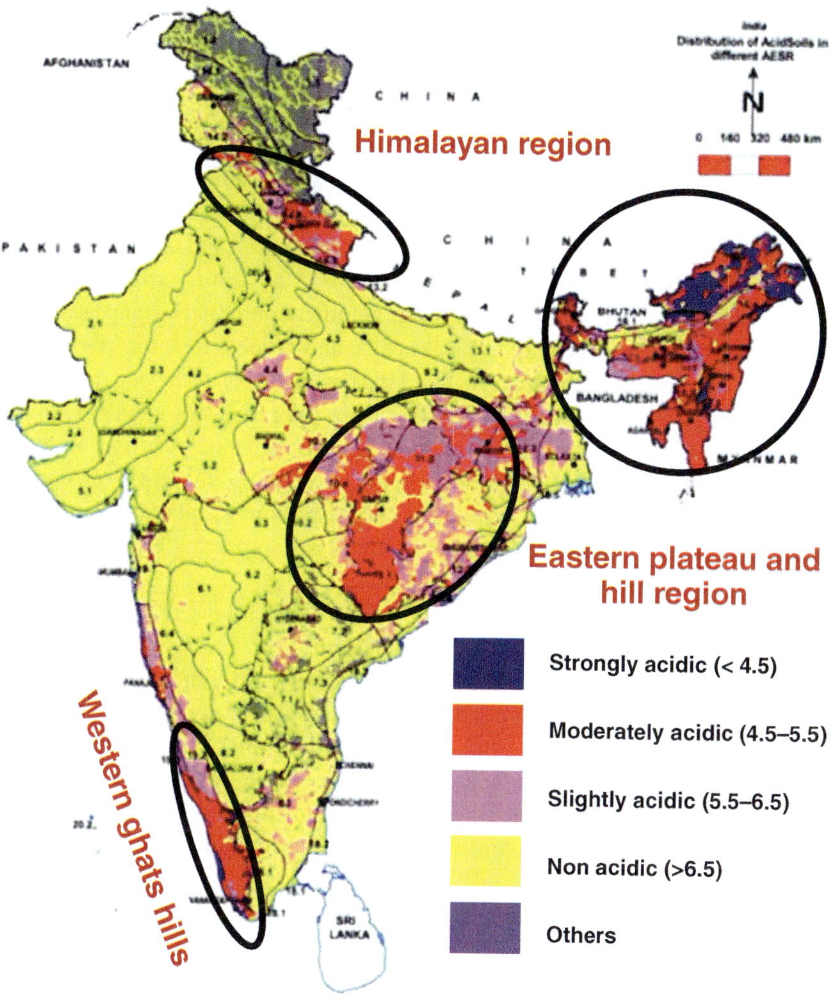

Figure 12.2 Spatial distribution of acid soils of India. Credit: Mandal et al. 2019/ Bhattacharyya.

under strong to moderate soil acidity. High rainfall and steep slopes of that region lead to surface soil loss due to erosion and subsequent development of soil acidity.

Surface soil acidity primarily results from excessive leaching of soluble salts in areas with humid tropical climate. In contrast, subsoil acidity is due to high exchangeable aluminum and iron hydroxides in the interlayer of silicate clays. Strongly to moderately acidic soils developed in humid and per-humid areas of India such as the soils of Assam and other areas of the Northeastern Hill (NEH)

Region, West Bengal, Bihar, Odisha, Andhra Pradesh, Kerala, Madhya Pradesh, Karnataka, Maharashtra, and Tamil Nadu primarily because of excessive cation leaching due to high rainfall. The result is lower pH and loss of soil fertility. Efforts have been made to map the acid soils and their spatial distribution in India on a 1:250,000 scale (Figure 12.2). This mapping was done to assess the severity of degradation (Sehgal et al., 1996) and to establish a base to prioritize areas for ameliorating acid soils (Mandal et al., 2019) within all the states.

Soil acidity is one of the major causes of land degradation and soil quality reduction *vis-à-vis* soil health. Depending upon severity of the acidity, crop productivity can be reduced by 10%–80%. Along with soil acidification, waterlogging is another notable issue that adversely affects soil health and quality. In India, about 8.3% of net sown area (approximately 11.6 Mha) is often waterlogged (Planning Commission, 2011). Poor aeration due to waterlogging causes soil degradation with respect to soil structure, temperature, salinization, denitrification, and nutrient imbalance. Formation of acid sulfate soil is another negative effect of continuous waterlogging. In particular, the acid sulfate soils in Kerala and Sundarban Delta of West Bengal possess peculiar characteristics affecting crop production because of seasonal changes in redox potential.

Reclamation of acid and waterlogged soils of India is very important for the development of sustainable agriculture. This chapter focuses on the different aspects of acid soil and waterlogging, with special reference to acid sulfate soil, and their impacts on soil health and nutrient dynamics. Different mitigation approaches are discussed along with an evaluation of the gaps in current knowledge and technology and an overview of ecosystem services provided by the acid soil. Important observations and areas for future research to manage soil acidity and waterlogged conditions for better soil health and crop production are also discussed.

Processes of Acid Soil Formation

Pedogenic acid soil forming processes, including laterization and podzolization (Singh and Chandran, 2015), are presented in Figure 12.3. Laterization is a weathering process that increases the proportion of iron and aluminum oxides in the solum and forms laterite (Oxisols) and lateritic (Ultisols) soils. Laterization generally occurs in tropical rain forests with basic broadleaf vegetation under a warm-humid environment with a pronounced dry season. The largest areas covered by acid soils in India are laterites and various latosolic soils. Podzolization generally occurs on granites or gneisses dominated by acidic parent materials (Schaetzl & Anderson, 2009); under humid and cool climatic conditions where precipitation is greater than evapotranspiration; and in nutrient-poor, acid-generating conifer

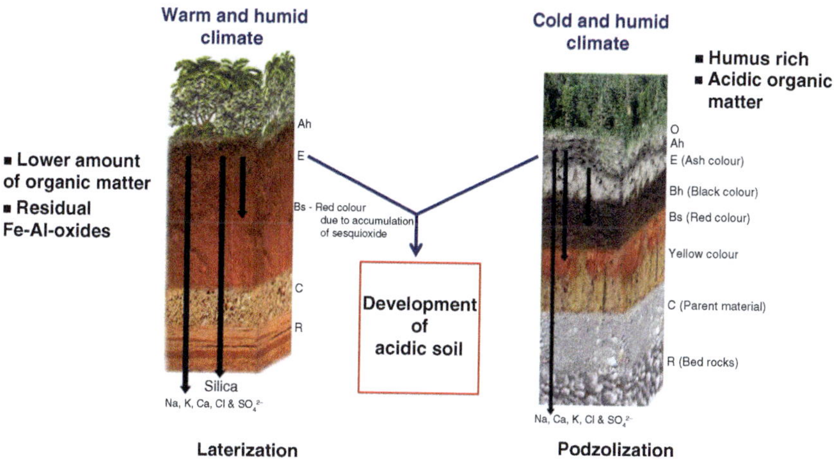

Figure 12.3 Formation of acid soil by pedogenic processes. Credit: Adapted and modified from Singh and Chandran, 2015.

forests with vegetation that is slowly degradable (Jordanova, 2017). Podzolization forms spodosols, which refer to the red, brown, or black subsurface horizon below a light-colored, near-surface layer (A horizon).

Non-pedogenic factors that accelerate the process of soil acidification are related to the intensification of agriculture and climatic factors such as acid rain (Figure 12.4).

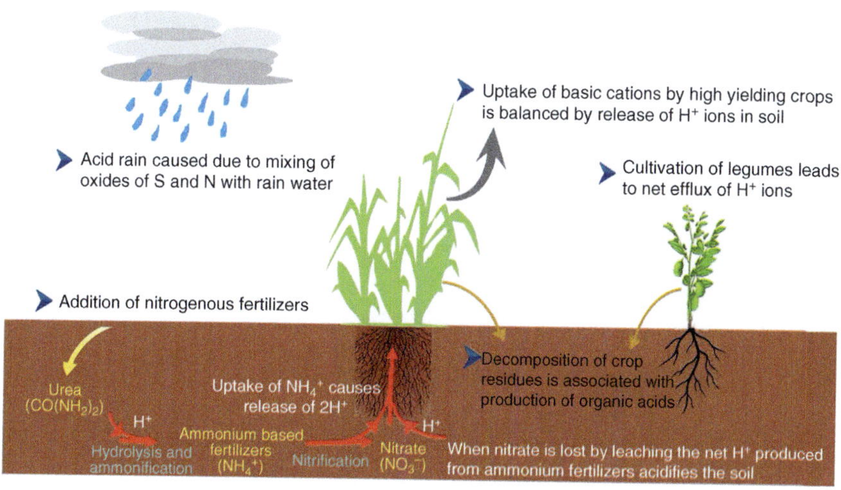

Figure 12.4 Non-pedogenic factors for development of soil acidity.

Soil Acidity and Soil Health

Given the diversity of soil in nature, a proper soil health assessment requires quantifying and combining multiple indicators into an index. Different approaches and methods are followed for determination of the soil quality index (SQI), such as additive and weighted datasets of total dataset and minimum dataset (MDS) and two scoring methods (linear and non-linear), which results in SQI ranges (Lenka et al., 2022; Mahajan et al., 2021). In India, fewer studies have used MDS-based linear weighted and non-linear weighted SQI assessment of acid soils than in other countries. Thus, assessment of SQI in acidic soils using MDS, linear scoring, and weighted approach of the soil quality indexing could save time and costs. Reviews of acid soils in India showed that integrated nutrient management (INM) exhibited a variation of SQI from 0.105 to 0.398, which is much lower than forest soils (0.562) and slightly improved in conservation agriculture system (Shah et al., 2022; Shao et al., 2020). A detailed review of 62 publications covering a total of 65 indicator sets has generated 10, 11, and 4 broad categories of physical, chemical, and biological properties that are frequently used for quantification of soil quality across the world (Figure 12.5; Bünemann et al., 2018).

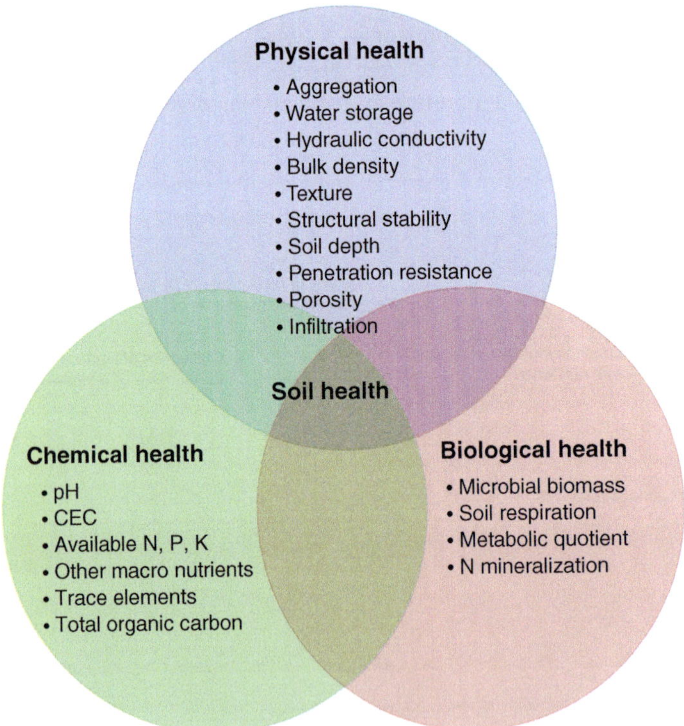

Figure 12.5 Soil physical, chemical, and biological indicators frequently used to assess soil health.

Impact of Soil Acidity on Soil Health

Acidification of soil decreases soil pH and deteriorates the soil health and plant growth. Soil acidity interferes with most of the soil's physical, chemical, and biological properties and consequently has a direct impact on soil health and quality. This section explains how soil acidity influences important soil properties and affects the soil health and quality.

Physical Soil Health

As discussed in earlier sections, acid soils may be generated through pedogenic processes and other external manmade or natural activities. Pedogenically developed lateritic acid soils are generally rich in poorly crystalline iron and aluminum hydroxides, which act as potential aggregating agents (Duiker et al., 2003). However, once the hydrated colloidal iron oxides become dehydrated, they form an irreversibly hard brick-like structure that is more useful for construction than for crop production (Sunil & Krishnappa, 2012). Sometimes, the sesquioxide-rich lateritic acid soils form very tough impenetrable crusts that severely decrease the water infiltration rate and trigger erosional loss of the productive topsoil layer (Ibrahim et al., 2020). Additionally, such acid soils are rich in low-activity clays especially kaolinite, which is non-expanding and has poor moisture-holding capacity. Overall, lateritic acid soils possess poor water infiltration, high soil erosion characteristics, and low available water storage capacity, which collectively contribute to poor physical soil health and limit the scope of crop production.

Apart from the topsoil, pedogenically developed acid soils also have poor physical health in the subsurface layers. Plinthite, which occurs as dark-red redox concretions of pedogenic origin in the subsurface layers, may irreversibly change to iron-rich hardpans and/or massive soil structures with irregular aggregates under repeated wetting and drying cycles. The hardpan or massive structure impedes subsurface water percolation and/or root penetration, significantly affecting the physical soil health and normal crop production. In some cases, the highly acidic condition of spodosols completely destroys the clay minerals and makes the soil sandy in texture (Retallack, 2021). Acid soils, developed by continuous application of NH_4^+-N fertilizer for a long period of time and excessive accumulation of NH_4^+ ions on the exchange sites, may adversely affect soil aggregation and break the soil structure, water storage and hydraulic conductivity, and ultimately lead to poorer soil physical health and crop production (Völker et al., 1995).

Chemical Soil Health

The chemical soil health mainly depends on the soil pH, cation exchange capacity (CEC), soil organic carbon content, and the availability of macro- and micronutrients. Soil pH plays a crucial role in governing the availability of essential nutrients to plants (Figure 12.6). Soils with low pH exhibit poor availability of the major plant nutrients (nitrogen, phosphorus, potassium, sulfur, calcium, and magnesium)

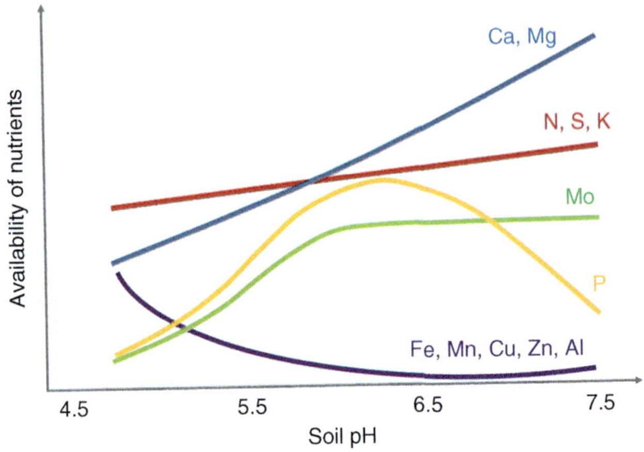

Figure 12.6 Relationship between soil pH and nutrient availability in acid soils.

and the trace element molybdenum and may be responsible for limiting the growth and development of plant biomass (both above and below ground) through interference in many essential metabolic functions. Low soil pH decreases the CEC because of excessive H^+ ions and/or solubilized Al^{3+} concentration in the soil solution and decreases the pH-dependent negative charges both on the clay and organic matter surfaces (phenolic OH and/or carboxylic COOH groups). Thus, a high Al^{3+} ion saturation in acidic soils lowers the amounts of exchangeable basic cations such as Ca^{2+} and Mg^{2+}, among others, that are essential for plant growth and development. Moreover, acid soils developed in high rainfall areas are inherently low in basic cations, including Na^+, K^+, Ca^{2+}, and Mg^{2+}, because those are removed from the soil profile owing to intense leaching. Lower availability of these basic cations adversely affects the chemical health of soil, expressed later in terms of their insufficiency for normal plant growth.

Although soil acidity leads to build up of soil organic matter (SOM) owing to slower microbial decomposition and higher fungal population, the availability of nutrients (carbon, nitrogen, and sulfur) through its dissolved components (amino acids, monomeric sugars, etc.) is considerably less under the low pH environment owing to the poor dissociation of the acidic functional groups (Andersson et al., 2000; Baldock, 2007). In low pH conditions, greater bond formation between positively charged organic constituents and negatively charged clay surfaces may also contribute to the poor dissolution of organic constituents containing carbon, nitrogen, and sulfur, affecting their availability (Curtin et al., 1998). In addition to poor availability of basic cations, nitrogen, sulfur, and phosphorus availability is also significantly reduced in acid soils by virtue of their strong adsorption to native iron and aluminum oxides and hydroxides, precipitation as insoluble secondary minerals (strengite, variscite, etc.), and/or very tight attachment through stronger Al/Fe-O-P bonds at clay surfaces. Similarly, acid soils with poor CEC also bear less exchangeable potassium in the clay colloidal surfaces and hence, K^+ ions are more prone to leaching and becomes less available than in neutral to alkaline soils. Therefore, deficiency of phosphorus and potassium in strongly acidic soils may significantly affect crop production as a result of poor chemical soil health.

The solubility, mobility, and hence, the availability of trace elements in soil depends on their partition between the solid and liquid phases by means of precipitation-dissolution reactions (Forstner, 1995; Rieuwerts et al., 1998). Under acidic soil conditions, cationic trace elements (Fe^{2+}, Mn^{2+}, Zn^{2+}, etc.) are generally more soluble owing to their high desorption but low adsorption (Bradl, 2004) and may become toxic to the plants. The depth-wise distribution of different micronutrients in the acidic Terai soil of West Bengal was found to be significantly influenced by the different forms of soil acidity (Pati & Mukhopadhyay, 2011). Boron is most soluble under acid conditions and is present as boric acid (H_3BO_3), which is prone to leaching loss owing to high rainfall. Therefore, boron deficiency is quite common in coarse-textured acidic soil and in soils with low organic matter content (Das et al., 2019; Mandal et al., 2004). The problem of molybdenum availability to plants growing in acid soils is widespread and becomes more severe, because molybdenum is both a trace element and highly insoluble in acid soils. In soils with low to medium organic matter content, copper rarely presents any deficiency problems, although in acid sandy soils deficiencies because of leaching losses of copper are possible. Therefore, soil acidity leads to lower solubility of labile SOM constituents, resulting in poor availability of nitrogen and sulfur through their dissolution; lower availability of inorganic phosphorus, potassium, other secondary nutrients, and some micronutrients; and toxic levels of some trace elements. These factors collectively deteriorate the soil health and quality.

Biological Soil Health

Among the biological soil heath indicators, soil respiration per unit of microbial biomass or the metabolic quotient (qCO_2) acts as a robust index to show the organic substrate utilization efficiency by soil microbes under specific conditions. In acidic soil, qCO_2 is 2.5 times greater than in the neutral soils (Blagodatskaya & Anderson, 1998), indicating less carbon availability per unit microbial biomass production and hence, poor utilization. Lower soil pH than the microbial internal cell pH (usually ~6.0) in acidic soils demands higher energy for maintaining the integrity of microbial cells and therefore reduces total microbial biomass produced (Anderson, 2003). The phenomena result in suppression of microbial growth, reproduction (especially of bacteria and fungi), and extracellular enzyme activities and eventually the important biological processes (such as nutrient cycling), which significantly deteriorates the biological health of the soil. Soil acidity also significantly reduces the organic matter decomposition rate, with corresponding decreases in net nitrogen mineralization and net nitrification compared with neutral and alkaline soils (Khalil et al., 2005). Hence, while some soils rich in total nitrogen could potentially be productive through significant N mineralization and nitrification, soil acidity may block such avenues and induce poor soil biological health, thus limiting the plant nitrogen uptake.

Many soil enzymes such as alkaline phosphatase (pH optima ~8.5–11) and organophosphorus hydrolase, along with some proteolytic and oxidative enzymes (pH optima ~7–9), exhibit optimum function at higher soil pH. Thus, the lower soil pH in acidic soil significantly hampers their activity. As a result, there is very little conversion of organic phosphorus and nitrogen from phosphate-rich and protein-like organic constituents to their inorganic forms and poor soil health. Soil acidity significantly hinders the root nodulation of the legume plants and the biological nitrogen fixation (Quilliam et al., 2013). Overall, soil acidity adversely affects the microbial population, its metabolic quotient, and extracellular enzyme activities, which has a detrimental impact, primarily on SOM mineralization and the nutrient cycling, and ultimately undermines soil health.

Approaches for Managing Acid Soils

Several approaches have been adopted to manage acid soil (Figure 12.6) for improving soil health and achieving better crop growth to develop sustainable agriculture production. Those approaches are discussed in this section.

Liming

Application of liming materials (i.e., oxides, hydroxides, carbonates, and silicates of calcium or calcium–magnesium mixtures) is an important intervention to restore soil acidity. Liming materials undergo chemical reactions in soil and shift soil pH from acidic to near neutral levels and thus improve the availability of nutrients to plants (Bhat et al., 2010; Das & Saha, 2014; Badole et al., 2015). According to Badole et al. (2015), neutralization of soil acidity by liming is highest in Entisols followed by Inceptisols and Alfisols. Liming is also beneficial to offset the effect of aluminum toxicity (Elisa et al., 2016), which would otherwise adversely affect crop productivity (Seguel et al., 2013). Lime application significantly increases crop yield, improves soil organic carbon status and soil respiration, and reduces greenhouse gas emission (Wang et al., 2021). Along with liming, INM with chemical fertilizers and organics was found to significantly improve maize and groundnut productivity in acidic soils of northeast India

(Dey et al., 2022; Kumar et al., 2012). However, frequent application of liming material is required to neutralize subsurface soil acidity, which would otherwise limit plant growth, development, and yield (Carmeis Filho et al., 2017).

Alternative Low Cost Liming Materials

Alternative cheap sources of liming materials such as sludge from paper mill by-products, basic slag from the steel industry, ferrochrome powder (trade name Bhusakti), and TISCO slag powder (trade name IRL clay) are largely available and can be used to reclaim acid soils. Several case studies for enhancing crop productivity in acid soils of eastern India using low-cost liming materials other than conventional agronomic and chemical calcite ($CaCO_3$) treatment have been conducted and reviewed. Ameliorating acid soils of eastern India with paper mill sludge at 50% lime requirement and 100% NPK produced higher net return in rainfed acidic uplands during the rainy season (Kar et al., 2010). It was also found that basic slag (applied at one-fifth of lime requirement) was more beneficial in improving nutrient availability in red and lateritic acid soils, compared with calcite low-cost liming material (Bhat et al., 2010; Figure 12.7). Application of basic slag or lime at 2 t/ha along with the recommended dose of NPK fertilizers significantly enhanced the yield and nutrient uptake of rice as well as the available nutrient status in acid soils of Karnataka (Mamatha et al., 2018). The presence of Ca^{2+} in waste slag produced from different steel companies maintained high alkalinity in soil for a longer period, thus those materials can be used to ameliorate soil acidity. The alkaline pH of those materials helped to reduce the mobility of heavy metals in soil solution and transformed them from being available to having a more stable form by strong binding mechanisms, thus improving the overall quality of acid soil (Ghisman et al., 2022). Application of silicon-rich amendment such as fly ash

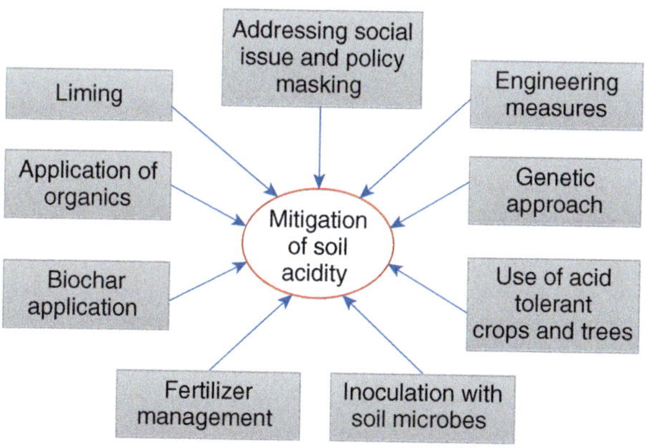

Figure 12.7 Different approaches to mitigate soil acidity.

and steel slag has also been found to help neutralize soil (Gu et al., 2011). The yellow gypsum (steel industry by-product) has been commercially used in vast areas of south India to ameliorate acid soils (Laxmanarayanan et al., 2022). Phosphogypsum, a by-product of diammonium phosphate factories, can also ameliorate subsoil acidity in light-textured soils as the calcium in it leaches faster than that in lime (Majumdar et al., 2022).

Application of Organic Amendment

In general, continuous addition of organic manure improves soil structure and porosity, reduces bulk density (BD), and increases carbon sequestration and nutrient availability (Purakayastha et al., 2019; Das et al., 2021), thereby contributing to the sustainability of crop production. Combined application (one-fifth of lime requirement) of farmyard manure or poultry manure with basic slag decreases soil acidity and increases nutrient availability in acid soil (Bhat et al., 2010; Figure 12.8). Long-term use of organic fertilizer alone decreases soil acidity, and integrated use of organic and mineral fertilizers increases crop yield in acid soils (Wang et al., 2019). The addition of crop residue in acidic Alfisol of northwest India significantly increased soil pH and decreased exchangeable Al^{3+} concentration (Panda et al., 2017). The pH of acid soil is increased owing to adsorption of organic anions and subsequent release of OH^- ions from mineral surfaces (Hue, 1992). The organic manure itself contains carbonyl groups that adsorb protons in acid soil and increase soil pH. Application of cattle manure was also found to increase soil pH by 15%–25% owing to the buffering capacity from

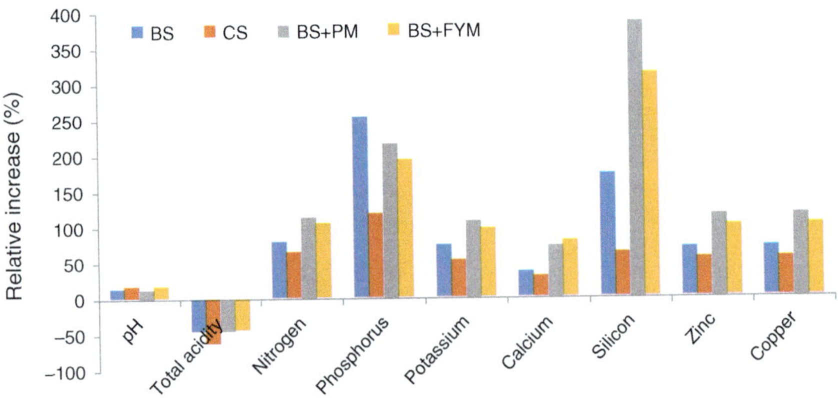

Figure 12.8 Effect of liming material on relative increase of pH, total acidity, and available nutrient in soil. credit: Bhat et al. 2010/with permission of ELSEVIER.

bicarbonates and organic acids present in the manure (Whalen et al., 2000). Application of organic manure with lime decreases the cadmium content in acid soil as well as its uptake by plants (Kumarpandit et al., 2017). In addition, treatment of acid soils with plant residue composts, urban waste compost, manures, and coal-derived organic products increases soil pH, decreases aluminum saturation, and improves conditions for plant growth.

Biochar Application

Biochar is a carbon-rich material obtained by pyrolysis of biomass. Application of biochar to improve soil fertility and carbon status is gaining remarkable importance. Pyrolysis of plant products generally produces alkaline biochar (Yuan et al., 2011), which has effectively been used to remediate acid soil (Fidel et al., 2017). Biochar acts as the most reactive material in acid soils owing to the presence of several alkaline functional groups, such as carbonyls (COO^-), carbonates (CO^{-3}), and phosphates (PO_4^{-3}), among others. Thus, addition of biochar in acidic soils neutralizes soil acidity and increases its pH (Dai et al., 2017). The effectiveness of biochar in ameliorating soil acidity positively correlates with its alkalinity (Shi et al., 2019).

In addition, the large surface area, high surface charge density, porosity, and presence of abundant functional groups allow biochar to adsorb and immobilize Al^{3+}, reduce its toxicity, and increase soil pH (Geng et al., 2022; Qian & Chen, 2013). Beyond ameliorating acid soil, biochar also improves CEC, enhances water retention capacity, reduces soil BD, and ultimately increases crop yields (Das et al., 2020). Biochar also leads to carbon sequestration with increased nutrient availability (potassium, calcium, and magnesium) and microbial diversity in soil, which helps to alleviate oxidative stress and promote crop growth and the overall quality of acid soil (Figure 12.9; Geng et al., 2022). However, pyrolysis temperature affects the impact of biochar on soil properties in acid soils.

Fertilizer Management

The proper management of applied mineral fertilizers by reducing the nitrogen fertilizer input and improving use efficiency of applied N (NUE) to decrease nitrate leaching is a promising way to mitigate soil acidification (Zeng et al., 2017). Improving NUE by proper management of the crop rhizosphere zone and synchronizing crop demand with the supply of nitrogen input can arrest nitrate leaching to some extent. Use of nitrate-based fertilizers has less capacity to acidify soils compared with ammonium-based fertilizers. Therefore, basic fertilizers such as sodium nitrate, basic slag, and so forth reduce the acidity in soils. Application

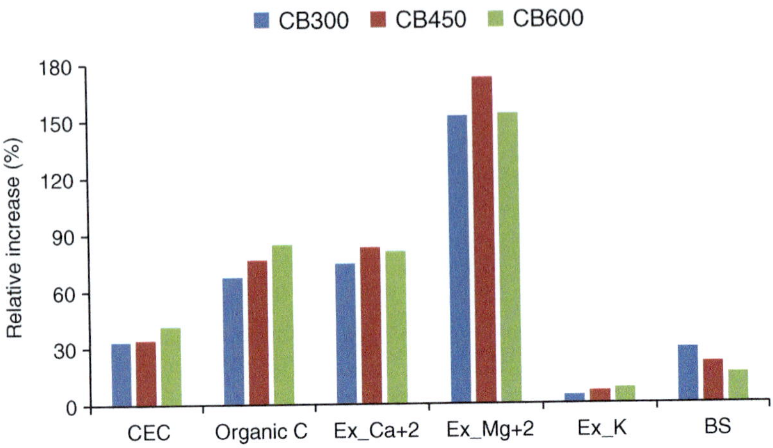

Figure 12.9 Effect of cow dung biochar pyrolyzed at 300 °C (CB300), 450 °C (CB450), and 600 °C (CB600) temperature on relative increase of different soil parameters over the control in an acid soil. Credit: Geng et al. 2022/with permission of ELSEVIER.

of phosphate fertilization is very beneficial for sustaining crop production in acid soils. Addition of rock phosphate decreases exchangeable aluminum, increases soil pH and available phosphorus content, and acts as a promising amendment for tropical acid soils (Utomo, 1995).

Integrated use of mineral fertilizer with enriched compost, biofertilizers, and organic manures has a beneficial effect in acid soils (Patra et al., 2021). Integrated use of mineral fertilizer with either farmyard manure or liming material significantly increases crop yield and improves soil organic carbon and total nitrogen status in an acid soil (Manna et al., 2007; Figure 12.10). Continuous application of inorganic fertilizers, farmyard manure, lime, and biofertilizers significantly improves crop yield and organic carbon status in acidic soils of the NEH ecosystem (Saha et al., 2010). Application of nano-fertilizers increases use efficiency of different nutrients, owing to the slow release and super absorbent property of the fertilizers, and improves the overall quality of soil (Khanam et al., 2022). Therefore, it represents a promising technology for sustainable crop production and overall development of acid soils.

Inoculation with Soil Microbes

Inoculation of soil and seed with different acid-tolerant soil microbes can help the host plant adapt to an acidic soil environment. Soil inoculation with arbuscular mycorrhizal fungi (AM) is a fast and promising tool for alleviating the problems of soil acidity, namely, the detrimental effect of low pH and Al toxicity

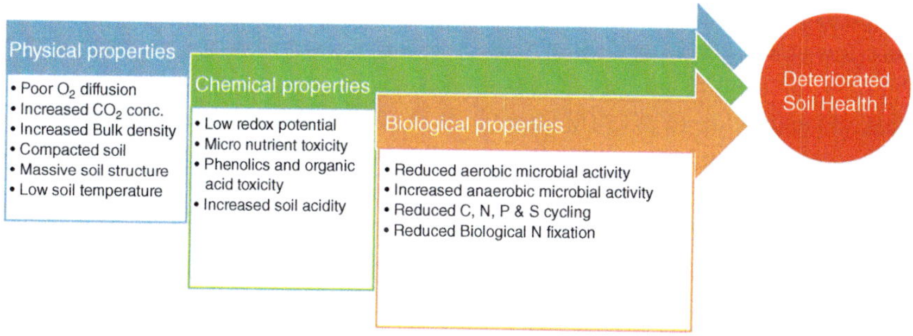

Figure 12.10 Effect of waterlogging on soil health.

on crop performance in acid soils (Marschner, 1995; Rouphael et al., 2015). Plants with AM inoculation were found to perform better under low pH and aluminum toxicity conditions; produced higher yields and improved fruit quality by increasing total soluble-solids content, phosphorus, and iron concentrations; and had increased nutritional status of both the macro- and micronutrients in leaves of host plants compared with non-inoculated ones (Rouphael et al., 2015). Plant growth-promoting microbes could potentially enhance sustainable plant growth and also help host plants adapt to extreme changes in soil pH (Msimbira & Smith, 2020). Rhizobia can produce extracellular polysaccharides (Gopalakrishnan et al., 2015) such as glutathione, a tripeptide that enhances tolerance for acidic pH (Wang et al., 2019). Rhizobia can also accumulate high levels of potassium and phosphorus and sustain them in very low soil pH (Watkin et al., 2003). Seed inoculation with acid-tolerant strains of phosphorus-solubilizing bacteria is beneficial for creating a greater rhizosphere effect to improve phosphorus nutrition of crops and boost crop yield in low-productivity hill soil (Pal, 1998).

Acid-Tolerant Crops and Trees

Both submerged and short duration upland rice may be able to tolerate soil acidity. Other than rice, millets, buckwheat, rye, groundnut, chickpea, sweet potato, potato, yam, and watermelon are some acid-loving crops that can tolerate 4.5–5.5 soil pH and are considered highly tolerant to soil acidity. Some moderately tolerant (pH 5.5–6.8) crops are barley, wheat, corn, maize, soybean, sunflower, linseed, sesame, lentil, beans, carrot, cucumber, brinjal, pea, chili, radish, tomato, squash, garlic, and colocasia (Ganeshamurthy et al., 2016; Majumdar et al., 2022; Ray et al., 2015). Some fruit crops, such as citrus, gooseberry, peach, watermelon, apple, strawberry, passion fruit, papaya, guava, litchi, kiwi, pineapple mango, and plum, as well as tea, can tolerate 4.5–7.0 pH and can be grown in strong to moderate acid soils (Ray et al., 2015). In addition, fruit crops such as strawberry, gooseberry, bael, elephant apple, plum, pineapple, orange, litchi, passion fruit, and jackfruit can tolerate soil acidity up to pH 5 (Majumdar et al., 2022). According to Panwar et al. (2011), soil fertility index and soil evaluation factors in acidic soil of Assam are both higher under forest and plantation crops than under agricultural crops. Some of the best known acid-tolerant tree species are *Albizia* sp., *Alnus nepalensis*, *Erythrina* sp., *Gmelina arborea*, and *Sesbania* sp., among others. In association with mycorrhizae and rhizobia, many of these tree species have great potential for growth and production on acid soils. These tree species also have capacity to ameliorate aluminum toxicity and to have increased crop yields in acid soils.

Genetic Approach

Adoption of acid-tolerant crop varieties is another approach to mitigate the impact of soil acidity. Traditional breeding methods, such as backcrossing, intercrossing, and others, can be used in production of acid-tolerant cereal varieties that can significantly decrease the impact of soil acidity on plants (Bian et al., 2015; Kang et al., 2010). Acid-tolerant wheat and barley varieties were reported decades ago by Foy et al. (1965). The submergence-tolerant rice variety IR53650 developed by the International Rice Research Institute and japonica-indica rice varieties with high-yielding potential can tolerate low pH and are suitable for acid sulfate soils (Kang et al., 2007, 2010). The identified crop varieties Smruti (groundnut) and UPAS 120 (pigeon pea) in Odisha are relatively tolerant to soil acidity ($pH < 5.5$; ICAR, 2012). The advanced molecular techniques can explain mechanisms and detect aluminum tolerance genes, and thus, they offer better options for marker-assisted selection in genetic breeding (Bian et al., 2013). The aluminum tolerance genes, such as *AtBCB*, *parB*, *NtPOX*, and *NtGDI1*, in transgenic *Arabidopsis* plants resulted in better relative root growth under an aluminum toxicity environment (Ezaki et al., 2000). The secretion of organic anions such as citrate and malate from plant roots plays an important role in excluding and detoxifying aluminum in acid soils (Kochian et al., 2004). A new malting barley variety that tolerates aluminum toxicity and low pH has been developed by selecting suitable tolerant alleles (Bian et al., 2015) by the marker-assisted selection technique, and it can be used in acid soil for sustainable production. Some acid-tolerant crop varieties in India are listed in the Table 12.1.

Engineering Measures

Conservation measures such as ridge and furrow tillage, conservation tillage, contour and strip cropping, bench terracing, raising of field bunds for in situ rain water harvesting, runoff water harvesting and recycling, compaction of surface layers with heavy rollers/heavy tractors, chiseling of soils with surface compact

Table 12.1 Acid tolerant (pH < 5.5) crop varieties.

State	Crops	Varieties
Assam	Rapeseed Summer green gram	Varuna, Sonmukhi K851, Sonmugu
Himachal Pradesh	Soybean Gobhi sarson	Bragg, Pb1, Harasoya ONK1, HPN3
Jharkhand	Black gram	KU 301
Kerala	Vegetable cowpea Cowpea (bush type)	Bhagyalakshmi V 16
Meghalaya	French bean	HUR 15
Odisha	Groundnut Pigeon pea	Smruti Upas-120
West Bengal	Mustard Wheat	Sanjukta, Pusa Bold K 9107, PBW 343

Note. Adapted from Mandal et al. (2019).

layers to a depth of 30–45 cm at 60- to 120-cm intervals are some important measures to mitigate soil acidity.

Addressing Social Issue and Policy Making

Most farmers are unaware of the ill effects of soil acidity until and unless there is drastic yield reduction. Therefore, providing proper training and disseminating knowledge to the farmers might help them manage acid soil in a better way. In addition, strong policies regarding the indiscriminate use of mineral fertilizers, periodic estimation of soil pH, use of improved varieties of seeds, and so forth, could open a new domain in managing soil acidity. The ambitious soil health card program of the Government of India used at mass scale in farmers' fields, soil pH reports, and pH correction through INM have reduced chemical fertilizer use by 10%–15%. Standard lime (< 60 mesh size) quality control and subsidy are long-pending issues in policy-making to tackle productivity loss in acid soils. Subsidies for alternate cheap liming sources (e.g., paper mill sludge) is another agenda of universal government policy.

Ecosystem Services of Acid Soil

Although acid soils are considered as degraded soils, they provide various ecosystem services for overall development of mankind. The eastern Himalayan region (EHR), which includes northeastern states and hilly parts of West Bengal, has acid soils and is one of the largest tea-growing areas in the world. A unique agroforestry system of tea under a canopy of trees covers 0.5 Mha of land in the EHR (Nath et al., 2021). The northeastern region alone contributes 16% of world's total tea production and nearly half of the total tea production of India as well as the major tea exporter of the country (Deka, 2010).

Aside from growing tea, the acid soil of the EHR is also suitable for production of coffee, rubber, different fruit crops, and floriculture. The previous forest area of the EHR was converted to rubber and areca plantation, and 84% of the EHR area is suitable for agro-forestry intervention, with 23% high and 71% good ecological suitability (Nath et al., 2018).

The hot tropical climate in the states of Kerala, Goa, Karnataka, and Tamil Nadu and in some parts of Assam and Tripura has resulted in the development of strongly to moderately acidic Inceptisols, Alfisols, and Mollisols. Furthermore, weathering under this climate will ultimately convert this soil into Ultisols with considerable amounts of layer silicate minerals (Pal et al., 2012a,b). However, the calcium-rich zeolites present in soils of Western Ghat and Konkan region prevent the total transformation of smectite into kaolinite and arrest the formation of base-poor Ultisols (< 35%). Naturally, these acid soils are loaded with organic carbon (>1.0%), sufficient water availability (10–15%), moderate to high shrink-swell character (coefficient of linear extensibility > 0.06), high base saturation (>50%), and high exchangeable calcium–magnesium ratio (>2.0), and they respond well to different management interventions. Therefore, despite being acidic in nature, these soils have improved physical, chemical, and mineralogical properties (Pal et al., 2018) and cannot be considered as degraded soil. These soils support abundant forest vegetation, suitable for production of rubber, coffee, tea, spices, and other horticultural crops such as mango and cereals (Kumar & Takeuchi, 2009; Pal et al., 2018). However, periodic monitoring and liming are needed to maintain the adequate base saturation and calcium–magnesium ratio.

Waterlogging and Soil Health

Globally, flooding is one of the most damaging abiotic stresses, affecting 17 million km^2 of land surface annually (Kaur et al., 2020), and continuous waterlogging due to flooding adversely affects the crop growth (Lone et al., 2018). According to the Intergovernmental Panel on Climate Change (IPCC, 2014), global climate change causes waterlogging events to be more frequent, severe, and unpredictable. Waterlogging affects 16% of the soils in the United States, 10% of the agricultural lands of Russia, and irrigated crop production areas of India, Pakistan, Bangladesh, and China (Food and Agriculture Organization [FAO], 2015). Globally, 10%–12% of the agricultural area is affected by waterlogging or severe soil drainage constraints (Shabala, 2011). Agricultural intervention in waterlogged areas is challenging as it causes land degradation and makes the soil unproductive. Soil waterlogging determines the direction and rate of chemical, geochemical, and biological reactions in soil and may indirectly influence soil physical, chemical, and microbiological properties (Fahmi et al., 2015). The changes in soil properties caused by waterlogging are presented in Figure 12.10. Lowland rice in India is still challenging for second crops (winter) because in most months (July–December) of the year, the soils remain waterlogged. Such lowlands cover significant areas of the eastern and northeastern states of India. Some areas might be converted to wetland (a type of peat soils) but drying up or growth of water hyacinth causes ecologic imbalance, as well as loss of biodiversity and ecosystem services in soil. Therefore, these waterlogged soils require appropriate management if they are not suitable for rice cultivation.

Soil Physical Properties

Soil waterlogging leads to swelling of soil colloids, increased BD, and reduced aggregate stability and permeability of soil. The increased BD in waterlogged soil is associated with the destruction of soil aggregates and the soil structure, which leads to soil compaction (Reddy & DeLaune, 2008). Long-term submergence leads to hardpan formation in subsurface layers and thus decreases the water infiltration and root hydraulic conductivity. Decreased root hydraulic conductivity has been reported to decelerate the leaf expansion in plants owing to induced stomatal closure and reduced leaf water potential, which negatively affects the photosynthetic food production, and hence, the yield of the plants (Else et al., 2008). Waterlogging also lowers the soil temperature, adversely affects the plant growth, and deteriorates soil health, especially for winter season crops (Zhen et al., 2020). The decreased aggregate stability in waterlogged soil is due to the depletion of O_2 and subsequent use of various elements for redox, which contributes to decreased aggregation. Hence, in continuously flooded areas, such as marshlands and rice fields, aggregate stability rarely improves (Rangasami et al., 2022). The depletion of O_2 increases the CO_2 concentration and poor soil aeration adversely affects the soil-water-plant relationship, thereby creating ecological imbalance and deteriorating soil health.

Soil Chemical Properties

Oxygen-deficient anaerobic waterlogged soil causes toxicity because of excess production of CO_2 and methane, which reduces the soil–atmosphere gaseous exchange. Secondary metabolites, such as phenolics and volatile fatty acids produced in submerged soil, cause injuries to the sensitive root tip (Goud et al., 2022).

Increased production of ethylene, acetaldehyde, and reactive oxygen species in plants under continuous submergence decreases the chlorophyll content, reduces photosynthetic capacity, and ultimately causes a decline in crop productivity. Waterlogging generally tends to neutralize the soil pH and to increase phosphorus, sulfur, and potassium availability in soil (Hemati Matin and Jalali, 2017). However, hypoxic or anoxic conditions and lower temperatures in waterlogged soil inhibit the nitrogen fixation mechanism and causes nitrogen deficiency. Waterlogging decreases nitrification increases denitrification, and ultimately reduces crop productivity. Waterlogging also lowers the redox potential and transforms the Fe^{3+} to Fe^{2+} and Mn^{4+} to Mn^{2+}, which increases their solubility and makes them toxic to plant roots. In addition, waterlogging causes an insufficient supply of sulfur owing to reduction of SO_4^{2-} and leads to hydrogen sulfide (H_2S) toxicity in iron-deficient soils (Bloomfield, 1969), thus causing deterioration of soil health.

Soil Biological Properties

Microbes and their enzymatic activities play vital roles in soils' biological health as they control the decomposition of SOM and the mineralization of carbon, nitrogen, phosphorus, and sulfur, and hence, their biogeochemical cycles. However, waterlogging conditions in soil restrict the free diffusion of atmospheric O_2 to soil and lower its soil concentration by ~10,000 times, which directly affects the microbial growth and activity (Pan et al., 2021). Reduced dehydrogenase and peroxidase enzyme activities act as indicators of cell respiration and biological SOM oxidation through hydrogen transfer (Tian & Shi, 2014; Xu et al., 2016). Additionally, prolonged soil submergence significantly drops the β-glucosidase, urease, acid and/or alkaline phosphatase, and aryl sulfatase enzyme activities, which are key players in carbon, nitrogen, phosphorus, and sulfur cycling in soil, respectively (Liu et al., 2015). Therefore, waterlogging directly hampers the nutrient cycling through decelerated nutrient mineralization and overall effects on the biological soil health.

Many reports have indicated that waterlogging strongly affects the obligate aerobic ammonia oxidizing and nitrifying soil communities, substantially affecting the native nitrogen-supplying capacity of the soil (Nguyen et al., 2018a, 2018b; Jiang et al., 2013). Apart from nutrient cycling, waterlogging has also been proved to be detrimental to rhizobium bacteria activity and biological nitrogen fixation potential of legume plants. Poor O_2 availability and substantially low soil temperature under prolonged submergence hampers the active nodulation process in roots and decreases the acetylene reduction activity responsible for nitrogen fixation (Jung et al., 2008); hence, the biological nitrogen fixation activity is affected, contributing to poor biological health of soil. Prolonged waterlogging may also sometimes result in conditions favoring proliferation of anaerobic microbes that generate volatile gases such as H_2S, CH_4, N_2O, PH_3, and others from decomposition of SOM. These gases may directly affect the plants through their toxicity (e.g., H_2S toxicity in rice) and/or may act as a means of gaseous loss of nutrients (e.g., nitrogen loss through denitrification) from soil, contributing to poor soil health. Overall, waterlogging significantly affects the growth and activity of the microbial communities in soil that directly interferes with nutrient cycling and associated important biological soil processes, ultimately leading to poor biological health of the soil.

Mitigation of Waterlogged Soils

Many soil management and crop management practices can mitigate the stresses associated with waterlogging. Important soil management practices that may be recommended in areas prone to water stagnation are discussed in this section.

Soil Management Practices

Surface drainage implemented by constructing channels such as spoon-drains, W-drains, and interceptor drains provides for safe removal of excess water from agricultural fields. Such techniques have been used very frequently to alleviate the conditions of waterlogging (Cox et al., 1994; Ayars & Evans, 2015).

Subsurface drainage removes excess water that has accumulated in the subsurface layers owing to the formation of hardpan, and it ensures a suitable environment in the root region (Xian et al., 2017). Subsurface drainage can be done through tile drains, a grid of perforated pipes, or mole drains, which are pipe-like structures based on the cohesion of clay particles (Filipović et al., 2014).

Strategic deep tillage and subsoil manuring are two more techniques that can be used to remove excess water from fields. Deep tillage loosens compacted subsurface soil layers, particularly in heavy-textured soils of humid regions where clay migration from surface to subsurface layers is prominent. Deep tillage with a deep ripper, spader, rotary, moldboard plow, or disk plow improves the drainage of stagnant water in subsoil, and it provides an environment for adequate root growth and productivity in no-till systems (Kuhwald et al., 2017; Scanlan & Davies, 2019). Subsoil manuring with high rates of nitrogen-rich organic materials following the deep tillage operations may further help to maintain the porous subsurface drainage condition in those soils.

In raised bed systems, crops are gown in raised beds, which enables avoiding waterlogged conditions (Hussain et al., 2018). Moreover, shallow drains between the beds allow excess stagnant water to easily flow out and ensure normal physical soil conditions for crop growth.

Controlling extensive unsystematic trafficking by farm machinery/vehicles that otherwise build hardpans below the plow layer, disrupt the soil aggregation, and facilitate the occurrence of waterlogging may prove to be an important soil management practice (Hamza & Anderson, 2005).

Crop Management Practices

Bio-drainage involves planting certain trees species, such as *Eucalyptus*, *Accacia*, and *Zyzyphus*, within the boundaries of agricultural fields. The presence of trees helps lower the groundwater table below the critical zone owing to their high-water absorption and transpiration rates. Inclusion of perennial deep-rooted herbaceous legumes such as lucerne, clovers, and others, is also able to extract high amounts of water, dry the soil to greater depths, and reduce the occurrence of waterlogging (Nichols, 2018).

Early and higher sowing and the use of vigorous agricultural crops may increase the crop water use, and thus helps to decrease the incidence of waterlogging (Ali et al., 2018; Manik et al., 2019).

Nutrient application based on deep placement of nitrogen fertilizers, application of slow and controlled release fertilizers, such as sulfur-coated urea and nitrification and urease inhibitors, may prove fruitful in mitigating the effect of water stress and increase nitrogen uptake in plants.

Plant growth regulators, such as auxins and cytokinins, can be applied at suitable growth stages. These regulators may mitigate waterlogging damage of plants and help to improve plant growth under waterlogged stress conditions (Wu et al., 2018).

Finally, the use of tolerant species and varieties can help to offset the impact of waterlogging. Introduction of waterlogging-tolerant plant varieties into the cultivation practices simply avoids waterlogging stresses (Tewari & Mishra, 2018; Wani et al., 2018).

Waterlogging and Soil Acidity

Continuous submergence creates a reduced environment, which decreases the activity of microorganisms (Nguyen et al., 2018b) and leads to incomplete oxidation of SOM and the consequent production of volatile organic acids. The higher concentration of dissolved CO_2 and volatile organic acids in stagnated water shifts the soil pH to more acidic and decreases the availability of phosphorus due to precipitation and subsequent formation of insoluble phosphate compounds in an acidic environment.

Acid Sulfate Soil

In addition to the direct effects on different soil health characteristics, prolonged waterlogging and subsequent drying-up conditions are responsible for the development of soil acidity in tropical coastal wetland areas of India (Sen, 2017). The term *acid sulfate soil* can be used to refer to soils with a pH < 4.0 (1:2.5 in 0.01 M $CaCl_2$ solution), having spots of jarosite or a soluble $SO_4^{-}S$ content more than 0.05% within 0.60 m of the soil depth (Dent, 1986). Acid sulfate soils can also be characterized by a very low pH (<3.5) and the presence of sulfidic materials (oxidizable sulfur compounds, such as pyrite) and/or a sulfuric horizon (Shamshuddin et al., 2014). The development of acid sulfate soils commonly occurs in coastal and estuarine regions with a sulfidic horizon under anoxic reducing conditions beneath the water table on account of microbial reduction of sulfates to sulfides (S^2) by anaerobic microbes, such as *Desulfovibrio* sp. and *Desulfotomaculum* sp. in the presence of ferrous ions in soil (Dent & Pons, 1995; Michael, 2013). Globally, ~17–24 Mha of acid sulfate soils is found in tropics and subtropics (Michael et al., 2016), out of which Asia, Africa, Australia, Latin America, and North America constitutes 6.5, 4.5, 3.0, 3.0, and 0.1 Mha, respectively (Simpson & Pedini, 1985). The area of acid sulfate soils in India is approximately 390,000 ha. These soils are found in Kerala, Orissa, West Bengal, Andhra Pradesh, and Tamil Nadu. Acid sulfate soils can be either potential (i.e., acid sulfate soil with sulfidic material), active (i.e., acid sulfate soil with sulfuric horizon), or postactive (Fanning, 2012; Michael et al., 2016). Because of their pale-yellow color and infertile nature, they were referred to as *Kattakali* meaning Cat Clays by Dutch farmers and *Maibolt* by Germans (Sarangi et al., 2022). In the Ganges Delta, these soils are locally called *Koimuromati* or *Koshmati* and *Kari* in the Kuttanad region of Kerala (Sarangi et al., 2022). In the West Bengal part of the Ganges Delta, soils with a very low pH (<4.5) are estimated to cover 0.04 Mha (Bandyopadhyay et al., 2003), whereas in Kerela, the total area under acid sulfate soil is estimated to be about 0.014 Mha (Indira & Covilakom, 2013).

These acid sulfate soils under anoxic conditions are rather harmless to crops and ecosystems, but soon after their exposure to air, they become very harmful (Dent & Pons, 1995; Ghosh et al., 2019; Sarangi et al., 2022). The reason for this change is the thermodynamically unstable nature of pyrites (FeS_2) in the presence of air or oxygen. When these soils are exposed for cultivation, oxidation of the pyrites or other oxidizable sulfidic materials occurs, forming free sulfuric acid (H_2SO_4), which results in severe soil acidity (pH < 4; Ghosh et al., 2019).

$$FeS_2 + 15/4\,O_2 + 7H_2O \rightarrow Fe(OH)_3 + 2SO_4^{2-} + 4H^+(\text{acid})$$

Such low soil pH causes solubilization of soil matrices, which in turn releases potentially toxic soil constituents (iron, aluminum, and other heavy metals).

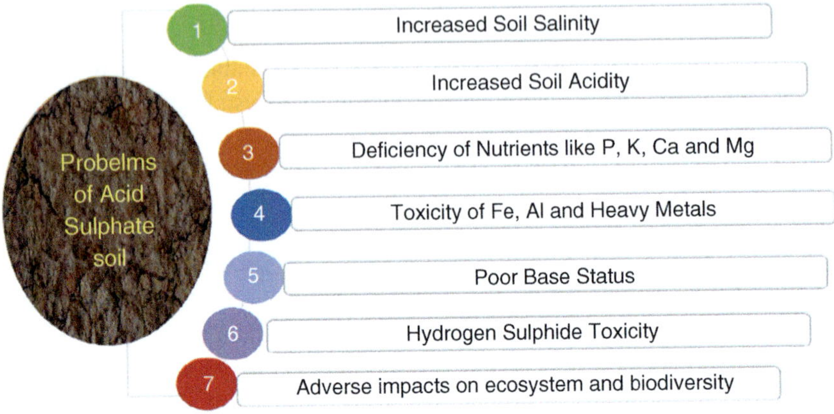

Figure 12.11 Soil-related problems of acid sulfate soils.

Upon accumulation in soils, these constituents have adverse impacts on crops and ecosystems (Michael, 2013; Sen & Bal, 2009). Major soil-related problems associated with acid sulfate soils are listed in Figure 12.11.

Reclamation of Acid Sulfate Soils

The reclamation of acid sulfate soils is necessary for better uses of these underutilized lands. The measures adopted for management of acid sulfate soils are mainly governed by factors such as the extent and severity of acidity developed upon the drainage, depth of nonacid topsoil covering potential acid sulfate soil horizon below, depth of water table during the dry season, and so forth (Burman et al., 2010). Maintaining a high-water table or continuous submergence and selecting suitable tolerant crop varieties for these areas are generally used as preventive and adaptive measures, respectively. In addition, some practical measures of reclamation of acid sulfate soils are explained in the following sections.

Chemical Measures

Use of materials such as limestone, rock phosphate, super phosphate, and manganese dioxide tends to increase pH and improve phosphorus content, while precipitating excess iron and aluminum. The application of liming materials in small quantities at short intervals has been considered more useful as the influence of liming on potential acidity is generally short lived (Thampatti et al., 1998). Application of burnt lime shell, dolomite, and rice husk ash in two splits at basal and 30 days after sowing have been reported to improve rice yield in acid sulfate soils of Kuttanad areas (Devi, 2021). In addition, the use of locally available basic slag, oyster shell (rich in calcium) may also be considered a cheap and better option of ameliorating acid sulfate soils (Sarangi et al., 2022).

Engineering Measures

Use of hydraulic suppression and the treatment of acid drainage water with hydrated lime ($Ca(OH)_2$) increases buffering and precipitates iron and aluminum compounds within the site. Hydraulic separation is suitable for coarse-textured sandy soils containing FeS_2, and hydro-cycloning may be used to hydraulically separate the sulfides from the sandy materials (Osman, 2018).

Use of INM and Inclusion of Organics

The application of lime, rock phosphate, and green manure was reported to be highly beneficial for the amelioration of acid sulfate soils in Sundarbans and led to significant improvements in the growth and yield of *kharif* (summer season) and *rabi* (winter season) crops (Rahman et al., 2005). In addition, INM has also been reported to address hidden deficiencies and nutrient imbalances and to improve and sustain the physical, chemical, and biological parameters of soil (Sarangi et al., 2022). Use of organic matter such as rice hull and poultry manure can reduce the problems of phosphorus fixation, and use of Lucerne hay has been reported to improve soil pH as well as reduce the redox potential in acid sulfate soils (Michael et al., 2016).

Microbial Remediation

Organic matter undergoes microbial degradation under anerobic or flooded conditions. During this process, electrons are generated and then accepted by a sequence of electron acceptors such as nitrate and manganese and iron oxides and sulfate, resulting in the reformation of pyrite with an increase in pH (Bucka et al., 2018). Inoculations with phosphorus-solubilizing bacteria and nitrogen-fixing bacteria produce low molecular weight organic acids or phytohormones, which can chelate aluminum ions. This chelation in turn reduces aluminum toxicity in acid sulfate soils, and when nitrogen-fixing and phosphorus-solubilizing bacteria were applied along with rice husk biochar ($pH = 10$), soil pH increased from 3.89 to 5.36, with significant improvement in rice yield (Panhwar et al., 2020).

Wetland Rice Cultivation Under Submergence

Rice is a semi-aquatic crop, suitable for growing under waterlogged conditions owing to the presence of special aerenchyma tissues. However, care must be taken to maintain the pH of soil solution above 5 by adding certain amendments. Below pH 5, iron precipitates as $FeOH_3$ and coats the root surface, restricting its uptake. Toxic effects due to H_2S gas can also be managed in rice fields by the occasional stirring of the rice soil in between rows to release H_2S in the air and reduce crop damage

(Chaitanya et al., 2017; Sarangi & Islam, 2019). The presence of jarosites above 40 cm of soil depth can also be suitably taken up for rice cultivation (Sarangi et al., 2022).

Conclusions and Way Forward

Soil acidity and waterlogged conditions adversely affect the soil health and hinder the goal of sustainable crop production. The development of acid sulfate soil is the result of prolonged waterlogging followed by drainage. Crop production with surface and subsurface drainage systems, periodic deep tillage and subsoil manuring, raised bed systems, use of bio-drainage facilitating trees, and so forth offers promising strategies to overcome the detrimental impacts of waterlogging on soil health. Application of low-cost liming materials other than the conventional one, addition of biochar and organic amendments, periodic inoculation with soil microbes, and other measures offer potential soil management strategies that help mitigate the detrimental effects of soil acidity and produce sustainable crop yields.

However, to understand the potential and the constraints of acid soils in India, it is necessary to locate these areas in a high-resolution, large-scale map (1:10,000 scale) and characterize them scientifically for land use planning in revised 60 AESRs and 20 agro-ecological regions. The acidic Alfisols, Inceptisols, Mollisols, and Ultisols are not at an alarming stage in India with respect to degradation in terms of soil acidity. In addition, the acidic soils of hot tropical climate should not be considered as chemically degraded soils owing to their high fertility status. However, those soils require improved management practices to maintain their nutrient status and make efficient use of soil water to sustain crop productivity at an enhanced level. Presently, no subsidy is given for land treatment with liming. National Rainfed Area Authority (NRAA), National Land Use and Conservation Board (NLCB), National Wastelands Development Board (NWDB), National Afforestation and Eco-development Board (NAEB), Krishi Vigyan Kendras (KVKs), nongovernmental organizations (NGOs), and various extension agencies can help in implementing the integrated management programs in conjunction with a map of acid soils across different AESRs. These measures can act greatly facilitate meeting the food security needs of the burgeoning Indian population. In addition, emphasis should be given to the following points for sustainability of acid soils:

- Acid soils of Alfisol in India respond to liming amelioration, whereas other soil orders particularly soils of northwest Indian Himalayan states (Uttarakhand, Himachal Pradesh, and Jammu & Kashmir) may not respond. Scant research has been done to address this issue, and few researchers have assessed the importance of pH-dependent charges or zero point of charge of these acid soils.

- Acid soils of India should be characterized based on the extent of calcium and phosphorus deficiency and the toxicity of heavy metals. They should be mapped based on agro-ecoregions for which an action plan needs to be developed.
- In large areas of acid soils, zinc deficiency is reported despite zinc solubility increasing with soil acidity. Application of di-ammonium phosphate as the nitrogen and phosphorus source might induce such deficiency in acid soil. However, appropriate research work is needed to understand the soil chemistry underlying zinc deficiency.
- Liming materials used for the amelioration of acid soils should be properly characterized and their specifications should be entered in the fertilizer control order.
- The basic chemistry underlying the mechanisms of yellow gypsum for ameliorating acid soil also represents an interesting topic of research.
- Periodic monitoring of geo-reference benchmark soils for soil acidity development can also be a research agenda in developing digital soil map techniques.

References

Ali, N., Dayal, A., Thomas, N., Lal, G., & Gupta, J. (2018). Effect of different sowing time on seed vigour parameters of wheat (*Triticum aestivum* L.) varieties. *International Journal of Pure and Applied Bioscience*, *6*, 1532–1538.

Anderson, T. H. (2003). Microbial eco-physiological indicators to assess soil quality. *Agriculture, Ecosystems & Environment*, *98*(1–3), 285–293.

Andersson, S., Nilsson, S. I., & Saetre, P. (2000). Leaching of dissolved organic carbon (DOC) and dissolved organic nitrogen (DON) in mor humus as affected by temperature and pH. *Soil Biology and Biochemistry*, *32*(1), 1–10.

Ayars, J. E., & Evans, R. G. (2015). Subsurface drainage—What's next? *Irrigation and Drainage*, *64*(3), 378–392.

Badole, S., Datta, A., Basak, N., Seth, A., Padhan, D., & Mandal, B. (2015). Liming influences forms of acidity in soils belonging to different orders under subtropical India. *Communications in Soil Science and Plant Analysis*, *46*(16), 2079–2094.

Baldock, J. A. (2007). Composition and cycling of organic carbon in soil. In P. Marschner & Z. Rengel (Eds.), *Soil biology: Nutrient cycling in terrestrial ecosystems* (Vol. 10, pp. 1–35). Springer-Verlag.

Bandyopadhyay, B. K., Maji, B., Sen, H. S., & Tyagi, N. K. (2003). *Coastal soils of West Bengal—Their nature, distribution and characteristics* (Bulletin 1/2003). Regional Research Station, CSSRI.

Bhat, J. A., Kundu, M. C., Hazra, G. C., Santra, G. H., & Mandal, B. (2010). Rehabilitating acid soils for increasing crop productivity through low-cost liming material. *Science of the Total Environment*, *408*(20), 4346–4353.

Bian, M., Jin, X., Broughton, S., Zhang, X. Q., Zhou, G., Zhou, M., Zhang, G., Sun, D., & Li, C. (2015). A new allele of acid soil tolerance gene from a malting barley variety. *BMC Genetics, 16*(1), Article 92

Bian, M., Zhou, M., Sun, D., & Li, C. (2013). Molecular approaches unravel the mechanism of acid soil tolerance in plants. *The Crop Journal, 1*(2), 91–104.

Blagodatskaya, E. V., & Anderson, T. H. (1998). Interactive effects of pH and substrate quality on the fungal-to-bacterial ratio and qCO_2 of microbial communities in forest soils. *Soil Biology and Biochemistry, 30*(10–11), 1269–1274.

Bloomfield, C. (1969). Sulphate reduction in waterlogged soils. *Journal of Soil Science, 20*(1), 207–221.

Bradl, H. B. (2004). Adsorption of heavy metal ions on soils and soils constituents. *Journal of Colloid and Interface Science, 277*(1), 1–18.

Bucka, F., Kölbl, A., Marschner, P., Fitzpatrick, R., Mosley, L., & Kögel-Knabner, I. (2018). Remediation of acid sulphate soils by addition of organic matter. *EGU General Assembly Conference, 20*, Article 2378

Bünemann, E. K., Bongiorno, G., Bai, Z., Creamer, R. E., De Deyn, G., De Goede, R., Fleskens, L., Geissen, V., Kuyper, T. W., Mäder, P., Pulleman, M., Sukkel, W., van Groenigen, J. W., & Brussaard, L. (2018). Soil quality—A critical review. *Soil Biology and Biochemistry, 120*, 105–125.

Burman, D., Bandyopadhyay, B. K., & Mahanta, K. K. (2010). Management of acid sulphate soil of coastal Sunderban region: Observations under on-farm trial. *Journal of Indian Society of Coastal Agricultural Research, 28*, 8–11.

Carmeis Filho, A. C. A., Crusciol, C. A. C., & Castilhos, A. M. (2017). Liming demand and plant growth improvements for an oxisol under long-term no-till cropping. *The Journal of Agricultural Science, 155*(7), 1093–1112.

Chaitanya, A. K., Badole, S., Gupta, A. K., & Pal, B. (2017). Prevention, reclamation and management of acid and acid sulphate soils. In S. K. Gupta & M. R. Goyal (Eds.), *Soil salinity management in agriculture* (pp. 337–363). Apple Academic Press.

Cox, J. W., McFarlane, D. J., & Skaggs, R. W. (1994). Field-evaluation of DRAINMOD for predicting waterlogging intensity and drain performance in South-Western Australia. *Soil Research, 32*(4), 653–671.

Curtin, D., Campbell, C. A., & Jalil, A. (1998). Effects of acidity on mineralization: pH-dependence of organic matter mineralization in weakly acidic soils. *Soil Biology and Biochemistry, 30*(1), 57–64.

Dai, Z., Zhang, X., Tang, C., Muhammad, N., Wu, J., Brookes, P. C., & Xu, J. (2017). Potential role of biochars in decreasing soil acidification—A critical review. *Science of the Total Environment, 581*, 601–611.

Das, R., Ghosh, A., Das, S., Basak, N., Singh, R., Priyanka, & Datta, A. (2021). Soil carbon sequestration for soil quality improvement and climate change mitigation. In D. Pant, A. K. Nadda, K. K. Pant, & A. K. Agarwal (Eds.), *Advances in carbon capture and utilization* (pp. 57–81). Springer Nature.

Das, R., Mandal, B., Sarkar, D., Pradhan, A. K., Datta, A., Padhan, D., Seth, A., Kumar, R., De, N., Mishra, V. N., Polara, K. B., Sharma, S., Thakur, N. P., Kachroo, D., Ray, M., Sharma, A., Patel, K. P., Garnayak, L. M., & Narkhede, W. N. (2019). Boron availability in soils and its nutrition of crops under long-term fertility experiments in India. *Geoderma*, *351*, 116–129.

Das, S. K., Ghosh, G. K., & Avasthe, R. (2020). Application of biochar in agriculture and environment, and its safety issues. *Biomass Conversion and Biorefinery*, *13*, 1359–1369.

Das, R., & Saha, D. (2014). Effect of liming on the changes of different forms of potassium in an acid soil treated with N and K fertilizers. *Journal of the Indian Chemical Society*, *91*(9), 1619–1625.

Deka, A. (2010). *Plantation development in north east India: Constraints and strategies. National Conference on Economic Development of Assam*, 1–10.

Dent, D. L. (1986). *Acid sulphate soils: A baseline for research and development.* International Institute for Land Reclamation and Improvement/ILRI.

Dent, D. L., & Pons, L. J. (1995). A world perspective on acid sulphate soils. *Geoderma*, *67*(3–4), 263–276.

Devi, V. (2021). Acidity amelioration practices for improving primary and secondary soil nutrient availability under rice cultivation in acid sulphate (Vaikom Kari) soils of Kuttanad. *Journal of Crop and Weed*, *17*(1), 27–33.

Dey, D., Kundu, M. C., & Sen, D. (2022). Productivity of groundnut as influenced by integrated use of lime, organics, inorganic fertilizers, and biofertilizers in acidic soil of Tripura. *International Journal of Plant & Soil Science*, *34*(2), 1–5.

Duiker, S. W., Rhoton, F. E., Torrent, J., Smeck, N. E., & Lal, R. (2003). Iron (hydr) oxide crystallinity effects on soil aggregation. *Soil Science Society of America Journal*, *67*(2), 606–611.

Elisa, A. A., Ninomiya, S., Shamshuddin, J., & Roslan, I. (2016). Alleviating aluminum toxicity in an acid sulfate soil from Peninsular Malaysia by calcium silicate application. *Solid Earth*, *7*(2), 367–374.

Else, M. A., Coupland, D., Dutton, L., & Jackson, M. B. (2008). Decreased root hydraulic conductivity reduces leaf water potential, initiates stomatal closure and slows leaf expansion in flooded plants of castor oil (*Ricinus communis*) despite diminished delivery of ABA from the roots to shoots in xylem sap. *Physiologia Plantarum*, *111*(1), 46–54.

Ezaki, B., Gardner, R. C., Ezaki, Y., & Matsumoto, H. (2000). Expression of aluminum-induced genes in transgenic Arabidopsis plants can ameliorate aluminum stress and/or oxidative stress. *Plant Physiology*, *122*(3), 657–666.

Fahmi, A., Susilawati, A., & Rachman, A. (2015). Influence of height waterlogging on soil physical properties of potential and actual acid sulphate soils. *Journal of Tropical Soils*, *19*(2), 67–73.

Fanning, D. S. (2012). Acid sulfate soils. In S. E. Jorgensen (Ed.), *Encyclopedia of environmental management* (pp. 26–30). Taylor and Francis.

Fidel, R. B., Laird, D. A., Thompson, M. L., & Lawrinenko, M. (2017). Characterization and quantification of biochar alkalinity. *Chemosphere*, *167*, 367–373.

Filipović, V., Mallmann, F. J. K., Coquet, Y., & Šimůnek, J. (2014). Numerical simulation of water flow in tile and mole drainage systems. *Agricultural Water Management*, *146*, 105–114.

Food and Agriculture Organization [FAO]. (2015). Status of the World's Soil Resources. Food and Agriculture Organization of the United Nations. Available at: http://www.fao.org/3/a-bc600e.pdf

Forstner, U. (1995). Land contamination by metals: Global scope and magnitude of problem. In H. E. Allen, C. P. Huang, G. W. Bailey, & A. R. Bowers (Eds.), *Metal speciation and contamination of soil* (pp. 1–33). CRC Press.

Foy, C. D., Armiger, W. H., Briggle, L. W., & Reid, D. A. (1965). Differential aluminum tolerance of wheat and barley varieties in acid soils 1. *Agronomy Journal*, *57*(5), 413–417.

Ganeshamurthy, A. N., Kalaivanan, D., & Satisha, G. C. (2016). Management of vegetable crops in acid soils of India. In K. V. Peter (Ed.), *Innovations in horticultural sciences* (pp. 559–584). New India Publishing Agency.

Geng, N., Kang, X., Yan, X., Yin, N., Wang, H., Pan, H., Yang, Q., Lou, Y., & Zhuge, Y. (2022). Biochar mitigation of soil acidification and carbon sequestration is influenced by materials and temperature. *Ecotoxicology and Environmental Safety*, *232*, Article 113241

Ghisman, V., Muresan, A. C., Buruiana, D. L., & Axente, E. R. (2022). Waste slag benefits for correction of soil acidity. *Scientific Reports*, *12*(1), Article 16042

Ghosh, S., Bakshi, M., Mitra, S., Mahanty, S., Ram, S. S., Banerjee, S., Chakraborty, A., Sudarshan, M., Bhattacharyya, S., & Chaudhuri, P. (2019). Elemental geochemistry in acid sulphate soils—A case study from reclaimed islands of Indian Sundarban. *Marine Pollution Bulletin*, *138*, 501–510.

Gopalakrishnan, S., Sathya, A., Vijayabharathi, R., Varshney, R. K., Gowda, C. L., & Krishnamurthy, L. (2015). Plant growth promoting rhizobia: Challenges and opportunities. *3. Biotech*, *5*(4), 355–377.

Goud, L. E., Singh, J., & Kumar, P. (2022). Climate change and their impact on global food production. In A. Kumar, J. Singh, & L. F. R. Ferreira (Eds.), *Microbiome under changing climate: Implications and solutions* (pp. 415–436). Woodhead Publishing.

Gu, H. H., Qiu, H., Tian, T., Zhan, S. S., Chaney, R. L., Wang, S. Z., Tang, Y. T., Morel, J. L., & Qiu, R. L. (2011). Mitigation effects of silicon rich amendments on heavy metal accumulation in rice (*Oryza sativa* L.) planted on multi-metal contaminated acidic soil. *Chemosphere*, *83*(9), 1234–1240.

Hamza, M. A., & Anderson, W. K. (2005). Soil compaction in cropping systems: A review of the nature, causes and possible solutions. *Soil and Tillage Research*, *82*(2), 121–145.

Hemati Matin, N., & Jalali, M. (2017). The effect of waterlogging on electrochemical properties and soluble nutrients in paddy soils. *Paddy and Water Environment, 15*(2), 443–455.

Hue, N. V. (1992). Correcting soil acidity of a highly weathered ultisol with chicken manure and sewage sludge. *Communications in Soil Science and Plant Analysis, 23*(3–4), 241–264.

Hussain, I., Sohail, M., Tanveer, S. K., & Muneer, M. (2018). Impact of planting density and growth habit of genotypes on wheat yield under raised bed planting method. *Science, 37*(3), 158–162.

Ibrahim, N. A., Magindran, T., & Shahrin, M. I. (2020). Sustainable use of laterite soil as compressed cement stabilized earth block for low cost housing construction. In *IOP Conference Series: Materials Science and Engineering* (Vol. 849, No. 1, pp. 012027). IOP Publishing.

ICAR. (2012). DARE/ICAR Annual report, (2011–2012).Available at https://icar.org.in/annual-report

Indira, B. V. N., & Covilakom, M. T. K. (2013). Characterization of acidity in acid sulphate soils of Kerala. *Journal of Life Sciences, 7*(8), 907.

Intergovernmental Panel on Climate Change. (2014). *Climate change 2014: Synthesis report. Contribution of working groups I, II and III to the fifth assessment report of the Intergovernmental Panel on Climate Change.* https://doi.org/10.1017/CBO9781107415416

Jiang, X., Liu, W., Liu, Q., Jia, Z., Wright, A. L., & Cao, Z. (2013). Soil N mineralization, nitrification, and dynamic changes in abundance of ammonia-oxidizing bacteria and archaea along a 2000 year chronosequence of rice cultivation. *Plant and the Soil, 365*(1), 59–68.

Jordanova, N. (2017). The magnetism of soils distinguished by iron/aluminum chemistry: Planosols, Pozdols, Andosols, Ferralsols, and Gleysols. In N. Jordanova (Ed.), *Soil magnetism* (pp. 139–220). Academic Press.

Jung, G., Matsunami, T., Oki, Y., & Kokubun, M. (2008). Effects of waterlogging on nitrogen fixation and photosynthesis in supernodulating soybean cultivar Kanto 100. *Plant Production Science, 11*(3), 291–297.

Kang, D. J., Futakuchi, K., Dumnoenngam, S., & Ishii, R. (2007). High-yielding performance of a new rice variety, IR53650 in mildly improved acid sulfate soil conditions. *Plant Production Science, 10*(1), 64–67.

Kang, D. J., Seo, Y. J., Lee, B. K., Vijarnsorn, P., & Ishii, R. (2010). Identification and crop performance of acid sulfate soil-tolerant rice varieties. *Journal of Crop Science and Biotechnology, 13*(2), 75–81.

Kar, G., Sahoo, N., & Kumar, A. (2010). *Integrated approach of crop diversification, soil and rainwater management for enhancing agricultural productivity of rainfed acidic land* (DWM Bulletin No. 47). Directorate of Water Management.

Kaur, G., Singh, G., Motavalli, P. P., Nelson, K. A., Orlowski, J. M., & Golden, B. R. (2020). Impacts and management strategies for crop production in waterlogged or flooded soils: A review. *Agronomy Journal, 112*(3), 1475–1501.

Khalil, M. I., Hossain, M. B., & Schmidhalter, U. (2005). Carbon and nitrogen mineralization in different upland soils of the subtropics treated with organic materials. *Soil Biology and Biochemistry, 37*(8), 1507–1518.

Khanam, R., Kulsum, P. G. P. S., Das, R., & Biswas, S. (2022). Nanofertilizers for sustainable crop production: A perspective in small farm agriculture. In A. Rakshit, S. Chakraborty, M. Parihar, V. S. Meena, P. K. Mishra, & H. B. Singh (Eds.), *Innovation in small-farm agriculture* (pp. 171–182). CRC Press.

Kochian, L. V., Hoekenga, O. A., & Pineros, M. A. (2004). How do crop plants tolerate acid soils? Mechanisms of aluminum tolerance and phosphorous efficiency. *Annual Review of Plant Biology, 55*(1), 459–493.

Kuhwald, M., Blaschek, M., Brunotte, J., & Duttmann, R. (2017). Comparing soil physical properties from continuous conventional tillage with long-term reduced tillage affected by one-time inversion. *Soil Use and Management, 33*(4), 611–619.

Kumar, B. M., & Takeuchi, K. (2009). Agroforestry in the Western Ghats of peninsular India and the satoyama landscapes of Japan: A comparison of two sustainable land use systems. *Sustainability Science, 4*, 215–232.

Kumar, M., Hazarika, S., Choudhury, B. U., Ramesh, T., Verma, B. C., & Bordoloi, L. J. (2012). Liming and integrated nutrient management for enhancing maize productivity on acidic soils of northeast India. *Indian Journal of Hill Farming, 25*(1), 36–38.

Kumarpandit, T., Kumarnaik, S., Patra, P. K., Dey, N., Patra, P. K., & Das, D. K. (2017). Influence of organic manure and lime on cadmium mobility in soil and uptake by spinach (*Spinacia oleracea* L.). *Communications in Soil Science and Plant Analysis, 48*(4), 357–369.

Laxmanarayanan, M., Dhumgond, P., Shruti, J. B. C. R., Sarkar, S., & Nagabovanalli, B.,. P. (2022). Influence of yellow gypsum on nutrient uptake and yield of groundnut in different acid soils of Southern India. *Scientific Reports, 12*(1), Article 5604

Lenka, N. K., Meena, B. P., Lal, R., Khandagle, A., Lenka, S., & Shirale, A. O. (2022). Comparing four indexing approaches to define soil quality in an intensively cropped region of northern India. *Frontiers in Environmental Science, 10*, 287.

Liu, Z., Zhou, W., Lv, J., He, P., Liang, G., & Jin, H. (2015). A simple evaluation of soil quality of waterlogged purple paddy soils with different productivities. *PLoS One, 10*(5), Article e0127690

Lone, A. A., Khan, M. H., Dar, Z. A., & Wani, S. H. (2018). Breeding strategies for improving growth and yield under waterlogging conditions in maize: A review. *Maydica, 61*(1), Article 11

Mahajan, G. R., Das, B., Manivannan, S., Manjunath, B. L., Verma, R. R., Desai, S., . . . Mulla, H. (2021). Soil and water conservation measures improve soil carbon sequestration and soil quality under cashews. *International Journal of Sediment Research*, *36*(2), 190–206.

Maji, A. K., Obi Reddy, G. P., & Sarkar, D. (2012). *Acid soils of India—Their extent and spatial distribution*. NBSS Publication.

Majumdar, S., Wanniang, S. K., & Behera, U. (2022). *Acid soil management in north eastern hill region of India*. Indian Council of Agricultural Research.

Mamatha, D. M. M., Gowda, R. C., & Shivakumara, M. N. (2018). Effect of basic slag on yield, nutrient status and uptake by paddy in acid soils of Karnataka, India. *International Journal of Current Microbiology and Applied Sciences*, *7*, 2286–2292.

Mandal, B., Ghosh, S., & Chattopadhyay, A. P. (2004). Distribution of extractable boron content in acidic soils of West Bengal in relation to soil properties. *Indian Journal of Agricultural Sciences*, *74*(12), 658–662.

Mandal, D. K., Mandal, C., Prasad, J., & Bhattacharyya, T. (2019). Acid soils in agro-ecological sub-regions of India: A revisit. *Indian Journal of Fertilisers*, *15*(10), 1156–1166.

Manik, S. M., Pengilley, G., Dean, G., Field, B., Shabala, S., & Zhou, M. (2019). Soil and crop management practices to minimize the impact of waterlogging on crop productivity. *Frontiers in Plant Science*, Article 140.

Manna, M. C., Swarup, A., Wanjari, R. H., Mishra, B., & Shahi, D. K. (2007). Long-term fertilization, manure and liming effects on soil organic matter and crop yields. *Soil and Tillage Research*, *94*(2), 397–409.

Marschner, H. (1995). Mineral nutrition of higher plants. *Annals of Botany*, *78*(4), 527–528.

Michael, P. S. (2013). Ecological impacts and management of acid sulphate soil: A review. *Asian Journal of Water, Environment and Pollution*, *10*(4), 13–24.

Michael, P. S., Fitzpatrick, R. W., & Reid, R. J. (2016). The importance of soil carbon and nitrogen for amelioration of acid sulphate soils. *Soil Use and Management*, *32*(1), 97–105.

Msimbira, L. A., & Smith, D. L. (2020). The roles of plant growth promoting microbes in enhancing plant tolerance to acidity and alkalinity stresses. *Frontiers in Sustainable Food Systems*, *4*, Article 106

Nath, A. J., Brahma, B., Sileshi, G. W., & Das, A. K. (2018). Impact of land use changes on the storage of soil organic carbon in active and recalcitrant pools in a humid tropical region of India. *Science of the Total Environment*, *624*, 908–917.

Nath, A. J., Kumar, R., Devi, N. B., Rocky, P., Giri, K., Sahoo, U. K., Bajpai, R. K., Sahu, N., & Pandey, R. (2021). Agroforestry land suitability analysis in the eastern Indian Himalayan region. *Environmental. Challenges*, *4*, Article 100199

Nguyen, L. T., Osanai, Y., Anderson, I. C., Bange, M. P., Braunack, M., Tissue, D. T., & Singh, B. K. (2018a). Impacts of waterlogging on soil nitrification and ammonia-oxidizing communities in farming system. *Plant and Soil*, *426*(1), 299–311.

Nguyen, L. T., Osanai, Y., Anderson, I. C., Bange, M. P., Tissue, D. T., & Singh, B. K. (2018b). Flooding and prolonged drought have differential legacy impacts on soil nitrogen cycling, microbial communities and plant productivity. *Plant and Soil, 431*(1), 371–387.

Nichols, P. (2018). Yanco Subterranean clover. Department of Primary Industries and Regional Development (DPIRD).

Osman, K. T. (2018). Acid soils and acid sulfate soils. In *Management of soil problems* (pp. 299–332). Springer.

Pal, D. K., Bhattacharyya, T., Chandran, P., & Ray, S. K. (2012a). Linking minerals to selected soil bulk properties and climate change: A review. *Clay Research, 31*(1), 38–69.

Pal, D. K., Bhattacharyya, T., & Lal, S. (2018). *Sustained mango cultivation in the ferruginous soils of Konkan and the Western Ghats: Soil acidity and pedological enigma*. International Mango Conference, Vengurle, India.

Pal, D. K., Bhattacharyya, T., Sinha, R., Srivastava, P., Dasgupta, A. S., Chandran, P., Ray, S. K., & Nimje, A. (2012b). Clay minerals record from Late Quaternary drill cores of the Ganga Plains and their implications for provenance and climate change in the Himalayan Foreland. *Palaeogeography, Palaeoclimatology, Palaeoecology, 356*, 27–37.

Pal, S. S. (1998). Interactions of an acid tolerant strain of phosphate solubilizing bacteria with a few acid tolerant crops. *Plant and Soil, 198*(2), 169–177.

Pan, J., Sharif, R., Xu, X., & Chen, X. (2021). Mechanisms of waterlogging tolerance in plants: Research progress and prospects. *Frontiers. Plant Science, 11*, Article 627331

Panda, A., Jha, P., Kumar, A., & Bajpai, R. K. (2017). Impact of organic residues application as a soil amendment on response in soil acidity. *International Journal of Chemical Studies, 5*(1), 105–109.

Panhwar, Q. A., Naher, U. A., Shamshuddin, J., & Ismail, M. R. (2020). Effects of biochar and ground magnesium limestone application, with or without bio-fertilizer addition, on biochemical properties of an acid sulfate soil and rice yield. *Agronomy, 10*(8), Article 1100

Panwar, P., Pal, S., Reza, S. K., & Sharma, B. (2011). Soil fertility index, soil evaluation factor, and microbial indices under different land uses in acidic soil of humid subtropical India. *Communications in Soil Science and Plant Analysis, 42*(22), 2724–2737.

Pati, R., & Mukhopadhyay, D. (2011). Distribution of cationic micronutrients in some acid soils of West Bengal. *Journal of the Indian Society of Soil Science, 59*(2), 125–133.

Patra, A., Sharma, V. K., Nath, D. J., Ghosh, A., Purakayastha, T. J., Barman, M., Kumar, S., Chobhe, K. A., Anil, A. S., & Rekwar, R. K. (2021). Impact of soil acidity influenced by long-term integrated use of enriched compost, biofertilizers, and fertilizer on soil microbial activity and biomass in rice under acidic soil. *Journal of Soil Science and Plant Nutrition, 21*(1), 756–767.

Planning Commission. (2011). *Mid-term Appraisal Eleventh Five Year Plan, 2007–2012*. Oxford University Press.

Purakayastha, T. J., Das, R., Kumari, S., Shivay, Y. S., Biswas, S., Kumar, D., & Chakrabarti, B. (2019). Impact of continuous organic manuring on mechanisms and processes of the stabilisation of soil organic C under rice–wheat cropping system. *Soil Research*, *58*(1), 73–83.

Qian, L., & Chen, B. (2013). Dual role of biochars as adsorbents for aluminum: The effects of oxygen-containing organic components and the scattering of silicate particles. *Environmental Science & Technology*, *47*(15), 8759–8768.

Quilliam, R. S., DeLuca, T. H., & Jones, D. L. (2013). Biochar application reduces nodulation but increases nitrogenase activity in clover. *Plant and Soil*, *366*, 83–92.

Rahman, M. A., Chikushi, J., Duxbury, J. M., Meisner, C. A., Lauen, J. G., & Yasunaga, E. (2005). Chemical control of soil environment by lime and nutrients to improve the productivity of acidic alluvial soils under rice-wheat cropping system in Bangladesh. *Environmental Control in Biology*, *43*(4), 259–266.

Rangasami, S. S., Murugaragavan, R., Rakesh, S. S., Prabu, P. C., Ramesh, A. R., & Elangovan, R. (2022). Chemistry of flooded soils: A review. *The Pharma Innovation Journal*, *11*(2), 2534–2539.

Ray, S. K., Chatterjee, D., Saha, S., Kamble, K. H., Deka, B. C., Kendra, V., & Jharnapani, M. (2015). *A guide for soil nutrient management with special reference to Wokha*. ICAR Research Complex for NEH Region, Nagaland Centre.

Reddy, K. R., & DeLaune, R. D. (2008). *Biogeochemistry of wetlands: Science and applications*. CRC press.

Retallack, G. J. (2021). Soil, soil processes, and Paleosols. In D. Alderton & S. A. Elias (Eds.), *Encyclopedia of geology* (2nd ed., Vol. 2, pp. 690–707). Academic Press.

Rieuwerts, J. S., Thornton, I., Farago, M. E., & Ashmore, M. R. (1998). Factors influencing metal bioavailability in soils: Preliminary investigations for the development of a critical loads approach for metals. *Chemical Speciation & Bioavailability*, *10*(2), 61–75.

Rouphael, Y., Cardarelli, M., & Colla, G. (2015). Role of arbuscular mycorrhizal fungi in alleviating the adverse effects of acidity and aluminium toxicity in zucchini squash. *Scientia Horticulturae*, *188*, 97–105.

Saha, R., Mishra, V. K., Majumdar, B., Laxminarayana, K., & Ghosh, P. K. (2010). Effect of integrated nutrient management on soil physical properties and crop productivity under a maize (*Zea mays*)–mustard (*Brassica campestris*) cropping sequence in acidic soils of Northeast India. *Communications in Soil Science and Plant Analysis*, *41*(18), 2187–2200.

Sarangi, S. K., & Islam, M. R. (2019). Advances in agronomic and related management options for Sundarbans. In H. S. Sen (Ed.), *The Sundarbans: A disaster-prone eco-region* (pp. 225–260). Coastal Research Library. Springer.

Sarangi, S. K., Mainuddin, M., & Maji, B. (2022). Problems, management, and prospects of acid sulphate soils in the Ganges Delta. *Soil Systems*, *6*(4), Article 95

Scanlan, C. A., & Davies, S. L. (2019). Soil mixing and redistribution by strategic deep tillage in a sandy soil. *Soil and Tillage Research*, *185*, 139–145.

Schaetzl, R., & Anderson, A. (2009). *Soils*. Cambridge University Press.

Seguel, A., Cumming, J. R., Klugh-Stewart, K., Cornejo, P., & Borie, F. (2013). The role of arbuscular mycorrhizas in decreasing aluminium phytotoxicity in acidic soils: A review. *Mycorrhiza*, *23*(3), 167–183.

Sehgal, J. L., Mandal, D. K., & Mandal, C. (1996). Agro ecological sub-region map of India. NBSS & LUP (ICAR).

Sen, H. S. (2017). Coastal ecosystems: Risk factors for development and threats due to climate change. In S. K. Gupta & M. R. Goyal (Eds.), *Soil salinity management in agriculture* (pp. 63–95). Apple Academic Press.

Sen, H. S., & Bal, A. R. (2009). Plants facing nutritional disorders in coastal saline soils and management options. In G. Singh, A. Qadar, N. P. S. Yaduvanshi, & P. Dey (Eds.), *Enhancing nutrient use efficiency in problem soils* (pp. 52–83). Central Soil Salinity Research Institute.

Shabala, S. (2011). Physiological and cellular aspects of phytotoxicity tolerance in plants: The role of membrane transporters and implications for crop breeding for waterlogging tolerance. *New Phytologist*, *190*, 289–298.

Shah, T. I., Shah, A. M., Bangroo, S. A., Sharma, M. P., Aezum, A. M., Kirmani, N. A., Lone, A. H., Jeelani, M. I., Rai, A. P., Wani, F. J., & Ahmad, L. (2022). Soil quality index as affected by integrated nutrient management in the Himalayan foothills. *Agronomy*, *12*(8), Article 1870

Shamshuddin, J., Azura, A. E., Shazana, M. A. R. S., Fauziah, C. I., Panhwar, Q. A., & Naher, U. A. (2014). Properties and management of acid sulfate soils in southeast Asia for sustainable cultivation of rice, oil palm, and cocoa. *Advances in Agronomy*, *124*, 91–142.

Shao, G., Ai, J., Sun, Q., Hou, L., & Dong, Y. (2020). Soil quality assessment under different forest types in the Mount Tai, central eastern China. *Ecological Indicators*, *115*, Article 106439

Shi, R. Y., Li, J. Y., Ni, N. I., & Xu, R. K. (2019). Understanding the biochar's role in ameliorating soil acidity. *Journal of Integrative Agriculture*, *18*(7), 1508–1517.

Simpson, H., & Pedini, P. (1985). Brackish water aquaculture in the tropics: The problem of acid sulfate soil environment. *Applied Geochemistry*, *19*, 1837–1853.

Singh, S. K., & Chandran, P. (2015). Soil genesis and classification. In R. K. Rattan, J. C. Katyal, B. S. Dwivedi, A. K. Sarkar, T. Bhattacharyya, J. C. Tarafdar, & S. S. Kukal (Eds.), *Soil science—An introduction* (pp. 57–96). Indian Society of Soil Science.

Sunil, B. M., & Krishnappa, H. (2012). Effect of drying on the index properties of lateritic soils. *Geotechnical and Geological Engineering*, *30*, 869–879.

Tewari, S., & Mishra, A. (2018). Flooding stress in plants and approaches to overcome. In P. Ahmad, M. A. Ahanger, V. P. Singh, D. K. Tripathi, P. Alam, & M. N. Alyemeni (Eds.), *Plant metabolites and regulation under environmental stress* (pp. 355–366). Academic Press.

Thampatti, K. C., Cherian, S., & Iyer, M. S. (1998). Influence of liming materials on soil acidity characteristics of an acid sulphate soil. *Journal of the Indian Society of Soil Science, 46*(2), 296–299.

Tian, L., & Shi, W. (2014). Soil peroxidase regulates organic matter decomposition through improving the accessibility of reducing sugars and amino acids. *Biology and Fertility of Soils, 50*(5), 785–794.

Utomo, M. (1995). Effect of rock phosphate on soil properties and apparent phosphorus recovery in acid soil of Sumatra. *Plant and Soil, 171*(1), 199–202.

Völker, G., Benecke, P., & Friedrich, J. (1995). Does acidification affect soil structure? In B. A. Stewart & K. H. Hartge (Eds.), *Soil structure: Its development and function* (pp. 351). Lewis Publishers, CRC Press.

Wang, H., Xu, J., Liu, X., Zhang, D., Li, L., Li, W., & Sheng, L. (2019). Effects of long-term application of organic fertilizer on improving organic matter content and retarding acidity in red soil from China. *Soil and Tillage Research, 195*, Article 104382

Wang, Y., Yao, Z., Zhan, Y., Zheng, X., Zhou, M., Yan, G., Wang, L., Werner, C., & Butterbach-Bahl, K. (2021). Potential benefits of liming to acid soils on climate change mitigation and food security. *Global Change Biology, 27*(12), 2807–2821.

Wani, S. H., Choudhary, M., Kumar, P., Akram, N. A., Surekha, C., Ahmad, P., & Gosal, S. S. (2018). Marker-assisted breeding for abiotic stress tolerance in crop plants. In S. S. Gosal & S. H. Wani (Eds.), *Biotechnologies of crop improvement, Volume 3* (pp. 1–23). Springer.

Watkin, E. L., O'Hara, G. W., & Glenn, A. R. (2003). Physiological responses to acid stress of an acid-soil tolerant and an acid-soil sensitive strain of *Rhizobium leguminosarum* biovar *trifolii*. *Soil Biology and Biochemistry, 35*(4), 621–624.

Whalen, J. K., Chang, C., Clayton, G. W., & Carefoot, J. P. (2000). Cattle manure amendments can increase the pH of acid soils. *Soil Science Society of America Journal, 64*(3), 962–966.

Wu, H., Xiang, J., Chen, H. Z., Zhang, Y. P., Zhang, Y. K., & Zhu, F. (2018). Effects of exogenous growth regulators on plant elongation and carbohydrate consumption of rice seedlings under submergence. *The Journal of Applied Ecology, 29*(1), 149–157.

Xian, C., Qi, Z., Tan, C. S., & Zhang, T. Q. (2017). Modeling hourly subsurface drainage using steady-state and transient methods. *Journal of Hydrology, 550*, 516–526.

Xu, B., Cheng, Y., Zou, X., & Zhang, X. (2016). Ethanol content in plants of *Brassica napus* L. correlated with waterlogging tolerance index and regulated by lactate dehydrogenase and citrate synthase. *Acta Physiologiae Plantarum, 38*(3), Article 81

Yuan, J. H., Xu, R. K., & Zhang, H. (2011). The forms of alkalis in the biochar produced from crop residues at different temperatures. *Bioresource Technology*, *102*(3), 3488–3497.

Zeng, M., de Vries, W., Bonten, L. T., Zhu, Q., Hao, T., Liu, X., Xu, M., Shi, X., Zhang, F., & Shen, J. (2017). Model-based analysis of the long-term effects of fertilization management on cropland soil acidification. *Environmental Science & Technology*, *51*(7), 3843–3851.

Zhen, B., Li, H., Niu, Q., Qiu, H., Tian, G., Lu, H., & Zhou, X. (2020). Effects of combined high temperature and waterlogging stress at booting stage on root anatomy of rice (*Oryza sativa* L.). *Water*, *12*(9), Article 2524

13

Conservation Agriculture Strategies for Enhancing Soil Health

Amit K. Dash, Abir Dey, Ranjan Bhattacharyya, T. K. Das, Arti Bhatia, and Shrila Das

Chapter Overview

In "The Nature and Properties of Soils," 14th edition, Brady and Weil (2013) defined soil health as the "ability of soil to self-regulate, maintain stability, demonstrate resilience, and exhibit an absence of stress symptoms, functioning effectively as an ecosystem." It refers to the biological integrity of the soil ecosystem, highlighting the equilibrium among soil organisms and between these organisms and their environment. Despite positive effects of conservation agriculture (CA), detailed study of the effects of CA on enhancing soil organic carbon sequestration and improving soil health under different cropping systems in India is limited. Information on CA impacts on greenhouse gas emission reduction in different agro-ecological regions of India is also limited. This chapter outlines the effects of CA on soil health and carbon sequestration and greenhouse gas emission mitigation, mainly in the irrigated arable lands of the Indo-Gangetic Plains (the food bowl of India). In addition, the chapter critically analyzes one of the soil organic carbon sequestration mechanisms: carbon stabilization through soil aggregation, especially inside microaggregates within macroaggregates in the Indo-Gangetic Plains.

Soil Health Series: Volume 4 Soil Health and Sustainability in India, First Edition.
Edited by Ranjan Bhattacharyya.

Introduction

The intensive system of cultivation, which involves imbalanced use of chemical fertilizers, intensive tillage, and nutrient-exhaustive varieties, has led to rapid deterioration of soil health. The major concerns of soil health are the degradation of fertile arable soils by soil erosion, soil organic carbon (SOC) depletion, and subsoil compaction, which is accelerated by intensive tillage operation. In the past two decades, rising production costs caused by price hikes in energy resources (diesel, electricity), labor, and other inputs have made the conventional tillage (CT) system economically less productive/feasible (Parihar et al., 2016). Conservation agriculture (CA) is essentially a production system involving (a) minimum disturbance to the soil surface through no or reduced tillage, (b) permanent covering of soil surface through retaining crop residues or growing cover crops, and (c) adoption of crop sequencing/rotations including agroforestry in spatial and temporal scale. Adoption of CA is gradually increasing in the tropics, subtropics, and temperate regions of the world, both under rainfed and irrigated ecologies. CA is also hailed as a panacea for restoration of yield and farm income in degraded agro-ecologies in arid and semi-arid tropics (Jat et al., 2012). CA is associated with a multitude of benefits related to soil health, such as improvement in quality and quantity of SOC (Dey et al., 2020), better soil structure (Parihar et al., 2020), favorable conditions for beneficial soil microbes (Kumawat et al., 2018), enhanced nutrient supply capacity (Bhattacharyya et al., 2015; Choudhary et al., 2019), and minimized carbon footprint owing to less emission from heavy machinery (Parihar et al., 2018). CA is a soil management practice that reduces soil erosion and degradation, which helps in the preservation of natural resources while optimizing agriculture productivity (FAO, 2015). It fosters biodiversity and natural biological processes in both surface and subsurface environments, leading to increased efficiency in water and nutrient utilization, thereby enhancing and maintaining crop productivity. These CA practices are frequently noted for their ability to increase SOC storage, while also reducing carbon emissions from agricultural inputs such as fertilizers and on-farm fuels (Lu et al., 2018). Therefore, CA has an untapped potential for improvement of soil health without compromising crop productivity. This chapter discusses the impact of CA on soil health, SOC sequestration, and greenhouse gas emission reduction, with a special emphasis on Indian subcontinental conditions.

Indicators of Soil Health

The present approach of soil health assessment is concerned with identification of specific indicators that best represent the soil function under consideration (Larson & Pierce, 1994). The indicators can be characterized as measurable soil attributes and processes. Selection of an indicator to assess soil health always depends on its scientific relevance to the specific soil function, sensitivity toward changes in management practices, and speed and accuracy of measurement under field condition, and it must also have relatively low sampling error (Carter et al., 1997). In many cases, some crucial soil attributes were found to be difficult or expensive to measure. Such cases highlight the importance of an indirect measure involving the pedotransfer function of related soil attributes that can be used as alternative method for computation of such indicators. A single soil attribute is unlikely to represent a specific soil function, while computation of all related soil attributes is not reasonable because of the high cost and time involvement. Hence, it is more practical to choose a set of minimum indicators (minimum data set) that will yield a comprehensive assessment of various soil processes important to soil quality. Different soil functions may have a different set of indicators, although one indicator may represent several soil functions. A simple example of a minimum data set from Das et al. (2021) is illustrated in Table 13.1.

Table 13.1 Minimum Data Set of Physical, Chemical, and Biological Properties Addressing Different Soil Functions.

Soil functions	Indicators
Physical stability	Mean weight diameter
	Bulk density
Water relations	Hydraulic conductivity
	Soil organic carbon
Nutrients cycling	Total nitrogen
	Total carbon
	Total organic carbon
	Oxidizable carbon
	Available phosphorus
	Available potassium
	Available micronutrients
	Microbial biomass carbon
	Dehydrogenase activity
Resistance and resilience	Soil organic carbon
	Bulk density
	Hydraulic conductivity
Biodiversity and habitat	Microbial biomass carbon
	Dehydrogenase activity

(*Source:* Das et al., 2021/with permission of Elsevier)

Impact of CA on Soil Attributes

CA has proven beneficial over CT practices under different soils, climates, and crop management scenarios across the country (Dey et al., 2016, 2018; Parihar et al., 2016, 2020; Bhattacharyya et al., 2019; Choudhary et al., 2019; Das et al., 2021).

Physical Attributes

Physical attributes of soil/indices act as important indicators for soil quality evaluation and other ecological processes within an agro-ecosystem. Key soil physical health indices include soil bulk density, soil compaction, soil moisture and temperature, aggregate size, distribution and stability, and hydraulic properties. Soil compaction affects the plant root behavior in soil through affecting seedling emergence, penetration resistance, root growth, soil aeration, and water infiltration, among other factors (Parihar et al., 2020).

Zero tillage (ZT) without residue retention increases bulk density in upper soil layers (Bhattacharyya et al., 2019, Dey et al., 2020) owing to the absence of repeated mechanical churning, which is responsible for a decrease in macroporosity and thus total porosity. These phenomena were prevalent in the initial years of CA adoption, especially in rice–wheat systems (Naresh et al., 2012; Bhattacharya et al., 2020). However, adoption of a complete CA package of practices (ZT with crop residue retention) does not increase soil bulk density because of the inherent lower density of soil organic matter (SOM) derived from crop residues (Choudhary et al., 2019; Parihar et al., 2020). In the long run, CA displays a significant reduction in bulk density by virtue of better aggregation and accumulation of a substantial amount of organic matter (Parihar et al., 2016; Dey et al., 2020; Table 13.2). CA promotes root growth and proliferation, which also contributes to decrease in bulk density. Crop diversification plays a key role in modifying bulk density and soil physical attributes by virtue of the different root growth, density, and architecture of the various crops. Owing to higher root density of maize compared with rice, bulk density of a maize–wheat– Mungbam cropping system was found to be significantly lower than a rice–wheat system (Patra et al., 2019). Penetration resistance influences plant root growth and water movement in soil depending on soil type, bulk density, and soil water content. CA decreased penetration resistance up to 30% compared with CT in maize-based systems in an Inceptisol (Parihar et al., 2016).

ZT has a strong influence on soil structure through reduced soil disturbance and maintenance of permanent residue cover. Adoption of CA improves the soil aggregation (Bhattacharyya et al., 2019; Choudhary et al., 2019; Bhattacharya et al., 2020; Parihar et al., 2020; Dash et al., 2023). Aggregates consist of primary

Table 13.2 Effect of Tillage and Residue Management on Soil Bulk Density Under Different Soil Types, Soil Depths, Crops, and Management Length.

Soil type	Tillage	Crop	Duration (years)	Depth (cm)	Bulk density (Mg m^{-3})	Reference
Sandy loam (New Delhi)	CT CA	Maize–wheat–mungbean	4	0–70	NS	Parihar et al., 2020
Clay loam	CT ZT ZT + R	Rice-wheat	6	0–15	Tillage-NS +R = 1.33 −R = 1.42	Dey et al., 2020
				15–30	Tillage-NS +R = 1.39 −R = 1.50	
Sandy loam	CA CT PB	Maize–wheat–mungbean Maize–chickpea–sesbania Maize–mustard–mungbean Maize–maize–sesbania	7	0–10	ZT = 1.46 PB = 1.46 CT = 1.53	Parihar et al., 2016
				10–20	ZT = 1.53 PB = 1.54 CT = 1.65	
				20–30	ZT = 1.55 PB = 1.57 CT = 1.65	
				30–60	NS	

(Continued)

Table 13.2 (Continued)

Soil type	Tillage	Crop	Duration (years)	Depth (cm)	Bulk density (Mg m^{-3})	Reference
Sandy loam	CT	Cotton–wheat (2008–2011)	4	0–5	ZT = 1.52	Bhattacharyya et al., 2019
	ZT	Maize–wheat–mungbean (2011–2012)			CT = 1.48	
	CT + R			5–15	ZT = 1.56	
	ZT + R				CT = 1.51	
				15–45	NS	
					+R < −R	
Clay loam	CT	Rice–wheat	3	0–10	Rice-Wheat	Choudhary et al., 2018a
	ZT	Maize–wheat			CT-R = 1.47	
	CT + R				ZT + R = 1.35	
	ZT + R				Maize-Wheat	
					CT-R = 1.38	
					ZT + R = 1.34	
Loam	CT	Rice–wheat (RW)	8	0–5	Tillage- NS	Patra et al., 2019
	NT + R	Maize–wheat–mungbean (MWMb)			NT-MWMB = 1.38	
					CT-RW = 1.60	

Note. CT = conventional till; CA = conservation agriculture; NT = no-till; PB = Permanent bed; ZT = zero till; +R = residue retention; −R = residue removal; NS = not significant.

mineral particles bound together by organic binding agents (Haynes & Swift, 1990). At the atomic level, negatively charged colloid surfaces, organic molecules, positively charged ions, and water molecules interact to cause flocculation. Microscopically, clay and organic matter serve as bridges between sand and silt particles, forming soil micro-aggregates (Ghosh et al., 2016). Macroscopically, macro-aggregates are formed through binding of these microaggregates with fungal hyphae, plant roots, and other stabilizing agents (Oades, 1984; Dash et al., 2023), and soil aggregate stability is enhanced through modifications of soil moisture, organic matter, microbial activity (Choudhary et al., 2019), and soil fauna population.

Macro-aggregates in soil form when micro-aggregates bind with fresh organic matter to create coarse intra-aggregate particulate organic matter (Six et al., 2000; Table 13.3). For macro-aggregate stability, physical protection is crucial, while micro-aggregates primarily benefit from biochemical recalcitrance (Dash et al., 2023). By minimizing mechanical disruption, CA thereby impedes aggregate breakdown and the formation of new micro-aggregates (Dash et al., 2023). The presence of crop residue enhances macro-aggregate formation, while reducing the proportion of micro-aggregates. Conversely, the absence of crop residue impedes the formation of macro-aggregates. Macro-aggregates naturally contain higher levels of carbon compared with micro-aggregates, which form following carbon loss owing to mechanical disturbances that break down macro-aggregates (Table 13.4; Modak et al., 2019). In a separate study, ZT and permanent bed treatments showed an increase in saturated hydraulic conductivity (Ks) by 11%–14%

Table 13.3 Effect of Crop Residue Retention and Nutrient Management on Aggregate Percentage and Aggregate Associated C (g kg^{-1} Soil) After 6 Years of Maize–Wheat–Mungbean Cropping Under Conservation Agriculture (Dash et al., 2023/Springer Nature).

	MA (%)		MI (%)		MA-TC (g kg^{-1} soil)		MI-TC (g kg^{-1} soil)	
Treatments	**0–5 cm**	**5–15 cm**	**0–5 cm**	**5–15 cm**	**0–5 cm**	**5–15 cm**	**0–5 cm**	**5–15 cm**
Crop residue loads								
CR_0	51.21	54.03	37.86	34.37	9.41	8.39	5.37	4.29
CR_2	58.91	58.44	30.32	28.47	12.06	9.38	6.44	5.28
CR_4	64.29	65.05	28.02	25.99	13.94	11.58	7.49	6.59
Mean	58.13	59.17	32.06	29.58	11.80	9.78	6.43	5.38
LSD ($p = 0.05$)	5.27	6.21	5.04	4.41	1.79	1.37	0.91	0.86

Note. MA = Macro-aggregates; MA-TC = Macro-aggregate-associated C; MI = Micro-aggregates; MI-TC = Micro-aggregate associated C; CR_0 = No crop residue retention; CR_2 = 2 t ha^{-1} crop residue retention; CR_4 = 4 t ha^{-1} crop residue retention; NS = non-significant.

Table 13.4 Effect of Tillage and Residue Management on Aggregate-Associated Carbon Under a Soybean–Wheat System (Modak et al., 2019/with permission of Elsevier).

	Macroaggregate-associated SOC (g kg^{-1})		Microaggregate-associated SOC (g kg^{-1})	
Treatment	**0–5 cm**	**5–15 cm**	**0–5 cm**	**5–15 cm**
Tillage management				
Zero tillage	10.5a	7.55a	9.43a	7.27a
Conventional tillage	8.34b	6.55b	7.61b	6.14b
Residue management				
No residue	6.66b	5.10d	5.59d	4.69d
Wheat residue alone	9.78a	6.75c	8.55c	6.30c
Soybean residue alone	10.5a	7.71b	9.47b	7.58b
Wheat and soybean residue	10.8a	8.64a	10.4a	8.23a

Note. Values followed by a different letter within a column are significantly different ($P < 0.05$).

compared with CT at soil depths of 0–45 cm. Additionally, penetrometer resistance was approximately 9%–30% lower in permanent beds (PB) and ZT treatments compared with CT at soil depths of 0–30 cm (Parihar et al., 2020).

Chemical Attributes

Chemical properties such as soil pH, electrical conductivity, SOC content, and macro- and micronutrient levels are used in soil health assessments. Among these, SOC is the most important parameter in evaluating soil health, generally correlating positively with soil nutrient-supplying capacity, crop yield, and nutrient uptake. Adoption of CA increases the organic carbon content in soil (Bhattacharyya et al., 2015; Parihar et al., 2016, 2019; Dey et al., 2020; Saraswat et al., 2023). Under CA, lability of SOC increases significantly compared with CT, as reflected in the higher values of lability index or carbon management index (Jat et al., 2019; Modak et al., 2019; Parihar et al., 2019; Dey et al., 2020). The enhancement in the labile SOC under CA in turn promotes nutrient cycling and improves the nutrient-supplying capacity of the soil (Choudhary et al., 2019). CA also enhances nonlabile SOC, indicating effective carbon sequestration (Dey et al., 2020). Adoption of CA reduces the soil pH as compared with CT (Sinha et al., 2019), especially in alkaline soils. Higher organic matter content and an increase in the release of hydrogen ions and weak organic acids during the decomposition of organic matter lowers the pH in plots with a lot of retained residue (Dash et al., 2023). Enhanced root exudation, from vigorous root systems under CA, further promotes the phenomenon. An excess amount of SOM or humus buffers the pH by high CEC.

The CA-based management enriches soils with organic matter (Bhattacharyya et al., 2012c, 2015; Dey et al., 2016, 2018), which in turn liberates plant nutrients through biological transformations, especially nitrogen, phosphorus, sulfur, and micronutrients (Table 13.5) (Choudhary et al., 2019). Bhattacharya et al., (2020) found that the in-surface soil layer available nitrogen content increases significantly by 18.3%, 30.8%, and 70.2% in 3-year CA, 6-year CA, and 9-year CA, respectively, compared with CT. Parihar et al. (2018) found that CA-based practices significantly increased the mineral-nitrogen fractions (15%–36% NO_3, 16%–35% NH_4, 0–30 cm depth) over the CT-based practices.

The nature and amount of retained crop residues govern nutrient cycling (Saraswat et al., 2023). The C/N ratio of retained residue together with its lignin, polyphenols, and soluble carbon contents has a major role to play in governing soil nitrogen availability (Moretto et al., 2001). During the decomposition of cereal residue with a larger C/N ratio, microbes assimilate nitrogen, leading to nitrogen immobilization under CA, especially in the initial years. Enhanced soil aggregation under CA promotes the accumulation and protection of nitrogen along with carbon within aggregates, which may limit immediate nitrogen availability (Dey et al., 2018). Better moisture conservation facilitates higher nutrient uptake and plant growth, in turn resulting in better nitrogen use efficiency in CA than in conventionally managed ecosystems (Sapkota et al., 2014). Another study reported significant increases in soil available nitrogen ranging from 23% to 50% in the topsoil (0–5 cm) and from 64% to 98% at depths of 5–15 cm under CA practices. Similar improvements were observed in available potassium levels (Roy et al., 2022). Enhanced SOM under CA liberates organic anions during decomposition, which can reduce phosphorus fixation

Table 13.5 Changes in Surface Soil (0–5 cm) Fertility (mg kg^{-1}) Caused by Long-Term Tillage Modifications (Choudhary et al., 2019).

Treatment	NO_3-N	NH_4-N	P	K	S	DTPA-Fe	DTPA-Zn	DTPA-Mn	DTPA-Cu
Tillage practices									
Permanent bed	28.3	15.9	14.3	93.0	17.6	9.46	5.12	3.10	3.09
Zero tillage	27.1	15.1	13.4	87.1	15.4	9.05	5.05	2.97	2.99
Conventional tillage	19.6	10.5	10.1	83.9	9.3	7.67	4.52	2.52	2.64
LSD ($P = 0.05$)	3.5	0.8	0.7	3.1	2.4	0.76	0.4	0.24	0.22

by competing for phosphorus fixation sites (Palm et al., 2014). Soil structure improvement owing to less disturbance and residue retention helps in phosphorus availability. Improved aggregation reduces the phosphorus fixation through decreasing the relative soil surface area available for phosphorus fixation (Zhang et al., 2003; Dash et al., 2023). Abundant crop residue retention also acts as a source of organic acids responsible for phosphorus solubilization, apart from its role as a phosphorus source. Haokip et al. (2020) reported that permanent bed planting with crop residue retention enhanced the soil microbial biomass phosphorus, Olsen phosphorus, and labile and moderately labile phosphorus. Phosphatase activity in soil increases under CA (Parihar et al., 2016; Choudhary et al., 2019), which drives the mineralization of organic phosphorus and transforms it into plant available inorganic phosphorus. Crop residue acts as a direct source of available potassium in soil. CA practices have been reported to modify the different pools, especially the non-exchangeable pools. Choudhary et al. (2019) reported ~13%, 23%, 23%, and 17% improvement in the availability of zinc, iron, manganese, and copper, respectively, under permanent beds compared with CT (Table 13.6).

SOC Enhancement by CA

CA reduces soil erosion and degradation, which ultimately helps in the preservation of natural resources, while optimizing agriculture productivity (FAO, 2015). Even though the Indo-Gangetic plains (IGP) of India have a huge potential for CA, only 1.50 Mha of the region is currently under CA practices (Das et al., 2018). CA has shown its effectiveness in enhancing soil carbon sequestration, improving overall soil quality, mitigating soil erosion, and promoting biodiversity (Bhattacharyya et al., 2012a, 2012b, 2013; Das et al., 2014, 2016, 2018, 2020). CA is responsible for the build-up of SOC, which depends on the type and quantity of crop residues reverting to soil (Table 13.6). The dynamics of SOC have garnered scientific interest owing to the potential impact of changes in land cover, land use, and management practices on global carbon storage and fluxes (Sharma et al., 2014)

Impacts of Long-Term CA Practices on SOC Pools and Carbon Dynamics

The SOC significantly influences the physical, chemical, and biological properties of soil. It serves as a critical indicator of soil quality and plays a vital role in binding agents essential for the formation of soil aggregates (Biswas et al., 2021). The decomposition of SOM increases with higher tillage intensity and reduces soil aggregation. To comprehend the effects of management practices on SOC dynamics, it is essential to fractionate SOC into distinct components. These fractions are highly diverse with contrasting turnover times, microbial decomposition, and so forth (Bhattacharyya et al., 2012a). Labile organic carbon fractions are recognized as sensitive indicators of the impacts of land use change on soil quality and health

Table 13.6 Impact of Conservation Agriculture on SOC and Other Nutrient Availability in Different Agro-Ecosystems of India.

Cropping system	Study site	Parameters	Increase/decrease due to adoption of CA	Reference
Lentil –finger millet	Himalayan subtemperate region	SOC stock	~8% increase (0–15 cm)	Bhattacharyya et al. (2012c)
Rice–wheat	IGP, subtropical region	SOC stock Crop productivity	13% increase (0–15 cm) 13% increase in system productivity	Bhattacharyya et al. (2015)
Maize–wheat	Himalayan sub temperate region	SOC stock Crop productivity	31% increase (0–15 cm) 17% increase in system productivity	Ghosh et al. (2016)
Rice-mustard	IGP, subtropical region	SOC Mineralizable N Olsen- P Ammonium acetate- K DTPA- Zn Hot water B DTPA- Fe	22% increase 17% increase 49% increase 49% increase 14% increase 18% increase 51% increase in 0–5 cm soil	Das et al. (2021)
Maize–wheat	IGP, subtropical region	Total soil N	23% increase in 0–5 cm soil	Bhattacharyya et al. (2018)
Rice-mustard	IGP, subtropical region	Labile C	50% increase in 0–5 cm soil	Das et al. (2020)

(*Continued*)

Table 13.6 (Continued)

Cropping system	Study site	Parameters	Increase/decrease due to adoption of CA	Reference
Maize–wheat	IGP, subtropical region	SOC concentration	19% increase in 0–30 cm soil	Das et al. (2018)
Maize-wheat-green gram	IGP, subtropical region	Total soil N	11% increase in 0–5 cm soil	Bhattacharyya et al. (2019)
		Nitrogen use efficiency	No change	
Maize–wheat	IGP, subtropical region	SOC stock	11% increase in 0–15 cm soil	Nath & Rattan (2017)
Rice–wheat–mungbean	Eastern Indo-Gangetic Plain	SOC stock	11% increase in OC stock in the 0–30 cm soil layer	Mondal et al. (2020)
Maize + pigeon pea	Vertisol located in Central India, the Research Farm of ICAR-Indian Institute of Soil Science, Bhopal, India	TOC	23.1% increase in TOC stock in the 0–10 cm soil layer	Naorem et al. (2023)
Finger millet (*Eleusine coracana* L. Garten) rotated with pigeon pea (*Cajanus cajan* L. Millspaugh) or horse gram (*Macrotyloma uniflorum* Lam. Verdc)	Karnataka in southern India, Oxic Haplustalfs, subtropical semi-arid climate	TOC	27% increase in TOC stock in the 0–20 cm soil layer	Prasad et al. (2016)

Note. DTPA = diethylenetriamine pentaacetate; IGP = Indo-Gangetic plains; SOC = soil organic carbon; TOC = total organic carbon.

(Nandan et al., 2019), emphasizing that the ratio of labile SOC to total SOC is crucial for SOC sequestration and overall soil health (Bhattacharyya et al., 2012a).

Guo et al. (2015) found that in the short term, no-tillage (NT) practices significantly increased microbial biomass carbon by 20% compared with CT, while having no impact on total organic carbon, dissolved organic carbon, easily oxidizable carbon, and the SOC of aggregates. A study on the long-term effects of NT in China revealed higher SOC concentrations in NT plots compared with CT plots, but only within the 0–10 cm soil layer (Liu et al., 2013). An increase in total organic carbon in the top 0.2 m of soil depth was recorded by Nandan et al. (2019) in ZT crop establishment treatments over CT treatments and the fractions increased by 21%, 16%, 13%, and 7% for the very labile C fraction, labile C fraction, nonlabile C fraction, and less-labile C fraction, respectively. On analysis of the SOC pools after 12 years in NT dryland ecosystems, it was found that microbial biomass carbon (active C) and particulate organic matter carbon (POM-C; slow C pool) increased as cropping intensity increased, depending on potential evapotranspiration. In comparison with the perennial grassland use system, the carbon pool increased in cropping systems. The highest carbon levels were observed in the continuous cropping system, which contained 91%, 78%, and 90% of the carbon amounts found in the perennial grass system within the active, slow, and passive carbon pools, respectively. Bhattacharyya et al. (2012b) found that in the Indian Himalayas, ZT plots exhibited approximately 11% higher concentrations of total SOC and 12% more POM-C compared with CT plots in the 0–5 cm soil layer. Tillage did not affect carbon pools in subsurface layers. Additionally, irrigation positively influenced SOC content in the 0–15 cm layer. While ZT positively affected the labile pools of SOC, it did not affect the recalcitrant pool.

Das et al. (2020) observed significantly higher very labile and labile carbon concentrations in the CA-based rice–mustard system by approximately 50.6% and 47.7%, respectively, at the 0–5 cm soil depth, compared with the conventional rice–mustard system. It was also found that higher water stable macro-aggregates, macro-aggregates to micro-aggregates ratio, and aggregate carbon content are seen under ZT crop establishment methods over CT.

Impacts of Long-Term CA Practices on Aggregate-Associated Carbon Stabilization

Tropical countries such as India have a very rapid carbon cycle, so SOC content is lower even with continuous application of good-quality organic sources. Soil stabilization is required for long-term carbon storage (Purakayastha et al. 2016). Physical, chemical, biochemical, and thermal stabilizations are common. Owing to their lower biological degradability, macro- and micro-aggregates protect carbon physically. Tisdall and Oades (1982) proposed that micro-aggregates are formed free and gradually become macro-aggregate constituents, but Oades

(1984) reported that carbon is first assimilated in macro-aggregates and then forms micro-aggregate nuclei. Physical stabilization of carbon is estimated by measuring the free light fraction, intra-aggregate POM, and mineral-associated fractions (Six et al., 1999).

The arrangement of soil particles into aggregates and the associated pore systems play a pivotal role in soil sustainability and are also important in plant growth, carbon sequestration, and other areas (Lehmann et al., 2017). The long-term stabilization of SOC can be achieved by soil aggregates through physical encapsulation, which protects it from decomposition by various decomposing agents (Six et al., 2000). Within aggregates, the position of SOC also affects the rate of decomposition (Christensen, 1996). The process of aggregate formation is considerably less in CT as tilling often interrupts the process (Six et al., 2002), which also affects the distribution of SOM in the profile (Carter, 1992). CA-based practices such as PB + residue treatments led to the highest share of recalcitrant carbon owing to less soil disturbance and greater protection of carbon through more stable aggregates, which indicates the potential of CA-based practices to stabilize organic carbon in soil (Bhattacharyya et al., 2012a; Dey, 2017). SOC plays an important role in forming and stabilizing the soil structure, enhancing soil physical properties, and nutrient recycling (Bhattacharyya et al., 2013).

A study conducted by Xin et al. (2015) revealed that soil aggregate stability and its associated organic carbon are strongly influenced by the tillage practices. In the 0–5 cm soil layer, tillage once in 2 years with residue, tillage once in 4 years with residue, and NT with residue enhanced the concentration of macro-aggregate-associated organic carbon by 14%, 56%, and 83%, respectively, compared with tillage every year with residue. In comparison with CT in a rice–wheat system, CA practices showed 9.89% higher SOC in the 0–50 cm soil depth, and a significant fraction of SOC was accumulated in the surface soil (28.3%) at 0–5 cm depth of the NT system compared with the CT system (Ghimire et al., 2012). Modak et al. (2019) discovered that zero-tilled plots contained approximately 26% and 15% more macro-aggregate-associated carbon compared with CT plots in the 0–5 cm and subsequent depth (5–15 cm), respectively. In the macro-aggregates, the levels of recalcitrant carbon were (a) around 57%, 76%, and 63% higher in wheat residue + stubble retention, stubble retention, and wheat residue plots, respectively, compared with no residue plots at 0–5 cm depth; and (b) 85%, 59%, and 40% higher in the same respective treatments at 5–15 cm depth. A similar trend was observed for micro-aggregate-associated recalcitrant carbon.

Glomalin as Affected by CA

Along their mycelial network, arbuscular mycorrhizal (AM) fungi synthesize glomalin, a glycoprotein, as a physical support molecule (Driver et al., 2005). The AM fungi get 20% of the total photosynthate produced in this way (Wang et al., 2017).

Carbon dioxide (CO_2) intake and photosynthesis are boosted by AM fungus growth stimulation, which in turn raises fungal biomass and glomalin (Rillig & Steinberg, 2002). Lovelock et al. (2004) defined glomalin as a glue-like sticky substance with the capacity to bind soil particles. Mycelia produced by AM fungi contribute to the consolidation of soil particles (Wright & Upadhyaya, 1996). Due to glomalin's poorly defined structure, glomalin-related soil proteins (GRSPs) are reported instead of pure glomalin when its concentration is measured. AM fungi producing glomalin can play an important role in atmospheric CO_2 sequestration because GRSP is a secure vault for SOC. Tillage can alter the soil environment and thereby affect the diversity of the AM fungal community. Many researchers have suggested that adoption of ZT or reducing mechanical disturbance of the soil can enhance the soil stability through a positive effect on AMF (Singh et al., 2018; Modak et al., 2019). Additionally, CA aids in enhancing soil quality in terms of increasing microbial size, diversity, and activities through which it encourages aggregation and carbon storage and turnover. The soil enzyme activities have been considered as appropriate indicators of microbial activities because they respond rapidly to changes in soil management. Modak et al. (2019) reported that residue retention had a significant impact on GRSP concentration. An increase in GRSPs in the surface layer by 15% and 16% was observed in bulk soil and macroaggregates, respectively, in plots under ZT compared with CT. Singh et al. (2018) observed that under rainfed conditions, plots with retained residue had better glomalin content in the soil compared with plots where residue was removed and plots with *Leucaena* added.

From Table 13.7, we can assess if the best CA practice (MBR + DSR − ZTW + R R − ZTMB + wheat residue) can replace 1 Mha of the existing farmers' practice in the IGP. If so, then total SOC could increase by ~19×10^6 Mg ha^{-1} yr^{-1} in the 0–5 cm soil layer, the system productivity would increase by ~1.22×10^6 Mg ha^{-1} yr^{-1} with a net profitability of ₹37407×10^6 annually in the IGP. Similarly, if the best CA practice can replace 0.5 Mha of the existing maize–wheat system of the farmers' practice, then then total SOC could increase by ~0.11×10^6 Mg C ha^{-1} yr^{-1}, total soil nitrogen would increase by 0.013×10^6 Mg N ha^{-1} yr^{-1}, and water use efficiency by 16×10^6 m^3 ha^{-1}.

Biological Attributes

Soil microorganisms are pivotal in transforming organic matter, cycling nutrients, and facilitating energy flow. Enzymes produced by these microorganisms are crucial for the decomposition of SOM. Consequently, soil microbial biomass and enzymatic activities are recognized as sensitive and valuable indicators of soil quality (Choudhary et al., 2018b). As agricultural management practices strongly influence microbial biomass and enzymatic activity, microbial diversity is negatively

Table 13.7 Impacts of Conservation Agriculture (CA) on Productivity, Soil Organic Carbon (SOC)/Total Soil Nitrogen (TSN), Water Use Efficiency, and Profitability Gain in the IGP.

Cropping system	Gain in SOC/TSN in the best CA treatment over farmers' practice in the 0–5 cm layer	Gain in system productivity in the best CA treatment over farmers' practice ($Mg\ ha^{-1}\ yr^{-1}$)	Gain in system water use efficiency in the best CA treatment over farmers' practice ($m^3\ ha^{-1}$)	Gain in system profitability in the best CA treatment over farmers' practice (₹ $ha^{-1}\ yr^{-1}$)
Rice–mustard	0.70 $Mg\ ha^{-1}$ SOC in 7 years	4.14 $Mg\ ha^{-1}\ yr^{-1}$ rice equivalent yield	55	₹53,000
Maize–wheat	1.05 $Mg\ ha^{-1}$ SOC in five years and 0.13 $Mg\ ha^{-1}$ TSN in five years	1.2 $Mg\ ha^{-1}\ yr^{-1}$ wheat equivalent yield	32	₹19,200
Cotton-wheat	1.12 $Mg\ ha^{-1}$ SOC in eight years	2.63 $Mg\ ha^{-1}\ yr^{-1}$ wheat equivalent yield	41	₹36,108
Rice-wheat	0.57 $Mg\ ha^{-1}$ SOC in three years	1.22 $Mg\ ha^{-1}\ yr^{-1}$ rice equivalent yield	—	₹37,407
Pigeonpea-wheat	1.19 $Mg\ ha^{-1}$ SOC in nine years	1.3 $Mg\ ha^{-1}\ yr^{-1}$ rice equivalent yield	30	—

correlated with tillage intensity. Choudhary et al. (2018a) reported a higher microbial population under CA in maize followed by a wheat rotation system in comparison with CT. Greater food source availability supplied through residue retention sustains higher microbial biomass in soil. The third principle of CA (i.e., crop diversification) positively affects microbial diversity. Diversification of crops increases the heterogeneity and biochemical complexity of residues and so these qualitative substrates build up a more diverse decomposer community.

The SOM acts as a source and substrate for microbial enzymes. In CA, the practices of residue recycling, ZT, and crop rotation significantly enhance soil enzyme activities. Residue recycling involves the return of crop residues to the soil, enriching it with organic matter and nutrients. ZT minimizes soil disturbance, preserving soil structure and microbial habitats. Crop rotation diversifies the plant species grown in a field over time, promoting a variety of root exudates and microbial communities. Together, these practices contribute to higher soil enzyme activities, which are essential for soil health and fertility (Kumawat et al., 2018; Parihar et al., 2016, 2020; Table 13.8). Residue-retained ZT has been shown to increase phosphatase activity (Parihar et al., 2016; Choudhary et al., 2018a; Choudhary et al., 2019). CA practices also significantly enhance dehydrogenase activity compared with CT, likely owing to increased microbial activity resulting from higher SOC levels. β-Glucosidase activity, critical for the carbon cycle, particularly in converting cellulose to glucose by microbes, was found to be 20%–27% more in the case of ZT and plant biomass retention compared with traditionally tilled plots (Parihar et al., 2020).

Table 13.8 Effect of Crop Residues Retention (CRR) on Soil Biological Properties Under Conservation Agriculture (Kumawat et al., 2018/Indian Council of Agricultural Research).

	Dehydrogenase activity (μg TPF g^{-1} dry soil h^{-1})		Microbial biomass carbon (mg kg^{-1})		Microbial biomass phosphorus (mg kg^{-1})	
Treatment	Rhizospheric soil	0–5 cm	Rhizospheric soil	0–5 cm	Rhizospheric soil	0–5 cm
No-CRR	21.0^{c}	19.9^{c}	443^{b}	353^{c}	45.7^{c}	37.7^{c}
25% CRR	20.8^{c}	21.3^{ab}	448^{b}	367^{bc}	55.2^{b}	40.6^{bc}
50% CRR	22.9^{b}	20.6^{bc}	454^{b}	374^{ab}	56.8^{b}	44.8^{ab}
75% CRR	24.3^{a}	21.5^{a}	476^{a}	388^{a}	61.2^{a}	48.1^{a}
SEM (±)	0.35	0.34	5.98	7.77	3.84	2.10
LSD ($P \leq 0.05$)	0.86	0.83	14.64	19.01	4.89	6.68

Note. Values followed by a different letter within a column are significantly different ($P < 0.05$).

Soil Quality Index Affected by CA

Indiscriminate use of inorganic fertilizers, coupled with high intensity tillage operation and complete removal of crop residues, has a detrimental effect on soil health and also accelerates the risk of soil erosion (Lal, 2015). Therefore, interest in adoption of different resource conservation technologies is rapidly increasing worldwide, including in India. Proper implementation of CA brings numerous improvements in physical, chemical, and biological attributes of soil, as discussed in the previous section, which is also reflected in the soil quality index (SQI). Recent studies on CA predominantly focused on the changes in soil attributes, but systematic studies regarding the impact of CA on SQI are very limited, especially in the Indian subcontinent (Table 13.9). In a recent study, it was found that adoption of 10 years of CA significantly enhanced the SQI more than use of CT in a rice–mustard system in the plains of Ganga (Das et al., 2021). The authors also observed that saturated hydraulic conductivity, soil pH, available phosphorus, available potassium, and total nitrogen are sensitive indicators for assessment of SQI, with the maximum variation in SQI value resulting from hydraulic conductivity. Parihar et al. (2020) attempted an assessment of SQI in a maize–wheat–mungbean system under CA. The principal component (PC) analysis revealed that the first two PCs with an eigenvalue greater than one can explain 76.3% of the variation in the total data set (Table 13.10), and all three quality indicators, namely, mean weight diameter (MWD) from PC1 and total SOC and available K from PC2, were selected in the minimum data set. CA was associated with significantly higher values of SQI, compared with CT (Table 13.10). Similarly, several other researchers reported higher SQI values under CA-based practices compared with CT across the country (Sharma et al., 2013; Sharma et al., 2016; Choudhary et al., 2018b).

Conclusion

CA is pivotal for soil health restoration and improvement. Adopting CA scientifically leads to improvement in different soil properties, including physical, chemical, and biological properties. To ensure the full potential benefit of CA, the three principles of CA need to be implemented carefully. Incomplete adoption of the CA package of practices is often found to be detrimental to soil health. The SQI is a comprehensive index that can assess the impact of any agronomic practice on overall soil health. For a truthful and realistic representation, a large data set should be considered for the calculation of SQI. Researchers across the country have reported improvement in SQI results upon implementation of CA. Therefore,

Table 13.9 Soil Quality Index Under Different Long-Term Conservation Agriculture Experiment.

Soil	Duration of CA	Depth (cm)	Cropping system	Minimum data set	Equation	SQI value	References
Inceptisol, New Delhi	4 years	0–30	Maize–wheat–mungbean	Total SOC, MWD, available K	SQI = 0.883 × MWD + 0.117 × total SOC + 0.117 × available K	0.82–1.01	Parihar et al., 2020
Inceptisol, New Delhi	10 years	0–5	Rice–mustard	K_S, soil pH, available P, total N	SQI = 0.686 × K_S + 0.137 × pH + 00.098 × Available P + 0.079 × Total N	0.596–1.0	Das et al., 2021
		5–15	Rice–mustard	K_S, soil pH, available K	SQI = 0.881 × K_S + 0.103 × Available P + 0.086 × pH	0.46–1.0	
Vertisol, Hyderabad	6 years	0–15	Soybean based	Organic C, MBC, available Zn, MWD, available B	—	1.21–1.65	Sharma et al., 2016
Aridisol, Hissar	7 years	0–15	Pearl millet-mungbean	Dehydrogenase activity, Exchangeable Mg, Available N, available Mn, BD, MBC	—	1.50–1.74	Sharma et al., 2013

Note. BD = bulk density; K_S = saturated hydraulic conductivity; MBC = microbial biomass carbon; MWD = mean weight diameter; SQI = soil quality index.

Table 13.10 Principal Component (PC) Analysis of 14 Soil Quality Indicators as Influenced by Conservation Agriculture (Parihar et al., 2020/with permission of Elsevier).

Soil attributes	PC1	PC2
Eigen value	3.07	1.12
Percentage of variance	67.36	8.94
Cumulative variance percentage	67.36	76.30
Eigen vectors		
Bulk density	−0.149	−0.416
Penetration resistance	−0.255	−0.148
Total water stable aggregate	0.286	0.064
Mean weight diameter	0.299	0.061
Hydraulic conductivity	0.276	0.001
Total soil organic carbon	0.262	−0.430
Available nitrogen	0.268	0.360
Available phosphorus	0.297	0.196
Available potassium	0.230	0.464
Microbial biomass carbon	0.285	−0.245
Fluorescein diacetate activity	0.284	−0.309
Dehydrogenase	0.279	−0.126
β-Glucosidase	0.297	−0.178
Alkaline phosphatase	0.242	−0.163

this package of practices needs to be encouraged and popularized among farmers, toward achieving a sustainable cropping system. To meet the needs of a growing global population, an additional ~1 billion tonnes of food must be produced by 2050. Simultaneously, efforts are needed to restore degraded soils and ecosystems, reduce net anthropogenic emissions, and enhance environmental quality. Given the constraints of limited resources, there is an urgent imperative to adopt and promote conservation-effective agricultural practices. The sustainable agricultural production goals can be achieved by CA-based crop production systems owing to their effects on promoting soil health, enhancing water and nutrient efficiency, and mitigating environmental impacts. Additionally, these production systems provide diverse provisioning (food production, water-use efficiency), regulating (soil-moisture conservation, soil aggregation, groundwater regulation, energy use, waste decomposition and detoxification, soil-erosion prevention, carbon sequestration, climate regulation), and supporting (nutrient accumulation and cycling, biodiversity conservation, primary production) ecosystem services that are crucial for improving the efficiency of natural resource use (soil, water, air, and fuel) and for achieving environmental and food security objectives in line with the United Nations Sustainable Development Goals (SDGs). CA can directly contribute to achieving several SDGs, including SDG 1 (No Poverty), SDG 2 (Zero Hunger), SDG 12 (Responsible Consumption and Production), SDG 13 (Climate Action), and SDG 15 (Life on Land).

References

Bhattacharya, P., Maity, P. P., Mowrer, J., Maity, A., Ray, M., Das, S., Chakrabarti, B., Ghosh, T., & Krishnan, P. (2020). Assessment of soil health parameters and application of the sustainability index to fields under conservation agriculture for 3, 6, and 9 years in India. *Heliyon*, *6*, 05640.

Bhattacharyya, P., Roy, K. S., Neogi, S., Adhya, T. K., Rao, K. S., & Manna, M. C. (2012a). Effects of rice straw and nitrogen fertilization on greenhouse gas emissions and carbon storage in tropical flooded soil planted with rice. *Soil and Tillage Research*, *124*, 119–130.

Bhattacharyya, P., Roy, K. S., Neogi, S., Chakravorti, S. P., Behera, K. S., Das, K. M., Bardhan, S., & Rao, K. S. (2012b). Effect of long-term application of organic amendment on C storage in relation to global warming potential and biological activities in tropical flooded soil planted to rice. *Nutrient Cycling in Agroecosystems*, *94*(2–3), 273–285.

Bhattacharyya, R., Bhatia, A., Das, T. K., Lata, S., Kumar, A., Tomer, R., & Biswas, A. K. (2018). Aggregate-associated N and global warming potential of conservation agriculture-based cropping of maize-wheat system in the north-western Indo-Gangetic Plains. *Soil and Tillage Research*, *182*, 66–77.

Bhattacharyya, R., Das, T. K., Das, S., Dey, A., Patra, A. K., Agnihotri, R., Ghosh, A., & Sharma, A. R. (2019). Four years of conservation agriculture affects topsoil aggregate-associated ^{15}nitrogen but not the ^{15}nitrogen use efficiency by wheat in a semi-arid climate. *Geoderma*, *337*, 333–340.

Bhattacharyya, R., Das, T. K., Pramanik, P., Ganeshan, V., Saad, A. A., & Sharma, A. R. (2013). Impacts of conservation agriculture on soil aggregation and aggregate-associated N under an irrigated agroecosystem of the Indo-Gangetic Plains. *Nutrient Cycling and Agro-Ecosystem*, *96*, 185–202.

Bhattacharyya, R., Das, T. K., Sudhishri, S., Dudhwal, A., Sharma, A. R., Bhatia, A., & Singh, G. (2015). Conservation agriculture effects on soil organic carbon accumulation and crop productivity under a rice-wheat cropping system in the western Indo-Gangetic Plains. *European Journal of Agronomy*, *70*, 11–21.

Bhattacharyya, R., Tuti, M. D., Kundu, S., Bisht, J. K., & Bhatt, J. C. (2012c). Conservation tillage impacts on soil aggregation and carbon pools in a sandy clay loam soil of the Indian Himalayas. *Soil Science Society of America Journal*, *76*, 617–627.

Biswas, S., Das, T. K., Bhattacharyya, R., Das, S., & Dwivedi, B. S. (2021). Biological soil quality and seasonal variation on enzyme activities under conservation agriculture-based rice–mustard system in the Indo-Gangetic Plains of India. *Soil Research*, *60*(6), 470–484.

Brady, N. C., & Weil, R. C. (2013). *The Nature and Properties of Soil* (14th ed.). Dorling Kindersley.

Carter, M. R. (1992). Influence of reduced tillage systems on organic matter, microbial biomass, macro-aggregate distribution and structural stability of the surface soil in a humid climate. *Soil and Tillage Research*, *23*, 361–372.

Carter, M. R., Gregorich, E. G., Anderson, D. W., Doran, J. W., Janzen, H. H., & Pierce, F. J. (1997). Concepts of soil quality and their significance. In E. G. Gregorich & M. R. Carter (Eds.), *Soil quality for crop production and ecosystem health* (Vol. 25, pp. 1–19). Elsevier.

Choudhary, M., Datta, A., Jat, H. S., Yadav, A. K., Gathala, M. K., Sapkota, T. B., Das, A. K., Sharma, P. C., Jat, M. L., Singh, R., & Ladha, J. K. (2018a). Changes in soil biology under conservation agriculture based sustainable intensification of cereal systems in Indo-Gangetic Plains. *Geoderma*, *313*, 193–204.

Choudhary, M., Rana, K. S., Meena, M. C., Bana, R. S., Jakhar, P., Ghasal, P. C., & Verma, R. K. (2019). Changes in physico-chemical and biological properties of soil under conservation agriculture based pearl millet–mustard cropping system in rainfed semi-arid region. *Archives of Agronomy and Soil Science*, *65*, 911–927.

Choudhary, M., Sharma, P. C., Jat, H. S., McDonald, A., Jat, M. L., Choudhary, S., & Garg, N. (2018b). Soil biological properties and fungal diversity under conservation agriculture in Indo-Gangetic Plains of India. *Journal of Soil Science and Plant Nutrition*, *18*, 1142–1156.

Christensen, B. T. (1996). Matching measurable soil organic matter fractions with conceptual pools in simulation models of carbon turnover: Revision of model structure. In D. S. Powlson, P. Smith, & J. V. Smith (Eds.), *Evaluation of soil organic matter models: Using existing long-term datasets* (pp. 143–159).

Das, A., Ramkrushna, G. I., Choudhury, B. U., Ngachan, S. V., Tripathi, A. K., Singh, R. K., & Munda, G. C. (2014). Conservation agriculture in rice and maize based cropping systems for enhancing crop and water productivity-participatory technology demonstration in north east India. *Indian Journal of Soil Conservation*, *41*, 196–203.

Das, S., Bhattacharyya, R., Das, T. K., Sharma, A. R., Dwivedi, B. S., Meena, M. C., Dey, A., Biswas, S., Aditya, K., Aggarwal, P., & Chaudhari, S. K. (2021). Soil quality indices in a conservation agriculture based rice-mustard cropping system in north-western Indo-Gangetic Plains. *Soil and Tillage Research*, *208*, 104914.

Das, T. K., Bandyopadhyay, K. K., Bhattacharyya, R., Sudhishri, S., Sharma, A. R., Behera, U. K., & Jat, M. L. (2016). Effects of conservation agriculture on crop productivity and water-use efficiency under an irrigated pigeonpea–wheat cropping system in the western Indo-Gangetic Plains. *The Journal of Agricultural Science*, *154*, 1327–1342.

Das, T. K., Nath, C. P., Das, S., Biswas, S., Bhattacharyya, R., Sudhishri, S., & Chaudhari, S. K. (2020). Conservation agriculture in rice-mustard cropping system for five years: Impacts on crop productivity, profitability, water-use efficiency, and soil properties. *Field Crops Research*, *250*, 107781.

Das, T. K., Saharawat, Y. S., Bhattacharyya, R., Sudhishri, S., Bandyopadhyay, K. K., Sharma, A. R., & Jat, M. L. (2018). Conservation agriculture effects on crop and water productivity, profitability and soil organic carbon accumulation under a maize-wheat cropping system in the north-western Indo-Gangetic Plains. *Field Crops Research*, *215*, 222–231.

Dash, A. K., Dwivedi, B. S., Dey, A., Meena, M. C., & Chakraborty, D. (2023). Temperature sensitivity of soil organic carbon as affected by crop residue and nutrient management options under conservation agriculture. *Journal of Soil Science and Plant Nutrition*, *23*, 4183–4197.

Dey, A. (2017). *Dynamics of soil organic carbon and nitrogen in conservation agriculture*. Ph.D. Thesis, Division of Soil Science and Agricultural Chemistry, Indian Agricultural Research Institute.

Dey, A., Dwivedi, B. S., Bhattacharyya, R., Datta, S. P., Meena, M. C., Das, T. K., & Singh, V. K. (2016). Conservation agriculture in a rice-wheat cropping system on an alluvial soil of north-western indo-Gangetic plains: Effect on soil carbon and nitrogen pools. *Journal of the Indian Society of Soil Science*, *64*, 246–254.

Dey, A., Dwivedi, B. S., Bhattacharyya, R., Datta, S. P., Meena, M. C., Jat, R. K., Gupta, R. K., Jat, M. L., Singh, V. K., Das, D., & Singh, R. G. (2020). Effect of conservation agriculture on soil organic and inorganic carbon sequestration and lability: A study from a rice–wheat cropping system on a calcareous soil of the eastern Indo-Gangetic Plains. *Soil Use and Management*, *36*, 429–438.

Dey, A., Dwivedi, B. S., Meena, M. C., & Datta, S. P. (2018). Dynamics of soil carbon and nitrogen under conservation agriculture in rice-wheat cropping system. *Indian Journal of Fertilisers*, *14*, 12–26.

Driver, J. D., Holben, W. E., & Rillig, M. C. (2005). Characterization of glomalin as a hyphal wall component of arbuscular mycorrhizal fungi. *Soil Biology and Biochemistry*, *37*, 101–106.

FAO. (2015). *Food and Agriculture Organization*. Conservation agriculture. http://www.fao.org/ag/ca/1a.html

Ghimire, R., Adhikari, K. R., Chen, Z. S., Shah, S. C., & Dahal, K. R. (2012). Soil organic carbon sequestration as affected by tillage, crop residue, and nitrogen application in rice–wheat rotation system. *Paddy and Water Environment*, *10*, 95–102.

Ghosh, A., Bhattacharyya, R., Dwivedi, B. S., Meena, M. C., Agarwal, B. K., Mahapatra, P., & Agnihorti, R. (2016). Temperature sensitivity of soil organic carbon decomposition as affected by long-term fertilization under a soybean-based cropping system in a sub-tropical Alfisol. *Agriculture, Ecosystems and Environment*, *233*, 202–213.

Guo, L. J., Zhang, Z. S., Wang, D. D., Li, C. F., & Cao, C. G. (2015). Effects of short-term conservation management practices on soil organic carbon fractions and microbial community composition under a rice-wheat rotation system. *Biology and Fertility of Soils*, *51*, 65–75.

Haokip, I. C., Dwivedi, B. S., Meena, M. C., Datta, S. P., Jat, H. S., Dey, A., & Tigga, P. (2020). Effect of conservation agriculture and nutrient management options on soil phosphorus fractions under maize-wheat cropping system. *Journal of the Indian Society of Soil Science*, *68*, 45–53.

Haynes, R. J., & Swift, R. S. (1990). Stability of soil aggregates in relation to organic constituents and soil water content. *Journal of Soil Science*, *41*, 73–83.

Jat, R. A., Wani, S. P., & Sahrawat, K. L. (2012). Conservation agriculture in the semi-arid tropics: Prospects and problems. *Advances in Agronomy*, *117*, 191–273.

Jat, S. L., Parihar, C. M., Dey, A., Nayak, H. S., Ghosh, A., Parihar, N., Goswami, A. K., & Singh, A. K. (2019). Dynamics and temperature sensitivity of soil organic carbon mineralization under medium-term conservation agriculture as affected by residue and nitrogen management options. *Soil and Tillage Research*, *190*, 175–185.

Kumawat, C., Sharma, V. K., Meena, M. C., Dwivedi, B. S., Barman, M., Kumar, S., Chobhe, K. A., & Dey, A. (2018). Effect of crop residue retention and phosphorus fertilization on P use efficiency of maize (*Zea mays*) and biological properties of soil under maize-wheat (*Triticum aestivum*) cropping system in an Inceptisol. *Indian Journal of Agricultural Sciences*, *88*, 1184–1189.

Lal, R. (2015). Sequestering carbon and increasing productivity by conservation agriculture. *Journal of Soil and Water Conservation*, *70*, 55A–62A.

Larson, W. E., & Pierce, F. J. (1994). The dynamics of soil quality as a measure of sustainable management. *Defining Soil Quality for a Sustainable Environment*, *35*, 37–51.

Lehmann, A., Zheng, W., & Rillig, M. C. (2017). Soil biota contributions to soil aggregation. *Nature Ecology & Evolution*, *1*, 1828–1835.

Liu, Y., Gao, M., Wu, W., Tanveer, S. K., Wen, X., & Liao, Y. (2013). The effects of conservation tillage practices on the soil water-holding capacity of a non-irrigated apple orchard in the loess plateau, China. *Soil and Tillage Research*, *130*, 7–12.

Lovelock, C. E., Wright, S. F., & Nichols, K. A. (2004). Using glomalin as an indicator for arbuscular mycorrhizal hyphal growth: An example from a tropical rain forest soil. *Soil Biology and Biochemistry*, *36*(6), 1009–1012.

Lu, X., Lu, X., & Liao, Y. (2018). Conservation tillage increases carbon sequestration of winter wheat-summer maize farmland on loess plateau in China. *PLoS One*, *13*, e0199846.

Modak, K., Ghosh, A., Bhattacharyya, R., Biswas, D. R., Das, T. K., Das, S., & Singh, G. (2019). Response of oxidative stability of aggregate-associated soil organic carbon and deep soil carbon sequestration to zero-tillage in subtropical India. *Soil and Tillage Research*, *195*, 104370.

Mondal, S., Poonia, S. P., Mishra, J. S., Bhagwati, P. B., Rao, K. K., Saurabh, K., Kumar, R., & Chakraborty, D. (2020). Short-term (5 years) impact of conservation agriculture on soil physical properties and organic carbon in a rice–wheat rotation in the Indo-Gangetic plains of Bihar. *European Journal of Soil Science*, *71*, 1076–1089.

Moretto, A. S., Distel, R. A., & Didoné, N. G. (2001). Decomposition and nutrient dynamic of leaf litter and roots from palatable and unpalatable grasses in a semi-arid grassland. *Applied Soil Ecology*, *18*, 31–37.

Nandan, R., Singh, V., Singh, S. S., Kumar, V., Hazra, K. K., Nath, C. P., & McDonald, A. (2019). Impact of conservation tillage in rice–based cropping systems on soil aggregation, carbon pools and nutrients. *Geoderma*, *340*, 104–114.

Naorem, A., Jayaraman, S., Sinha, N. K., Mohanty, M., Chaudhary, R. S., Hati, K. M., Mandal, A., Thakur, J. K., Patra, A. K., Srinivasarao, C., Chaudhari, S. K., Dalal, R. C., & Lal, R. (2023). Eight-year impacts of conservation agriculture on soil quality, carbon storage, and carbon emission footprint. *Soil and Tillage Research*, *232*, 105748.

Naresh, R. K., Singh, S. P., & Chauhan, P. (2012). Influence of conservation agriculture, permanent raised bed planting and residue management on soil quality and productivity in maize-wheat system in western Uttar Pradesh. *International Journal of Life Sciences Biotechnology and Pharma Research*, *1*, 27–33.

Nath, A. J., & Rattan, L. (2017). Effects of tillage practices and land use management on soil aggregates and soil organic carbon in the north Appalachian region, USA. *Pedosphere*, *27*, 172–176.

Oades, J. M. (1984). Soil organic matter and structural stability: Mechanisms and implications for management. *Plant and Soil*, *76*, 319–337.

Palm, C., Blanco-Canqui, H., DeClerck, F., Gatere, L., & Grace, P. (2014). Conservation agriculture and ecosystem services: An overview. *Agriculture, Ecosystems and Environment*, *187*, 87–105.

Parihar, C. M., Parihar, M. D., Sapkota, T. B., Nanwal, R. K., Singh, A. K., Jat, S. L., Nayak, H. S., Mahala, D. M., Singh, L. K., Kakraliya, S. K., & Stirling, C. M. (2018). Long-term impact of conservation agriculture and diversified maize rotations on carbon pools and stocks, mineral nitrogen fractions and nitrous oxide fluxes in Inceptisol of India. *Science of the Total Environment*, *640*, 1382–1392.

Parihar, C. M., Singh, A. K., Jat, S. L., Dey, A., Nayak, H. S., Mandal, B. N., Saharawat, Y. S., Jat, M. L., & Yadav, O. P. (2020). Soil quality and carbon sequestration under conservation agriculture with balanced nutrition in intensive cereal-based system. *Soil and Tillage Research*, *202*, 104653.

Parihar, C. M., Singh, A. K., Jat, S. L., Ghosh, A., Dey, A., Nayak, H. S., Parihar, M. D., Mahala, D. M., Yadav, R. K., Rai, V., & Satayanaryana, T. (2019). Dependence of temperature sensitivity of soil organic carbon decomposition on nutrient management options under conservation agriculture in a sub-tropical Inceptisol. *Soil and Tillage Research*, *190*, 50–60.

Parihar, C. M., Yadav, M. R., Jat, S. L., Singh, A. K., Kumar, B., Pradhan, S., Chakraborty, D., Jat, M. L., Jat, R. K., Saharawat, Y. S., & Yadav, O. P. (2016). Long term effect of conservation agriculture in maize rotations on total organic carbon, physical and biological properties of a sandy loam soil in north-western Indo-Gangetic Plains. *Soil and Tillage Research*, *161*, 116–128.

Patra, S., Julich, S., Feger, K. H., Jat, M. L., Jat, H., Sharma, P. C., & Schwärzel, K. (2019). Soil hydraulic response to conservation agriculture under irrigated intensive cereal-based cropping systems in a semiarid climate. *Soil and Tillage Research*, *192*, 151–163.

Prasad, J. V. N. S., Rao, C. S., Srinivas, K., Jyothi, C. N., Venkateswarlu, B., Ramachandrappa, B. K., . . . Mishra, P. K. (2016). Effect of ten years of reduced tillage and recycling of organic matter on crop yields, soil organic carbon and its fractions in Alfisols of semi arid tropics of southern India. *Soil and Tillage Research*, *156*, 131–139.

Purakayastha, T. J., Bhaduri, D., Aishwath, O. P., & Shivay, Y. S. (2016). Soil management strategies to enhance carbon sequestration potential of degraded lands. *Indian Journal of Agronomy*, *61*(4), 407–419.

Rillig, M. C., & Steinberg, P. D. (2002). Glomalin production by an arbuscular mycorrhizal fungus: A mechanism of habitat modification. *Soil Biology and Biochemistry*, *34*, 1371–1374.

Roy, D., Datta, A., Jat, H. S., Choudhary, M., Sharma, P. C., Singh, P. K., & Jat, M. L. (2022). Impact of long term conservation agriculture on soil quality under cereal based systems of north West India. *Geoderma*, *405*, 115391.

Sapkota, T. B., Majumdar, K., Jat, M. L., Kumar, A., Bishnoi, D. K., McDonald, A. J., & Pampolino, M. (2014). Precision nutrient management in conservation agriculture based wheat production of Northwest India: Profitability, nutrient use efficiency and environmental footprint. *Field Crops Research*, *155*, 233–244.

Saraswat, A., Ram, S., AbdelRahman, M. A., Raza, M. B., Golui, D., Hc, H., . . . Rahman, M. M. (2023). Combining fuzzy, multicriteria and mapping techniques to assess soil fertility for agricultural development: A case study of Firozabad District, Uttar Pradesh, India. *Land*, *12*(4), 860.

Sharma, K. L., Chandrika, D. S., Grace, J. K., Shankar, G. M., Sharma, S. K., Thakur, H. S., Jain, M. P., Sharma, R. A., Chary, G. R., Srinivas, K., & Gajbhiye, P. (2016). Soil quality assessment under restorative soil management practices in soybean (*Glycine max*) after six years in semi-arid tropical black lands of Central India. *Communications in Soil Science and Plant Analysis, 47*, 1465–1475.

Sharma, K. L., Grace, J. K., Raj, M., Mittal, S. B., Singh, J., Sharma, S. K., Sangwan, P. S., Sidhpuria, M. S., Vittal, K. P. R., Mishra, P. K., & Sankar, G. M. (2013). Improvement and assessment of soil quality under long-term conservation agricultural practices in hot, arid tropical Aridisol. *Communications in Soil Science and Plant Analysis, 44*, 1033–1055.

Sharma, V., Hussain, S., Sharma, K. R., & Arya, V. M. (2014). Labile carbon pools and soil organic carbon stocks in the foothill Himalayas under different land use systems. *Geoderma, 232*, 81–87.

Singh, G., Bhattacharyya, R., Das, T. K., Sharma, A. R., Ghosh, A., Das, S., & Jha, P. (2018). Crop rotation and residue management effects on soil enzyme activities, glomalin and aggregate stability under zero tillage in the Indo-Gangetic Plains. *Soil and Tillage Research, 184*, 291–300.

Sinha, A. K., Ghosh, A., Dhar, T., Bhattacharya, P. M., Mitra, B., Rakesh, S., Paneru, P., Shrestha, S. R., Manandhar, S., Beura, K., & Dutta, S. (2019). Trends in key soil parameters under conservation agriculture-based sustainable intensification farming practices in the eastern ganga Alluvial Plains. *Soil Research, 57*, 883–893.

Six, J., Elliott, E. T., & Paustian, K. (1999). Aggregate and soil organic matter dynamics under conventional and no-tillage systems. *Soil Science Society of America Journal, 63*, 1350–1358.

Six, J., Feller, C., Denef, K., Ogle, S., de Moraes Sa, J. C., & Albrecht, A. (2002). Soil organic matter, biota and aggregation in temperate and tropical soils—Effects of no-tillage. *Agronomie, 22*, 755–775.

Six, J. Á., Elliott, E. T., & Paustian, K. (2000). Soil macroaggregate turnover and microaggregate formation: A mechanism for C sequestration under no-tillage agriculture. *Soil Biology and Biochemistry, 32*, 2099–2103.

Tisdall, J. M., & Oades, J. M. (1982). Organic matter and water-stable aggregates in soils. *Journal of Soil Science, 33*, 141–163.

Wang, W., Shi, J., Xie, Q., Jiang, Y., Yu, N., & Wang, E. (2017). Nutrient exchange and regulation in arbuscular mycorrhizal symbiosis. *Molecular Plant, 10*, 1147–1158.

Wright, S. F., & Upadhyaya, A. (1996). Extraction of an abundant and unusual protein from soil and comparison with hyphal protein of arbuscular mycorrhizal fungi. *Soil Science, 161*(9), 575–586.

Xin, S. H. U., Zhu, A. N., Zhang, J. B., Yang, W. L., Xin, X. L., & Zhang, X. F. (2015). Changes in soil organic carbon and aggregate stability after conversion to conservation tillage for seven years in the Huang-Huai-Hai Plain of China. *Journal of Integrative Agriculture*, *14*, 1202–1211.

Zhang, M. K., He, Z. L., Calvert, D. V., Stoffella, P. J., Yang, X. E., & Li, Y. C. (2003). Phosphorus and heavy metal attachment and release in sandy soil aggregate fractions. *Soil Science Society of America Journal*, *67*, 1158–1167.

14

Managing Crop Residue for Soil Health Improvement in India

Dibyendu Sarkar, Siddhartha Mukherjee, G. Pratibha, and Sumanta Kundu

Chapter Overview

Crop biomass left in the field after the crop is harvested is known as crop residue, which include stalks, stubble, stovers, leaves, and husks. Most crop residue in India is used as fodder and fuel in a variety of ways. Since proper disposal of huge quantity of residues is a great challenge, farmers prefer burning them to clear the field and allow easy tillage for sowing of the next crop. However, the adoption of suitable in situ and ex situ (i.e., off-field) residue management strategies can be beneficial for improving soil health. For example, crop residue is a huge source of carbon and mineral nutrients. Under in situ management, crop residues are either retained on the soil surface or mixed into the soil during sowing or before sowing using appropriate machinary. Retention and/or incorporation of residues can be practiced efficiently through adoption of appropriate conservation agriculture practices. Large volumes of agricultural residues can also be recycled through off-field compost and biochar preparation. Adoption of sustainable crop residue management plans (in situ and/or ex situ) has favorable effects on soil physical, chemical, and biological properties and is considered a sustainable crop production option. In situ management of crop residues under conservation agriculture can be a more economic and energy-conserving strategy for improving soil health. Conventional tillage can efficiently recycle leguminous crop residues back into the soil, while the same tillage practices can effectively recycle cereal and sugarcane residues with a high C:N ratio. Development of need based residue management

Soil Health Series: Volume 4 Soil Health and Sustainability in India, First Edition.
Edited by Ranjan Bhattacharyya.

machinery; CA-based crop production technology; effective strategies to control weeds, insects, and pests; and utliizing bio-decomposers for efficient decomposition of cereal and sugarcane residues will drive future research to address the existing challenges of crop residue management by farmers for improving soil health in India.

Introduction

Crop plant biomass left in the field after the economic components of the crop are harvested is known as crop residue. Crop residues include stalks, stubbles, stovers, leaves, and husks, which are regarded as agricultural waste. Sugarcane bagasse and molasses are examples of processed crop residues. The annual global production of agricultural residues is 5,287 million Mg (Shinde et al., 2022), with India generating about 686 million Mg of crop residues per year (Kumar et al., 2023). Rice, wheat, maize, and sugarcane account for about 80% of the total residue production in India. A majority of this crop residue is used as fodder and fuel for various domestic and agricultural situations. Nevertheless, there is an annual surplus of 178 million Mg of unused residue in India.

Proper disposal of such a huge quantity of residues is a great challenge. Farmers prefer burning them, which clears the field economically and quickly and allows easy tillage and land preparation for sowing of the next crop. About 87 million Mg of crop residue is burned each year in India (Datta et al., 2020). The impact of residue burning on soil health is extremely detrimental owing to the loss of organic matter through CO_2 and conversion of nitrogen into NO_3^- (Kumar et al., 2019a). Although burning of crop residues temporarily increases the mineral nitrogen (NH_4^+) and plant available phosphorus content in soil, it has serious impacts on soil health, organic matter content, plant micro- and macronutrient content, and the soil microbial population (Raza et al., 2019). Residue burning also emits greenhouse gases such as CO_2, CH_4, CO, SO_2, nitrous oxides (NO, NO_x), hydrocarbons, volatile and semivolatile organic compounds, and particulate matter (Mittal et al., 2009). Thus, burning of crop residues has several negative impacts on the environment, including soil health degradation, air pollution, global warming, and contribution to climate change (Mathur & Srivastava, 2019).

Crop residues generated in Indian agriculture represent a huge source of organic carbon and mineral nutrients, and adoption of suitable residue management strategies, in situ and ex situ (i.e., off-field), could be beneficial for improving soil health and maintaining environmental sustainability. Recycling of crop residues to the soil can be done through in situ practices such as retention on the soil surface or incorporation into the soil during sowing or a few days before sowing of the next crop, given the availability of suitable machinery (Kumar et al., 2023). Retention and/or incorporation of residues in the agricultural production system can be performed efficiently through adoption of appropriate conservation agriculture (CA) technologies. Large volumes of agricultural residues can also be recycled through off-field compost and biochar preparation. Considering these options, the present chapter discusses various crop residue management issues, including strategies, effects, and challenges for soil health improvement in India.

Residue Management Options for Improving Soil Health

Since large quantities of residues generated from agriculture serve as a source of plant nutrients and carbon (Table 14.1), it is important to adopt judicious crop residue management techniques that will increase agricultural profitability and improve soil health, and in turn, maintain environmental sustainability. Any particular residue management option may not be viable for all regions owing to variability in agro-ecosystems, residue utilization pattern, and socio-economic status of farmers. Various in situ and off-field residue management options are compiled and discussed in the following sections.

In Situ Crop Residue Management

Crop residue management within the field is an effective option because it eliminates the cost of transportation, recycles plant nutrients back into the soil, and saves time. Under in situ management, crop residues are either retained on the soil surface or incorporated into the soil during sowing of the next crop or few days beforehand with suitable machinery. Recycling of crop residues into the soil serves many ecosystem functions, including supplying nutrient elements to plants and improving soil health (Sarkar et al., 2020).

Retention of Crop Residues on Soil Surface

One management approach involves spreading the whole quantity or a portion of the residues on the soil or keeping anchored stubble in the field after crop harvest and allowing it to decompose with time. Crop residue retention is an energy-saving and environmental conservation practice. Recent advances in farm machinery allow direct sowing of crops in residue-covered soil by opening narrow slits for placing seeds and fertilizers, keeping the rest of the soil surface undisturbed. A combine harvester with a straw management system, Happy Seeder, rotary disc drill, and strip-till drill are currently used for retaining crop residues on the soil surface (Kumar et al., 2023). Mulching of residues on the soil surface

Table 14.1 Important Physical and Chemical Properties of Some Crop Residues in India.

Parameters	Rice straw	Wheat straw	Maize straw	Soybean residue	Soybean leaf	Sugarcane straw	Cotton stubble	Red gram stubble
pH	7.60			7.40	6.68	7.1		
Moisture content (%)	11.4			78.3	72.4			
Total organic C (%)	39.2–51.7	40–55	41–47	18.8	20.2	20–45	41.9	42.9
C:N ratio	61.3–73.9			3.4	10.6	12.4	48.8	58.0
Nitrogen (%)	0.4–1.2	0.30–0.82	0.60–1.50	5.4	1.9	0.60–1.62	0.86	0.74
Phosphorus (%)	0.02–0.21	0.06–0.12	0.17–0.09			0.02–1.12	0.30	0.28
Potassium (%)	1.0–3.7	0.75–3.25	0.08–1.25			0.23–1.36	0.92	0.89
Calcium (%)	0.03–0.39	0.24–0.34						
Magnesium (%)	0.10–0.26	0.18						
Sulfur (%)	0.06–0.10	0.04–0.60	0.01–0.22			0.10–0.20		
Iron ($mg\,kg^{-1}$)	129.2	25–100						
Manganese ($mg\,kg^{-1}$)	560–1,090	10–82						
Zinc ($mg\,kg^{-1}$)	30–2,680	5–20						
Copper ($mg\,kg^{-1}$)	4.2–9.3	1.0–2.5						
Boron ($mg\,kg^{-1}$)	3.4–8.9	14–21						
Molybdenum ($mg\,kg^{-1}$)	2.5	0.09–0.18						

Note. Jusoh et al. (2013), Mandal et al. (2004), Goyal and Sindhu (2011), Dobermann and Fairhurst (2002), Nghi et al. (2020), Kumar et al. (2023), Singh et al. (2014), Maguyon-Detras et al. (2020), Bah et al. (2020), Wong et al. (2001), Canilha et al. (2012), Fortes et al. (2012), Carvalho et al. (2017), Solangi et al. (2018).

provides good soil health by moderating soil and crop canopy temperature and reducing evaporation losses and irrigation requirements during summer months (Sidhu et al., 2015). Surface retention of crop residues reduces soil erosion, conserves carbon in the soil, increases crop tolerance against terminal heat stress, and improves physical, chemical, and biological soil properties (Bimbraw, 2019; Lohan et al., 2018; Lohan & Sharma, 2012; Singh et al., 2009). Mulching of soil with crop residues obstructs germination and growth of weeds by providing benefits to a crop for nutrients and water. Surface retention of crop residues under zero-tillage (ZT) affects weed dynamics and favors restricting the weed seed bank to the top soil layers (Chauhan et al., 2006). However, crop residues on the soil surface can intercept >50% of applied pre-emergence herbicide, which results in poor efficacy of herbicides to control weeds in crop fields. Therefore, further research on efficient weed control practices, mechanical residue spreaders, and bio-decomposers for rapid mineralization of residues is required for improving crop yield and widespread adoption of residue retention practices among farmers.

Incorporation of Crop Residues into Soil

Incorporating or mixing crop residues into the soil is another option for in situ residue management. A zero-till seed drill, special drill chopper and spreader, super seeder, rotavator, disc harrows, moldboard plow, and cultivator are used for residue incorporation into soil. Incorporation of crop residues into the soil yields advantages such as improved soil health and soil nutrient balances (Mandal et al., 2004; Mukherjee et al., 2024; Sidhu & Beri, 2005). However, the availability of essential plant nutrients such as nitrogen, phosphorus, and potassium to crops is dependent on residue quality. Residues of cereal crops and sugarcane, which are associated with larger C:N ratios, are slowly decomposed and release mineralized nitrogen into the soil. According to Samra et al. (2003), nitrogen immobilization in the soil affects the yield of succeeding crops that are grown immediately after incorporation of cereal crop residues with a high C:N ratio. The management options to overcome such nitrogen immobilization issues are (a) incorporation of cereal crop residues into soil 10–40 days before crop sowing, and (b) application of an additional amount of 20–40 kg N ha^{-1} at sowing immediately after incorporation of cereal crop residues into soil. But the additional costs and time associated with these options make them less adoptable by the farmers, particularly in rice–wheat and sugarcane–wheat cropping systems, where time availability for wheat sowing is a constraint. Residue incorporation of pulses, vegetables, and oilseed crops, such as groundnut with low C:N ratios help rapid mineralization of nitrogen for immediate crop uptake. Incorporation of residues from leguminous crops through conventional as well as reduced tillage systems without any additional investment of time and money has the potential to improve soil health.

Conservation Agriculture

Conservation agriculture, which has been emphasized as a key route to sustainable intensification, is based on three principles: (a) minimal soil disturbance or no-tillage/ZT; (b) continuous soil cover with a crop, cover crop, or crop residues; and (c) crop rotation (FAO, 2019). Conservation agriculture technologies are adapted in different agro-ecological conditions of both rainfed and irrigated systems. This involves various resource conservation technologies, such as ZT cultivation, furrow-irrigated raised bed, direct-seeded rice, and crop diversification, which are practiced on only about 1.5 million ha in India including rice–wheat cropping in the Indo-Gangetic Plains (IGP; Jat et al., 2012). While practicing CA with permanent cover crops, sowing new crop seed in the presence of residues from the preceding crop is difficult. Development of an innovative zero-till seed drill has revolutionized sowing of wheat seed in the presence of rice residues in fields. Such a development can be effective in sowing fields with a rice residue load up to 10 Mg ha^{-1} (Kumar et al., 2013; Saharawat et al., 2010). Several research works conducted by the Indian Council of Agricultural Research, New Delhi; Punjab Agricultural University, Ludhiana; and International Maize and Wheat Improvement Centre – Borlaug Institute for South Asia (CIMMYT-BISA) efficiently demonstrated the use of Happy Seeder with a spreader, which is popularly known as the straw management system for sowing wheat with a reduced input requirement (Lohan et al., 2018; Sidhu et al., 2015). In eastern India, where crop residues have competing domestic uses, such as animal feed and roof thatching, Bidhan Chandra Krishi Viswavidyalaya, West Bengal, has shown that during rice harvest, a part of the stubble can be left in fields for restoration of soil health. The Central Research Institute for Dryland Agriculture, Hyderabad, recommended that under dryland agro-ecosystems where only a single crop is raised in a year, it is possible to grow a second crop with residual soil moisture by covering the soil with crop residues (Indian Agricultural Research Institute, 2012).

Zero-tillage has proved to be advantageous for sustaining both soil and crop health. The benefits of ZT over the conventional practice include better crop productivity; saving of costly inputs, such as labor, fuel, irrigation water, and nutrients; and reducing pest, disease, and weed infestation (Jat et al., 2009; Kumar et al., 2019b; Majumdar et al., 2017; Saharawat et al., 2010; Singh et al., 2016). Legume cover crops and short-duration fallows can fix substantial amounts of N_2 from air and improve soil fertility, which increases yield of the subsequent cereal crops and provides substantial biomass for retention/incorporation. Increased carbon accumulation in the soil surface layers, together with the presence of crop residue mulch, has important effects on improving the soil water supply for crop growth. Increased structural stability and protection of the soil surface from the direct impact of raindrops result in greater infiltration and capture of rainfall for crop growth. Enhanced activity of soil macrofauna, particularly earthworms, in the absence of tillage can alleviate excessive build-up of soil organic matter in the surface horizons (Singh et al., 2015). The detritivore earthworm *Lumbricus terrestris* was shown to reduce the biomass of *Fusarium culmorum* in wheat straw, thus compensating for the negative impacts of ZT from crop residue accumulation. A major benefit of CA is the control of soil erosion owing to maintenance of soil cover, greater infiltration, and reduced runoff.

Off-field Crop Residue Management

Crop residues, which are collected from fields after crop harvest, can be used for multiple purposes, such as agricultural and industrial applications. Manual collection of crop residues is less preferred because of the need for labor and the delay in sowing the next crop. Preferably, collection can be done with the help of baling machines, which collect crop residues and make cylindrical or square-shaped bales that can be stored for future utilization. Various off-field crop residue management options for improving nutrient supply to crops, reducing inorganic fertilizer requirements, and thereby enhancing soil health are discussed in the following sections.

Composting of Crop Residues

Composting is the eco-friendly microbial transformation of different biodegradable organic materials into a humus-rich product that can enhance soil health. This is one of the most popular and beneficial technologies for management of crop residues. Composting has long been used to recycle solid wastes so that the organic material can be easily degraded through microbial decomposition until it can be safely handled, stored, and applied to soil. Composting of agricultural and municipal solid waste has been prevalent since ancient times. By far the most important decomposers in composting are bacteria. They are also known as saprophytic organisms and belong to the group of chemoorganotrophic microorganisms, including species from the genera *Pseudomonas*, *Zymomonas*, *Xanthomonas*, and *Burkholderia* (Sánchez et al., 2017). *Actinomycetes* and fungi are also active in the composting process. Crop residues can be removed from the field; composted alone or with other organic materials, such as animal and household wastes; and then returned to soil as manure.

In India, crop residues are traditionally used for compost preparation. In this method, agricultural residues are used as bedding material for domestic animals and are then stacked in piles or put into pits to decompose over a long period. In animal sheds, straw absorbs urine, which enriches the nitrogen content in it. Composting of rice crop residues from 1 ha of land produces about 3 Mg of manure, which is rich in nutrients like farmyard manure. Vermicomposting is the composting process, which involves decomposition and subsequent stabilization of organic materials involving the combined activities of earthworms and microorganisms. Microbes are responsible for biochemical decomposition of organic matter, while earthworms act as important drivers of the process in conditioning the substrate and altering microbiological activity. Vermicomposting is done using windrows for composting. The agricultural byproducts, such as residues and animal droppings and urine, are first mixed and layered in long narrow piles, which are called windrows. A few important points to consider when preparing compost from crop residues are (a) substrate composition (C:N ratio and pH), (b) temperature, (c) moisture content, (d) aeration, and (e) inoculation (Harindintwali et al., 2020).

C:N Ratio and pH Composting of organic residues in a pile or pit requires carbon as an energy source and nitrogen to maintain and develop cells, proteins, enzymes, and hormones (Harindintwali et al., 2020). Microbes in a composting pile are active within a C:N ratio from organic wastes between 20:1 and 30:1. An optimal C:N ratio for the most efficient decomposition is about 25:1. Biodegradation of organic residues will decline if C:N ratios decrease to 10:1 or less or increase to 50:1 or more. With a very high C:N ratio, rice residues take a longer time to decompose with an accompanying loss of nitrogen as NH_3. So, to solve that issue, mixing with cow dung and an inorganic nitrogen source at a proper ratio should be followed (Vigneswaran et al., 2016).

pH influences microbial activity during composting of agricultural residues. During composting, bacteria and other microorganisms can withstand a wide range of pH (5–8), although pH values of 7–8 are ideal (Torres-Climent et al., 2015). Organic acids are produced during the early phases of composting, and these acids reduce pH to 4–5. Acidic conditions are favorable for the growth of fungi and the breakdown of lignin and cellulose. As composting progresses, organic acids are neutralized and the mature compost typically has pH values within 6–8 (Tognetti et al., 2007).

Temperature Composting is a self-heating process that normally passes through mesophilic (25–40 °C), thermophilic (40–80 °C), cooling, and maturation (10–40 °C) stages. Each of these stages is dominated by different microbial communities. Mesophilic microorganisms, such as bacteria of *Pseudomonadaceae*, *Enterobacteriaceae*, *Streptomycetaceae*, and *Erythrobacteraceae* families grow well in the temperature range of 15–35 °C and consume soluble compounds, such as sugar, amino acids, and lipids (Bernal et al., 2009). Biological degradation of organic residues takes place rapidly when the temperature is in the range of 32–60 °C. Temperatures above 60 °C kill many microorganisms, including human and plant pathogens, and limit the rate of decomposition of organic residues (Gerba &

Pepper, 2019). Generally, the optimum composting temperature is achieved by controlling structure, C:N ratio, and aeration of the pile. A thermophilic fungal consortia of *Aspergillus nidulans*, *Scytalidium thermophilum*, and *Humcola* sp. was proved to be very efficient in decomposing soybean trash and rice straw (Pandey et al., 2009). Composting of rice straw at low temperatures was enhanced by psychrotrophic microbes such as *Eupenicillium crustaceum*, *Paecilomyces* sp., *Bacillus atrophaeus*, and *Bacillus* sp. (Shukla et al., 2016).

Moisture Content Maintaining adequate moisture in the composting pile or pit is essential because microorganisms need water to grow and perform their activities during decomposition. Composting materials must not be too wet, which limits oxygen supply and favors anaerobic activity, or too dry, which reduces microbial degradation. For effective composting, moisture content in the range of 55%–70% ensures appropriate conditions for microorganisms to decompose and transform crop residues into organic matter (Goyal & Sindhu, 2011). The optimum moisture content for composting maize stalks, silage, oat straw, wood shavings, alfalfa hay, soybean straw, wheat straw, and leaves ranges from about 60% to 80% (Ahn et al., 2008).

Aeration Composting is usually performed under aerobic conditions. During composting, convenient aeration is required to enhance growth of adequate aerobic microbial communities and for improving the quality of the end product. Aeration is also helpful in reducing a high initial moisture content and controlling the formation of bad odors and the presence of flies in the composting materials (Shen et al., 2011). Forced aeration and compost agitation are the two methods used to incorporate O_2 into the composting material.

Each conventional composting method, such as windrow composting, aerated static pile composting, and in-vessel composting, has a particular mechanism for introducing O_2 into compost. The traditional windrow composting method largely depends on mechanical aeration. With the use of a compost turner, the composting materials are turned or agitated periodically to introduce O_2. Aerated static pile composting methods involve decomposition of biomass without mechanical manipulation. In this system, the composting mixture remains static and aeration is maintained at an appropriate level by pushing or pulling air through the composting pile. In-vessel composting is carried out within a container or vessel, and aeration is conducted by either agitation or forced aeration. Low aeration in a compost pile results in anaerobic conditions, which produce CH_4 and H_2S gas, while excessive aeration can slow the composting process because of excessive heat and loss of water and NH_3. Different studies have shown that optimal aeration depends on the composition of raw materials and the aeration methods. An aeration rate of 0.25 $L\,min^{-1}\,kg^{-1}$ was found to be effective for composting rice straw with dairy manure in continuous ventilation conditions (Li et al., 2008). Another study showed that emission of CH_4 and N_2O could be controlled during composting of straw and dry grasses with chicken manure at an aeration rate of 0.10 m^3 min^{-1} m^{-3} (Shen et al., 2011).

Inoculation Composting generally takes place owing to omnipresent aerobic microorganisms, which are responsible for recycling nutrients from organic residues and waste products from other organisms. However, inoculation of specific microorganisms for composting is beneficial for producing mature compost in a short time. For example, inoculation of bacteria isolates (e.g., *Bacillus shackletonii*, *Streptomyces thermovulgaris*, and *Ureibacillus thermosphaericus*) in a compost pile of lignocellulosic materials facilitated decomposition of lignin and improved the quality of the mature compost (Vargas-García et al., 2007). Inoculation with a thermophilic fungal consortium of *Aspergillus niger*, *Aspergillus flavus*, and *Trichoderma viride* significantly improved humification during composting of municipal organic solid waste (Awasthi et al., 2014). Therefore, inoculation of crop residues with selected microorganisms is a feasible option for enhancing the composting process and improving compost quality.

Use of Compost for Improving Soil Health The use of compost in soil is an environment-friendly option for recovery of fertility, controlling erosion and sedimentation, and supplying substances such as organic carbon, nutrients, phytohormones, antioxidants, and enzymes that enhance soil health (Goyal et al., 2009; Saha et al., 2007). Compost prepared from crop residues serves as a promising organic manure with a rich source of plant nutrients (Table 14.2) and is the best approach to reduce the usage of inorganic fertilizers. Nutrient elements are organically bound within compost and are slowly released in soil through microbial activity over a period of time, which ensures a continuous supply of nutrients for plant uptake (Batabyal et al., 2016; Sarkar et al., 2018). Integrated use of rice straw compost and an inorganic fertilizer in rice crops improved soil health by increasing carbon and nitrogen fractions and maintaining microbiological properties of soil in Haryana (Goyal et al., 2009). In an experiment conducted in Nagaland, Chatterjee et al. (2021) observed that application of vermicompost prepared from paddy straw and farm animal litter to green bell pepper enhanced soil health by increasing organic carbon and available plant nutrients.

Biochar Preparation from Crop Residues

Use of crop residues through conversion to biochar may be a promising option for improving soil quality and crop productivity in India. Biochar is a carbon-rich solid product created through thermal decomposition of plant biomass under a limited O_2 supply or an O_2-free environment at a relatively low

Table 14.2 Nutrient Contents in Composts Prepared From Crop Residues.

Compost	Total N	Total P	Total K	Total C	C:N ratio
			%		
Sugar cane plant residue compost	1.68	1.13	2.11	23.89	14.2
Sugar cane plants residue + cattle manure (50:50) compost	1.32	0.51	0.62	20.64	15.6
Paddy straw compost	1.15	0.062	0.134	39.7	34.50
Paddy straw + biogas slurry compost	1.72	0.148	0.138	37.8	21.90
Paddy straw + fungal culture compost	2.09	0.164	0.169	34.9	16.60
Paddy straw + cattle dung compost	2.17	0.159	0.146	36.9	17.00
Sugar cane trash + cow dung compost	1.08	0.54	1.23		32.7
Sugar cane trash + cow dung vermicompost	1.15	0.50	1.30		32.0
Mustard residue + cow dung compost	1.08	0.56	1.75		32.8
Mustard residue + cow dung vermicompost	1.21	0.54	1.72		29
Paddy straw + rabbit dung vermicompost	1.68	0.43	3.41	18.90	11.2
Paddy straw + poultry litter vermicompost	1.54	1.06	7.23	16.20	10.6
Paddy straw + cow dung vermicompost	2.24	0.34	1.51	29.3	13.1
Paddy straw + pig dung vermicompost	1.48	1.56	2.15	13.9	5.59
Cotton stubble vermicompost	2.00	1.08	0.99	30.71	15.3
Cotton stubble vermicompost + rock phosphate	2.10	1.32	0.98	30.41	14.4
Red gram stubble vermicompost	2.20	2.15	0.98	31.41	14.2
Red gram stubble vermicompost + rock phosphate	2.35	2.60	1.08	30.95	13.1

Note. Aparna et al. (2022), Khater (2015), Chatterjee et al. (2021), Goyal and Sindhu (2011), Bansal and Kapoor (2000).

temperature (<700 °C) in a process known as pyrolysis (Lehmann et al., 2006). The estimated biochar production potential from different crop residues in India is 162 million Mg per year, and the combined carbon sequestration potential of incorporating this biochar in soil is about 80 million Mg per year (Table 14.3). However, very little biochar is produced and utilized as soil amendment in Indian agriculture.

The ancient method of biochar production was the "pit" or "trench" method. The most successful approach for efficient biochar production is through slow pyrolysis. Under slow pyrolysis, a biochar yield between 25% and 35% can be

Table 14.3 Estimated Biochar Production and Carbon Sequestration Potential From Crop Residues in India.

Crop	Residue availability	Biochar production potential (estimated 35% conversion efficiency by slow pyrolysis)	C sequestration potential (estimated about 50% of initial C sequestrated)
		million Mg yr^{-1}	
Arhar	5.7	2.0	1.0
Bajra	15.8	5.5	2.8
Cotton	52.9	18.5	9.3
Groundnut (shelly stalk)	15.1	5.3	2.6
Jowar (cob, stalk, husk)	24.2	8.5	4.2
Maize (stalk, cob)	27.0	9.5	4.7
Mesta	1.7	0.6	0.3
Mustard (stalk, husk)	8.7	3.1	1.5
Rice	170.0	59.5	29.8
Soybean stalk	9.9	3.5	1.7
Sugarcane	12.1	4.2	2.1
Sunflower	1.4	0.5	0.3
Tapioca	4.0	1.4	0.7
Wheat	112.0	39.2	19.6
Til stalk	1.2	0.4	0.2
Coffee	1.6	0.6	0.3

Note. Venkatesh et al. 2018/Central Research Institute for Dryland Agriculture.

obtained; fast pyrolysis processes aim at production of bio-oil, and the amount of biochar produced is about 12% (Cheng et al., 2012). The cook stove, earth mound kilns, and drum kilns are traditionally used for biochar production in India (Srinivasarao et al., 2013). Many biochar production kilns are designed, developed, and used for making biochar from crop residues and forest biomass (Gangil & Wakudkar, 2013; Venkatesh et al., 2013) in India.

Effect of Biochar on Soil Health Addition of biochar to soil has beneficial effects in increasing organic carbon content, nutrient retention, and microbial activity, which in turn improves soil health (Mukherjee et al., 2014; Purakayastha et al., 2019; Srinivasarao et al., 2013; Venkatesh et al., 2022). Conversion of crop biomass carbon to biochar leads to the sequestration of about 50% of the initial carbon compared with 3% retained after burning and less than 10%–20% in biological decomposition (Lehmann et al., 2006). In comparison with other terrestrial carbon sequestration strategies (e.g., afforestation), carbon sequestration through biochar increases the residence time of carbon in soil (Pariyar et al., 2020). Biochar prepared from crop residues serves as a source of plant nutrients (Table 14.4). Purakayastha et al. (2015) showed that mixing maize stover biochar into a sandy loam soil increased available nitrogen and phosphorus; wheat straw biochar enhanced available potassium; and rice straw biochar, which was relatively labile, improved microbial activity. Application of biochar prepared from maize, redgram, and cotton stalks improved physical, chemical, and biological health of an Alfisol under rainfed conditions (Pandian et al., 2016). Recently, Sharma et al. (2021) observed that rice husk biochar application for growing knol khol in subtropical northwest India improved soil physical health by enhancing infiltration rate, aggregate stability, maximum water-holding capacity, and hydraulic conductivity of a sandy clay loam soil. Therefore, preparation of biochar from crop residues and its application to soils seems to be a sustainable strategy for soil health management.

Role of Using Crop Residues in Improving Soil Health

Management of crop residues has important implications in improving soil health and agricultural productivity and protecting the environment. Incorporation or retention of residues in soil has a favorable effect on soil physical, chemical, and biological properties, such as bulk density, porosity, water-holding capacity, pH, availability of nutrients, organic carbon, and improved microbial activities. Adoption of a sustainable crop residue management plan is regarded as one of the cleaner crop production options (Raza et al., 2019).

Effect on Physical Health

Addition of crop residues on the soil surface protects soil from the beating action of raindrops, which prevents soil crust formation and increases the water infiltration rate, resulting in better soil physical health with minimal soil erosion (Dutta et al., 2022). Bulk density is often lower in areas with the least soil disturbance owing to coverage of the soil surface with crop residues or their integration into soil. Straw incorporation helps to stabilize soil aggregates by increasing microbial cells and microbial secretions and the release of the decomposition products of dead microorganisms. Crop residues are lighter than soil mineral components and their decomposition is faster, which

Table 14.4 Some Properties of Biochar Prepared From Crop Residues[a].

Biochar made from	Bulk density	pH	EC	CEC	Carbon	Nitrogen	Phosphorus	Potassium
	$Mg\ m^{-3}$		$dS\ m^{-1}$	$Cmol\ kg^{-1}$			%	
Purakayastha et al. (2015)								
Maize stalk	0.52	10.7	3.84	63.5	66	1.31	1.76	0.20
Pearl millet stalk	0.60	10.6	10.90	57.6	64	1.10	1.60	0.22
Rice straw	0.46	8.6	7.68	97.3	60	0.95	1.52	0.05
Wheat straw	0.48	8.8	13.30	10.9	64	1.23	0.74	0.29
Pandian et al. (2016)								
Maize stalk	0.36–0.39	8.9–10.0	0.65–1.09		66–71	0.31–0.45	0.24–0.84	0.30–0.80
Cotton stalk	0.40–0.43	8.8–10.2	0.58–0.85		74–85	0.26–0.35	0.15–0.39	1.10–1.20
Redgram stalk	0.37–0.40	8.4–10.0	0.63–1.00		79–85	0.27–0.39	0.18–0.76	0.90–1.40
Sharma et al. (2021)								
Rice husk		8.9	0.49		54	0.09	0.15	0.50

[a] CEC = cation exchange capacity; EC = electrical conductivity.

promotes better soil aggregation and lower bulk density under a rice–wheat cropping system (Gathala et al., 2013; Singh et al., 2010). By improving infiltration, aeration, and microbial populations, ZT coupled with residue retention helps increase soil organic matter and thus decreases bulk density (Choudhary et al., 2018; Jat et al., 2009; Walia & Brar, 2006). Happy Seeder can lower tillage frequency, which causes the least soil disturbance (Prasad et al., 2020). Bandyopadhyay et al. (2010) showed that soils receiving rice straw plus inorganic NPK fertilizers for 21 years had more water-stable macro-aggregates and higher aggregate stability, mean weight diameter, and geometric mean diameter than the control without fertilizers under a rice–wheat cropping system in the hot, humid subtropical eastern India. Incorporation of leguminous crop resides has been shown to improve physical soil properties, such as water-holding capacity and soil permeability. Application of crop residues coupled with conservation tillage was reported to improve the soil aggregate and carbon storage in rice-based cropping systems of northwest India (Singh et al., 2020).

Covering soil with crop residues and stubble helps to increase soil moisture availability. It reduces evaporative loss of moisture by protecting the soil surface from direct sunlight and acting as a barrier to air passage over the surface compared with tilled soil without crop residue. Owing to increased soil porosity and physical aggregation with crop residue application, there is enhanced water infiltration and reduced surface runoff, which results in more plant-available water in the soil (Indoria et al., 2017; Upadhyay et al., 2022). Similarly, higher infiltration rates under ZT with residues can also reduce water runoff. Cultivation under ZT with residues helps to stabilize soil temperature variations during crop growth compared with residue removal. During early crop growth, soil temperature was higher with ZT plus residues than with conventional cultivation, while it was reduced during the later stages of wheat growth (Dwivedi & Thakur, 2000). During winter months, residue retention serves as an insulator against drops in soil temperature at night and reduces temperature variations between day and night, compared with conventional cultivation. Compared with residue removal, residue incorporation at soil depths of 5, 15, and 30 cm reduced soil temperature by 0.58, 0.66, and 0.74 °C, respectively (Korav et al., 2022).

Effect on Soil Chemical Health

Efficient crop residue management can effectively influence chemical soil properties such as pH, electrical conductivity, and cation exchange capacity as well as the transformation of different plant nutrients. The effect of residue management on important chemical properties are discussed in the following sections.

Soil pH

Retention of crop residues influences soil pH since a change in soil pH is related to the chemical composition of the residues and the soil properties. Soil properties that influence residue decomposition, such as texture, moisture content, temperature, available nitrogen, soil organic carbon, and initial pH, control the effect of residues on soil pH (Turmel et al., 2015). Residues high in ash and nitrogen content, such as residues of leguminous crops, may have a greater effect on soil pH compared with residues of wheat, which have lower nitrogen. Rice residues are expected to cause a decrease in soil pH through accumulation of CO_2 and organic acids during their decomposition in soil (Goswami et al., 2020). In contrast, Pathak et al. (2011) reported that rice straw supplied different bases such as hydroxyl (-OH), which maintained a neutral soil pH. However, the effect of residue retention/incorporation on pH is generally restricted to the topsoil layer; it still remains uncertain whether incorporation or surface retention of crop residues has an influence on pH in lower soil layers (Turmel et al., 2015).

Soil Organic Carbon

Increased soil organic carbon (SOC) stock is important to maintain good soil health with sustainable crop production (Majumder et al., 2008). Intensive agricultural practices can hamper growth of SOC stock, which leads to emission of greenhouse gasses and soil carbon loss. Low SOC levels aggravate the problem of soil erosion and particularly result in soil degradation problems such as decline in soil fertility and biodiversity. This outcome leads to lower soil quality and greater pressure on sustainable crop production and food security (Lal, 2007). Crop residue retention or incorporation can increase or maintain SOC; however, such an effect may be controlled by soil type, agroclimatic conditions, and management practices (Nandan et al., 2018; Singh et al., 2008b). Several studies reported a higher SOC content under no-tillage or minimum tillage with residue retention in soil, particularly at surface layers (Gangwar et al., 2006; Jat et al., 2020; Nayak et al., 2022). In contrast, tillage influences distribution of SOC in the soil profile by incorporating residues in soil, and thus, a subsequent increase in SOC content in deeper layers can also be observed (Yadvinder-Singh et al., 2015). Incorporation of rice residue leads to a huge addition of carbon to the soil and can maintain a long-term balance between addition of organic carbon and its loss through different pathways (Mandal et al., 2008). Ultimately, there would be either a net build-up or depletion of SOC stock. In fact, turnover of crop residue carbon that results in its stabilization into SOC is controlled by many factors, including climate, substrate biochemistry, and soil environment (Majumder et al., 2008; Yadav et al., 2019a). Only around 1.4%–9.0% (mean 4.0%) of the total carbon added through crop residues was found to be stabilized in SOC pools under long-term rice cultivation. A rice–rice cropping system can add 3.03–4.81 Mg C ha^{-1} annually in soil through crop residue carbon input under different fertilizer regimes and management practices (Mandal et al., 2008). Incorporation of rice straw increased SOC in both lowland and upland systems, but the effect was more prominent in the latter (Yadvinder-Singh et al., 2005). Application of crop residues can influence soil carbon fractions. Application of sorghum residues at different levels (2–6 Mg ha^{-1}) for 8 years significantly increased various labile carbon fractions and the lability index in rainfed Alfisols under a sorghum–cowpea cropping system. Similarly, application of partially decomposed rice straw manure (at 10 Mg ha^{-1} yr^{-1} on wet basis) with NPK fertilizers for 19 years increased SOC fractions compared with control or NPK in an irrigated Inceptisol under a rice–wheat cropping system (Majumder et al., 2008).

Nutrient Availability to Crops

Although large amounts of plant nutrients are removed in harvested grain, sizeable amounts remain in crop residue (Table 14.1). Addition of crop residue to fields can thus influence the dynamics of nutrients in soil and their availability to crops. The C:N ratio of crop residue controls the rate of nitrogen release and consequently nitrogen immobilization or mineralization (Kumar et al., 2023). Upon mineralization, nitrogen from residue may be taken up by plants, recycled by microbes for their growth, stabilized in complex soil organic matter, or lost from the soil plant system through gaseous loss of nitrogen (NH_3, N_2O, N_2, and NO) and nitrate (NO_3^-) leaching (Kumar & Goh, 1999). Addition of legume residues with a low C:N ratio can result in nitrogen mineralization, whereas cereal residues with a large C:N ratio lead to temporary locking of nitrogen in soil through immobilization during decomposition. Yadvinder-Singh et al. (2005) observed substantial immobilization of fertilizer nitrogen after 10 days of fertilizer application with rice straw incorporation at 0 and 10 days before fertilizer application. Patra et al. (2004) also claimed that surface retention of crop residue prevents nitrogen mineralization and increases nitrogen immobilization, denitrification, and NH_3 volatilization. Kushwaha et al. (2000) reported that in residue removal treatment, the nitrogen mineralization rate was maximum during the seedling stage of crops and then decreased as crops reached maturity. Incorporation of crop residue enhances the nitrogen content of wetland rice soils (Nandan et al., 2018; Tripathi et al., 2021). There might be depletion in nitrogen availability in soils over time without residue application in conventional cultivation over residue application in CA owing to greater nitrogen losses from the system.

Microbial decomposition of crop residues also regulates phosphorus availability in soil through mineralization from the residues. ZT with rice residue retention as mulch resulted in significantly higher total and available phosphorus than conventional tillage without any residue retention under a rice–wheat cropping on a sandy loam soil in Punjab, India (Sharma et al., 2022). Mishra et al. (2001) reported that about 80% of total potassium present in rice straw was released within 5 weeks after its incorporation into soil and 95% of the potassium from straw was mineralized by the end of 23 weeks. Higher concentrations of total, exchangeable, and NH_4OAc-extractable potassium were observed in plots receiving rice straw compared with those without residue additions (Yadav et al., 2019b). Residue incorporation could also recycle available potassium by 15% as well as direct the demand for an external supply of potassium from fertilizer sources (Singh et al., 2018). Crop residue influenced availability of micronutrients, such as zinc and iron for rice (Gupta et al., 2007; Yadvinder-Singh et al., 2005), and around 50%–80% of micronutrients (zinc, iron, copper, and manganese) taken up by rice and wheat crops can be recycled through crop residue incorporation. Yadvinder-Singh et al. (2005) reported that incorporation of crop residue on a long-term basis could increase the DTPA-extractable zinc, copper, iron, and manganese content in soil. Smitha et al. (2019) reported that application of cluster bean residues before transplanting of sacred basil could enhance SOC and the availability of macronutrients in soils.

Effect on Soil Biological Health

Sustainable management of crop residue has a significant impact on the regulation of soil microbial biomass and microbial diversity (Singh et al., 2021; Korav et al., 2022). Retention/incorporation of crop residue in topsoil layers enhances microbial activity (Baghel et al., 2020; Das et al., 2021; Mondal et al., 2021; Samui et al., 2020), which may be due to alteration of plant–soil microclimate, availability of carbonaceous material, increased water and nutrient availability, and regulation of soil temperature (Singh et al., 2021). Addition of crop residue was associated with a significant increase in microbial biomass carbon and activity of enzymes such as dehydrogenase and phosphatase in soil compared with residue burning (Chatterjee et al., 2018; Kumar et al., 2019a). From an experiment under a dryland rice–barley cropping system in an Inceptisol, Kushwaha et al. (2000) showed that residue addition with minimum tillage significantly enhanced soil microbial biomass carbon and microbial biomass nitrogen compared with conventional tillage with residue removal. Incorporation of residues from leguminous crops, such as cluster bean before sowing of the next crop was reported to increase soil microbial biomass and dehydrogenase activity in soil (Smitha et al., 2019). Residue management caused 4.9% and 15.4% increases in nematode population and biomass, respectively, as soil nematodes were positively influenced by residue management under long-term ZT (Bhagat & Gosal, 2018; Gaind & Nain, 2011). Residue retention on soil accumulates cellulose as a component of soil organic matter, and this energy-rich cellulose source is eco-functional through promoting increased adaptability of decomposers with cellulolytic enzymes, by which they can gain dominance over decomposers that depend on simpler carbon sources (Singh et al., 2021).

Challenges in Managing Crop Residues

The major challenges of crop residue management in improving soil health in India are discussed in the following sections.

Competitive Uses of Crop Residues

India generates about 686 million Mg of crop residues per year (Datta et al., 2020). Most of these residues (about 75%) is used as animal fodder, roof thatching, and domestic and industrial fuel. Even after consumption of crop residues for these purposes, there still remains a surplus of 178 million Mg, of which 87 million Mg is burned in India each year. In spite of the huge potential of using crop residues for improving soils, either directly through surface retention and incorporation or indirectly by converting the residues into compost or biochar, presently only ~3% is returned to Indian soils (Kumar et al., 2023). There is genuine potential for utilizing 62 million Mg of rice residues (based on the average rice residue load of 6 Mg ha^{-1}) and 3 million Mg of sugarcane trash (based on the average trash load of 10 Mg ha^{-1}) from 10.3 million hectares under rice–wheat (Kumar et al., 2018) and 0.3 million hectares under sugarcane–wheat (Shanthy et al., 2014) in the country as the major share of rice and sugarcane residues is burned for timely sowing of the succeeding wheat crop. Additionally, 23.9 and 9.3 million Mg of wheat and maize residue, respectively, can be returned back to the soil in the major residue-burning states of Uttar Pradesh, Punjab, and Haryana. Considering all crops, including rice, coarse cereals, pulses, oilseeds, and sugarcane, about 450 million Mg of residue can be recycled back to the soil in the whole country (Kumar et al., 2023).

Residue Burning in Fields

Judicious management of rice residues under intensively cultivated rice–wheat cropping in the IGP is challenging because farmers are concerned about timely sowing of wheat seed after rice harvest (Sidhu et al., 2007; Singh et al., 2020). Farmers of this region prefer to burn rice residues due to the very short time available for wheat sowing after the rice harvest, unavailability of proper residue-handling machines, and costly labor-intensive manual residue removal

(Kumar et al., 2022a). The burning of sugarcane trash is also a common practice in northern India because of its poor palatability as fodder and immediate removal required from the fields for sowing of the next crop. Farmers also derive benefits of controlling diseases, pests, and weeds by burning crop residues (Gadde et al., 2009). Residue burning emits greenhouse gases and potent radioactive compounds in the air, such as CO_2, CH_4, CO, SO_2, nitrous oxides (NO, NO_x), hydrocarbons, volatile and semivolatile organic compounds, particulate matter, and heavy metals (Mittal et al., 2009), which contribute to climate change and adversely affect environmental (air and soil) quality and human health (Mathur & Srivastava, 2019). Emissions from rice residue burning alone contributed about 65 Gg of CH_4, 1.7 Gg of N_2O, and 1,897 Gg of CO_2 during 2019 in India (Kumar et al., 2023). Another detrimental effect of burning crop residues is the loss of plant nutrients, which adversely affects soil quality (Gupta et al., 2004; Thind et al., 2019). It was estimated that 20%–100% of carbon, nitrogen, phosphorus, potassium, and other nutrients retained in various crop residues are lost during burning (Singh et al., 2008a; Singh et al., 2017). This results in a loss of about 400, 6, 3.5, 25, and 1.2 kg of carbon, nitrogen, phosphorus, potassium, and sulfur, respectively, for each 1 Mg of rice straw burned. Residue burning also increases soil temperature, which results in decreases in the macro- and microbial populations and the loss of SOC (Bhuvaneshwari et al., 2019; Gupta et al., 2004; Kaur et al., 2019). The impact of residue burning on soil health is detrimental because it causes the loss of organic carbon through CO_2 and the conversion of nitrogen into NO_3^- (Kumar et al., 2019a). According to a report from Punjab Agricultural University, 824 Gg of NPK is lost from the soil of Punjab every year owing to residue burning (Porichha et al., 2021). Although burning of crop residues temporarily increases the mineral nitrogen (NH_4^+) and plant available phosphorus content in soil (Mandal et al., 2004; Moharana et al., 2012; Pandey, 2019), it has a serious impact on soil health, organic matter content, plant micro- and macronutrient content, and soil microbial populations (Raza et al., 2019).

Problems Associated with Residue Incorporation in Soil

Recycling of cereal crops residues and sugarcane straw with higher C:N ratios through incorporation has been challenging because of microbial immobilization of mineral nitrogen in the soil. Such immobilization of nitrogen negatively influences productivity of the crops that are sown immediately after incorporation of these residues (Samra et al., 2003). In addition, increased emission of CH_4 and CO_2 from rice residues incorporated into fields is of great concern (Rath et al., 2005).

Unavailability of Suitable Equipment

Various machines are used for in situ crop residue management and sowing of succeeding crops. For managing crop residues, farmers chose machines by considering the local availability of resources, such as power supplies and fuel. Crop residue management using suitable machines under CA with adoption of reduced tillage conserves energy, time, and moisture for cereals, sugarcane, pulses, vegetables, and oilseed crops compared with residue incorporation under intensive tillage systems. The small, marginal, and medium farmers in India face challenges for adoption of in situ residue management because of the higher cost, higher power requirements, and lack of timely availability of crop-specific machines.

Insect, Pest, and Weed Problems

Stubbles and residues of previous crops on the soil surface can provide food and a favorable habitat for insects, pests, and rodents that cause damage to crops. A significant level of rodent problems and a higher burrow density are common under residue-retained conditions in wheat, which significantly reduces crop yield (Sharma & Singh, 2018; Singh et al., 2019). There are recent reports of increased incidence of army worm [*Mythimna (Pseudaletia) separata* (Wlk.)], pink stem borer (*Sesamia inferens*), mealybug [*Brevennia rehi* (Lindinger)], bandicoot rat [*Bandicota bengalensis* (Gray)], and *Fusarium* head blight in wheat crops cultivated with residue retention under CA-based rice–wheat and rice–wheat–mung bean systems (Kaur et al., 2021; Kumar et al., 2022b). Increased occurrence of insects and pests in ZT under CA are due to the availability of stable habitat for larvae and carryover of pests from infestation of the previous crop to the succeeding crop.

Crop residues also provide an environment that can be either inhibitive or favorable for germination and growth of weeds, either through a physical barrier for emergence or by allelopathic influences. In general, retention of previous crop residues on the soil surface favors weed emergence in partially covered soil. Although long-term ZT reduces growth of *Phalaris minor* in wheat due to less disturbance to the soil and insufficient exposure to light, it can also stimulate broad-leaved weeds such as *Rumex dentatus* and *Malva parviflora* (Chhokar et al., 2007). In addition, the efficacy of pre-emergence herbicides is questionable because of significant absorption of herbicides by straw of the previous crop. Therefore, inadequate control of insects, pests, and weeds in crops cultivated with in situ crop residue management can increase the cost of cultivation owing to reduced crop yield and profitability.

Conclusions

Crop residues serve as a huge source of organic carbon and mineral nutrients. Adoption of suitable in situ and ex situ residue management practices is beneficial for improving soil health and maintaining environmental sustainability. Under in situ management, crop residues are either retained on the soil surface or mixed into the soil during sowing of the next crop or a few days beforehand, provided suitable machinery is available. Retention and/or incorporation of residues can be practiced efficiently through adoption of appropriate CA technologies. Important ex situ crop residue management options for improving nutrient supply to crops, reducing inorganic fertilizer requirements, and enhancing soil health are composting and biochar preparation.

Judicious management of crop residues has important implications for improving soil health in India. Both in situ and ex situ management of residues has favorable effects on soil physical, chemical, and biological properties, such as bulk density, porosity, water-holding capacity, pH, availability of nutrients, organic carbon, and improved microbial activities. Adoption of sustainable crop residue management plans is considered one of the cleaner crop production options. Amid various challenges, in situ management of crop residues under CA can be a more economic and energy-conserving strategy for improving soil functioning. Conventional tillage can efficiently recycle leguminous crop residues back into the soil, while conservation tillage can recycle large amounts of cereal and sugarcane residues with a high C:N ratio in a single field operation with less energy involvement than the former. Development of location- and crop-specific residue management machinery; CA-based crop production technology; effective strategies to control weeds, insects, and pests under crop residue retention/incorporation; and bio-decomposers for fast decomposition of cereal and sugarcane crop residues can drive future research to address the existing challenges in accelerating crop residue management by farmers for improving soil health in India.

References

Ahn, H. K., Richard, T. L., & Glanville, T. D. (2008). Optimum moisture levels for biodegradation of mortality composting envelope materials. *Waste Management*, *28*(8), 1411–1416. https://doi.org/10.1016/j.wasman.2007.05.022

Aparna, K., Rekha, K. B., Vani, K. P., & Prakash, T. R. (2022). Nutrient dynamics of cotton and red gram residue compost. *Current Science*, *122*(6), 657–659.

Awasthi, M. K., Pandey, A. K., Khan, J., Bundela, P. S., Wong, J. W. C., & Selvam, A. (2014). Evaluation of thermophilic fungal consortium for organic municipal solid waste composting. *Bioresource Technology, 168*, 214–221. https://doi.org/10.1016/j.biortech.2014.01.048

Baghel, J. K., Das, T. K., Mukherjee, I., Nath, C. P., Bhattacharyya, R., Ghosh, S., & Raj, R. (2020). Impacts of conservation agriculture and herbicides on weeds, nematodes, herbicide residue and productivity in direct-seeded rice. *Soil and Tillage Research, 201*, 104634. https://doi.org/10.1016/j.still.2020.104634

Bah, H., Zhou, M., Kizito, S., Xiao, R., Raza, S. T., Dong, Z., & Zhu, B. (2020). Carbon balance under organic amendments in the wheat-maize cropping systems of sloppy upland soil. *Sustainability, 12*(7), Article 2747. https://doi.org/10.3390/su12072747

Bandyopadhyay, P. K., Saha, S., Mani, P. K., & Mandal, B. (2010). Effect of organic inputs on aggregate associated organic carbon concentration under long-term rice-wheat cropping system. *Geoderma, 154*(3–4), 379–386. https://doi.org/10.1016/j.geoderma.2009.11.011

Bansal, S., & Kapoor, K. K. (2000). Vermicomposting of crop residues and cattle dung with *Eisenia foetida*. *Bioresource Technology, 73*(2), 95–98. https://doi.org/10.1016/S0960-8524(99)00173-X

Batabyal, K., Mandal, B., Sarkar, D., Murmu, S., Tamang, A., Das, I., Hazra, G. C., & Chattopadhyay, P. S. (2016). Comprehensive assessment of nutrient management technologies for cauliflower production under subtropical conditions. *European Journal of Agronomy, 79*, 1–13. https://doi.org/10.1016/j.eja.2016.04.009

Bernal, M. P., Alburquerque, J. A., & Moral, R. (2009). Composting of animal manures and chemical criteria for compost maturity assessment. *A review. Bioresource Technology, 100*(22), 5444–5453. https://doi.org/10.1016/j.biortech.2008.11.027

Bhagat, P., & Gosal, S. K. (2018). Long term application of rice straw and nitrogen fertilizer affects soil health and microbial communities. *Chemical Science Review and Letters, 7*(26), 586–593.

Bhuvaneshwari, S., Hettiarachchi, H., & Meegoda, J. N. (2019). Crop residue burning in India: Policy challenges and potential solutions. *International Journal of Environmental Research and Public Health, 16*(5), 832. https://doi.org/10.3390/ijerph16050832

Bimbraw, A. S. (2019). Generation and impact of crop residue and its management. *Current Agriculture Research Journal, 7*(3), 304–309. https://doi.org/10.12944/CARJ.7.3.05

Canilha, L., Chandel, A. K., dos Santos Milessi, T. S., Antunes, F. A. F., da Costa Freitas, W. L., das Graças Almeida Felipe, M., & da Silva, S. S. (2012). Bioconversion of sugarcane biomass into ethanol: An overview about composition, pretreatment methods, detoxification of hydrolysates, enzymatic saccharification, and ethanol fermentation. *Journal of Biomedicine and Biotechnology*, *2012*, Article 989572. https://doi.org/10.1155/2012/989572

Carvalho, J. L. N., Nogueirol, R. C., Menandro, L. M. S., Bordonal, R. d. O., Borges, C. D., Cantarella, H., & Franco, H. C. J. (2017). Agronomic and environmental implications of sugarcane straw removal: A major review. *GCB Bioenergy*, *9*(7), 1181–1195. https://doi.org/10.1111/gcbb.12410

Chatterjee, D., Dutta, S. K., Kikon, Z. J., Kuotsu, R., Sarkar, D., Satapathy, B. S., & Deka, B. C. (2021). Recycling of agricultural wastes to vermicomposts: Characterization and application for clean and quality production of green bell pepper (*Capsicum annuum* L.). *Journal of Cleaner Production*, *315*, Article 128115. https://doi.org/10.1016/j.jclepro.2021.128115

Chatterjee, S., Bandyopadhyay, K. K., Pradhan, S., Singh, R., & Datta, S. P. (2018). Effects of irrigation, crop residue mulch and nitrogen management in maize (*Zea mays* L.) on soil carbon pools in a sandy loam soil of Indo-Gangetic Plain region. *Catena*, *165*, 207–216. https://doi.org/10.1016/j.catena.2018.02.005

Chauhan, B. S., Gill, G., & Preston, C. (2006). Influence of tillage systems on vertical distribution, seedling recruitment and persistence of rigid ryegrass (*Lolium rigidum*) seed bank. *Weed Science*, *54*(4), 669–676. https://doi.org/10.1614/WS-05-184R.1

Cheng, G., Li, Q., Qi, F., Xiao, B., Liu, S., Hu, Z., & He, P. (2012). Allothermal gasification of biomass using micron size biomass as external heat source. *Bioresource Technology*, *107*, 471–475. https://doi.org/10.1016/j.biortech.2011.12.074

Chhokar, R. S., Sharma, R. K., Jat, G. R., Pundir, A. K., & Gathala, M. K. (2007). Effect of tillage and herbicides on weeds and productivity of wheat under rice-wheat growing system. *Crop Protection*, *26*(11), 1689–1696. https://doi.org/10.1016/j.cropro.2007.01.010

Choudhary, M., Datta, A., Jat, H. S., Yadav, K. A., Gathala, M. K., Sapkota, T. B., Das, A. K., Sharma, P. C., Jat, M. L., Singh, R., & Ladha, J. K. (2018). Changes in soil biology under conservation agriculture based sustainable intensification of cereal systems in Indo-Gangetic Plains. *Geoderma*, *313*, 193–204. https://doi.org/10.1016/j.geoderma.2017.10.041

Das, T. K., Sourav, G., Das, A., Sen, S., Datta, D., Sonaka, G., Raj, R., Behera, B., Roy, A., Vyas, A. K., & Rana, D. S. (2021). Conservation agriculture impacts on productivity, resource-use efficiency and environmental sustainability: A holistic review. *Indian Journal of Agronomy*, *66*, S111–S127.

Datta, A., Emmanuel, M. A., Ram, N. K., & Dhingra, S. (2020). *Crop residue management: Solution to achieve better air quality*. The Energy and Resources Institute (TERI). https://www.teriin.org/sites/default/files/2020-01/crop-residue-management.pdf

Dobermann, A., & Fairhurst, T. H. (2002). Rice straw management. *Better Crops International*, *16*, 7–11.

Dutta, A., Patra, A., Hazra, K. K., Nath, C. P., Kumar, N., & Rakshit, A. (2022). A state of the art review in crop residue burning in India: Previous knowledge, present circumstances and future strategies. *Environmental Challenges*, *8*, Article 100581. https://doi.org/10.1016/j.envc.2022.100581

Dwivedi, D. K., & Thakur, S. S. (2000). Production potential of wheat (*Triticum aestivum* L.) crop as influenced by residual organics, direct and residual fertility levels under rice (*Oryza sativa*)-wheat cropping system. *Indian Journal of Agronomy*, *45*(4), 641–647.

FAO. (2019). Conservation agriculture. https://www.fao.org/conservation-agriculture/en/

Fortes, C., Trivelin, P. C. O., & Vitti, A. C. (2012). Long-term decomposition of sugarcane harvest residues in Sao Paulo state, Brazil. *Biomass and Bioenergy*, *42*, 189–198. https://doi.org/10.1016/j.biombioe.2012.03.011

Gadde, B., Bonnet, S., Menke, C., & Garivait, S. (2009). Air pollutant emissions from rice straw open field burning in India, Thailand and the Philippines. *Environmental Pollution*, *157*(5), 1554–1558. https://doi.org/10.1016/j.envpol.2009.01.004

Gaind, S., & Nain, L. (2011). Soil health in response to bio-augmented paddy straw compost. *World Journal of Agricultural Sciences*, *7*(4), 480–488.

Gangil, S., & Wakudkar, H. M. (2013). Generation of bio-char from crop residues. *International Journal of Emerging Technology and Advanced Engineering*, *3*(3), 566–570.

Gangwar, K. S., Singh, K. K., Sharma, S. K., & Tomar, O. K. (2006). Alternative tillage and crop residue management in wheat after rice in sandy loam soils of Indo-Gangetic plains. *Soil and Tillage Research*, *88*(1–2), 242–252. https://doi.org/10.1016/j.still.2005.06.015

Gathala, M. K., Kumar, V., Sharma, P. C., Saharawat, Y. S., Jat, H. S., Singh, M., Kumar, A., Jat, M. L., Humphreys, E., & Sharma, D. K. (2013). Optimizing intensive cereal-based cropping systems addressing current and future drivers of agricultural change in the north western Indo-Gangetic Plains of India. *Agriculture, Ecosystems & Environment*, *177*, 85–97. https://doi.org/10.1016/j.agee.2013.06.002

Gerba, C. P., & Pepper, I. L. (2019). Municipal wastewater treatment. In M. L. Brusseau, I. L. Pepper, & C. P. Gerba (Eds.), *Environmental and pollution science* (3rd ed., pp. 393–418). Academic Press. https://doi.org/10.1016/B978-0-12-814719-1.00022-7

Goswami, S. B., Mondal, R., & Mandi, S. K. (2020). Crop residue management options in rice-rice system: A review. *Archives of Agronomy and Soil Science*, *66*(9), 1218–1234. https://doi.org/10.1080/03650340.2019.1661994

Goyal, S., & Sindhu, S. S. (2011). Composting of rice straw using different inocula and analysis of compost quality. *Microbiology Journal*, *1*(4), 126–138. https://doi.org/10.3923/mj.2011.126.138

Goyal, S., Singh, D., Suneja, S., & Kapoor, K. K. (2009). Effect of rice straw compost on soil microbiological properties and yield of rice. *Indian Journal of Agricultural Research*, *43*(4), 263–268.

Gupta, P. K., Sahai, S., Singh, N., Dixit, C. K., Singh, D. P., Sharma, C., Tiwari, M. K., Gupta, R. K., & Garg, S. C. (2004). Residue burning in rice-wheat cropping system: Causes and implications. *Current Science*, *25*(12), 1713–1717.

Gupta, R. K., Ladha, J. K., Singh, J., Singh, G., & Pathak, H. (2007). Yield and phosphorus transformations in a rice-wheat system with crop residue and phosphorus management. *Soil Science Society of America Journal*, *71*(5), 1500–1507. https://doi.org/10.2136/sssaj2006.0325

Harindintwali, J. D., Zhou, J., & Yu, X. (2020). Lignocellulosic crop residue composting by cellulolytic nitrogen-fixing bacteria: A novel tool for environmental sustainability. *Science of the Total Environment*, *715*, Article 136912. https://doi.org/10.1016/j.scitotenv.2020.136912

Indian Agricultural Research Institute. (2012). *Crop residues management with conservation agriculture: Potential, constraints and policy needs.*

Indoria, A. K., Srinivasarao, C., Sharma, K. L., & Reddy, K. S. (2017). Conservation agriculture—A panacea to improve soil physical health. *Current Science*, *112*(1), 52–61.

Jat, H. S., Kumar, V., Datta, A., Choudhary, M., Kakraliya, S. K., Poonia, T., McDonald, A. J., Jat, M. L., & Sharma, P. C. (2020). Designing profitable, resource use efficient and environmentally sound cereal-based systems for the Western Indo-Gangetic plains. *Scientific Reports*, *10*, Article 19267. https://doi.org/10.1038/s41598-020-76035-z

Jat, M. L., Gathala, M. K., Ladha, J. K., Saharawat, Y. S., Jat, A. S., Kumar, V., Sharma, S. K., Kumar, V., & Gupta, R. K. (2009). Evaluation of precision land levelling and double zero-tillage systems in the rice-wheat rotation: Water use, productivity, profitability and soil physical properties. *Soil and Tillage Research*, *105*(1), 112–121. https://doi.org/10.1016/j.still.2009.06.003

Jat, M. L., Malik, R. K., Saharawat, Y. S., Gupta, R., Mal, B., & Paroda, R. (2012). *Regional dialogue on conservation agricultural in South Asia: Proceedings and recommendations.* Asia Pacific Association of Agricultural Research Institutions (APAARI), International Maize and Wheat Improvement Center (CIMMYT), Indian Council of Agricultural Research (ICAR). http://hdl.handle.net/10883/1319

Jusoh, M. L. C., Manaf, L. A., & Latiff, P. A. (2013). Composting of rice straw with effective microorganisms (EM) and its influence on compost quality. *Iranian Journal of Environmental Health Sciences & Engineering*, *10*(1), Article 17. https://doi.org/10.1186/1735-2746-10-17

Kaur, R., Bansal, M., Sharma, S., & Tallapragada, S. (2019). Impact of in situ rice crop residue burning on agricultural soil of district Bathinda, Punjab, India. *Rasāyan Journal of Chemistry*, *12*(2), 421–430. http://dx.doi.org/10.31788/RJC.2019.1225160

Kaur, R., Kaur, S., Deol, J. S., Sharma, R., Kaur, T., Brar, A. S., & Choudhary, O. P. (2021). Soil properties and weed dynamics in wheat as affected by rice residue management in the rice-wheat cropping system in south Asia: A review. *Plants*, *10*(5), Article 953. https://doi.org/10.3390/plants10050953

Khater, E. G. (2015). Some physical and chemical properties of compost. *International Journal of Waste Resources*, *5*(1), Article 172. https://doi.org/10.4172/2252-5211.1000172

Korav, S., Rajanna, G. A., Yadav, D. B., Paramesha, V., Mehta, C. M., Jha, P. K., Singh, S., & Singh, S. (2022). Impacts of mechanized crop residue management on rice-wheat cropping system—A review. *Sustainability*, *14*(23), Article 15641. https://doi.org/10.3390/su142315641

Kumar, A., Kushwaha, K. K., Singh, S., Shivay, Y. S., Meena, M. C., & Nain, L. (2019a). Effect of paddy straw burning on soil microbial dynamics in sandy loam soil of Indo-Gangetic plains. *Environmental Technology & Innovation*, *16*, Article 100469. https://doi.org/10.1016/j.eti.2019.100469

Kumar, K., & Goh, K. M. (1999). Crop residues and management practices: Effects on soil quality, soil nitrogen dynamics, crop yield, and nitrogen recovery. *Advances in Agronomy*, *68*, 197–319. https://doi.org/10.1016/S0065-2113(08)60846-9

Kumar, N., Chaudhary, A., Ahlawat, O. P., Naorem, A., Upadhyay, G., Chhokar, R. S., Gill, S. C., Khippal, A., Tripathi, S. C., & Singh, G. P. (2023). Crop residue management challenges, opportunities and way forward for sustainable food-energy security in India: A review. *Soil and Tillage Research*, *228*, Article 105641. https://doi.org/10.1016/j.still.2023.105641

Kumar, N., Chhokar, R. S., Meena, R. P., Kharub, A. S., Gill, S. C., Tripathi, S. C., Gupta, O. P., Mangrauthia, S. K., Sundaram, R. M., Sawant, C. P., & Gupta, A. (2022a). Challenges and opportunities in productivity and sustainability of rice cultivation system: A critical review in Indian perspective. *Cereal Research Communications*, *50*, 573–601. https://doi.org/10.1007/s42976-021-00214-5

Kumar, R., Choudhary, J. S., Mishra, J. S., Mondal, S., Poonia, S., Monobrullah, M., Hans, H., Verma, M., Kumar, U., Bhatt, B. P., & Malik, R. K. (2022b). Outburst of pest populations in rice-based cropping systems under conservation agricultural practices in the middle Indo-Gangetic Plains of South Asia. *Scientific Reports*, *12*, Article 3753. https://doi.org/10.1038/s41598-022-07760-w

Kumar, V., Gathala, M. K., Saharawat, Y. S., Parihar, C. M., Kumar, R., Kumar, R., & Nayak, H. S. (2019b). Impact of tillage and crop establishment methods on crop yields, profitability and soil physical properties in rice-wheat system of Indo-Gangetic Plains of India. *Soil Use and Management*, *35*(2), 303–313. https://doi.org/10.1111/sum.12473

Kumar, V., Jat, H. S., Sharma, P. C., Gathala, M. K., Malik, R. K., Kamboj, B. R., Yadav, A. K., Ladha, J. K., Raman, A., Sharma, D. K., & McDonald, A. (2018). Can productivity and profitability be enhanced in intensively managed cereal systems while reducing the environmental footprint of production? Assessing sustainable intensification options in the breadbasket of India. *Agriculture, Ecosystems & Environment*, *252*, 132–147. https://doi.org/10.1016/j.agee.2017.10.006

Kumar, V., Saharawat, Y. S., Gathala, M. K., Jat, A. S., Singh, S. K., Chaudhary, N., & Jat, M. L. (2013). Effect of different tillage and seeding methods on energy use efficiency and productivity of wheat in the Indo-Gangetic Plains. *Field Crops Research*, *142*, 1–8. https://doi.org/10.1016/j.fcr.2012.11.013

Kushwaha, C. P., Tripathi, S. K., & Singh, K. P. (2000). Variations in soil microbial biomass and N availability due to residue and tillage management in a dryland rice agroecosystem. *Soil and Tillage Research*, *56*(3–4), 153–166. https://doi.org/10.1016/S0167-1987(00)00135-5

Lal, R. (2007). Anthropogenic influences on world soils and implications to global food security. *Advances in Agronomy*, *93*, 69–93. https://doi.org/10.1016/S0065-2113(06)93002-8

Lehmann, J., Gaunt, J., & Rondon, M. (2006). Bio-char sequestration in terrestrial ecosystems—A review. *Mitigation and Adaptation Strategies for Global Change*, *11*, 403–427. https://doi.org/10.1007/s11027-005-9006-5

Li, X., Zhang, R., & Pang, Y. (2008). Characteristics of dairy manure composting with rice straw. *Bioresource Technology*, *99*(2), 359–367. https://doi.org/10.1016/j.biortech.2006.12.009

Lohan, S. K., Jat, H. S., Yadav, A. K., Sidhu, H. S., Jat, M. L., Choudhary, M., Peter, J. K., & Sharma, P. C. (2018). Burning issues of paddy residue management in north-west states of India. *Renewable and Sustainable Energy Reviews*, *81*(1), 693–706. https://doi.org/10.1016/j.rser.2017.08.057

Lohan, S. K., & Sharma, S. (2012). Present status of renewable energy resources in Jammu and Kashmir State of India. *Renewable and Sustainable Energy Reviews*, *16*(5), 3251–3258. https://doi.org/10.1016/j.rser.2012.02.011

Maguyon-Detras, M. C., Migo, M. V., Van Hung, N., & Gummert, M. (2020). Thermochemical conversion of rice straw. In M. Gummert, N. V. Hung, P. Chivenge, & B. Douthwaite (Eds.), *Sustainable rice straw management* (pp. 43–64). Springer. https://doi.org/10.1007/978-3-030-32373-8_4

Majumdar, K., Sanyal, S. K., Singh, V. K., Dutta, S. K., Satyanarayana, T., & Dwivedi, B. S. (2017). Potassium fertilizer management in Indian agriculture: Current trends and future needs. *Indian Journal Fertilizers*, *13*(5), 20–30.

Majumder, B., Mandal, B., Bandyopadhyay, P. K., Gangopadhyay, A., Mani, P. K., Kundu, A. L., & Mazumdar, D. (2008). Organic amendments influence soil organic carbon pools and rice-wheat productivity. *Soil Science Society of America Journal*, *72*(3), 775–785. https://doi.org/10.2136/sssaj2006.0378

Mandal, B., Majumder, B., Adhya, T. K., Bandyopadhyay, P. K., Gangopadhyay, A., Sarkar, D., Kundu, M. C., Choudhury, S. G., Hazra, G. C., Kundu, S., & Samantaray, R. N. (2008). Potential of double-cropped rice ecology to conserve organic carbon under subtropical climate. *Global Change Biology*, *14*(9), 2139–2151. https://doi.org/10.1111/j.1365-2486.2008.01627.x

Mandal, K. G., Misra, A. K., Hati, K. M., Bandyopadhyay, K. K., Ghosh, P. K., & Mohanty, M. (2004). Rice residue-management options and effects on soil properties and crop productivity. *Food, Agriculture & Environment, 2*(1), 224–231.

Mathur, R., & Srivastava, V. K. (2019). Crop residue burning: Effects on environment. In N. Shurpali, A. Agarwal, & V. Srivastava (Eds.), *Greenhouse gas emissions: Energy, environment, and sustainability* (pp. 127–140). Springer. https://doi.org/10.1007/978-981-13-3272-2_9

Mishra, B., Sharma, P. K., & Bronson, K. F. (2001). Decomposition of rice straw and mineralization of carbon, nitrogen, phosphorus and potassium in wheat field soil in western Uttar Pradesh. *Journal of the Indian Society of Soil Science, 49*(3), 419–424.

Mittal, S. K., Singh, N., Agarwal, R., Awasthi, A., & Gupta, P. K. (2009). Ambient air quality during wheat and rice crop stubble burning episodes in Patiala. *Atmospheric Environment, 43*(2), 238–244. https://doi.org/10.1016/j.atmosenv.2008.09.068

Moharana, P. C., Sharma, B. M., Biswas, D. R., Dwivedi, B. S., & Singh, R. V. (2012). Long-term effect of nutrient management on soil fertility and soil organic carbon pools under a 6-year-old pearl millet-wheat cropping system in an Inceptisol of subtropical India. *Field Crops Research, 136*, 32–41. https://doi.org/10.1016/j.fcr.2012.07.002

Mondal, M., Garai, S., Banerjee, H., Sarkar, S., & Kundu, R. (2021). Mulching and nitrogen management in peanut cultivation: An evaluation of productivity, energy trade-off, carbon footprint and profitability. *Energy, Ecology and Environment, 6*, 133–147. https://doi.org/10.1007/s40974-020-00189-9

Mukherjee, A., Lal, R., & Zimmerman, A. R. (2014). Effects of biochar and other amendments on the physical properties and greenhouse gas emissions of an artificially degraded soil. *Science of the Total Environment, 487*, 26–36. https://doi.org/10.1016/j.scitotenv.2014.03.141

Mukherjee, S., Sarkar, D., Mandal, B., Kanthal, S., Ghosh, S., Sahu, B., Singh, P., Dey, A., Jaison, M., Dutta, J., Dash, B., & Saha, N. (2024). Conservation agriculture influences soil nitrogen availability in the lower Indo-Gangetic Plains. *Plant and Soil*. https://doi.org/10.1007/s11104-024-06826-0

Nandan, R., Singh, V., Singh, S., Hazra, K., & Nath, P. C. (2018). Performance of crop residue management with different tillage and crop establishment practices on weed flora and crop productivity in rice-wheat cropping system of eastern Indo-Gangetic plains. *Journal of Crop and Weed, 14*(2), 65–71.

Nayak, H. S., Parihar, C. M., Mandal, B. N., Patra, K., Jat, S. L., Singh, R., Singh, V. K., Jat, M. L., Garnaik, S., Nayak, J., & Abdallah, A. M. (2022). Point placement of late vegetative stage nitrogen splits increase the productivity, N-use efficiency and profitability of tropical maize under decade long conservation agriculture. *European Journal of Agronomy, 133*, Article 126417. https://doi.org/10.1016/j.eja.2021.126417

Nghi, N. T., Romasanta, R. R., Hieu, N. V., Du, N. X., Ngan, N. V. C., Chivenge, P., & Hung, N. V. (2020). Rice straw-based composting. In N. Gummert, N. V. Hung, P. Chivenge, & B. Douthwaite (Eds.), *Sustainable rice straw management* (pp. 33–41). Springer. https://doi.org/10.1007/978-3-030-32373-8_3

Pandey, A. K., Gaind, S., Ali, A., & Nain, L. (2009). Effect of bio augmentation and nitrogen supplementation on composting of paddy straw. *Biodegradation*, *20*, 293–306. https://doi.org/10.1007/s10532-008-9221-3

Pandey, C. (2019). Management of crop residue for sustaining soil fertility and food grains production in India. *Acta Scientific Agriculture*, *3*(1), 188–195.

Pandian, K., Subramaniayan, P., Gnasekaran, P., & Chitraputhirapillai, S. (2016). Effect of biochar amendment on soil physical, chemical and biological properties and groundnut yield in rainfed Alfisol of semi-arid tropics. *Archives of Agronomy and Soil Science*, *62*(9), 1293–1310. https://doi.org/10.1080/03650340.2016.1139086

Pariyar, P., Kumari, K., Jain, M. K., & Jadhao, P. S. (2020). Evaluation of change in biochar properties derived from different feedstock and pyrolysis temperature for environmental and agricultural application. *Science of the Total Environment*, *713*, Article 136433. https://doi.org/10.1016/j.scitotenv.2019.136433

Pathak, H., Tewari, A. N., Sankhyan, S., Dubey, D. S., Mina, U., Singh, V. K., & Jain, N. (2011). Direct-seeded rice: Potential, performance and problems—A review. *Current Advances in Agricultural Sciences*, *3*(2), 77–88.

Patra, A. K., Chhonkar, P. K., & Khan, M. A. (2004). Nitrogen loss and wheat (*Triticum aestivum* L.) yields in response to zero tillage and sowing time in a semi-arid tropical environment. *Journal of Agronomy and Crop Science*, *190*(5), 324–331. https://doi.org/10.1111/j.1439-037X.2004.00112.x

Porichha, G. K., Hu, Y., Rao, K. T. V., & Xu, C. C. (2021). Crop residue management in India: Stubble burning vs. other utilizations including bioenergy. *Energies*, *14*(14), Article 4281. https://doi.org/10.3390/en14144281

Prasad, S., Kumar, S., Sheetal, K., & Venkatramanan, V. (2020). Global climate change and biofuels policy: Indian perspectives. In V. V. Shah (Ed.), *Global climate change and environmental policy: Agriculture perspectives* (pp. 207–226). Springer Nature. https://doi.org/10.1007/978-981-13-9570-3_6

Purakayastha, T. J., Bera, T., Bhaduri, D., Sarkar, B., Mandal, S., Wade, P., Kumari, S., Biswas, S., Menon, M., Pathak, H., & Tsang, D. C. W. (2019). A review on biochar modulated soil condition improvements and nutrient dynamics concerning crop yields: Pathways to climate change mitigation and global food security. *Chemosphere*, *227*, 345–365. https://doi.org/10.1016/j.chemosphere.2019.03.170

Purakayastha, T. J., Kumari, S., & Pathak, H. (2015). Characterization, stability, and microbial effects of four biochars produced from crop residues. *Geoderma*, *239*, 293–303. https://doi.org/10.1016/j.geoderma.2014.11.009

Rath, A. K., Ramakrishnan, B., Rao, V. R., & Sethunathan, N. (2005). Effects of rice-straw and phosphorus application on production and emission of methane from tropical rice soil. *Journal of Plant Nutrition and Soil Science, 168*(2), 248–254. https://doi.org/10.1002/jpln.200421604

Raza, M. H., Abid, M., Yan, T., Naqvi, S. A. A., Akhtar, S., & Faisal, M. (2019). Understanding farmers' intentions to adopt sustainable crop residue management practices: A structural equation modeling approach. *Journal of Cleaner Production, 227*, 613–623. https://doi.org/10.1016/j.jclepro.2019.04.244

Saha, S., Pandey, A. K., Gopinath, K. A., Bhattacharaya, R., Kundu, S., & Gupta, H. S. (2007). Nutritional quality of organic rice grown on organic composts. *Agronomy for Sustainable Development, 27*, 223–229. http://doi.org/10.1051/agro:2007002

Saharawat, Y. S., Singh, B., Malik, R. K., Ladha, J. K., Gathala, M., Jat, M. L., & Kumar, V. (2010). Evaluation of alternative tillage and crop establishment methods in a rice–wheat rotation in North Western IGP. *Field Crops Research, 116*(3), 260–267. https://doi.org/10.1016/j.fcr.2010.01.003

Samra, J. S., Singh, B., & Kumar, K. (2003). Managing crop residues in the rice-wheat system of the Indo-Gangetic Plain. In J. K. Ladha (Ed.), *Improving the productivity and sustainability of rice-wheat systems, issues and impact* (ASA Spec. Publ. 65 pp. 173–195). ASA, CSSA, SSSA.

Samui, I., Skalicky, M., Sarkar, S., Brahmachari, K., Sau, S., Ray, K., Hossain, A., Ghosh, A., Nanda, M. K., Bell, R. W., & Mainuddin, M. (2020). Yield response, nutritional quality and water productivity of tomato (*Solanum lycopersicum* L.) are influenced by drip irrigation and straw mulch in the coastal saline ecosystem of Ganges delta, India. *Sustainability, 12*(17), Article 6779. https://doi.org/10.3390/su12176779

Sánchez, Ó., Ospina, D. A., & Montoya, S. (2017). Compost supplementation with nutrients and microorganisms in composting process. *Waste Management, 69*, 136–153. https://doi.org/10.1016/j.wasman.2017.08.012

Sarkar, D., Baishya, L. K., Meitei, C. B., Naorem, G. C., Thokchom, R. C., Singh, J., Bhuvaneswari, S., Batabyal, K., Das, R., Padhan, D., Prakash, N., & Rahman, F. H. (2018). Can sustainability of maize-mustard cropping system be achieved through balanced nutrient management? *Field Crops Research, 225*, 9–21. https://doi.org/10.1016/j.fcr.2018.05.018

Sarkar, S., Skalicky, M., Hossain, A., Brestic, M., Saha, S., Garai, S., Ray, K., & Brahmachari, K. (2020). Management of crop residues for improving input use efficiency and agricultural sustainability. *Sustainability, 12*(23), Article 9808. https://doi.org/10.3390/su12239808

Shanthy, T. R., Ram, B., Venkatasubramanian, V., Karpagam, C., & Prathap, D. P. (2014). *Scientific sugarcane cultivation* (Extension Publication No. 227). *ICAR-Sugarcane Breeding Institute.*

Sharma, A., & Singh, R. (2018). Effect of abiotic factors on burrow density of Indian gerbil, *Tatera indica* (Hardwicke) (Rodentia: Muridae) in Punjab. *Journal of Entomology and Zoology Studies*, *6*(4), 1508–1513.

Sharma, P., Abrol, V., Sharma, V., Chaddha, S., Srinivasarao, C., Ganie, A. Q., Hefft, D. I., El-Sheikh, M. A., & Mansoor, S. (2021). Effectiveness of biochar and compost on improving soil hydro-physical properties, crop yield and monetary returns in inceptisol subtropics. *Saudi Journal of Biological Sciences*, *28*(12), 7539–7549. https://doi.org/10.1016/j.sjbs.2021.09.043

Sharma, S., Kaur, S., Choudhary, O. P., Singh, M., Al-Huqail, A. A., Ali, H. M., Kumar, R., & Siddiqui, M. H. (2022). Tillage, green manure and residue retention improves aggregate-associated phosphorus fractions under rice–wheat cropping. *Scientific Reports*, *12*, Article 7167. https://doi.org/10.1038/s41598-022-11106-x

Shen, Y., Ren, L., Li, G., Chen, T., & Guo, R. (2011). Influence of aeration on CH_4, N_2O and NH_3 emissions during aerobic composting of a chicken manure and high C/N waste mixture. *Waste Management*, *31*(1), 33–38. https://doi.org/10.1016/j.wasman.2010.08.019

Shinde, R., Shahi, D. K., Mahapatra, P., Naik, S. K., Thombare, N., & Singh, A. K. (2022). Potential of lignocellulose degrading microorganisms for agricultural residue decomposition in soil: A review. *Journal of Environmental Management*, *320*, Article 115843. https://doi.org/10.1016/j.jenvman.2022.115843

Shukla, L., Suman, A., & Verma, P. (2016). Syntrophic microbial system for ex-situ degradation of paddy straw at low temperature under controlled and natural environment. *Journal of Applied Biology & Biotechnology*, *4*(2), 30–37. https://doi.org/10.7324/JABB.2016.40205

Sidhu, B. S., & Beri, V. (2005). Experience with managing rice residues in intensive rice-wheat cropping system in Punjab. In I. P. Abrol, R. K. Gupta, & R. K. Malik (Eds.), *Conservation agriculture-status and prospects* (pp. 55–63). Centre for Advancement of Sustainable Agriculture.

Sidhu, H. S., Manpreet-Singh, Humphreys, E., Yadvinder-Singh, Balwinder-Singh, Dhillon, S. S., Blackwell, J., Bector, V., Malkeet-Singh, & Sarbjeet-Singh (2007). The Happy Seeder enables direct drilling of wheat into rice stubble. *Australian Journal of Experimental Agriculture*, *47*(7), 844–854. https://doi.org/10.1071/EA06225

Sidhu, H. S., Singh, M., Singh, Y., Blackwell, J., Lohan, S. K., Humphreys, E., Jat, M. L., Singh, V., & Singh, S. (2015). Development and evaluation of the Turbo Happy Seeder for sowing wheat into heavy rice residues in NW India. *Field Crops Research*, *184*, 201–212. https://doi.org/10.1016/j.fcr.2015.07.025

Singh, J. M., Singh, J., Kumar, H., Singh, S., Sachdeva, J., Kaur, B., Chopra, S., & Chand, P. (2019). Management of paddy straw in Punjab: An economic analysis of different techniques. *Indian Journal of Agricultural Economics, 74*(3), 301–310.

Singh, P., Heikkinen, J., Ketoja, E., Nuutinen, V., Palojärvi, A., Sheehy, J., Esala, M., Mitra, S., Alakukku, L., & Reginaet, K. (2015). Tillage and crop residue management methods had minor effects on the stock and stabilization of topsoil carbon in a 30-year field experiment. *Science of the Total Environment, 518–519*, 337–344. https://doi.org/10.1016/j.scitotenv.2015.03.027

Singh, P., Mukherjee, S., Saha, N., Biswas, S., & Mandal, B. (2021). Conservation agriculture in reshaping belowground microbial diversity. In A. Rakshit, S. Singh, P. Abhilash, & A. Biswas (Eds.), *Soil science: Fundamentals to recent advances* (pp. 141–173). Springer. https://doi.org/10.1007/978-981-16-0917-6_8

Singh, R., Behera, S., Yadav, Y. K., & Kumar, S. (2014). Potential of wheat straw for biogas production using thermophiles. In S. Kumar, A. K. Sarma, S. K. Tyagi, & Y. K. Yadav (Eds.), *Recent advances in bio-energy research* (pp. 242–249). Sardar Swaran Singh National Institute of Renewable Energy. https://doi.org/10.13140/RG.2.1.2286.3844

Singh, R., Yadav, D. B., Ravisankar, N., Yadav, A., & Singh, H. (2020). Crop residue management in rice-wheat cropping system for resource conservation and environmental protection in north-western India. *Environment, Development and Sustainability, 22*, 3871–3896. https://doi.org/10.1007/s10668-019-00370-z

Singh, R. P., Dhaliwal, H. S., Humphreys, E., Sidhu, H. S., Singh, M., Singh, Y., & Blackwell, J. (2008a). Economic assessment of the Happy Seeder for rice-wheat systems in Punjab, India. 52nd Annual Conference of the Australian Agricultural and Resource Economics Society (AARES), February 5–8, 2008, Canberra, Australia. https://doi.org/10.22004/ag.econ.5975

Singh, R. P., Verma, S. K., Kumar, S., & Lakara, K. (2017). Impact of tillage and herbicides on the dynamics of broad leaf weeds in wheat (*Triticum aestivum* L.). *International Journal of Agriculture, Environment and Biotechnology, 10*(6), 643–652. https://doi.org/10.5958/2230-732X.2017.00079.1

Singh, V. K., Dwivedi, B. S., Singh, S. K., Majumdar, K., Jat, M. L., Mishra, R. P., & Rani, M. (2016). Soil physical properties, yield trends and economics after five years of conservation agriculture based rice-maize system in north-western India. *Soil and Tillage Research, 155*, 133–148. https://doi.org/10.1016/j.still.2015.08.001

Singh, V. K., Dwivedi, B. S., Singh, S. K., Mishra, R. P., Shukla, A. K., Rathore, S. S., Shekhawat, K., Majumdar, K., & Jat, M. L. (2018). Effect of tillage and crop establishment, residue management and K fertilization on yield, K use efficiency and apparent K balance under rice-maize system in north-western India. *Field Crops Research, 224*, 1–12. https://doi.org/10.1016/j.fcr.2018.04.012

Singh, Y., Gupta, R. K., Singh, J., Singh, G., Singh, G., & Ladha, J. K. (2010). Placement effects on rice residue decomposition and nutrient dynamics on two soil types during wheat cropping in rice-wheat system in north-western India. *Nutrient Cycling in Agroecosystems, 88*, 471–480. https://doi.org/10.1007/s10705-010-9370-8

Singh, Y., Sidhu, H. S., Singh, M., Dhaliwal, H. S., Blackwell, J., Singh, R. P., Humphreys, L., Singla, N., Thind, H. S., Lohan, S. K., & Sran, D. S. (2009). *Happy seeder: A conservation agriculture technology for managing rice residues* [Technical Bulletin]. Department of Soils, Australia.

Singh, Y. H. B., Shan, Y. H., Beebout, S. E. J., Singh, Y., & Buresh, R. J. (2008b). Crop residue management for lowland rice-based cropping systems in Asia. *Advances in Agronomy, 98*, 117–199. https://doi.org/10.1016/S0065-2113(08)00203-4

Smitha, G. R., Basak, B. B., Thondaiman, V., & Saha, A. (2019). Nutrient management through organics, bio-fertilizers and crop residues improves growth, yield and quality of sacred basil (*Ocimum sanctum* Linn). *Industrial Crops and Products, 128*, 599–606. https://doi.org/10.1016/j.indcrop.2018.11.058

Solangi, S. H., Jakhrani, A. Q., Mukwana, K. C., Jatoi, A. R., & Luhur, M. R. (2018). Investigation of quantity, quality and energy content of indigenous sugarcane trash in Naoshehro Feroze District, Sindh. *Engineering Technology and Applied Science Research, 8*(6), 3609–3613. https://doi.org/10.48084/etasr.2411

Srinivasarao, C., Gopinath, K. A., Venkatesh, G., Dubey, A. K., Harsha, W., Purakayastha, T. J., Pathak, H., Pramod, J., Lakaria, B. L., Rajkhowa, D. J., Sandip, M., Jeyaraman, S., Venkateswarlu, B., & Sikka, A. K. (2013). *Use of biochar for soil health management and greenhouse gas mitigation in India: Potential and constraints.* Central Research Institute for Dryland Agriculture.

Thind, H. S., Sharma, S., Singh, Y., & Sidhu, H. S. (2019). Rice-wheat productivity and profitability with residue, tillage and green manure management. *Nutrient Cycling in Agroecosystems, 113*, 113–125. https://doi.org/10.1007/s10705-018-09967-8

Tognetti, C., Mazzarino, M. J., & Laos, F. (2007). Improving the quality of municipal organic waste compost. *Bioresource Technology, 98*(5), 1067–1076. https://doi.org/10.1016/j.biortech.2006.04.025

Torres-Climent, A., Martin-Mata, J., Marhuenda-Egea, F., Moral, R., Barber, X., Perez-Murcia, M. D., & Paredes, C. (2015). Composting of the solid phase of digestate from biogas production: Optimization of the moisture, C/N ratio, and pH conditions. *Communications in Soil Science and Plant Analysis, 46*(suppl 1), 197–207. https://doi.org/10.1080/00103624.2014.988591

Tripathi, S. C., Chander, S., Meena, R. P., Venkatesh, K., & Verma, A. (2021). Incorporation of rice residue and green gram cultivation saves nitrogen, improve soil health and sustainability of rice-wheat system. *Field Crops Research, 271*, Article 108248. https://doi.org/10.1016/j.fcr.2021.108248

Turmel, M. S., Speratti, A., Baudron, F., Verhulst, N., & Govaerts, B. (2015). Crop residue management and soil health: A systems analysis. *Agricultural Systems, 134*, 6–16. https://doi.org/10.1016/j.agsy.2014.05.009

Upadhyay, P. K., Sen, A., Singh, Y., Singh, R. K., Prasad, S. K., Sankar, A., Singh, V. K., Dutta, S. K., Kumar, R., Rathore, S. S., Shekhawat, K., Babu, S., Singh, R. K., Kumar, B., Dey, A., Rajanna, G. A., & Kulshekaran, R. (2022). Soil health, energy budget, and rice productivity as influenced by cow products application with fertilizers under South Asian Eastern Indo-Gangetic Plains Zone. *Frontiers in Agronomy*, *3*, Article 758572. https://doi.org/10.3389/fagro.2021.758572

Vargas-García, M. C., Suárez-Estrella, F., López, M. J., & Moreno, J. (2007). Effect of inoculation in composting processes: Modifications in lignocellulosic fraction. *Waste Management*, *27*(9), 1099–1107. https://doi.org/10.1016/j.wasman.2006.06.013

Venkatesh, G., Gopinath, K. A., Reddy, K. S., Reddy, B. S., Prabhakar, M., Srinivasarao, C., Kumari, V. V., & Singh, V. K. (2022). Characterization of biochar derived from crop residues for soil amendment, carbon sequestration and energy use. *Sustainability*, *14*(4), 2295. https://doi.org/10.3390/su14042295

Venkatesh, G., Gopinath, K. A., Reddy, K. S., Reddy, B. S., Prasad, J. V. N. S., Rao, G. R., Pratibha, G., Srinivasarao, C., Chary G. R., Prabhakar, M., Kumari V. V., Shankar, A. K., & Venkateswarlu, B. (2018). *Biochar Production and its use in rainfed Agriculture: Experiences from CRIDA* (CRIDA-NICRA Research Bulletin 02/2018). ICAR - Central Research Institute for Dryland Agriculture.

Venkatesh, G., Venkateswarlu, B., Gopinath, K. A., Srinivasarao, C., Korwar, G. R., & Reddy, B. S. (2013). *Low cost portable kiln for biochar production* (NICRA Technical Brochure). Central Research Institute for Dryland Agriculture.

Vigneswaran, S., Kandasamy, J., & Johir, M. A. H. (2016). Sustainable operation of composting in solid waste management. *Procedia Environmental Sciences*, *35*, 408–415. https://doi.org/10.1016/j.proenv.2016.07.022

Walia, U. S., & Brar, L. S. (2006). Effect of tillage and weed management on seed bank of *Phalaris minor* in wheat under rice-wheat sequence. *Indian Journal of Weed Science*, *38*(1&2), 104–107.

Wong, J. W. C., Mak, K. F., Chan, N. W., Lam, A., Fang, M., Zhou, L. X., Wu, Q. T., & Liao, X. D. (2001). Co-composting of soybean residues and leaves in Hong Kong. *Bioresource Technology*, *76*(2), 99–106. https://doi.org/10.1016/S0960-8524(00)00103-6

Yadav, G. S., Lal, R., Meena, R. S., Babu, S., Das, A., Bhowmik, S. N., Datta, M., Layak, J., & Saha, P. (2019a). Conservation tillage and nutrient management effects on productivity and soil carbon sequestration under double cropping of rice in north eastern region of India. *Ecological Indicators*, *105*, 303–315. https://doi.org/10.1016/j.ecolind.2017.08.071

Yadav, S. K., Benbi, D. K., & Toor, A. S. (2019b). Effect of long-term application of rice straw, farmyard manure and inorganic fertilizer on potassium dynamics in soil. *Archives of Agronomy and Soil Science*, *65*(3), 374–384. https://doi.org/10.1080/03650340.2018.1505040

Yadvinder-Singh, Bijay-Singh, & Timsina, J. (2005). Crop residue management for nutrient cycling and improving soil productivity in rice-based cropping systems in the tropics. *Advances in Agronomy*, *85*, 269–407. https://doi.org/10.1016/S0065-2113(04)85006-5

Yadvinder-Singh, Singh, M., Sidhu, H. S., Humphreys, E., Thind, H. S., Jat, M. L., Blackwell, J., & Singh, V. (2015). Nitrogen management for zero till wheat with surface retention of rice residues in north-west India. *Field Crops Research*, *184*, 183–191. https://doi.org/10.1016/j.fcr.2015.03.025

15

Recommendations for Long-Term Soil Health Enhancement in India

Kuntal M. Hati, Ranjan Bhattacharyya, Pramod Jha, J. Somasundaram, B. N. Ghosh, Avijit Ghosh, and Douglas L. Karlen

Chapter Overview

Soil degradation is a major problem in many parts of India. Therefore, improving soil health is essential for sustainable development of agriculture, bioenergy, and the infrastructure supporting a current population of more than 1.4 billion people. This chapter summarizes the practices discussed in previous chapters. All of these practices, as well as the research and technology transfer efforts needed to effectively implement them, are crucial for meeting the challenges posed by soil degradation in India.

Balanced and Integrated Nutrient Management

Imbalanced and inappropriate fertilizer application adversely affects soil health and limits the capacity of soil to produce at optimal and sustainable levels at spatial and temporal scales. Balanced application of inorganic fertilizer and organic amendments greatly influence the accumulation of organic matter in soil and also influence soil health. A balanced supply of nutrients promotes higher crop production, which stimulates better soil health. Integrated nutrient management (INM) involves the judicious use of organic and inorganic sources of nutrients to provide macro- and micronutrients to plants and thus improve crop yield and quality over the long run, thus making farming more sustainable and economically viable. Long-term INM experiments have clearly shown that fertilizer

Soil Health Series: Volume 4 Soil Health and Sustainability in India, First Edition.
Edited by Ranjan Bhattacharyya.

response can be significantly increased with a complete and balanced supply of nutrients along with organic manure. The soil organic carbon (SOC) stock and carbon sequestration rate are also significantly improved with the adoption of INM practices.

Improvement of Nutrient Use Efficiency

Efficient use of fertilizer nutrients is essential to achieve high crop yields and economic profitability, while minimizing the negative environmental impacts of agriculture. However, temporal and spatial variability of soil nutrients can restrict the effectiveness of broad-based fertilizer recommendations. Site-specific nutrient management (SSNM) can be effectively practiced to achieve high nutrient use efficiency (NUE) and economic profitability, while also leaving a lower environmental footprint. SSNM encompasses several practices, including using field-specific fertilizer nutrient management, recycling crop residues, and integrating legumes into the cropping cycle. To increase NUE, it is essential to strive for better soil health by using nutrient management technologies and practices adapted to different production systems. Some examples of modern tools and technologies include Nutrient Expert, Green Seeker, and other remote sensing tools. These tools can provide critical nutrient management support and increase productivity of resource-poor small and marginal farmers, who often cannot access or afford soil testing. Fertigation or application of water-soluble fertilizers with sprinkler/drip irrigation systems or through foliar spray can also help improve crop yield and the quality of fruits and vegetables. Water-soluble fertilizers used for fertigation generally contain a combination of all essential nutrients at ratios that suit various combinations of crops and site-specific water quality, soil fertility, and climatic conditions. Fertigation can markedly improve NUE, crop yield, and quality. Higher NUE increases system productivity, profitability, and soil carbon status, and it reduces the environmental footprint of agricultural activities by reducing nutrient losses to groundwater and surface-water bodies. Collectively, these practices also improve soil health and biodiversity.

Nano-Fertilizers and Novel Fertilizer Products

Nanotechnology has shown great potential in agriculture, particularly for improving crop water and fertilizer use efficiencies. By applying various synthetic nano-fertilizers and other nano-materials, it is possible to deliver nutrients to plants in a more efficient and timely manner. This method can increase the bioavailability of soil nutrients and lead to higher crop yields and a better soil environment. Research has been conducted on commercially available nano-particles of zinc oxide (ZnO), copper oxide (CuO), and iron oxide (FeO) to investigate their stability in an aqueous system and their assimilation by plants. A solution culture experiment with nano-particles of ZnO and soluble $ZnSO_4$, showed that ZnO nano-particles had a stimulating effect, resulting in more rapid growth of maize as compared with higher doses of zinc applied as $ZnSO_4$. However, it is important to note that the use of nano-materials in agriculture is still a relatively new practice and more research is needed to fully understand the potential benefits and risks to the environment and human health. It is crucial to ensure that any nano-materials used in agriculture are thoroughly tested and regulated to minimize any potential negative impacts.

A variety of novel fertilizer products that can promote better soil health are available. Some currently on the market are the following:

- Humic acid-based fertilizers: Humic acid is naturally present in soil and can improve soil structure, nutrient retention, and water-holding capacity. Humic acid-based fertilizers combine humic acid with other nutrients to provide a complete fertilizer product.
- Bio-fertilizers/microbial inoculants: Bio-fertilizers contain beneficial microorganisms such as bacteria and fungi that can improve soil health by promoting nutrient cycling, improving soil structure, and suppressing plant pathogens. Some microbial inoculants also contain nutrients such as nitrogen-fixing bacteria that can help reduce the need for synthetic fertilizers.
- Seaweed-based fertilizers: Seaweed contains a variety of nutrients that can benefit soil health, including potassium, magnesium, and trace minerals. Seaweed-based fertilizers can improve soil structure, water retention, and nutrient availability.
- Slow-release fertilizers: Slow-release fertilizers release nutrients over an extended period of time. The slow release reduces the risk of nutrient leaching, while ensuring that plants have a consistent and adequate supply of nutrients, thereby enhancing NUE. Some slow-release fertilizers also contain organic matter that can improve soil structure and water-holding capacity.

Overall, using a combination of novel fertilizer products may help promote better soil health by improving soil structure, nutrient availability, and water-holding capacity. It is important to choose the right fertilizer product for specific soil and crop needs and to follow application guidelines carefully to avoid overuse or misuse of these products.

Recycling of Residues and Farm Waste Through Composting

Crop residues and farm waste in India are generally removed from the field to keep the field clean and to make sowing new crops easier. Currently, the residues and waste are either burned in fields or used as feed for farm animals. Either way, the nutrients and organic carbon present in these residues are lost from the system, causing nutrient deficiency and consequent soil degradation. However, these byproducts of the production system, which are rich in organic matter, could be recycled through composting and used to enhance soil health. Composting is a natural process that breaks down organic materials into a nutrient-rich soil amendment. Application of compost prepared from recycled farm residues and wastes helps in building soil organic matter content, which is essential for soil health. Organic matter in soil acts as a sponge, holding water and nutrients in the soil and slowly releasing them to plants. It also helps to improve soil structure, which can increase water infiltration and reduce erosion. Composting can thus help reduce waste, improve soil quality, and increase crop productivity, making it an ideal solution for farmers looking to improve their yields while also reducing their environmental impact. With the support of government schemes, nongovernmental organizations, and private organizations, more farmers can be encouraged to adopt composting practices, which will help to reduce waste, improve soil fertility, and promote sustainable agriculture.

Biochar as a Soil Amendment for Soil Health Improvement

Biochar is a carbon-rich substance that is created by heating organic matter in the absence of oxygen. Soil health management interventions with soil amendments such as biochar are logical solutions for improving soil quality and consequently crop productivity. Biochar can help overcome soil problems such as acidity, salinity, drought, and low fertility, while also maintaining soil health and agricultural productivity. Application of biochar as a soil amendment helps to increase water retention and microbial growth within the soil. In addition, through its chemical characteristics that include providing macro-nutrients (nitrogen, phosphorus, potassium, calcium, and magnesium) and micro-nutrients (copper, iron, manganese, and zinc), biochar-based fertilizers provide a complete fertilizer product for plant growth in nutrient-poor soils, help reduce nutrient leaching and heavy metal contamination in crops, enhance crop growth and yield, and increase carbon sequestration. Biochar addition to hard-setting soil is reported to increase the carbon, nitrogen, phosphorus, and pH and decrease the soil strength. Biochar application to soil has thus emerged as a potential technology for sequestering carbon, reducing greenhouse gas emissions, and improving soil health.

Cover Cropping and Mulching

Cover cropping and mulching are effective practices for improving soil health. Cover cropping involves growing one or more plant species between the primary growing seasons to cover and protect the soil. Cover crops can help to prevent soil erosion, reduce weed growth, improve soil structure, and increase soil organic matter, which all help increase soil fertility and water-holding capacity. Mulch is a layer of organic material, such as leaves, straw, or wood chips, that is applied to the soil surface. It helps to retain moisture in the soil, suppress weed growth, and regulate soil temperature. As the mulch breaks down, it adds organic matter to the soil, which improves soil structure and fertility. Both cover cropping and mulching can also help to reduce the need for synthetic fertilizers and pesticides that can have negative impacts on soil health and the environment. Cover cropping and mulching are particularly important in areas where soil fertility is low and where drought and other environmental stresses are common. In India, cover cropping and mulching have been used for centuries to improve soil fertility and agricultural productivity. Some common cover crops used in India include legumes such as cowpea and soybean, as well as oilseeds such as mustard and sesame. Mulching is often done with rice straw, sugarcane trash, or other crop residues. By improving soil health through these low-cost and sustainable practices, farmers can increase productivity, profitability, and food security, while also reducing the environmental impact of agriculture.

Crop Rotation/Diversification with Inclusion of Legumes in Crop Cycle

Crop rotation with the inclusion of legumes in crop cycles can have a significant impact on soil health. These practices promote soil health by improving soil structural stability and reducing soil erosion and buildup of soil-borne diseases and pests. By alternating different crops in a field, nutrients that were depleted by the previous crop can be replenished in the soil, thus reducing the need for synthetic fertilizers and pesticides. Incorporating legumes into the crop cycle is beneficial for soil health owing to the ability of the legumes to fix nitrogen in the soil, reducing the need for application of synthetic fertilizers to the system. Using crop rotation and diversification, as well as incorporating legumes into the crop cycle, can help to maintain or improve soil health, which is essential for sustainable agriculture.

Conservation Agriculture

Intensive tillage degrades soil health because it breaks soil aggregates, which increases crop residue decomposition rates, accelerates SOC oxidation by increasing soil temperature, and reduces biodiversity among microbes by disturbing their habitats. Mono-cropping without legumes in the rotation further degrades soil health by affecting nutrient cycling and soil biodiversity. Conservation agriculture (CA) is a form of natural farming in which crops are grown with minimal soil disturbance, crop residues are retained on the surface, and the cropping system is diversified by including legumes and cover crops in the rotation. CA studies in India have shown a multitude of benefits regarding soil health. These benefits include improving the quality and quantity of SOC, stabilizing soil structure, creating more favorable conditions for beneficial soil microbes, enhancing nutrient supplying capacity, minimizing the carbon footprint owing to lower greenhouse gas emissions from machinery, and reducing erosion and the loss of fertile topsoil.

CA also offers several sustainable crop residue management options for a large part of the country where residues are currently being burned, causing air pollution, releasing a large volume of carbon dioxide to the atmosphere, and increasing the loss of valuable essential plant nutrients from the ecosystem. Adoption of suitable, site-specific CA practices for both irrigated and rainfed agro-ecosystems would be beneficial throughout the country. Stopping crop residue burning alone has great potential for soil health improvement, sustainability of the production system, and mitigation of the adverse impacts of climate change on agriculture. CA has an untapped potential for improving soil health without compromising crop productivity. However, challenges that need to be addressed through CA research include sowing of crops in heavy residue, especially when soils are moist during the rainy season. Addressing this challenge throughout India will require increasing the availability of appropriate sowing equipment in many agroecoregions; developing new management strategies that include, but do not solely depend on, herbicides to address challenges posed by shifts in weed dynamics; and teaching farmers the new skills required for implementing CA practices. Concerted efforts are needed to address these issues to achieve widespread adoption of CA in India.

Agro-Forestry Intervention for Improving Soil Health of Marginal Land

The use of agro-forestry (AF) interventions is one option for improving and managing degraded or marginal land that is unsuitable for regular or economically viable cultivation. AF is a land-use system in which woody perennials are grown in some arrangement or temporal sequence on the same land as agricultural crops and/or livestock. The system is self-sustaining because of the diversified components such as arable crops, forage species, tree components, and domestic animals. The tree component of AF protects the soil from erosion through an extensive root network and large canopy. It is also helpful for stabilizing gullies and preventing their spread. The grassy component involved in AF also mitigates soil erosion owing to thick ground cover and increases SOC content and structural stability of the soil, which in turn reduces land degradation. Including leguminous tree and shrub species further enriches the soil through biological nitrogen fixation along with the addition of organic matter through leaf fall, which also helps improve biological soil. As a whole, the AF system supports several functions that enhance biogeochemical nutrient cycling with active involvement of plants and microorganisms within the biosphere. Areas along field boundaries, farm roads, or canals that remain barren and severely affected by one or more types of land degradation are especially suitable for one or more components of AF. The practice can also improve biodiversity of cultivated farms and consequently enhance soil health of marginal lands as a whole. At the same time, AF produces a large amount of woody matter that, if retained over a long period, can be used.

Rainwater Harvesting and Recycling

Much of the rainfed semi-arid region of India does not receive adequate and/or well-distributed rainfall. About 75% to 85% of the total annual rainfall is received during the four monsoonal months of the year. Harvesting the excess rainfall received during those months and recycling it during rain-deficient or lean periods is essential for growing winter season crops and providing life-saving irrigation when rain is deficient for rainy season crops. Irrigating crops with stored rainwater not only increases system productivity but also adds biomass to the soil, which promotes soil health through sequestration of carbon in the soil profile. Rainwater can be harvested either in situ or ex situ, but it must be used judiciously with site-specific technologies for maximizing productivity. The aim of any efficient water management system (irrigated or rain-fed) is to maximize productivity per unit of water (transported, pumped, or retained in situ). Watershed-based crop planning optimizes harvesting of rainwater and its judicious use. Broad-bed and furrow, narrow-bed and furrow, ridge and furrow, and raised- and sunken-bed systems are potential land configuration systems that can improve surface drainage, enhance water infiltration, and facilitate safe drainage of excess water through grassed waterways into water-harvesting ponds for use during dry periods. Summer cultivation and subsoiling of soils can also help increase storage of rainwater in soil profile for higher water-use efficiency, productivity, and crop yield.

Controlling Runoff and Soil Loss for Soil Health Management

Reducing runoff and soil loss, particularly in undulating land areas, is imperative for maintaining healthy soil and preserving agricultural productivity. Excessive runoff and soil loss occur because of various factors, including excessive precipitation, wind, and human activities such as land-use change, tillage, and use of heavy machinery that causes soil compaction. Several combinations of surface land management and agronomic interventions can greatly help to control runoff and soil loss, while also enhancing soil health and promoting sustainable agriculture. Land surface management techniques such as laser land leveling, broad-bed and furrow systems, ridge and furrow systems, contour bunding, and so forth, can greatly help in preventing soil erosion and improving soil health. These practices can be combined with agronomic interventions to enhance the overall effectiveness of soil conservation.

Laser land leveling is a technique that uses precise laser-based monitoring to level the soil surface and reduce the slope gradient. This helps to minimize runoff and soil loss and provides more uniform distribution of irrigation water for attaining higher water-use efficiency. Broad-bed and furrow systems involve creating broad-beds and furrows in the field, which not only helps to reduce runoff but also facilitates better drainage and aeration of the soil. Similarly, ridge and furrow management involves creating ridges and furrows in the field to reduce runoff and soil erosion and to facilitate better water infiltration and distribution. Contour bunding involves creating bunds or small barriers along the contour lines of a field with a moderate to gentle slope to slow runoff and trap sediment, thereby preventing loss of fertile topsoil. This technique can also help improve soil moisture retention, which can be further enhanced by incorporating organic matter into the soil through agronomic interventions such as crop residue management, green manuring, and composting. Higher productivity increases the addition of above- and belowground biomass and thus facilitates improvement in soil health.

Structural measures such as terraces, grassed waterways, and sediment basins also reduce runoff and soil loss to a great extent. Agronomic interventions, including intercropping, cover cropping, crop rotation, contour farming, vegetative barriers, and INM, can also help enhance soil health by improving soil structure, nutrient content, and microbial activity. These practices not only improve soil health but also help reduce the need for chemical fertilizers, which can further contribute to environmental sustainability and mitigation of climate change. Managing water can help control runoff and soil loss. Techniques such as diverting water away from sensitive areas, installing drainage systems, and using irrigation techniques that minimize runoff can all help reduce soil erosion. Soil erosion and degradation caused by overgrazing can be managed through controlled and rotational grazing of pasture land, with livestock being frequently moved between different pastures.

Management of Soil Salinity and Sodicity to Improve Soil Health

Soil salinity and sodicity have detrimental effects on soil health, particularly soil physical health, leading to reduced plant growth and yield. Climate, rainfall pattern, soil type, irrigation water quality and its management, groundwater level and quality, and soil management practices influence salinity development. Arid and semi-arid environments, where the amount of rainfall is not sufficient to leach salts from surface soil, are conducive to salinity development. Faulty agricultural practices such as excessive use of basic fertilizers (e.g., sodium nitrate or basic slag), removal of native vegetation, burning of crop residues, excessive irrigation (particularly with saline groundwater), and growing shallow-rooted annual crops can accelerate the salinization process.

Inundation and submergence of low-lying coastal lines with brackish water are a major cause of soil salinization in the coastal belt. The salinity can be alleviated through adoption of good agronomic practices such as balanced fertilization, applying vermicompost or farmyard manure, green manuring, residue retention, farm waste recycling, efficient/judicious use of irrigation water, and cultivation of salt-tolerant cultivars. Surface soil salinization can be reduced to a certain extent through mulching, as it reduces the evaporation and consequently slows capillary rise of underground saline water. Sodicity can be mitigated by application of calcium-based amendments such as gypsum, phospho-gypsum, or $CaCO_3$ mobilizers, including elemental sulfur, sulfuric acid, pyrites, fly ash, aluminum chloride, agricultural industry by-products, and press mud, or through afforestation of sodic land with sodium-tolerant trees. Some recently developed synthetic polymers and nano-materials, acidified biochars, elemental sulfur, sulfuric acid, and flue gas desulfurization gypsum have shown great potential as an amendment alternative to replace gypsum. Salt-affected soil reclamation makes the resource congenial for plant growth, leading to higher productivity of both above- and belowground biomass. As a result, the SOC content increases, and other soil properties subsequently improve, which leads to an overall soil quality improvement in these marginal lands.

Acid Soil and Iron and Aluminum Oxide Toxicity Management

Soil acidification is a major problem in the coastal state of Kerala, hilly areas of the northeast and northwest Himalayan region, and the eastern plateau region with humid to per-humid climatic conditions. Excessive leaching of soluble salts or cations with high rainfall causes acidity in surface soils, while the subsoil acidity develops owing to the accumulation of exchangeable aluminum and iron hydroxides in the interlayer of silicate clays. Soil acidity negatively influences most soil physical, chemical, and biological properties, resulting in the deterioration of the soil health and quality. Low pH in acid soils causes aluminum and iron to become soluble and toxic to plants, thus limiting their growth and yield. Application of liming materials on a regular basis to bring the soils to near neutral reactivity and restore the productivity of the soils is imperative to overcome this constraint. Application of biochar, organic amendments, periodic inoculation with soil microbes, and low-cost liming materials other than conventional calcite or dolomite are the other potential soil management strategies useful for mitigating the detrimental effects of soil acidity and sustaining crop yields. Moderate to slightly acidic soils in hot, tropical climatic regions require additional practices such as improved crop varieties; use of acid-tolerant microbial inoculants; integrated use of NPK fertilizers, micronutrients, farmyard manure, and liming materials; planting of acid-tolerant tree species; and improvements in water management, land development, and soil conservation. Collective use of those practices can maintain the soil nutrient status and make efficient use of soil water, thus helping to sustain crop productivity at an enhanced level. Developing crop varieties that are tolerant to acid soils, including iron and aluminum oxide toxicity, can improve crop yields and reduce the need for additional soil management inputs. Overall, management of acid soils and iron and aluminum oxide toxicity in India requires a multipronged approach that includes application of soil amendments, organic matter management, integrated and balanced nutrient management, soil conservation, and plant breeding. Implementation of these strategies can help improve crop productivity and sustainably manage soil resources in the country.

Alleviation of Micro-nutrient Deficiency to Improve Soil Health

Single or multiple micro-nutrient deficiencies are widely reported in different parts of India. These deficiencies are due to imbalanced fertilizer application and little or no application of organic manures or recycling of crop residues. They cause yield loss and deterioration of nutrient quality in the produce. Addition of micronutrients based on their availability could check their deficiency and improve soil. Among the micro-nutrients, boron deficiency is extensive in the light soils of Kerala, Northern Bengal. These deficiencies can be alleviated by applying borax at 10 kg/ha.

Soil Health Management for Coastal Regions of India

Coastal regions of India cover an area of 10.78 million hectares, encompassing nine states, two island groups, and two union territories. Major challenges for farming faced in this zone are the buildup of soil salinity, intrusion of brackish water, scarcity of good-quality irrigation water, and waterlogged soils during the rainy season. Strategies for mitigating these farming challenges and improving soil health include efficient surface and subsurface drainage for the low-lying coastal belt; rainwater harvesting and its efficient recycling for winter season crops; adoption of land modification; need-based calcium supplementation for acidity neutralization; INM with locally available organic amendments (farmyard manure, green manure, vermicompost, smart city waste compost, press mud, etc.); and adoption of CA practices. Recycling of crop residues, green manuring, and application of organic manures would reduce soil salinity and alkalinity, while also improving soil physical health. Land management practices such as bunding and deep drainage can also be used to facilitate leaching of salts and increase crop productivity in foot-slopes of hilly coastal terrains. Adoption of the broad-bed and furrow system of land surface management can provide safe drainage of excess water and harvesting of the rainwater for recycling. Protective embankment of agricultural land, brick pitching of the earthen embankments, and planting of mangrove trees as wind breaks are also useful to check tidal inundation.

Soil Health Management for Eastern India

Major soil-related constraints for the eastern region of India include soil acidity, poor nutrient status, micronutrient deficiency, salinity in the coastal belt, water inundation in low-lying areas, and soil erosion from hilly terrains. Lack of crop diversification, mono-cropping, intensive cultivation, residue removal, excessive use of inorganic fertilizer, and low application of organic amendments have all had a negative impact on soil health. Managing soil health in this region is crucial for sustainability of the production system, as it allows the soil to efficiently perform its fundamental activities. Site-specific management of soil constraints is imperative for improving soil health and system productivity. Application of organic residues or organic amendments such as farmyard manure, compost, solid animal manures, biosolid, green manure, agro-industrial wastes, food-processing wastes, and biochar at sufficient rates, either alone or in combination with inorganic fertilizers for crop production, would improve soil health. Efficient use of bio-wastes (crop residues, weed biomass, forest litter, and animal dung) has great potential to improve soil health and crop productivity in northeastern India. Balancing nutrient supplies through integrated use of inorganic and organic sources could maintain or enhance soil health by improving biological, chemical, and physical soil properties. In the northeastern hill region, where high rainfall, moderate to steep slopes, and shallow soil depth prevail, AF systems are recommended in place of shifting cultivation for managing soil health and thereby improving SOC status, soil aggregation, and in situ soil moisture conservation, while reducing soil loss and erosion. Crop rotation and diversification with leguminous crops (e.g., black gram, green gram, groundnut, and soybean) can improve soil health compared with the existing cereal-based cropping practices in eastern India and the northeastern Himalayas. By strip-cropping cereals and erosion-resistant legumes, rainwater can be stored and runoff slowed in many areas. Growing groundnut as an intercrop with maize is another useful practice for soil conservation and improving soil health in the northeastern region. Crop rotation and diversification can increase crop yield, enhance carbon sequestration, balance soil fertility, and improve soil structure. It can also increase erosion resistance, biodiversity, biological activity, and overall soil health. Adoption of CA with judicious plant nutrient management and residue recycling can enhance soil health by sequestering carbon and nitrogen in soils and improving crop productivity for several cropping systems in eastern and northeastern India. Crop–livestock integration and controlled grazing can also improve system productivity and soil health in northeastern hilly terrain. Managing low soil pH and improving crop productivity on acid soils are also important for enhancing food security and soil health. Managing acid soils with lime raises soil pH, increases nutrient availability to plants, and reduces iron and aluminum toxicity. Lower doses of lime (i.e., one-tenth to one-fifth of the required lime) applied in furrows at the time of crop sowing can be very useful and cost effective for amelioration of acid soils.

Soil Health Management for Dryland and Arid Regions of India

About 51% of the farming communities, and therefore cultivated land, in India depend on rainfed agriculture. It contributes about 44% of the total food production and plays a critical role in country's food security. Most of the dryland agricultural areas in arid and semi-arid zones of the country have poor nutrient status, low water-holding capacity, high pH, low soil organic matter content, shallow soil depth, poor soil structure, and high susceptibility to erosion by wind and water. Collectively, these factors are responsible for low productivity in these zones. Soil erosion and land degradation owing to frequent floods and droughts are the other major causes for low productivity in dryland areas. Efficient management of limited water and soil resources are two of the major challenges for sustainable crop production in those areas. Poor soil conservation practices, lack of technology transfer, and low-yielding crop varieties all contribute to further degradation of soil fertility under dryland farming. Increasing soil water storage, sequestering carbon, adopting CA and organic/natural farming practices, mulching, using suitable and region-specific land configurations, and applying soil amendments including manures, tank silts, gypsum, and other materials such as biochar, polymers, zeolites, and bentonites can increase water and nutrient retention capacity of the soil and consequently improve their use efficiency. Improving water-use efficiency through sprinkler or drip irrigation combined with efficient nutrient management (i.e., foliar spraying of water-soluble fertilizers, nano-fertilizers, SSNM), and conjunctive use of chemical fertilizers and locally available organic manures can bridge current yield gaps in dryland areas and improve soil health.

Summary and Conclusion

Long-term soil health enhancement in India will require a variety of soil and plant management strategies tailored for the distinctly different agro-ecoregions. By adopting sustainable, site-specific soil management measures, farmers can improve their soil health, reduce their dependence on chemical fertilizers and pesticides, and ensure sustainable agricultural production. It is our hope and desire that this volume of the Soil Science Society of America's Soil Health Series provides ideas and inspiration for solving this critical global challenge.

Printed and bound by CPI Group (UK) Ltd, Croydon, CR0 4YY

16/09/2026

14963422-0002